P9-DMH-668

"The definitive work on the West's water crisis"
—*Newsweek*

"Intelligent, provocative, and compulsively readable"
—*Chicago Sun-Times*

"A savagely witty history of America's reckless depletion of its water resources . . . *Cadillac Desert* is a much-needed blast of cold reason on a hot topic of debate."
—*Newsday*

"The scale of this book is as staggering as that of Hoover Dam. Beautifully written and meticulously researched, it spans our century-long effort to moisten the arid West. . . . Anyone thinking of moving west of the hundredth meridian should read this book before they call their real estate agent."
—*St. Louis Post-Dispatch*

"A masterful account . . . among the best-read, most influential environmental books published by an American since *Silent Spring*"
—*San Francisco Examiner*

"A highly partisan, wonderfully readable portrayal of the damming, diverting and dirtying of western rivers"
—*Washington Post Book World*

"One of the triumphs of the year" —*London Observer*

"A revealing, absorbing, often amusing and alarming report on where billions of [taxpayers'] dollars have gone—and where a lot more are going . . . [Reisner] has put the story together in trenchant form."
—*The New York Times Book Review*

"Well-written history and analysis, thoroughly researched and abundantly clear in its message" —*Los Angeles Times Book Review*

"Pugnacious . . . Well-documented . . . A compelling, cautionary tale, one that should be required reading for the federal bureaucracy"
—*Chicago Tribune*

"Sometimes startling, always perceptive . . . Written in a style that pops and sparkles like a fresh mountain stream"
—*Smithsonian*

"*Cadillac Desert* is a book that no American concerned about the nation's future can afford to miss. . . . A fascinating, stubbornly logical, expert analysis of the problems the American West has had in trying to capture and control enough water to supply its agriculture and growing population"
—*The Grand Rapids Press*

"Thoughtful and sprightly . . . Reisner's book deserves to be widely read by political leaders, as well as environmentalists and just about anyone interested in water policies. . . . After reading *Cadillac Desert*, it is hard to be indifferent about the importance of water."
—*Christian Science Monitor*

"Fascinating . . . Reisner has an uncanny gift for mingling narrative with quotes, historical references, and even suspense."
—*Oakland Tribune*

"A magnificent piece of investigative journalism . . . Reisner documents how the unholy rivalry between the Bureau of Reclamation and the Army Corps of Engineers led to the construction of dozens of wasteful, useless dams in the past sixty years."
—*Palo Alto Times Tribune*

"An incredible story . . . This timely and important book should be required reading for all citizens."
—*Publishers Weekly*

PENGUIN BOOKS

CADILLAC DESERT

Marc Reisner was staff writer of the Natural Resources Defense Council from 1972 to 1979. In 1979 he received an Alicia Patterson Journalism Fellowship to investigate water resources in the West. In addition to *Cadillac Desert*, which was nominated for a National Book Critics Circle award, Reisner is also author of *Game Wars: The Undercover Pursuit of Wildlife Poachers* (Viking, 1991). He lives in San Francisco with his wife, Lawrie Mott, and two young daughters.

OZYMANDIAS

I met a traveller from an antique land
Who said: Two vast and trunkless legs of stone
Stand in the desert . . . Near them, on the sand,
Half sunk, a shattered visage lies, whose frown,
And wrinkled lip, and sneer of cold command,
Tell that its sculptor well those passions read
Which yet survive, stamped on these lifeless things,
The hand that mocked them, and the heart that fed:
And on the pedestal these words appear:
"My name is Ozymandias, king of kings:
Look on my works, ye Mighty, and despair!"
Nothing beside remains. Round the decay
Of that colossal wreck, boundless and bare
The lone and level sands stretch far away.

PERCY BYSSHE SHELLEY

MARC REISNER

CADILLAC DESERT

The American West and Its Disappearing Water

REVISED AND UPDATED

PENGUIN BOOKS

For Konrad and Else Reisner

PENGUIN BOOKS
Published by the Penguin Group
Penguin Books USA Inc., 375 Hudson Street,
New York, New York 10014, U.S.A.
Penguin Books Ltd, 27 Wrights Lane, London W8 5TZ, England
Penguin Books Australia Ltd, Ringwood, Victoria, Australia
Penguin Books Canada Ltd,10 Alcorn Avenue,
Toronto, Ontario, Canada M4V 3B2
Penguin Books (N.Z.) Ltd, 182–190 Wairau Road,
Auckland 10, New Zealand

Penguin Books Ltd, Registered Offices:
Harmondsworth, Middlesex, England

First published in the United States of America by
Viking Penguin Inc. 1986
Published in Penguin Books 1987
This revised and updated edition published in Penguin Books 1993

5   7   9   10   8   6

Copyright © Marc Reisner, 1986, 1993
Maps copyright © Viking Penguin Inc., 1986
All rights reserved

Grateful acknowledgment is made for permission to reprint an excerpt from
"Talking Columbia," words and music by Woody Guthrie. TRO—© Copyright
1961 and 1963 Ludlow Music, Inc., New York, N.Y. Used by permission.

LIBRARY OF CONGRESS CATALOGING IN PUBLICATION DATA
Reisner, Marc
Cadillac desert
Reprint. Originally published, New York, N.Y., U.S.A.
Viking, 1986.
Bibliography.
Includes index.
1. Irrigation—Government policy—West (U.S.)—History.
2. Water resources development—Government policy—West
(U.S.)—History. 3. Corruption (in politics)—West (U.S.)
—History. I. Title.
[HD1739.A17R45 1987] 333.91'00978 87.7602
ISBN 0 14 01. 17824 4 (revised edition)

Printed in the United States of America
Set in Aster
Maps by David Lindroth

Except in the United States of America, this book is sold subject to the condition that
it shall not, by way of trade or otherwise, be lent, re-sold, hired out, or otherwise
circulated without the publisher's prior consent in any form of binding or cover other
than that in which it is published and without a similar condition including this
condition being imposed on the subsequent purchaser.

CONTENTS

Illustrations follow pages 182 and 374.

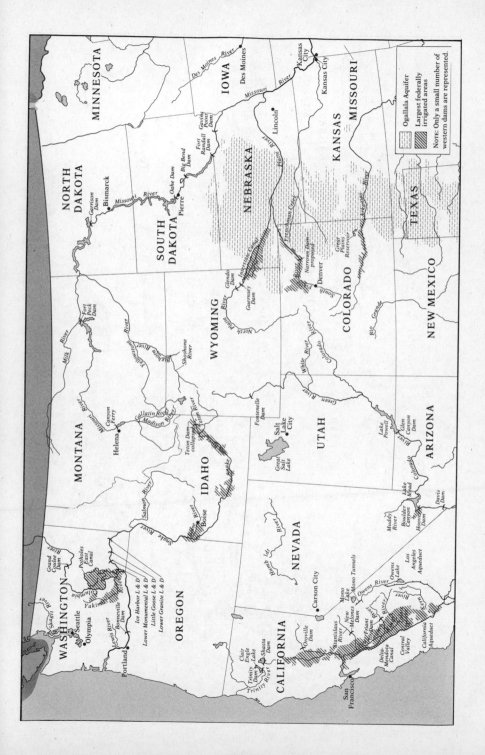

NOTE: Only a small number of western dams are represented.

Ogallala Aquifer

Largest federally irrigated areas

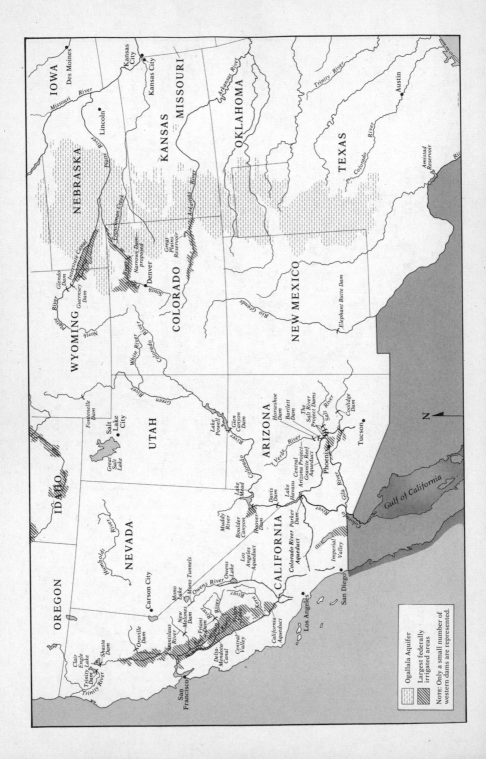

OREGON

IDAHO

Trinity River
Clair Engle
Trinity Lake
Shasta Dam
San Francisco
Stanislaus River
New Melones Dam
Oroville Dam
Friant Dam
Delta Mendota Canal
Central Valley
California Aqueduct
Los Angeles
Kern River
Owens River
Mono Tunnels
Mono Lake
Owens Lake
Los Angeles Aqueduct
Carson City
Humboldt River

NEVADA

San Diego

CALIFORNIA

Imperial Valley
Colorado River Aqueduct
Parker Dam
Davis Dam
Lake Havasu
Boulder Canyon
Hoover Dam
Muddy River
Lake Mead

UTAH

Salt Lake City
Great Salt Lake
Fontenelle Dam

WYOMING

North Platte River
White River
Green River
Colorado River

Lake Powell
Glen Canyon Dam

ARIZONA

Verde River
Horseshoe Dam
Bartlett Dam
The Salt River Project Dams
Salt River
Phoenix
Central Arizona Project—Granite Reef Aqueduct
Gila River
Coolidge Dam
Tucson

Colorado River
Gulf of California

N

NEW MEXICO

Rio Grande
Elephant Butte Dam

TEXAS

Colorado River
Amistad Reservoir
Austin
Rio

Phil River
Glendo Dam
Guernsey Dam
Kortes Dam
Seminoe Dam
Pathfinder Dam
North Platte River
Interstate Canal
Frenchman Creek
Narrows Dam proposed
Denver
South Platte River

COLORADO

Great Plains Reservoir
Arkansas River

IOWA

Missouri River
Des Moines

NEBRASKA

Platte River
Lincoln

KANSAS

Kansas City
Kansas City

MISSOURI

Arkansas River

OKLAHOMA

Trinity River

Ogallala Aquifer

Largest federally irrigated areas

NOTE: Only a small number of western dams are represented.

INTRODUCTION

A Semidesert
with a Desert Heart

Oone late November night in 1980 I was flying over the state of
Utah on my way back to California. I had an aisle seat, and
since I believe that anyone who flies in an airplane and doesn't
spend most of his time looking out the window wastes his money, I
walked back to the rear door of the airplane and stood for a long time
at the door's tiny aperture, squinting out at Utah.

Two days earlier, a fierce early blizzard had gone through the
Rocky Mountain states. In its wake, the air was pellucid. The frozen
fire of a winter's moon poured cold light on the desert below. Six inches
away from the tip of my nose the temperature was, according to the
pilot, minus sixty-five, and seven miles below it was four above zero.
But here we were, two hundred highly inventive creatures safe and
comfortable inside a fat winged cylinder racing toward the Great Basin
of North America, dozing, drinking, chattering, oblivious to the frigid
emptiness outside.

Emptiness. There was nothing down there on the earth—no towns,
no light, no signs of civilization at all. Barren mountains rose duskily
from the desert floor; isolated mesas and buttes broke the wind-
haunted distance. You couldn't see much in the moonlight, but obviously there were no forests, no pastures, no lakes, no rivers; there
was no fruited plain. I counted the minutes between clusters of lights.
Six, eight, nine, eleven—going nine miles a minute, that was a lot of
uninhabited distance in a crowded century, a lot of emptiness amid
a civilization whose success was achieved on the pretension that natural obstacles do not exist.

Then the landscape heaved upward. We were crossing a high, thin cordillera of mountains, their tops already covered with snow. The Wasatch Range. As suddenly as the mountains appeared, they fell away, and a vast gridiron of lights appeared out of nowhere. It was clustered thickly under the aircraft and trailed off toward the south, erupting in ganglionic clots that winked and shimmered in the night. Salt Lake City, Orem, Draper, Provo: we were over most of the population of Utah.

That thin avenue of civilization pressed against the Wasatches, intimidated by a fierce desert on three sides, was a poignant sight. More startling than its existence was the fact that it had been there only 134 years, since Brigham Young led his band of social outcasts to the old bed of a drying desert sea and proclaimed, "This is the place!" *This* was the place? Someone in that first group must have felt that Young had become unhinged by two thousand horribly arduous miles. Nonetheless, within hours of ending their ordeal, the Mormons were digging shovels into the earth beside the streams draining the Wasatch Range, leading canals into the surrounding desert which they would convert to fields that would nourish them. Without realizing it, they were laying the foundation of the most ambitious desert civilization the world has seen. In the New World, Indians had dabbled with irrigation, and the Spanish had improved their techniques, but the Mormons attacked the desert full-bore, flooded it, subverted its dreadful indifference—moralized it—until they had made a Mesopotamia in America between the valleys of the Green River and the middle Snake. Fifty-six years after the first earth was turned beside City Creek, the Mormons had six million acres under full or partial irrigation in several states. In that year—1902—the United States government launched its own irrigation program, based on Mormon experience, guided by Mormon laws, run largely by Mormons. The agency responsible for it, the U.S. Bureau of Reclamation, would build the highest and largest dams in the world on rivers few believed could be controlled—the Colorado, the Sacramento, the Columbia, the lower Snake—and run aqueducts for hundreds of miles across deserts and over mountains and through the Continental Divide in order to irrigate more millions of acres and provide water and power to a population equal to that of Italy. Thanks to irrigation, thanks to the Bureau—an agency few people know—states such as California, Arizona, and Idaho became populous and wealthy; millions settled in regions where nature, left alone, would have countenanced thousands at best; great valleys and hemispherical basins metamorphosed from desert blond to semitropic green.

On the other hand, what has it all amounted to?

Stare for a while at a LANDSAT photograph of the West, and you will see the answer: not all that much. Most of the West is still untrammeled, unirrigated, depopulate in the extreme. Modern Utah, where large-scale irrigation has been going on longer than anywhere else, has 3 percent of its land area under cultivation. California has twelve hundred major dams, the two biggest irrigation projects on earth, and more irrigated acreage than any other state, but its irrigated acreage is not much larger than Vermont. Except for the population centers of the Pacific Coast and the occasional desert metropolis—El Paso, Albuquerque, Tucson, Denver—you can drive a thousand miles in the West and encounter fewer towns than you would crossing New Hampshire. Westerners call what they have established out here a civilization, but it would be more accurate to call it a beachhead. And if history is any guide, the odds that we can sustain it would have to be regarded as low. Only one desert civilization, out of dozens that grew up in antiquity, has survived uninterrupted into modern times. And Egypt's approach to irrigation was fundamentally different from all the rest.

If you begin at the Pacific rim and move inland, you will find large cities, many towns, and prosperous-looking farms until you cross the Sierra Nevada and the Cascades, which block the seasonal weather fronts moving in from the Pacific and wring out their moisture in snows and drenching rains. On the east side of the Sierra-Cascade crest, moisture drops immediately—from as much as 150 inches of precipitation on the western slope to as little as four inches on the eastern—and it doesn't increase much, except at higher elevations, until you have crossed the hundredth meridian, which bisects the Dakotas and Nebraska and Kansas down to Abilene, Texas, and divides the country into its two most significant halves—the one receiving at least twenty inches of precipitation a year, the other generally receiving less. Any place with less than twenty inches of rainfall is hostile terrain to a farmer depending solely on the sky, and a place that receives seven inches or less—as Phoenix, El Paso, and Reno do—is arguably no place to inhabit at all. Everything depends on the manipulation of water—on capturing it behind dams, storing it, and rerouting it in concrete rivers over distances of hundreds of miles. Were it not for a century and a half of messianic effort toward that end, the West as we know it would not exist.

The word "messianic" is not used casually. Confronted by the desert, the first thing Americans want to do is change it. People say that they "love" the desert, but few of them love it enough to live there. I mean in the real desert, not in a make-believe city like Phoenix with exotic palms and golf-course lawns and a five-hundred-foot fountain and an artificial surf. Most people "love" the desert by driving through it in air-conditioned cars, "experiencing" its grandeur. That may be some kind of experience, but it is living in a fool's paradise. To *really* experience the desert you have to march right into its white bowl of sky and shape-contorting heat with your mind on your canteen as if it were your last gallon of gas and you were being chased by a carload of escaped murderers. You have to imagine what it would be like to drink blood from a lizard or, in the grip of dementia, claw bare-handed through sand and rock for the vestigial moisture beneath a dry wash.

Trees, because of their moisture requirements, are our physiological counterparts in the kingdom of plants. Throughout most of the West they begin to appear high up on mountainsides, usually at five or six thousand feet, or else they huddle like cows along occasional streambeds. Higher up the rain falls, but the soil is miserable, the weather is extreme, and human efforts are under siege. Lower down, in the valleys and on the plains, the weather, the soil, and the terrain are more welcoming, but it is almost invariably too dry. A drought lasting three weeks can terrorize an eastern farmer; a drought of five months is, to a California farmer, a normal state of affairs. (The lettuce farmers of the Imperial Valley don't even *like* rain; it is so hot in the summer it wilts the leaves.) The Napa Valley of California receives as much Godwater—a term for rain in the arid West—as Illinois, but almost all of it falls from November to March; a weather front between May and September rates as much press attention as a meteor shower. In Nevada you see rainclouds, formed by orographic updrafts over the mountains, almost every day. But rainclouds in the desert seldom mean rain, because the heat reflected off the earth and the ravenous dryness can vaporize a shower in midair, leaving the blackest-looking cumulonimbus trailing a few pathetic ribbons of moisture that disappear before reaching the ground. And if rain does manage to fall to earth, there is nothing to hold it, so it races off in evanescent brown torrents, evaporating, running to nowhere.

One does not really conquer a place like this. One inhabits it like an occupying army and makes, at best, an uneasy truce with it. New England was completely forested in 1620 and nearly deforested 150 years later; Arkansas saw nine million acres of marsh and swamp forest converted to farms. Through such Promethean effort, the eastern half of the continent was radically made over, for better or worse. The West never can be. The only way to make the region over is to irrigate it. But there is too little water to begin with, and water in rivers is phenomenally expensive to move. And even if you succeeded in moving every drop, it wouldn't make much of a difference. John Wesley Powell, the first person who clearly understood this, figured that if you evenly distributed all the surface water flowing between the Columbia River and the Gulf of Mexico, you would *still* have a desert almost indistinguishable from the one that is there today. Powell failed to appreciate the vast amount of water sitting in underground aquifers, a legacy of the Ice Ages and their glacial melt, but even this water, which has turned the western plains and large portions of California and Arizona green, will be mostly gone within a hundred years—a resource squandered as quickly as oil.

At first, no one listened to Powell when he said the overwhelming portion of the West could never be transformed. People figured that when the region was settled, rainfall would magically increase, that it would "follow the plow." In the late 1800s, such theories amounted to Biblical dogma. When they proved catastrophically wrong, Powell's irrigation ideas were finally embraced and pursued with near fanaticism, until the most gigantic dams were being built on the most minuscule foundations of economic rationality and need. Greening the desert became a kind of Christian ideal. In May of 1957, a very distinguished Texas historian, Walter Prescott Webb, wrote an article for *Harper's* entitled "The American West, Perpetual Mirage," in which he called the West "a semidesert with a desert heart" and said it had too dark a soul to be truly converted. The greatest national folly we could commit, Webb argued, would be to exhaust the Treasury trying to make over the West in the image of Illinois—a folly which, by then, had taken on the appearance of national policy. The editors of *Harper's* were soon up to their knees in a flood of vitriolic mail from westerners condemning Webb as an infidel, a heretic, a doomsayer.

Desert, semidesert, call it what you will. The point is that despite heroic efforts and many billions of dollars, all we have managed to do in the arid West is turn a Missouri-size section green—and that conversion has been wrought mainly with nonrenewable groundwater. But a goal of many westerners and of their federal archangels, the Bureau of Reclamation and Corps of Engineers, has long been to double, triple, quadruple the amount of desert that has been civilized and farmed, and now these same people say that the future of a hungry world depends on it, even if it means importing water from as far away as Alaska. What they seem not to understand is how difficult it will be just to hang on to the beachhead they have made. Such a surfeit of ambition stems, of course, from the remarkable record of success we have had in reclaiming the American desert. But the same could have been said about any number of desert civilizations throughout history—Assyria, Carthage, Mesopotamia; the Inca, the Aztec, the Hohokam—before they collapsed.

And it may not even have been drought that did them in. It may have been salt.

The Colorado River rises high in the Rockies, a trickle of frigid snowmelt bubbling down the west face of Longs Peak, and begins its fifteen-hundred-mile, twelve-thousand-foot descent to the Gulf of California. Up there, amid mountain fastnesses, its waters are sweet. The river swells quickly, taking in the runoff of most of western Colorado, and before long becomes a substantial torrent churning violently through red canyons down the long west slope of the range. Not far from Utah, at the threshold of the Great Basin, the rapids die into riffles and the Colorado River becomes, for a stretch of forty miles, calm and sedate. It has entered the Grand Valley, a small oasis of orchards and cows looking utterly out of place in a landscape where it appears to have rained once, about half a million years ago. The oasis is man-made and depends entirely on the river. Canals divert a good share of the flow and spread it over fields, and when the water percolates through the soil and returns to the river it passes through thick deposits of mineral salts, a common phenomenon in the West. As the water leaves the river, its salinity content is around two hundred parts per million; when it returns, the salinity content is sixty-five hundred parts per million.

The Colorado takes in the Gunnison River, whose waters have also filtered repeatedly through irrigated, saline earth, and disappears into the canyonlands of Utah. Near the northernmost tentacle of Lake Powell, where the river backs up for nearly two hundred miles behind Glen Canyon Dam, it receives its major tributary, the Green River. The land along the upper Green is heavily irrigated, and so is the land beside its two major tributaries, the Yampa and the White. Some of *their* tributaries, which come out of the Piceance Basin, are saltier than the ocean. In Lake Powell, the water spreads, exposing vast surface acreage to the sun, which evaporates several feet each year, leaving all the salts behind. Released by Glen Canyon Dam, the Colorado takes in the Little Colorado, Kanab Creek, the Muddy, and one of the more misnamed rivers on earth, the Virgin. It pools again in Lake Mead, again in Lake Mojave, and again in Lake Havasu; it takes in the Gila River and its oft-used tributaries, the Salt and the Verde, all turbid with alkaline leachate. A third of its flow then goes to California, where some of it irrigates the Imperial Valley and the rest allows Los Angeles and San Diego to exist. By then, the water is so salty that restaurants often serve it with a slice of lemon. If you pour it on certain plants, they will die.

Along the Gila River in Arizona, the last tributary of the Colorado, is a small agricultural basin which Spaniards and Indians tried to irrigate as early as the sixteenth century. It has poor drainage—the soil is underlain by impermeable clays—so the irrigation water rose right up to the root zones of the crops. With each irrigation, it became saltier, and before long everything that was planted died. The Spaniards finally left, and the desert took the basin back; for a quarter of a millennium, it remained desert. Then, in the 1940s, the Bureau of Reclamation reclaimed it again, building the Welton-Mohawk Project and adding an expensive drainage system to collect the sumpwater and carry it away. Just above the Mexican border, the drain empties into the Colorado River.

In 1963, the Bureau closed the gates of Glen Canyon Dam. As Lake Powell filled, the flow of fresh water below it was greatly reduced. At the same time, the Welton-Mohawk drain was pouring water with a salinity content of sixty-three hundred parts per million directly into the Colorado. The salinity of the river—what was left of it—soared to fifteen hundred parts per million at the Mexican border. The most important agricultural region in all of Mexico lies right below the border, utterly dependent on the Colorado River; we were giving the farmers slow liquid death to pour over their fields.

The Mexicans complained bitterly, to no avail. By treaty, we had promised them a million and a half acre-feet of water. But we hadn't promised them *usable* water. By 1973, Mexico was in a state of apoplexy. The ruin of its irrigated agricultural lands along the lower Colorado was the biggest issue in the campaign of presidential candidate Luis Echeverría, who was elected by a wide margin in that year. Still, the United States continued to do nothing. But 1973 also saw the arrival of OPEC. Some new geologic soundings in the Bay of Campeche indicated that Mexico might soon become one of the greatest oil-exporting nations in the world. When Echeverría threatened to drag the United States before the World Court at The Hague, Richard Nixon sent his negotiators down to work out a salinity-control treaty. It was signed within a few months.

Once we agreed to give Mexico water of tolerable quality, we had to decide how to do it. Congress's solution was to authorize a desalination plant ten times larger than any in existence that will clean up the Colorado River just as it enters Mexico. What it will cost nobody knows; the official estimate in 1985 was $300 million, not counting the 40,000 kilowatts of electricity required to run it. Having done that, Congress wrote what amounts to a blank check for a welter of engineered solutions farther upriver, whose exact nature is still under debate. Those could cost another $600 million, probably more. One could easily achieve the same results by buying out the few thousand acres of alkaline and poorly drained land that contribute most to the problem, but there, once again, one runs up against the holiness of the blooming desert. Western Congressmen, in the 1970s, were perfectly willing to watch New York City collapse when it was threatened with bankruptcy and financial ruin. After all, New York was a profligate and sinful place and probably deserved such a fate. But they were not willing to see one acre of irrigated land succumb to the forces of nature, regardless of cost. So they authorized probably $1 billion worth of engineered solutions to the Colorado salinity problem in order that a few hundred upstream farmers could go on irrigating and poisoning the river. The Yuma Plant will remove the Colorado's salt—actually just enough of it to fulfill our treaty obligations to Mexico—at a cost of around $300 per acre-foot of water. The upriver irrigators buy the same amount from the Bureau for three dollars and fifty cents.

Nowhere is the salinity problem more serious than in the San Joaquin Valley of California, the most productive farming region in the entire world. There you have a shallow and impermeable clay layer, the residual bottom of an ancient sea, underlying a million or so acres of fabulously profitable land. During the irrigation season, temperatures in the valley fluctuate between 90 and 110 degrees; the good water evaporates as if the sky were a sponge, the junk water goes down, and the problem gets worse and worse. Very little of the water seeps through the Corcoran Clay, so it rises back up into the root zones—in places, the clay is only a few feet down—waterlogs the land, and kills the crops. A few thousand acres have already gone out of production—you can see the salt on the ground like a dusting of snow. In the next few decades, as irrigation continues, that figure is expected to increase almost exponentially. To build a drainage system for the valley—a giant network of underground pipes and surface canals that would intercept the junk water and carry it off—could cost as much as a small country's GNP. In 1985, the Secretary of the Interior put forth a figure of $5 billion for the Westlands region, and Westlands is only half the problem. Where would the drainwater go? The Westlands' drainwater, temporarily stored in a huge sump which was christened a wildlife preserve, has been killing thousands of migrating waterfowl; the water contains not just salts but selenium, pesticides, and God knows what else. There is one logical terminus: San Francisco Bay. As far as northern Californians are concerned, the farmers stole all this water from them; now they want to ship it back full of crud.

As is the case with most western states, California's very existence is premised on epic liberties taken with water—mostly water that fell as rain on the north and was diverted to the south, thus precipitating the state's longest-running political wars. With the exception of a few of the rivers draining the remote North Coast, virtually every drop of water in the state is put to some economic use before being allowed to return to the sea. Very little of this water is used by people, however. Most of it is used for irrigation—80 percent of it, to be exact. That is a low percentage, by western standards. In Arizona, 87 percent of the water consumed goes to irrigation; in Colorado and New Mexico, the figure is almost as high. In Kansas, Nevada, Nebraska, North Dakota, South Dakota, and Idaho—in all of those states, irrigation accounts for nearly all of the water that is consumptively used.

By the late 1970s, there were 1,251 major reservoirs in California, and every significant river—save one—had been dammed at least once. The Stanislaus River is dammed fourteen times on its short run to the sea. California has some of the biggest reservoirs in the country; its rivers, seasonally swollen by the huge Sierra snowpack, carry ten times the runoff of Colorado's. And yet all of those rivers and reservoirs satisfy only 60 percent of the demand. The rest of the water comes from under the ground. The rivers are infinitely renewable, at least until the reservoirs silt up or the climate changes. But a lot of the water being pumped out of the ground is as nonrenewable as oil.

Early in the century, before the federal government got into the business of building dams, most of the water used for irrigation in California was groundwater. The farmers in the Central Valley (which comprises both the Sacramento and the San Joaquin) pumped it out so relentlessly that by the 1930s the state's biggest industry was threatened with collapse. The growers, by then, had such a stranglehold on the legislature that they convinced it, in the depths of the Depression, to authorize a huge water project—by far the largest in the world— to rescue them from their own greed. When the bonds to finance the project could not be sold, Franklin Delano Roosevelt picked up the unfinished task. Today, the Central Valley Project is still the most mind-boggling public works project on five continents, and in the 1960s the state built its own project, nearly as large. Together, the California Water Project and the Central Valley Project have captured enough water to supply eight cities the size of New York. But the projects brought into production far more land than they had water to supply, so the growers had to supplement their surface water with tens of thousands of wells. As a result, the groundwater overdraft, instead of being alleviated, has gotten worse.

In the San Joaquin Valley, pumping now exceeds natural replenishment by more than half a trillion gallons a year. By the end of the century it could rise to a trillion gallons—a mining operation that, in sheer volume, beggars the exhaustion of oil. How long it can go on, no one knows. It depends on a lot of things, such as the price of food and the cost of energy and the question whether, as carbon dioxide changes the world's climate, California will become drier. (It is expected to become much drier.) But it is one reason you hear talk about redirecting the Eel and the Klamath and the Columbia and, someday, the Yukon River.

The problem in California is that there is absolutely no regulation over groundwater pumping, and, from the looks of things, there won't be any for many years to come. The farmers loathe the idea, and in California "the farmers" are the likes of Exxon, Tenneco, and Getty Oil. Out on the high plains, the problem is of a different nature. There, the pumping of groundwater is regulated. But the states have all decided to regulate their groundwater out of existence.

The vanishing groundwater in Texas, Kansas, Colorado, Oklahoma, New Mexico, and Nebraska is all part of the Ogallala aquifer, which holds two distinctions: one of being the largest discrete aquifer in the world, the other of being the fastest-disappearing aquifer in the world. The rate of withdrawal over natural replenishment is now roughly equivalent to the flow of the Colorado River. This was the region called the Dust Bowl, the one devastated by the Great Drought; that was back before anyone knew there was so much water underfoot, and before the invention of the centrifugal pump. The prospect that a region so plagued by catastrophe could become rich and fertile was far too tantalizing to resist; the more irrigation, everyone thought, the better. The states knew the groundwater couldn't last forever (even if the farmers thought it would), so, like the Saudis with their oil, they had to decide how long to make it last. A reasonable period, they decided, was twenty-five to fifty years.

"What are you going to do with all that water?" asks Felix Sparks, the former head of the Colorado Water Conservation Board. "Are you just going to leave it in the ground?" Not necessarily, one could reply, but fifty years or a little longer is an awfully short period in which to exhaust the providence of half a million years, to consume as much nonrenewable water as there is in Lake Huron. "Well," says Sparks, "when we use it up, we'll just have to get more water from somewhere else."

Stephen Reynolds, Sparks's former counterpart in New Mexico—as state engineer, the man in charge of water, he may have been the most powerful person in the state—says much the same thing: "We made a conscious decision to mine out our share of the Ogallala in a period of twenty-five to forty years." In the portions of New Mexico that overlie the Ogallala, according to Reynolds, some farmers withdraw as much as five feet of water a year, while nature puts back a quarter of an inch. What will happen to the economy of Reynolds's state when its major agricultural region turns to dust? "Agriculture uses about 90 percent of our water, and produces around 20 percent of the state's income, so it wouldn't necessarily be a knockout economic blow," he answers. "Of course, you are talking about drastic changes in the whole life and culture of a very big region encompassing seven states.

"On the other hand," says Reynolds, half-hopefully, "we may decide as a matter of national policy that all this agriculture is too important to lose. We can always decide to build some more water projects."

More water projects. During the first and only term of his presidency, Jimmy Carter decided that the age of water projects had come to a deserved end. As a result, he drafted a "hit list" on which were a couple of dozen big dams and irrigation projects, east and west, which he vowed not to fund. Carter was merely stunned by the reaction from the East; he was blown over backward by the reaction from the West. Of about two hundred western members of Congress, there weren't more than a dozen who dared to support him. One of the projects would return five cents in economic benefits for every taxpayer dollar invested; one offered irrigation farmers subsidies worth more than $1 million each; another, a huge dam on a middling California river, would cost more than Hoover, Shasta, Glen Canyon, Bonneville, and Grand Coulee combined. But Carter's hit list had as much to do with his one-term presidency as Iran.

Like millions of easterners who wonder how such projects get built, Jimmy Carter had never spent much time in the West. He had never driven across the country and watched the landscape turn from green to brown at the hundredth meridian, the threshold of what was once called the Great American Desert—but which is still wet compared to the vast ultramontane basins beyond. In southern Louisiana, water is the central fact of existence, and a whole culture and set of values have grown up around it. In the West, lack of water is the central fact of existence, and a whole culture and set of values have grown up around it. In the East, to "waste" water is to consume it needlessly or excessively. In the West, to waste water is *not* to consume it—to let it flow unimpeded and undiverted down rivers. Use of water is, by definition, "beneficial" use—the term is right in the law—even if it goes to Fountain Hills, Arizona, and is shot five hundred feet into 115-degree skies; even if it is sold, at vastly subsidized rates, to farmers irrigating crops in the desert which their counterparts in Mississippi or Arkansas are, at that very moment, being paid not to grow. To easterners, "conservation" of water usually means protecting rivers from development; in the West, it means building dams.

More water projects. In the West, nearly everyone is for them. Politicians of every stripe have sacrificed their most sacred principles on the altar of water development. Barry Goldwater, scourge of welfare and champion of free enterprise, was a lifelong supporter of the Central Arizona Project, which comes as close to socialism as anything this country has ever done (the main difference being that those who are subsidized are well-off, even rich). Former Governor Jerry Brown of California attended the funeral of E. F. Schumacher, the English economist who wrote *Small Is Beautiful*, then flew back home to lobby for a water project that would cost more than it did to put a man on the moon. Alan Cranston, once the leading liberal in the U.S. Senate, the champion of the poor and the oppressed, successfully lobbied to legalize illegal sales of subsidized water to giant corporate farms, thus denying water—and farms—to thousands of the poor and oppressed.

In the West, it is said, water flows uphill toward money. And it literally does, as it leaps three thousand feet across the Tehachapi Mountains in gigantic siphons to slake the thirst of Los Angeles, as it is shoved a thousand feet out of Colorado River canyons to water Phoenix and Palm Springs and the irrigated lands around them. It goes 444 miles (the distance from Boston to Washington) by aqueduct from the Feather River to south of L.A. It goes in man-made rivers, in siphons, in tunnels. In a hundred years, actually less, God's riverine handiwork in the West has been stood on its head. A number of rivers have been nearly dried up. One now flows backward. Some flow through mountains into other rivers' beds. There are huge reservoirs where there was once desert; there is desert, or cropland, where there were once huge shallow swamps and lakes.

It still isn't enough.

———

In 1971, the Bureau of Reclamation released a plan to divert six million acre-feet from the lower Mississippi River and create a river in reverse, pumping the water up a staircase of reservoirs to the high plains in order to save the irrigation economy of West Texas and eastern New Mexico, utterly dependent on groundwater, from collapse. Since the distance the water would have to travel is a thousand miles, and the elevation gain four thousand feet, and since six million acre-feet of water weigh roughly 16.5 trillion pounds, a lot of energy would be required to pump it. The Bureau figured that six nuclear plants would do, and calculated the cost of the power at one mill per kilowatt-hour, a tiny fraction of what it costs today. The whole package came to $20 billion, in 1971 dollars; the benefit-cost ratio would have been .27 to 1. For each dollar invested, twenty-seven cents in economic productivity would be returned. "That's kind of discouraging," says Stephen Reynolds. "But when you consider our balance-of-payments deficits, you have to remember that we send $100 billion out of this country each year just to pay for imported oil. The main thing we export is food. The Ogallala region produces a very large share of our agricultural exports."

More water projects. In the early 1960s, the Ralph M. Parsons Corporation, a giant engineering firm based in Pasadena, California, released a plan to capture much of the floor of the Yukon and Tanana rivers and divert it two thousand miles to the Southwest through the Rocky Mountain Trench. The proposal, called the North American Water and Power Alliance, wasn't highly regarded by Canada, which was the key to the "alliance," but in the West it was passionately received. Ten years later, as environmentalism and inflation both took root, NAWAPA seemed destined for permanent oblivion. But then OPEC raised the price of oil 1,600 percent, and Three Mile Island looked as if it might seal fission's doom. California was hit by the worst drought in its history; had it lasted one more year, its citizens might have begun migrating back east, their mattresses strapped to the tops of their Porsches and BMWs. All of a sudden the hollowness of our triumph over nature hit home with striking effect. With hydroelectricity now regarded by many as salvation, and with nearly half the irrigated farmland in the West facing some kind of doom—drought, salt, or both combined—NAWAPA, in the early 1980s, began to twitch again. The cost estimate (phony, of course) had doubled, from $100 billion to $200 billion, but by then we were spending that much in a single year on defense. The project could produce 100,000 megawatts of electricity; it could rescue California, the high plains, and Arizona and still have enough water left to turn half of Nevada green. The new

Romans were now saying that it wasn't a matter of *whether* NAWAPA would be built, but when.

Perhaps they are right. Perhaps, despite the fifty thousand major dams we have built in America; despite the fact that federal irrigation has, for the most part, been a horribly bad investment in free-market terms; despite the fact that the number of free-flowing rivers that remain in the West can be counted on two hands; perhaps, despite all of this, the grand adventure of playing God with our waters will go on. Perhaps it will be consummated on a scale of which our forebears could scarcely dream. By encouraging millions of people to leave the frigid Northeast, we could save a lot of imported oil; by doubling our agricultural exports, we could pay for the oil we import today. As the ancient, leaking water systems and infrastructure of the great eastern cities continue to decay, we may see an East-West alliance develop: you give us our water projects, we'll give you yours. Perhaps, in some future haunted by scarcity, the unthinkable may be thinkable after all.

In the West, of course, where water is concerned, logic and reason have never figured prominently in the scheme of things. As long as we maintain a civilization in a semidesert with a desert heart, the yearning to civilize more of it will always be there. It is an instinct that followed close on the heels of food, sleep, and sex, predating the Bible by thousands of years. The instinct, if nothing else, is bound to persist.

The lights of Salt Lake City began to fade, an evanescent shimmer on the rear horizon. A few more minutes and the landscape was again a black void. We were crossing the Great Basin, the arid heart of the American West. The pilot announced that the next glow of civilization would be Reno, some six hundred miles away. I remembered two things about Reno. The annual precipitation there is seven inches, an amount that Florida and Louisiana and Virginia have received in a day. But even though gambling and prostitution are legal around Reno, water metering, out of principle, was for a long time against the law.

CHAPTER ONE

A Country of Illusion

The American West was explored by white men half a century before the first colonists set foot on Virginia's beaches, but it went virtually uninhabited by whites for another three hundred years. In 1539, Don Francisco Vásquez de Coronado, a nobleman who had married rich and been appointed governor of Guadalajara by the Spanish king, set out on horseback from Mexico with a couple of hundred men, driving into the uncharted north. Coronado was a far kinder conquistador than his ruthless contemporaries Pizarro and De Soto, but he was equally obsessed with gold. His objective was a place called Cibola, seven cities where, legend had it, houses and streets were veneered with gold and silver. All he found, somewhere in northwestern Arizona, were some savage people living in earthen hovels, perhaps descendants of the great Hohokam culture, which had thrived in central Arizona until about 1400, when it mysteriously disappeared. Crestfallen, but afraid of disgracing the Spanish crown, Coronado pushed on. Tusayan, Cicuye, Tiguex, Quivira—no gold. His fruitless expedition took him from the baking desert canyons of south-central Arizona up to the cool ponderosa highlands of the Mogollon Rim, then down again into the vast, flat, treeless plains of West Texas and Oklahoma and Kansas. He returned, miraculously, a couple of years later, having lost half his men and some of his sanity when his horse stepped on his skull as he was exercising it. Since the climate of the American West is often compared, by those who don't know better, with that of Spain, it is instructive to quote part of the letter

Coronado wrote to Viceroy Mendoza as he was recovering along the Rio Grande:

> After traveling seventy-seven days from Tiguex over these barren lands, our Lord willed that I should arrive in the province called Quivira [Kansas], to which the guides El Turco and the other savage were taking me. They had pictured it as having stone houses many stories high; not only are there none of stone, but, on the contrary, they are of grass, and the people are savage like all I have seen and passed up to that place. They have no woven fabrics, nor cotton with which to make them. All they have is tanned skins of the cattle they kill, for the herds are near the place where they live, a fair-sized river. [The Indian guides' reward for their misleading travelogue was to be garroted to death.] . . .
> The natives gave me a piece of copper which an Indian chief wore suspended from his neck. I am sending it to the viceroy of New Spain, for I have not seen any other metal. . . . I have done everything within my power to serve you, as your faithful sergeant and vassal, and to discover some country where God our Lord might be served by extending your royal patrimony. . . . The best country I have discovered is this Tiguex River [the Rio Grande] and the settlement where I am now camping. But they are not suitable for colonizing, for, besides being four hundred leagues from the North Sea and more than two hundred from the South Sea, thus prohibiting all intercourse, the land is so cold, as I have informed Your Majesty, that it seems impossible for anyone to spend the winter here, since there is no firewood, nor any clothing with which the men may keep themselves warm, except the skins which the natives wear. . . .

The greatest irony of Coronado's adventure was that he must have passed within a few miles of the gold and silver lodes at Tombstone and Tubac, Arizona. A few of his party, on a side excursion, discovered the Grand Canyon, but they were unimpressed by its beauty, and guessed the width of the Colorado River far below them at eight feet or so. The Rio Grande, which would later sustain the only appreciable Spanish settlements outside of California, didn't impress them, either. When he returned to Guadalajara, Coronado was put on trial for inept leadership, which, though an utterly unfounded charge, was enough to discourage would-be successors who might have discovered the precious metals that would have induced Spain to lay a far stronger claim on the New World. His expedition also lost a few horses, which found their way into the hands of the native Americans. The two dominant tribes of the Southwest, the Apache and Comanche, soon evolved into the best horsemen who ever lived, and their ferocity toward incursionists made them formidable adversaries of the Spaniards who tried to settle the region later.

The Spanish did make a more than desultory try at establishing a civilization in California, which was more to their liking than the remainder of the West. (And, in fact, the huge California land grants doled out by the king established a pattern of giant fiefdoms that persists there to this day.) But they never found gold in California, so the territory didn't seem worth a fight. Challenged by the first American expeditionary force in 1842, Mexico ceded the entire territory six years later—just a few months before a man named James Marshall was to discover a malleable yellow rock in the tailrace of Sutter's Mill on the American River above Sacramento.

In 1803, the United States of America consisted of sixteen states along the Atlantic Seaboard, three-quarters of whose area were still untrammeled wilderness, and a vast unmapped tract across the Appalachian Mountains—which would metamorphose, more quickly than anyone might have expected, into the likes of Cleveland and Detroit. In that same year, the new First Consul of France, Napoleon Bonaparte, sat in Paris wrestling with a question: what to conquer? France had recently acquired a million square miles of terrain in North America from Spain—Spain having gotten it originally from France—and the prospect of a huge colonial empire in the New World was tempting. On the other hand, here was Europe—settled, tamed, productive— waiting for civilized dominion by the French. For what would history remember him better—the conquest of Russia or the conquest of buffalo?

The new President of the United States was Thomas Jefferson, an ardent Francophile, but, above all, a practical man. Jefferson knew better than anyone that a French presence in the New World could only be considered a threat. Jefferson was also exceedingly clever, and he was not above a little ruse. "The day that France takes possession of Louisiana," he wrote in a message to his ministers in Paris, "we must marry ourselves to the British fleet and nation." Having said that, Jefferson, through the offices of a Franco-American gunpowder manufacturer named du Pont de Nemours, then inaugurated a hallowed presidential tradition known as the intentional leak. Reading the "intercepted" message, Napoleon lost his half-formed resolve to create an empire on two continents. The result was the Louisiana Purchase.

Napoleon had no idea what he had sold for $15 million, and Jefferson had no idea what he had bought. For fifteen years, however, he had been trying to send an expedition to the unknown country west of the Mississippi River, and now, for the first time, he was able to persuade Congress to put up the money. In 1804, Jefferson's personal secretary, a private, moody, and sensitive young man named Meriwether Lewis, together with a bluff and uncomplicated army captain named William Clark, left St. Louis with a party of fifty men. Poling, tugging, and, at times, literally carrying a fifty-foot bateau up the whipsawing braided channels of the Missouri River, they arrived at the villages of the Mandan tribe, in what has come to be North Dakota, in the early winter. When the ice broke in the spring, some of the party returned to St. Louis with the boat. The thirty-one others, accompanied by a Shoshone Indian girl named Sacajawea, who had been captured and enslaved by the Mandans, and her newborn baby, continued westward on horseback and on foot. Guided by Sacajawea— whose usefulness as an interpreter was only a small part of the Lewis and Clark expedition's fabulous luck—they pressed across the plains to the beginning of the true Missouri at Three Forks, Montana. From there, they struggled over the Continental Divide and found the Salmon River, whose alternative name, the River of No Return, is an indication of the experiences they had trying to follow it. In despair, the party gave up and turned northward, finding the Clearwater River, which offered them an easier path westward. The Clearwater led them to the Snake, and the Snake led them to the Columbia—a huge anomaly of a river in the pale desert east of the Cascades. Entering the Columbia gorge, they made an almost instantaneous transition from arid grasslands to rain forest as the river sliced through the Cascade Range—a type of transition utterly fantastic to an easterner. From there, it was a short hop to the Pacific, where the party spent the winter, fattening on seafood. In August of 1806, they were back in St. Louis.

The country Lewis and Clark saw amazed, appalled, and enchanted them. Above all, it bewildered them. They had seen the western plains at their wettest—in the springtime of an apparently wet year—but still there were few rivers, and full ones were fewer. The sky was so immense it swallowed the landscape, but the land swallowed up the provenance of the sky. There was game—at times a ludicrous abundance of it—but there were no trees. To an easterner, no trees meant no possibility of agriculture. If the potential wealth of the land could be judged by the layers of fat on its inhabitants, it was worthwhile to note that the only fat Indians seen by Lewis and Clark were those on the Pacific Coast, sating themselves on salmon and clams. Reading their journals, one gets the impression that Lewis and Clark simply didn't know what to think. They had never seen a landscape like this, never guessed one could even exist. Each "fertile prairie" and "happy prospect" is counterweighted by a "forbidding plain." Louisiana, though penetrated, remained an enigma.

The explorers who followed Lewis and Clark were more certain of their impressions. In the same year the expedition returned, General Zebulon Montgomery Pike crossed the plains on a more southerly course, through what was to become Kansas and Colorado. There he saw "tracts of many leagues . . . where not a speck of vegetable matter existed" and dismissed the whole country as an arid waste. "These vast plains of the western hemisphere may become in time as celebrated as the sandy deserts of Africa," wrote Pike. Major Stephen Long, who followed Pike a decade later, had a similar impression. Long referred to the whole territory between the Mississippi and the Rocky Mountains as the Great American Desert—a phrase and an image that held for almost half a century. The desert might have sat there even longer in the public mind, ineradicable and fixed, had not a member of the Lewis and Clark expedition by the name of John Colter noticed, in the rivers and streams tumbling out of the Rocky Mountains, a plenitude of beaver.

The settlement of the American West owed itself, as much as anything, to a hat. The hat was made of beaver felt, and, during the 1820s and 1830s, no dedicated follower of fashion would settle for anything less. Demand was great enough, and beavers east of the Mississippi were scarce enough, that a cured plew could fetch $6 to $10—at the time, a week's wages. If one was reckless, adventurous, mildly to strongly sociopathic, and used to living by one's wits, it was enough money to make the ride across the plains and winters spent amid the hostile Blackfoot and Crow worth the danger and travail. The mountain men never numbered more than a few hundred, but their names—Bridger, Jackson, Carson, Colter, Bent, Walker, Ogden, Sublette—are writ large all over the American West. Supreme outdoorsmen, they could read important facts in the angle and depth of a bear track; they could hide from the Blackfoot in an icy stream, breathing through a hollow stem, and live out a sudden blizzard in the warm corpse of an eviscerated mountain sheep. As trappers, they were equally proficient—so proficient that within a few years of their arrival in the Rocky Mountain territory, the beavers had already begun to thin out. But that was all the more reason for the more restless of them, especially those backed by eastern money, to go off exploring unknown parts for more beaver streams. And no explorer in the continent's history was more compulsive and indefatigable than Jedediah Smith.

In 1822, when he joined the Rocky Mountain Fur Trading Company, Smith was twenty-two years old, and had never seen the other side of the Rockies. Within two years, however, he was in charge of an exploratory party of trappers heading into utterly unfamiliar territory along the Green River. They found beaver there in fabulous numbers, and Smith, feeling unneeded, decided to see what lay off to the north and west. With six others, he set a course across the Great Basin toward Great Salt Lake. The landscape was more desolate than anything they had seen. If the Great American Desert was on the other side of the mountains, then what would you call this? Game was pitifully scarce. The herds of buffalo had vanished, and the only creatures appearing in numbers were rattlesnakes and jackrabbits. The few human beings encountered were numbingly primitive. They built no lodges, used the crudest tools, made no art. They subsisted, from all appearances, on roots and insects; a live gecko made a fine repast. Mark Twain, encountering some of the last of the wild Digger Indians half a century later, called them "the wretchedest type of mankind I have ever seen." But they were, as Twain noted, merely a reflection of the landscape they found themselves in.

Smith's party skirted Great Salt Lake and continued westward, becoming the first whites, and probably the first humans, to cross the Bonneville Salt Flats—a hundred miles of horrifyingly barren terrain. They then struck across what is now eastern Oregon, eventually reaching a British fort near the Columbia River. Sensing something less than a generous welcome (the British still wanted at least a piece of this subcontinent), the party turned around, and was back on the Green River by July of 1825, in time for the trappers' first rendezvous.

The rendezvous was the first all-male ritual in the non-Indian West—a kind of Baghdad bazaar leavened by fighting, fornication, and adventure stories that would have seemed outlandish if they hadn't, for the most part, been true. Trappers arrived from hundreds of miles around with their pelts, which they traded for whiskey sold by St. Louis entrepreneurs at $25 the gallon, for ammunition, and for staples such as squaws. There was usually carnage, inhibited mainly by the water the traders had added to the whiskey. At the Green River rendezvous, however, Smith and two of his partners, David Jackson and William Sublette, forsook the festivities for serious business. They had decided to take over the Missouri Fur Trading Company from its owner, General William Ashley, who had amassed a substantial fortune in an astonishingly short time. When the deal was consummated, Smith was given the assignment he coveted—to be in charge of finding new sources of pelts.

Within days of returning from Oregon, Smith was already heading out with a party of fourteen men from Cache Valley, Utah, in search of virgin beaver streams. They followed the languid Sevier River through the red-and-blond deserts of southwestern Utah, then jumped across to the Virgin River, which led them to the Colorado above the present site of Hoover Dam. Unknowingly, they were breaking the Mormon Outlet Trail, by which the secrets of successful irrigation would migrate to California and Arizona and be applied with such ambition that, within a scant century and a half, there would be proposals to import irrigation water from Alaska along the same route. By the time they reached the Colorado River, winter was already near; they had trapped only a few beaver, and didn't feel like turning back. Anxious to find warmth and food, Smith decided to lead the party across the Mojave Desert toward the ocean coast. "A complete barrens" was his description, "a country of starvation." After several exhausting days (they had to carry all their water), the explorers sighted two tall ranges to the west. They crossed the pass between them and found themselves in the Los Angeles Basin, at Mission San Gabriel Archangel in Spanish California. The padres' reception was friendly, but the Spanish governor's was not. Ever since hearing about the expedition of "Capitán Merrie Weather," his attitude toward Yankees had tilted toward paranoia. Exiled from the basin, Smith led his party up the San Joaquin Valley and into the Sierra Nevada, where, along the Stanislaus River, they found beaver in urban concentrations. After a few weeks of trapping, Smith loaded hundreds of plews on horses, selected his two toughest men, and set off across the spine of the Sierra Nevada into what is now Nevada.

Of all the routes across the Great Basin, the one he chose is the longest and driest. U.S. Highway 6 now runs parallel and slightly south; the trip is so desolate and frightening that many motorists will not take it, even in an air-conditioned car loaded with water jugs; they go north, along Interstate 80, which stays reassuringly in sight of the Humboldt River. In six hundred miles of travel, Smith's party crossed three small inconstant streams. That they survived at all is a miracle. "My arrival caused a considerable bustle in camp," he wrote in his diary after arriving in time for the second rendezvous on the Bear

River in Utah. "A small cannon, brought up from St. Louis, was loaded and fired for a salute. . . . Myself and party had been given up for lost."

Two weeks after the rendezvous, Smith was, incredibly, on the way to California again, anxious to relieve the men who had remained on the Stanislaus and to trap out the beaver of the Sierra Nevada before someone else discovered them. His route was pretty much the same as the time before. While crossing the Colorado, however, his party was ambushed by a band of Mojave Indians; nine of the nineteen men survived, among them Smith. Fleeing across the desert, they finally reached southern California, where Smith left three wounded men to recover. The rest of the party then joined the trappers they had left the year before. (How they managed to find each other is a subject Smith passes over lightly in his diary.) Both groups, by now, were bereft of supplies. Selecting his two friendliest surviving men, Smith rode across the Central Valley to the missions at Santa Clara and San Jose to barter plews for food, medicine, clothing, and ammunition. As soon as the members of the party were sighted, they were dragged off to jail in Monterey. Bail was set at $30,000, an amount calculated to ensure that they would remain there at the governor's whim. Smith's luck, however, seemed to ricochet between the abominable and sublime; a wealthy sea captain from New England, who was holding over in Monterey, was so impressed by Smith's courage that he arranged to post the entire amount.

Freed but banished forever from California, Smith gathered the remnants of his expedition, and they wandered up the Sacramento Valley, trapping as they went. It was by then the middle of winter, and the snowpack in the Sierra was twelve feet deep; crossing the range was out of the question. Smith decided to venture back toward the ocean. Crossing the Yolla Bolly and Trinity mountains, the party found itself in a rain forest dominated by a gigantic species of conifer they had never seen. Reaching the Pacific near the mouth of the river that now bears Smith's name, they slogged northward through country which can receive a hundred inches of rain during six winter months. At the mouth of the Umpqua River, they stopped to rest. Smith went off to reconnoiter in an improvised canoe. While he was gone, a band of the Umpqua tribe stole into camp and murdered all but three of the men. Fleeing through the tangled forest beneath giant trees, two of the survivors found Smith, and they raced off together in the direction of Fort Vancouver on the Columbia River. They arrived there in August of 1828, emaciated and in shock. Their last surviving companion straggled in after them; he had found his way alone.

The British, by then well established in Oregon, considered the attack ominous enough to demand a reprisal. An expedition was dispatched for the Umpqua Valley, where the marauding band was cornered; thirty-nine horses and Smith's seven hundred beaver pelts were seized. Although the British were still smarting from the War of 1812, the commander refused to let Smith compensate him for his trouble; instead, he paid him $3,200 for the horses and pelts. He also offered the Americans a long rest at the fort, since it would take most of the winter for them to tell all their tales. In the spring of 1829, the assembled force of Fort Vancouver watched in disbelief as Smith and Arthur Black, the last of the four survivors who still retained their nerve, strode confidently through the gates and up the Columbia River, en route to the June rendezvous. "They are sporting with life or courting danger to madness," remarked the commander, who never went out with fewer than forty men. Within twelve weeks, Smith and Black were back among their companions in Jackson Hole.

After six years of hair-raising adventures, Jedediah Smith decided to relax and devote a season to tranquil pursuits—trapping beaver on icy mountain streams in territory claimed by Indians and grizzly bears—and then returned to St. Louis to see what opportunity lay there. But civilization stank in his nostrils, and wilderness coursed through his blood. After a brief stay in the frontier capital, Smith was back on the Santa Fe Trail, guiding pioneers westward. It was there, at the age of thirty, that his life came to an abrupt end, a Comanche tomahawk embedded in his skull. He is memorialized today across a region the size of Europe, though modern explorers in a Prowler or a Winnebago may not realize that half a dozen Smith Rivers and a landscape of Smith Parks, Passes, Peaks, and Valleys in eleven states are mostly named after the same Smith.

The "useful" role ascribed to the mountain men is that they opened the door to settlement of the West. It might be more accurate, however, to say that they slammed it shut. The terrors they endured were hardly apt to draw settlers, and their written accounts of the region had to lie heavy on a settler's mind: plains so arid that they could barely support bunchgrass; deserts that were fiercely hot and fiercely cold; streams that flooded a few weeks each year and went dry the rest; forests with trees so large it might take days to bring one down; Indians, grizzly bears, wolves, and grasshopper plagues; hail followed by drought followed by hail; no gold. You could live off the land in better years, but the life of a trapper, a hunter, a fortune seeker—the only type of life that seemed possible in the West—was not what the vast majority of Americans sought.

There were those who believed, in the 1830s, that the Louisiana

Purchase had been a waste of $15 million—that the whole billion acres would remain as empty as Mongolia or the Sahara. And then, just a generation later, there were those who believed a billion people were destined to settle there. It seemed there was only one person in the whole United States with the wisdom, the scientific detachment, and the explorer's insight to dissect both myths and find the truth that lay buried within.

John Wesley Powell belonged to a subspecies of American which flourished briefly during the nineteenth century and went extinct with the end of the frontier. It was an estimable company, one that included the likes of Mark Twain, John Muir, Abraham Lincoln, William Dean Howells, and Hamlin Garland. They were genuine Renaissance men, though their circumstances were vastly different from those of Jefferson or Benjamin Franklin. The founding fathers, the most notable among them, were urban gentlemen or gentlemen farmers who grew up in a society that, though it sought to keep Europe and its mannerisms at arm's length, had a fair amount in common with the Old World. They lived in very civilized style, even if they lived at the edge of a frontier. Powell, Howells, Lincoln, and the others were children of the real frontier. Most grew up on subsistence farms hacked out of ancient forests or grafted onto tallgrass prairie; they lacked formal education, breeding, and refinement. Schooled by teachers who knew barely more than they did, chained to the rigors of farm life, they got their education from borrowed books devoured by the embers of a fireplace or surreptitiously smuggled into the fields. What they lacked in worldliness and schooling, however, they more than made up in vitality, originality, and circumambient intelligence. John Wesley Powell may be one of the lesser-known of this group, but he stood alone in the variety of his interests and the indefatigability of his pursuits.

Powell's father was a poor itinerant preacher who transplanted his family westward behind the breaking wave of the frontier. As a boy in the 1840s, Powell moved from Chillicothe, Ohio, to Walworth County, Wisconsin, to Bonus Prairie, Illinois. Nothing was paved, little was fenced; the forests were full of cougars and the streams full of fish. To Powell, the frontier was a rapturous experience. Like John Muir, he got a vagabond's education, rambling cross-country in order to become intimate with forests and fauna, with hydrology and weather. In the summer of 1855, Powell struck out for four months

and walked across Wisconsin. Two years later he floated down the
Ohio River from Pittsburgh to St. Louis. A few months later, he was
gathering fossils in interior Missouri. The next spring he was rowing
alone down the Illinois River and up the Mississippi and the Des
Moines River to the middle of Iowa, then a wilderness. Between his
peregrinations Powell picked up some frantic education—Greek,
Latin, botany, a bit of philosophy—at Wheaton, Oberlin, and Illinois
College, but he never graduated and he never stayed long. Powell
learned on the run.

When the Civil War broke out, Powell enlisted on the Union side,
fought bravely, and came out a major, a confidant of Ulysses Grant,
and minus an arm, which was removed by a steel ball at the Battle
of Shiloh. To Powell, the loss of an arm was merely a nuisance, though
the raw nerve endings in his amputated stump kept him in pain for
the rest of his life. After the war he tried a stint at teaching, first at
Illinois Wesleyan and then at Illinois State, but it didn't satisfy him.
He helped found the Illinois Museum of Natural History, and was an
obvious candidate for the position of curator, but decided that this,
too, was too dull an avenue with too visible an end. Powell, like the
mountain men, was compulsively drawn to the frontier. In the United
States of the late 1860s, there was but one place where the frontier
was still nearly intact.

By 1869, the population of New York City had surpassed one million.
The city had built a great water-supply aqueduct to the Croton River
and was imagining its future subway system. Chicago, founded thirty
years earlier, was already a big sprawling industrial town. The millionaires of San Francisco were building their palatial mansions on
Nob Hill. New England was deforested, farms and settlements were
spilling onto the prairie. However, on maps of the United States published in that year a substantial area remained a complete blank, and
was marked "unexplored."

The region overlay parts of what is now Colorado, Utah, Arizona,
New Mexico, and Nevada. It was about the size of France, and through
the middle of it ran the Colorado River. That was about all that was
known about it, except that the topography was awesome and the
rainfall scarce. The region was known as the Plateau Province, and
parties heading westward tended to avoid it at all costs.

Some of the Franciscan friars, who were as tough as anyone in the
Old West, had wandered through it on the Old Spanish Trail. Other-
wise, the Mormon Outlet Trail skirted the region to the west, the
California and Oregon trails swung northward, and the El Paso-Yuma

Trail went south. From a distance, one could see multicolored and multistoried mesas and cliffs, saurian ridges, and occasionally a distant snowcapped peak. There were accounts of canyons that began without reason and were suddenly a thousand feet deep, eroded more by wind than by water. A distance that a bird could cover in an hour might require a week to negotiate. The days were hot and the nights were often frigid, owing to the region's high interior vastness, and water was almost impossible to find. Lacking wings, there was only one good way to explore it: by boat.

On the 24th of May, 1869, the Powell Geographic Expedition set out on the Green River from the town of Green River, Wyoming, in four wooden dories: the *Maid of the Canyon*, the *Kitty Clyde's Sister*, the *Emma Dean*, and the *No Name*. For a scientific expedition, it was an odd group. Powell, the leader, was the closest thing to a scientist. He had brought along his brother Walter—moody, sarcastic, morose, one of the thousands of psychiatric casualties of the Civil War. The rest of the party was made up mostly of mountain men: O. G. Howland, his brother Seneca, Bill Dunn, Billy Hawkins, and Jack Sumner, all of whom had been collected by Powell en route to Green River. He had also invited a beet-faced Englishman named Frank Goodman, who had been patrolling the frontier towns looking for adventure, and Andy Hall, an eighteen-year-old roustabout whose casual skill as an oarsman had impressed Powell when he saw him playing with a boat on the Green River. There was also George Bradley, a tough guy whom Powell had met by accident at Fort Bridger and who had agreed to come along in exchange for a discharge from the army, which Powell managed to obtain for him.

For sixty miles out of the town of Green River, the river was sandy-bottomed and amiable. There were riffles, but nothing that could legitimately be called a rapid. The boatmen played in the currents, acquiring a feel for moving water; the others admired the scenery. As they neared the Uinta Mountains, they went into a sandstone canyon colored in marvelous hues, which Powell, who had a knack for naming things, called Flaming Gorge. The river bore southward until it came up against the flanks of the range, then turned eastward and entered Red Canyon.

In Red Canyon, the expedition got its first lesson in how a few feet of drop per mile can turn a quiet river into something startling. Several of the rapids frightened them into racing for shore and lining or portaging, an awful strain with several thousand pounds of boats, supplies, and gear. After a while, however, even the bigger rapids were not so menacing anymore—if, compared to what was about to come, one could call them big.

Beyond Flaming Gorge the landscape opened up into Brown's Park, but soon the river gathered imperceptible momentum and the canyon ramparts closed around them like a pair of jaws. A maelstrom followed. Huge scissoring waves leaped between naked boulders; the river plunged into devouring holes. The awestruck Andy Hall recited an alliterative verse he had learned as a Scottish schoolboy, "The Cataract of Lodore," by the English Romantic poet Robert Southey. Over Powell's objection—he did not like using a European name—the stretch became the Canyon of Lodore.

As they approached the first big rapid in the canyon, the *No Name* was sucked in by the accelerating current before anyone had a chance to scout. "I pass around a great crag just in time to see the boat strike a rock and rebounding from the shock career and fill the open compartment with water," wrote Powell in his serialized journal of the trip. "Two of the men lost their oars, she swings around, and is carried down at a rapid rate broadside on for quite a few yards and strikes amidships on another rock with great force, is broken quite in two, and then men are thrown into the river, the larger part of the boat floating buoyantly. They soon seize it and down the river they drift for a few hundred yards to a second rapid filled with huge boulders where the boat strikes again and is dashed to pieces and the men and fragments are soon carried beyond my sight."

The three crew members survived, but most of the extra clothes, the barometers, and several weeks' worth of food were gone. The next day the party found the stern of the boat intact, still holding the barometers, some flour, and a barrel of whiskey that Powell, who was something of a prig, did not realize had been smuggled aboard. When they finally floated out of Lodore Canyon into the sunlit beauty of Echo Park, Powell wrote in his journal that despite "a chapter of disaster and toil . . . the canyon of Lodore was not devoid of scenic interest, even beyond the power of the pen to tell." And O. G. Howland, who nearly lost his life in Disaster Falls, wrote haughtily that "a calm, smooth stream is a horror we all detest now."

Desolation Canyon. Gray Canyon. They were now in territory even Indians hadn't seen. The landscape closed in and opened up. Labyrinth Canyon. Stillwater Canyon. They shot a buck and scared a bighorn lamb off a cliff, their first fresh meat in weeks. Powell, climbing a cliff with his one arm, got himself rimmed and required rescue by Bradley, who got above him, dangled his long johns, and pulled Powell up.

The country grew drier and more desolate. Fantastic mesas loomed in the distance, banded like shells. The Grand Mesa, to the east, the largest mesa in the world, rose to eleven thousand feet from desert badlands into an alpine landscape of forests and lakes. Wind-eroded shiprocks loomed over the rubblized beds of prehistoric seas. Battlements of sandstone rose in the distance like ruins of empire. Deep in uncharted territory the Colorado River, then known as the Grand, rushed in quietly from the northeast, carrying the snowmelt of Longs Peak and most of western Colorado. The river's volume had now doubled, but still it remained quite placid. Was it conceivable that they were near the end of its run? Powell was tempted to believe so, but knew better. There were four thousand feet of elevation loss ahead. On the 21st, after a short stop to rest and reseal the boats, they were on the water again, which was high, roiled, and the color of cocoa. In a few miles they came to a canyon, frothing with rapids. They lined or portaged wherever they could, ran if they had no alternative. Soon they were between vertical walls and the river was roaring mud. Cataracts launched them downriver before they had time to think; waves like mud huts threw them eight feet into the air. The scouts would venture ahead if there was room enough to walk, and return ashenfaced. The canyon relented a little at times, so they could portage, but the river did not. In one day, they made three-quarters of a mile in Cataract Canyon, portaging everything they saw.

During the daytime, the temperature would reach 106 degrees; at night the men shivered in their dank drawers. Some became edgy, prone to violent outbursts. Bradley's incendiary moods lasted through most of a day, and he would run almost anything rather than portage. Powell's instinctive caution infuriated Bradley, as did his indefatigable specimen gathering, surveying, and consignment of everything to notes. The pace was maddeningly uneven: they would do eight miles in a day, then a mere mile or two. Two months' worth of food remained, most of it musty bread, dried apples, spoiled bacon, and coffee. Once, Billy Hawkins got up in the middle of dinner, walked to the boats, and pulled out the sextant. He said he was trying to find the latitude and longitude of the nearest pie.

On the 23rd of July they passed a foul-smelling little stream coming in from the west; they called it the Dirty Devil. The big river quieted. The hunters took off up the cliffsides and returned with a couple of desert bighorn sheep, which were devoured with sybaritic abandon. The sheep were an omen. For the next several days, they floated on a brisk but serene river through a canyon such as no one had seen. Instead of the pitiless angular black-burned walls of Cataract Canyon, they were now enveloped by rounded pink-and-salmon-colored sandstone, undulating ahead of them in soft contours. There were huge arched chasms, arcadian glens hung with maidenhair ferns, zebra-striped walls, opalescent green fractures irrigated by secret springs. Groping for a name that would properly convey their sense of both awe and relief, Powell decided on Glen Canyon. On August 1 and 2, the party camped in Music Temple.

By the 5th of August, they were down to fifteen pounds of rancid bacon, several bags of matted flour, a small store of dried apples, and a large quantity of coffee. Other than that they would have to try to live off the land, but the land was mostly vertical and the game, which had never been plentiful, had all but disappeared. They met the Escalante River, draining unknown territory in Utah, then the San Juan, carrying in snowmelt from southwestern Colorado.

The river on which they were floating was made up now of most of the mentionable runoff of the far Southwest. They were in country that no white person had ever seen, riding the runoff of a region the size of Iraq, and they approached each blind bend in the river with a mixture of anticipation and terror. Soon the soft sandstone of Glen Canyon was replaced by the fabulous coloration of Marble Canyon. Then, on August 14, the hard black rock of Cataract Canyon reemerged from the crust of the earth. "The river enters the gneiss!" wrote Powell. Downriver, they heard what sounded like an avalanche.

Soap Creek Rapids, Badger Creek Rapids, Crystal Creek Rapids, Lava Falls. Nearly all of the time, the creeks that plunge down the ravines of the Grand Canyon will barely float a walnut shell, but the flash floods resulting from a desert downpour can dislodge boulders as big as a jitney bus. Tumbled by gravity, the boulders carom into the main river and sit there, creating a dam, which doesn't so much stop the river as make it mad. Except for the rapids of the Susitna, the Niagara, and perhaps a couple of rivers in Canada, the modern Colorado's rapids are the biggest on the continent. Before the dams were built, however, the Colorado's rapids were *really* big. At Lava Falls, where huge chunks of basalt dumped in the main river create a thirty-foot drop, waves at flood stage were as high as three-story houses. There was a cycling wave at the bottom that, every few seconds, would burst apart with the retort of a sixteen-inch gun, drenching anyone on either bank of the river—two hundred feet apart. To run Lava Falls today, in a thirty-foot Hypalon raft, wrapped in a Mae West life jacket, vaguely secure in the knowledge that a rescue helicopter sits on the canyon rim, is a lesson in panic. The Powell expedition was running most of the canyon's rapids in a fifteen-foot pilot boat made of pine and a couple of twenty-one-foot dories made of oak—with the rudest of life jackets, without hope of rescue, without a single human being within hundreds of miles. And Powell himself was running them strapped to a captain's chair, gesticulating wildly with his one arm.

The river twisted madly. It swung north, then headed south, then back north, then east—east!—then back south. Even Powell, constantly consulting sextant and compass, felt flummoxed. The rapids, meanwhile, had grown so powerful that the boats received a terrible battering from the force of the waves alone, and had to be recaulked every day. As they ran out of food and out of caulk, Powell realized that the men were also beginning to run out of will. There was mutiny in their whisperings.

August 25. They had come thirty-five miles, including a portage around a spellbinding rapid where a boulder dam of hardened lava turned the river into the aftermath of Vesuvius. (That, as it turned out, had been Lava Falls.) There were still no Grand Wash Cliffs, which would signal the confluence with the Virgin River and the end of their ordeal. They saw, for the first time in weeks, some traces of Indian habitation, but obviously no one had lived there in years. Occasionally they caught a glimpse of trees on the canyon rim, five thousand feet above. They were in the deepest canyon any of them had ever seen.

August 26. They came on an Indian garden full of fresh squash. With starvation imminent, they stole a dozen gourds and ate them ravenously. "We are three-quarters of a mile in the depths of the earth," wrote Powell. "And the great river shrinks into insignificance, as it dashes its angry waves against the walls and cliffs, that rise to the world above; they are but puny ripples and we but pigmies, running up and down the sands or lost among the boulders. . . . But," he added hopefully, "a few more days like this and we are out of prison."

August 27. The river, which had been tending toward the west, veered again toward the south. The hated Precambrian granite, which had dropped below the riverbed, surfaced again. Immediately came a rapid which they decided to portage. At eleven o'clock in the morning, they came to the worst rapids yet.

"The billows are huge," wrote Bradley. "The spectacle is appalling." It was, Jack Sumner wrote, a "hell of foam." The rapids was bookended by cliffs; there was no way to portage and no way to line. There wasn't even a decent way to scout.

After the party had had a meal of fried flour patties and coffee, O. G. Howland asked Powell to go for a walk with him. The major knew what was coming. It saddened him that if there was to be mutiny, the leader would be Howland. He was a mountain man by nature and experience, but, after Powell, still the most literate and scientific-minded of the group. Nonetheless, Howland had been plagued by bad luck; it was he who had steered the *No Name* to its destruction in Lodore Canyon; he who had twice lost maps and notes in swampings. He had tested fate enough. In the morning, Howland told Powell, he and his brother Seneca, together with Bill Dunn, were going to aban-don the boats and climb out of the canyon.

Powell did not sleep that night. He took reading after reading with his sextant until he was as positive as he dared be that they were within fifty miles of Grand Wash Cliffs. At the most, they ought to be four days from civilization, with the only remaining obstacle in view a wild twenty-second ride through a terrific rapid. Powell woke How-land in the middle of the night and poured out his conviction, but it was too late. His immediate reaction was two laconic sentences in his journal, but later he offered this version of what took place:

> We have another short talk about the morrow, but for me there is no sleep. All night long, I pace up and down a little path, on a few yards of sand beach, along by the river. Is it wise to go on? I go to the boats again, to look at our rations. I feel satisfied that we can get over the danger immediately before us; what there may be below I know not. From our outlook yesterday, on the cliffs, the cañon seemed to make another great bend to the south, and this, from our experience heretofore, means more and higher granite walls. I am not sure that we can climb out of the cañon here, and, when at the top of the wall, I know enough of the country to be certain that it is a desert of rock and sand, between this and the nearest Mormon town, which, on the most direct line, must be seventy-five miles away. True, the last rains have been favorable to us, should we go out, for the probabilities are that we shall find water still standing in holes, and, at one time, I almost conclude to leave the river. But for years I have been contemplating this trip. To leave the exploration unfinished, to say that there is a part of the cañon which I cannot explore, having already almost accomplished it, is more than I am willing to acknowledge, and I determine to go on.

August 28. Breakfast was as "solemn as a funeral." Afterward, Powell asked all of the men, for the last time, whether they planned to go ahead or climb out. The Howlands and Bill Dunn still intended to walk out; the rest would remain. The party gave the three some guns and offered them their equal share of the remaining rations. They accepted the guns. "Some tears are shed," Powell wrote. "It is rather a solemn parting; each party thinks the other is taking the dangerous course." Billy Hawkins stole away and laid some biscuits on a rock the mutineers would pass on their way up the cliffs. "They are as fine fellows as I ever had the good fortune to meet," declared taciturn George Bradley, blinking away a tear.

As the others rowed cautiously toward the monster rapids in their two boats, the Howland brothers and Bill Dunn had already begun climbing up one of the canyon arroyos. Powell felt himself torn between watching them and the approaching rapids. They plunged down the first drop. The hydraulic wave at the bottom inundated them, but the water was so swift that they were out of it before the boat could fill. They were launched atop a pillow of water covering a rock, slid off, then rode out a landscape of haystacks. As the *Maid of the Canyon* circulated quietly in the whirlpool at rapids' end, *Kitty Clyde's Sister* wallowed up alongside. The roar of the rapids was almost submerged by the men's ecstatic shouts. They grabbed rifles and fired volley after volley into the air to show their erstwhile companions that it could be done. Unable to see around the bend in the river or to walk back up, they waited in the eddy for nearly two hours, hoping the others would rejoin them, but they never did.

A few miles below Separation Rapid, the party came to another rapid, Lava Cliffs, which, were it not now under the waters of Lake Mead, would perhaps be the biggest on the river. In a style so much like the man himself—exact and fastidious, yet felicitous and engaging—Powell wrote down what happened there:

[O]n [the] northern side of the canyon [is] a bold escarpment that seems to be a hundred feet high. We can climb it and walk along its summit to a point where we are just at the head of the fall. Here the basalt is broken down again, so it seems to us, and I direct the men to take a line to the top of the cliff and let the boats down along the wall. One man remains in the boat to keep her clear of the rocks and prevent her line from being caught on the projecting angles. I climb the cliff and pass along to a point just over the fall and descend by broken rocks, and find that the break of the fall is above the break of the wall, so that we cannot land, and that still below the river is very bad, and that there is no possibility of a portage. Without waiting further to examine and determine what shall be done, I hasten

back to the top of the cliff to stop the boats from coming down.
When I arrive I find the men have let one of them down to the
head of the fall. She is in swift water and they are not able to
pull her back; nor are they able to go on with the line, as it is
not long enough to reach the higher part of the cliff which is
just before them; so they take a bight around a crag. I send two
men back for the other line. The boat is in very swift water,
and Bradley is standing in the open compartment, holding out
his oar to prevent her from striking against the foot of the cliff.
Now she shoots out into the stream and up as far as the line
will permit, and then, wheeling, drives headlong against the
rock, and then out and back again, now straining on the line,
now striking against the rock. As soon as the second line is
brought, we pass it down to him; but his attention is all taken
up with his own situation, and he does not see that we are
passing him the line. I stand on a projecting rock, waving my
hat to gain his attention, for my voice is drowned by the roaring
of the falls. Just at this moment I see him take his knife from
its sheath and step forward to cut the line. He has evidently
decided that it is better to go over with the boat as it is than
to wait for her to be broken to pieces. As he leans over, the boat
sheers again into the stream, the stem-post breaks away and
she is loose. With perfect composure Bradley seizes the great
scull oar, places it in the stern rowlock, and pulls with all his
power (and he is an athlete) to turn the bow of the boat down
stream, for he wishes to go bow down, rather than to drift
broad-side on. One, two strokes he makes, and a third just as
she goes over, and the boat is fairly turned, and she goes down
almost beyond our sight, though we are more than a hundred
feet above the river. Then she comes up again on a great wave,
and down and up, then around behind some great rocks, and
is lost in the mad, white foam below. We stand frozen with
fear, for we see no boat. Bradley is gone! so it seems. But now,
away below, we see something coming out of the waves. It is
evidently a boat. A moment more, and we see Bradley standing
on deck, swinging his hat to show that he is all right. But he
is in a whirlpool. We have the stem-post of his boat attached
to the line. How badly she may be disabled we know not. I
direct Sumner and [Walter] Powell to pass along the cliff and
see if they can reach him from below. Hawkins, Hall, and myself
run to the other boat, jump aboard, push out, and away we go
over the falls. A wave rolls over us and our boat is unmanage-

able. Another great wave strikes us, and the boat rolls over, and tumbles and tosses, I know not how. All I know is that Bradley is picking us up. We soon have all right again, and row to the cliff and wait until Sumner and Powell can come. After a difficult climb they reach us. We run two or three miles farther and turn again to the northwest, continuing until night, when we have run out of the granite once more.

August 30. At the confluence of the Colorado and the Virgin River, three Mormons and an Indian helper are seine-netting fish. They have been there for weeks, under orders from Brigham Young to watch for the Powell expedition. Since the members of the expedition have already been reported dead several times in the newspapers, the Mormons are really on the lookout for corpses and wreckage; they hope to salvage whatever journals and maps have survived in order that they might learn something about the unexplored portion of the region where they have banished themselves. Late in the morning, one of them flings a glance upriver and freezes. There are two boats coming down, and, unless they are ghosts, the people inside them seem to be alive.

It had taken three months and six days for the expedition to travel from Green River to Grand Wash Cliffs. Though wilder water than the Colorado is routinely run today, few river runners would dispute that the Powell expedition accomplished the most impressive feat of perilous river exploration in history. But the expedition ended, as fate would have it, on an ironically tragic note. While Powell and those who stayed with him were being fed and pumped for information by the Mormons, the Howland brothers and Bill Dunn were lying dead on the rim of the Grand Canyon, murdered by a band of Shivwits Indians. Later there were rumors that they had molested a Shivwits girl, but the Indian wars were raging and they may have been killed simply for taking the band by surprise. That the Shivwits shot Powell's companions full of holes contains a cold irony, for years later, after Powell had sat around many campfires with them, the Shivwits tribe would come to regard the one-armed major as their most faithful white friend.

When John Wesley Powell first left Council Bluffs, Iowa, in 1867, bound for Denver and the valley of the Green River, the region he crossed was virtually empty. It was like modern interior Alaska, after removing Fairbanks. Indians were more common than whites, and buffalo were much more prevalent than Indians. By the time he reached the ninety-eighth meridian, about two-fifths of the way across Nebraska, the light dusting of settlers' towns and farms had thinned out to nothing. Before him were another five hundred miles of virgin plains, almost uninhabited by whites; then there was Denver, a rowdy little town that owed its existence mainly to furs and gold, and not much else until one got to Salt Lake and California.

On each successive trip west the changes took away Powell's breath. The breaking wave of settlement was eating up half a meridian a year; from one season to the next, settlements were thirty miles farther out. By the late 1870s, the hundredth meridian had been fatefully crossed. There were homes sprouting in central Nebraska, miles from water, trees, and neighbors, their occupants living in sod dugouts suggestive of termite mounds. Farms began to grow up around Denver, where a type of agriculture thoroughly alien to America's farmers—irrigation—was being experimented with. (Horace Greeley, the publisher of the New York *Herald Tribune*—the publisher whose words "Go west, young man" galvanized the nineteenth century—was mainly responsible for this; he had dispatched his agricultural editor, Nathan Meeker, to a spot north of Denver to found a utopian irrigation colony which, not surprisingly, became Greeley, Colorado. The colony appeared to be a success, even forgetting the large annual contribution from Greeley.) On their way across the plains, travelers could see huge rolling clouds of dust on the southern horizon, caused by cattle drives from Texas to railheads at Dodge and Kansas City. The plains were being dug up; the buffalo were being annihilated to starve the Indians and make way for cows; the vanishing tribes were being herded like cattle onto reservations.

This enormous gush of humanity pouring into a region still marked on some maps as the Great American Desert was encouraged by wishful thinking, by salesmanship, that most American of motivating forces, and, most of all, by natural caprice. For a number of years after 1865, a long humid cycle brought uninterrupted above-average rainfall to the plains. Guides leading wagon trains to Oregon reported that western Nebraska, usually blond from drought or black from prairie fires, had turned opalescent green. Late in the 1870s, the boundary of the Great American Desert appeared to have retreated westward across the Rockies to the threshold of the Great Basin. Such a spectacular climatic transformation was not about to be dismissed as a fluke, not by a people who thought themselves handpicked by God to occupy a wild continent. A new school of meteorology was founded to explain it. Its unspoken principle was divine intervention, and its motto was "Rain Follows the Plow." Since the rains coincided with

the headlong westward advance of settlement, the two must somehow
be related. Professor Cyrus Thomas, a noted climatologist, was a leading proponent. "Since the territory [of Colorado] has begun to be
settled," he announced in declamatory tones, "towns and cities built
up, farms cultivated, mines opened, and roads made and travelled,
there has been a gradual increase in moisture. . . . I therefore give it
as my firm conviction that this increase is of a permanent nature, and
not periodical, and that it has commenced within eight years past,
and that it is in some way connected to the settlement of the country,
and that as population increases the moisture will increase." Ferdinand V. Hayden, who was Thomas's boss and one of the most famous
geographers and geologists of his time, also subscribed to the theory.
(Hayden happened to be a notable rival of John Wesley Powell, who
believed otherwise.) The exact explanations varied. Plowing the land
exposed the soil's moisture to the sky. Newly planted trees enhanced
rainfall. The smoke from trains caused it. Vibrations in the air created
by all the commotion helped clouds to form. Dynamiting the air became a popular means of inducing rain to fall. Even the Secretary of
Agriculture came out for a demonstration in Texas. "The result," he
reported, "was—a loud noise!"

The notion that settlement was changing the climate on the flat,
loamy, treeless plains rang irresistibly true to the subsistence farmer
from the East who spent more time clearing his land of rocks and
stumps than plowing and harvesting. Hamlin Garland, the writer, was
the son of such a subsistence farmer, a man hounded out of Wisconsin
by trees and hills. "More and more," Garland was to remember, "[my
father] resented the stumps and ridges which interrupted his plow.
Much of his quarter section remained unbroken. There were ditches
to be dug and young oaks to be uprooted in the forest. . . . [B]itterly
he resented his uptilted, horse-killing fields, and his complaining
words sank so deep in the minds of his sons that for years thereafter
they were unable to look upon any rise of ground as an object to be
admired."

The Irish potato famine, a bad drought in the Ohio Valley, the
reflexive restlessness which, Alexis de Tocqueville thought, set Americans apart from the Europeans they had recently been—all of these,
too, were behind the flood. When Hamlin Garland's family settled in
Iowa, they had no neighbors within sight. A year later, they were
surrounded, fencepost to fencepost. "All the wild things died or hurried
away, never to return," wrote Garland mournfully. "The tender plants,
the sweet flowers, the fragrant fruits, the busy insects . . . prairie

wolves [that] lurked in the grass and swales . . . all of the swarming
lives which had been native here for countless centuries were utterly
destroyed." If poor immigrants arrived in Iowa and found land too
expensive, they could either return East and look for some hardscrab-
ble farm they could afford—in West Virginia, perhaps, or New Hampshire—or continue on to Nebraska. Since rain was bound to follow
the plow, they went to Nebraska. Merchants in St. Louis and other
railhead cities, who dreamed of markets expanding in three directions
at once, became cheerleaders for the New Meteorology. So did land
speculators, who figured that even if it was nonsense, they could buy
out the burned-out homesteaders for a pittance and convert their farms
to rangeland. But nothing did away with the Great American Desert
quite as effectively as the railroads.

In 1867, the Kansas Pacific did not reach the Pacific—few of the railroads which veiled themselves in oceanic mists ever did—but it did
reach as far as Abilene, Kansas. The Atchison, Topeka, and Santa Fe
Railroad was already to La Junta, Colorado, and branching south to
Santa Fe. The Union Pacific made Cheyenne, and two years later it
met the Central Pacific at Promontory, Utah, spanning the continent.
The Southern Pacific linked Texas to San Francisco. The Northern
Pacific hitched Montana to Duluth. The initial result of such unparalleled expansion was an ocean of debt. The federal government had
arranged the loans, but what was a loan worth if you didn't see how
you could raise the income to pay it back? Of course, there was a way
for the government to help with that problem: after all, it did own
plenty of land.
    During the four decades following the Civil War, 183 million acres
went out of the public domain into railroad ownership. To call it a
bonanza is to understate the matter significantly. The railroad land
grants were a gift the size of California plus the major part of Montana.
The deeded lands usually paralleled the railroad's track; reproduced
on maps, they resembled jet streams flowing in reverse. Anyone who
bought land from the railroads would be utterly dependent on them
for getting his harvests to eastern markets and receiving supplies in
return. When the time came to set rates, the railroads could charge
pretty much what they pleased. But first they had to seduce the settlers
who were still content to battle stumps in Kentucky or endure peonage
in Germany and Ireland. J. J. Hill, the founder of the Great Northern,
said as much himself. "You can lay track through the Garden of Eden,"
he told an acquaintance. "But why bother if the only inhabitants are

Adam and Eve?" The upswing in precipitation, and the crypto-science that explained it, were exactly what was needed. From there it became a job for advertising.

The creative juices flowed. A publicist working for the Rio Grande and Western Railroad noticed, while gazing at a map of the territory of Deseret—now Utah—a faint resemblance to the cradle of civilization. The Rio Grande and Western promptly published a map of Deseret that contained an inset map of Palestine ("The Promised Land!"), calling attention to their "striking similarity." "Follow prairie dogs and Mormons," went a pamphlet of the Burlington line, "and you will find good land." (It failed to mention that prairie dogs, which build their homes underground, cannot do so in wet or soggy ground, and therefore loathe any place receiving a decent amount of rain.) A Northern Pacific circular proclaimed, with no evident sense of shame, that not a single case of illness had been recorded in Montana during the previous year, except for indigestion caused by overeating.

Many of the railroads published their own newspapers, full of so-called testimonials from alleged Kansas farmers who were raising a hundred bushels of corn to the acre, from settlers who had traded rags for riches in five years. "Why emigrate to Kansas?" asked a testimonial in *Western Trail*, the Rock Island Railroad's gazette. "Because it is the garden spot of the world. Because it will grow anything that any other country will grow, and with less work. Because it rains here more than in any other place, and at just the right time." The railroads were careful to conceal their ties with the land-sales companies they owned, and with the journalists to whom they gave free passage and free meals, if not paychecks. One such journalist, Frederick Goddard, produced a popular publication entitled *Where to Emigrate and Why*. The Laramie Plains of Wyoming, he said, were a good place, "as ready today for the plow and spade as the fertile prairies of Illinois." (The Laramie Plains are five thousand feet higher than Illinois; the growing season is at least fifty days shorter; there is about a third as much rain.) Western Nebraska was also a delight. A few patches of drift sand, perhaps, but calling it a desert was preposterous. By drift sand, Goddard may have meant the Sand Hills, a fifteen-thousand-square-mile expanse of thirsty dunes which, to this day, remains mostly uninhabited and unfarmed.

"The utmost care has been exercised to admit nothing . . . that cannot be depended upon as correct." "All claims may be fully sustained, upon investigation." "If hard work doesn't agree with you, or you can't get on without luxuries, stay where you are. If you don't have enough capital to equip and stock a farm, if you are susceptible to homesickness, if you do not have pluck and perseverance, stay where you are." At a time when a five-course dinner in a fancy restaurant cost $1.25, the Union Pacific and the Burlington spent $1 million on advertising for Nebraska alone. Even so, sooner or later the railroads were bound to run out of settlers—long before they ran out of land. Then it became a problem of moving the more intrepid ones westward so that others could fill their places. The strategy used most often had to do with the effects of western climate on health. In 1871, the Union Pacific described the climate throughout eastern Kansas as "genial and healthy." With irresistible logic, the railroad asked, "What doth it profit a man to buy a farm . . . if he and his family lose their health?" That was enough to bring pioneers from the malarial swamps of Louisiana. Eleven years later, when eastern Kansas was filling up with settlers and five million acres of Union Pacific land remained unsold at the other end of the state, the climate in eastern Kansas suddenly turned unhealthy. For their own benefit, the railroad began advising settlers to "get to the higher elevations of the state."

Meanwhile, in Europe, an enormous harvest of souls was waiting to be converted. Western railroad agents frequently showed up in port cities, where they held court under striped awnings and dazzled groups of murmuring listeners with claims they wouldn't dare utter in the States. Swedes, who seemed to have a tendency toward homesickness, were promised a free passage back to Europe if they returned to port with a small quota of relatives in tow. The steamship companies, which were having trouble filling their expensive ships—partly because they had a chronic inclination to explode—were happy to cooperate. When a new ship docked in New York harbor, the mob of land-sales agents rushing aboard was like a migration in reverse. The terms of sale— 10 percent down, 7 percent interest, interest alone required for the first three years—could have been regarded as usurious, since deflation was the chronic economic ailment of the time. But terms like this were not to be found in Europe. Neither, for that matter, was land.

The number-one allies of the railroads in their efforts to bring settlers to the West were the politicians, newspaper editors, and territorial jingoists who were already there. No one excelled William Gilpin in this role. Gilpin, who had been a member of John C. Frémont's expedition to Oregon in 1843, was the prototypical nineteenth-century Renaissance man of the American West: soldier, philosopher, orator, lawyer, geographer, governor, author, windbag, and booby. In an essay—"Geopolitics with Dew on It"—published in *Harper's* magazine in 1943, Bernard DeVoto called Gilpin's thinking typical of what passed, in nineteenth-century America, for science: "a priori, deduced, generalized, falsely systematized, and therefore wrong." He might have added "dotty." Imagining himself in space, Gilpin saw the North American continent as a "vast amphitheater, opening toward heaven"—an enormous continent-wide bowl formed by the Rockies and the Appalachian ridges which was ready, as far as Gilpin was concerned, "to receive and fuse harmoniously whatever enters within its rim." A capitalist-expansionist mystic as only the nineteenth century could offer up, Gilpin thundered to a meeting of the Fenian Brotherhood in Denver, "What an immense geography has been revealed! What infinite hives of population and laboratories of industry have been set in motion! . . . North America is known to our own people. Its concave form and homogeneous structure are revealed."

The hives of population of which Gilpin spoke were the 1,310,000,000 people who, he was convinced, could fit comfortably within his continental bowl—and because they *could* fit, then it was weakness of will to settle for anything less. Obviously, a desert had no place in such a galvanic vision. "The PLAINS are not *deserts*," Gilpin shouted in one of his books, which was modestly titled *The Continental Railway, Compacting and Fusing Together All the World's Continents*, "but the OPPOSITE, and the cardinal basis for the future empire now erecting itself upon the North American continent." Empire was a passion with Gilpin, as it was with his mentor, Senator Thomas Hart Benton of Missouri. Benton, in addition to being the father of John C. Frémont's wife, was the father of Manifest Destiny, which was to become the rationalization for those excesses that its companion doctrine, Social Darwinism, could not excuse.

While Benton sat in Missouri flogging pioneers westward, Gilpin stood in Colorado welcoming them and shrieking for more. And there was no scarcity of Bentons and Gilpins in the states between. Kansas's Board of Agriculture was reporting a statewide average of 44.17 inches of precipitation in 1888 and 43.99 inches in 1889. It has never rained that much in Kansas since. There was also a Kansas Bureau of Immigration, which announced that the climate in Kansas was, without exception, the most desirable in the United States. Summer might linger into November, and then "at the close of February we are reminded by a soft gentle breeze from the South, that winter is gone." At the same time, a story began to circulate among disillusioned settlers about a mule standing in a field of Kansas corn. It grew so hot that all the corn around him began to pop, and mistaking it for a blizzard, he froze to death.

Nebraska had its Bureau of Immigration, too, which specialized in isothermal belts. These were longitudinal and latitudinal bands within which, by natural laws, the most advanced muscular and mental development, as well as the most heroic achievements of invention and creative genius, were invariably produced. The most significant isothermal belt in America ran right through Nebraska. As evidence, you had only to look at Colorado, which was farther south and west and full of dirty Spaniards and Indians. Coloradans, of course, shrugged off this type of thing: they were busy describing their own miracles.

Capitalists, newspaper editors, lonely pioneers, local emperors of Gilpin's ilk—all had a stake in retreating deserts. But they were not the only ones. Abolitionists, for example, did, too. In the 1850s, when Kansas seemed likely to be the next state admitted to the Union, something approaching warfare broke out between those who would have made it a free state and those who would have tolerated slavery. Horace Greeley, an avowed abolitionist with considerable interest in the West, found the climate in Kansas wonderful and the rainfall abundant. In such a state, Greeley said in his influential editorials, a 160-acre homestead could produce an ample living. A plantation, of course, demanded more land—but if Kansas was full of yeoman farmers working 160-acre plots, plantations and slaves were not likely to intrude.

One hundred and sixty acres. If anything unifies the story of the American West—its past and its present, its successes and its dreadful mistakes—it is this mythical allotment of land. Its origins are found in the original Homestead Act of 1862, which settled on such an amount— a half-mile square, more often referred to as a quarter section—as the ideal acreage for a Jeffersonian utopia of small farmers. The idea was to carve millions of quarter sections out of the public domain, sell them cheaply to restless Americans and arriving immigrants, and, by letting them try to scratch a living out of them, develop the nation's resources and build up its character.

In the West, the Homestead Act had several later incarnations. The Desert Lands Act, the Timber Culture Act, and the Timber and Stone Act were the principal ones. Neither Congress nor the General Land Office, which was responsible for administering the acts, could ever comprehend that the relative success of the land program east of the Mississippi River had less to do with the perseverance of the settlers or the wisdom of the legislation than with the forgiving nature of the climate. In the East, virtually every acre received enough rainfall, except during years of extraordinary drought, to grow most anything that didn't mind the soil and the temperature. (Unlike much of the West, which suffers through months of habitual drought, the East gets precipitation year-round; in the spring and early summer, when crops need water most, much of the East is exceptionally wet.) Since the growing season, except in the extreme north, was at least five months long, even an ignorant or lazy farmer could raise *some* kind of crop.

In the West, even if you believed that the rainfall was magically increasing, you still had to contend with high altitudes (the western plains, the Snake River Valley, and most of the irrigable lands in the Great Basin would float over the tops of all but the highest Appalachian Mountains) and, as a result, chronic frost danger even in May and September. Then there were the relentless winds, hailstones bigger than oranges, tornadoes, and breathtaking thunderstorms. There were sandy lands that would not retain moisture and poorly drained lands that retained too much; there were alkaline lands that poisoned crops.

The General Land Office bureaucrats sat in Washington pretending that such conditions did not exist. Their job, as they perceived it, was to fill little squares with people. They extended no credit, provided no water, offered no services. And the permutations of the Homestead Act that found their way into the western versions of the law sometimes *added* to the farmers' burdens. Under the Timber Culture Act, for example, you had to plant one-quarter of your quarter section with trees, a stipulation inserted because it was thought that trees increased the rainfall. In West Texas, where, meteorologically speaking, all that is predictable is the wind, you would have to spend most of your time replanting your fallen-down trees. Under the Desert Lands Act, which applied to land so arid even the government realized that farming was hopeless without irrigation, you had to demonstrate "proof of irrigation" before you could own the land. Unless you owned reasonably flat land immediately adjacent to a relatively constant stream which did not, as most western rivers do for much of their length, flow in a canyon, complying with the Desert Lands Act was almost out of the question. A mutual irrigation effort by the inhabitants of a valley was, perhaps, a possibility. That was what the Mormons had done, but they were a close-knit society linked by a common faith and a history of persecution.

The members of Congress who wrote the legislation, the land office agents who doled out land, and the newspaper editors who celebrated the settlers' heroism had, in a great many cases, never laid eyes on the land or the region that enclosed it. They were unaware that in Utah, Wyoming, and Montana—to pick three of the colder and drier states—there was not a single quarter section on which a farmer could subsist, even with luck, without irrigation, because an unirrigated quarter section was enough land for about five cows. The Indians accepted things as they were; that is why they were mostly nomadic, wandering toward greener grass and fuller herds and flowing water. If whites were going to insist on living there—fixed, settled, mortgaged, fenced—the best they could do with the land was graze it. But in those three states, an economical grazing unit was, say, twenty-five hundred to five thousand acres, depending on the circumstances. To amass that much land you had to cheat—on a magnificent scale. If you didn't, you had to overgraze the land and ruin it, and many millions of acres were damaged or ruined in exactly this way. Many settlers were tasting property ownership for the first time in their lives, and all they had in common was greed.

Speculation. Water monopoly. Land monopoly. Erosion. Corruption. Catastrophe. By 1876, after several trips across the plains and through the Rocky Mountain states, John Wesley Powell was pretty well convinced that those would be the fruits of a western land policy based on wishful thinking, willfulness, and lousy science. And by then everything he predicted was happening, especially land monopoly, water monopoly, graft, and fraud.

Homesteads fronting on streams went like oranges aboard a scurvy-ridden ship. The doctrine of riparian rights, which had been unthinkingly imported from the East, made it possible to monopolize the water in a stream if you owned the land alongside it. But if the stream was anything larger than a creek, only the person who owned land upstream, where it was still small, could manage to build a dam or barrage to guarantee a summer flow; then he could divert all he wanted, leaving his downstream neighbors with a bed of dry rocks. Riparian doctrine alone, therefore, made it possible for a tiny handful of landowners to monopolize the few manageable rivers of the West. When their neighbors saw their predicament and sold out, they could monopolize the best land, too.

As for the Desert Lands Act and the Timber and Stone Act, they could not have promoted land monopoly and corruption more efficiently if they had been expressly designed for that purpose. A typical irrigation scene under the Desert Lands Act went as follows: A beneficiary hauled a hogshead of water and a witness to his barren land, dumped the water on the land, paid the witness $20, and brought him to the land office, where the witness swore he had seen the land irrigated. Then, with borrowed identification and different names, another land application was filed, and the scene was repeated. If you could pull it off six or seven times, you had yourself a ranch. Foreign sailors arriving in San Francisco were offered a few dollars, a jug of whiskey, and an evening in a whorehouse in exchange for filing a land claim under the Timber and Stone Act. Before shipping out, the sailors abdicated title; there were no restrictions on transfer of ownership. Whole redwood forests were acquired in such a manner.

Then there was the Swamplands Act, or Swamp and Overflow Act—a Desert Lands Act of the bulrushes. If there was federal land that overflowed enough so that you could traverse it at times in a flat-bottomed boat, and you promised to reclaim it (which is to say, dike and drain it), it was yours. Henry Miller, a mythical figure in the history of California land fraud, acquired a large part of his 1,090,000-acre empire under this act. According to legend, he bought himself a boat, hired some witnesses, put the boat and witnesses over county-size tracts near the San Joaquin River where it rains, on the average, about eight or nine inches a year. The land became his. The sanitized version of the story, the one told by Miller's descendants, has him benefiting more from luck than from ruse. During the winter of 1861 and 1862, most of California got three times its normal precipitation, and the usually semiarid Central Valley became a shallow sea the size of Lake Ontario. But the only difference in this version is that Miller didn't need a wagon for his boat; he still had no business acquiring hundreds of thousands of acres of the public domain, yet he managed it with ease.

One of the unforeseen results of the homestead legislation was a high rate of employment among builders of birdhouses. In most instances, you were required to display an "erected domicile" on your land. The Congress, after all, was much too smart to give people land without requiring them to live on it. In a number of instances, the erected domicile was a birdhouse, put there to satisfy a paid witness with a tender conscience. It is quite possible that the greatest opportunity offered by the homestead legislation in the West was the opportunity to earn a little honest graft. By conservative estimates, 95 percent of the final proofs under the Desert Lands Act were fraudulent. "Whole townships have been entered under this law in the interest of one person or firm," thundered Binger Hermann, a commissioner of the General Land Office, about the Timber and Stone Act. Not long afterward, Hermann himself was fired for allowing unrestricted fraud.

Mark Twain might have written it off to the human condition, but Powell, who subscribed to a more benevolent view of humanity, wrote it off to the conditions of the desert and the failure to understand them. Americans were making a Procrustean effort to turn half a continent into something they were used to. It was a doomed effort. Even worse, it was unscientific.

The document that Powell hoped would bring the country to its senses was called *A Report on the Lands of the Arid Region of the United States, with a More Detailed Account of the Lands of Utah.* Published in 1876, the volume was seven years in preparation—though Powell took time out for a second expedition down the Colorado, in 1871, and for his usual plethora of intermittent pursuits. Powell's *Report* is remarkably brief, a scant two hundred pages in all. Unlike many of his rivals, such as the bombastic Ferdinand V. Hayden, Powell was more interested in being right than in being long. But his portrait of the American West has revolutionary implications even today.

At the beginning, Powell reconfirmed his view, which he had already submitted to an unbelieving Congress, that two-fifths of the United States has a climate that generally cannot support farming without irrigation. On top of that, irrigation could reclaim only a fraction of it. "When all the waters running in the streams found in this region are conducted on the land," Powell said, "there will be but a small portion of the country redeemed, varying in the different territories perhaps from *one to three percent"* (emphasis added). Powell regarded the theory that increased rainfall accompanied human settlement as bunk, but, typically, he disposed of it in a sympathetic and felicitous way: "If it be true that increase of the water supply is due to increase in precipitation, as many have supposed, the fact is not cheering to the agriculturalist of the arid region. . . . Any sudden great change [in climate] is ephemeral, and usually such changes go in cycles, and the opposite or compensating change may reasonably be anticipated. . . . [W]e shall have to expect a speedy return to extreme aridity, in which case a large portion of the agricultural industries of these now growing up would be destroyed."

The whole problem with the Homestead Acts, Powell went on, was that they were blind to reality. In the West, a 160-acre *irrigated* farm was too *large*, while a 160-acre *unirrigated* farm was too *small.* Most western valley soil was fertile, and a good crop was a near certainty once irrigation water was applied; in the milder regions the growing season was very long and two crops were possible, so one could often subsist on eighty irrigated acres or less. That, in fact, was about all the irrigated land one family could be expected to work. Remove the irrigation water, however, and things were drastically different. Then even a whole section was too small a piece of land. Under most circumstances, Powell claimed, no one could make a living through dryland ranching on fewer than 2,560 acres—four full sections. And even with that much land, a settler's prospects would be dicey in times of drought, because the land might lie utterly bare. Therefore, every pasturage farm should ideally have a water right sufficient to irrigate twenty acres or so during emergencies.

Having thrown over the preeminent myths about agriculture in the American West, Powell went on to the truly revolutionary part of his report. Under riparian water law, to give everyone a water right for twenty irrigated acres was impossible if you gave everyone a neat little square of land. Some squares would contain much greater stream footage than others, and their owners would have too much water compared with the others. The property boundaries would therefore have to be gerrymandered to give everyone a sufficient piece of the stream. That was one way you could help avert the monopolization of water. Another way was to insist that people *use* their water rights, not hold on to them in the hope that cities would grow up and one could make a killing someday selling water to them. An unused water right should revert—let us say after five years—to the public trust so someone else could claim it.

Doing all this, Powell reasoned, might help assure that water would be used equitably, but not necessarily efficiently. Ideally, to get through drier months and times of drought, you needed a reservoir in a good location—at a low altitude, and on the main branch of a stream. That way you could get more efficient storage of water—a dam only twice as large, but lower down, might capture five times as much water as a smaller one upstream. Also, you could then irrigate the lower valley lands, which usually have better soil and a longer growing season. In any event, an on-stream storage reservoir was, from the point of view of irrigation, preferable to small shallow ponds filled with diverted streamwater, the typical irrigation reservoirs of his day; the ponds evaporated much greater amounts of water and displaced valuable cropland.

But who, Powell asked, was building on-stream reservoirs? Practically no one. Homesteaders couldn't build them at all, let alone build them right, nor could groups of homesteaders—unless perhaps they were Mormons. Such dams required amounts of capital and commitment that were beyond the limits of aggregations of self-interested mortals. Private companies probably couldn't build good irrigation projects, either, nor even states. Sooner or later, the federal government would have to get into the irrigation business or watch its efforts to settle the West degenerate into failure and chaos. Once it realized that, it would have to undertake a careful survey of the soil characteristics so as not to waste a lot of money irrigating inferior land with drainage problems. And (he implied rather than stated) the government ought to put J. W. Powell in charge; the General Land Office, which would otherwise be responsible, was, as anyone could see, "a gigantic illustration of the evils of badly directed scientific work."

Having gone this far, Powell figured he might as well go the whole route. Fences, for example, bothered him. What was the sense of every rancher enclosing his land with a barbed-wire fence? Fenced lands tended to be unevenly grazed, and fences were obvious hazards to cattle in winter storms. Fencing was also a waste of time and money, especially in a region where rainfall could skid from twenty to six inches in successive years and someone was lucky to survive at all, let alone survive while constantly repairing and replacing fences. Individually fenced lands were a waste of resources, too; it takes a lot more tin, Powell reasoned, to make five eight-ounce cans than to make one forty-ounce can. The sensible thing was for farms to be clustered together and the individually owned lands treated as a commons, an *ejido*, with a single fence around the perimeter.

States bothered Powell, too. Their borders were too often nonsensical. They followed rivers for convenience, then struck out in a straight line, bisecting mountain ranges, cutting watersheds in half. Boxing out landscapes, sneering at natural reality, they were wholly arbitrary and, therefore, stupid. In the West, where the one thing that really mattered was water, states should logically be formed around watersheds. Each major river, from the glacial drip at its headwaters to the delta at its mouth, should be a state or semistate. The great state of Upper Platte River. Will the Senator from the state of Rio Grande yield? To divide the West any other way was to sow the future with rivalries, jealousies, and bitter squabbles whose fruits would contribute solely to the nourishment of lawyers.

While Powell knew that his plan for settling the American West would be considered revolutionary, he saw a precedent. After all, what was the difference between a cooperative irrigation district and a New England barn-raising? One was informal, the other organized and legalized, but otherwise they were the same thing. Communal pasturelands might be a gross affront to America's preoccupation with private property rights, but they were common in Europe. In the East, where inland navigation was as important as irrigation was in the West, you already had a strong federal presence in the Corps of Engineers. If anything was revolutionary, it was trying to graft English common law and the principles and habits of wet-zone agriculture onto a desert landscape. There was not a desert civilization in the world where that had been tried—and most of those civilizations had withered even after following sensible rules.

Powell was advocating cooperation, reason, science, an equitable sharing of the natural wealth, and—implicitly if not explicitly—a return to the Jeffersonian ideal. He wanted the West settled slowly, cautiously, in a manner that would work. If it was done intelligently instead of in a mad, unplanned rush, the settlement of the West could help defuse the dangerous conditions building in the squalid industrial cities of the East. If it was done wrong, the migration west might go right into reverse.

The nation at large, however, was in no mood for any such thing. It was avid for imperial expansion, and the majority of its citizens wanted to get rich. New immigrants were arriving, dozens of boatloads a day, with that motive burning in their brains. To them America was not so much a democratic utopia as a gold mine. If monopolists reigned here, they could accept that; someday *they* would be monopolists, too. Forty years earlier, Alexis de Tocqueville had captured the raw new country's soul: "To clear, to till, and to transform the vast uninhabited continent which is his domain, the American requires the daily support of an energetic passion; that passion can only be the love of wealth; the passion for wealth is therefore not reprobated in America, and, provided it does not go beyond the bounds assigned to it for public security, it is held in honor." In Powell's day, that passion for wealth had if anything grown more intense. A pseudoscientific dogma, Social Darwinism, had been invented to give predatory behavior a good name. Darwin could not be taught in the schools; but a perversion of Darwin could be practiced in real life.

The unpeopled West, naturally, was where a great many immigrants hoped to find their fortunes. They didn't want to hear that the West was dry. Few had ever seen a desert, and the East was so much like Europe that they imagined the West would be, too. A tiny bit semiarid, perhaps, like Italy. But a desert? Never! They didn't want to hear of communal paturelands—they had left those behind, in Europe, in order that they could become the emperors of Wyoming. They didn't want the federal government parceling out water and otherwise meddling in their affairs; that was another European tradition they had left an ocean away. Agricultural fortunes were being made in California by rampant capitalists like Henry Miller, acreages the size of European principalities were being amassed in Texas, in Montana. If the federal government controlled the water, it could also control the land, and then the United States might become a nation of small farmers after all—which was exactly what most Americans *didn't* want. For this was the late nineteenth century, when, as Henry Adams wrote, "the majority at last declared itself, once and for all, in favor of the capitalistic system with all its necessary machinery . . . the whole mechanical consolidation of force . . . ruthlessly . . . created monopolies capable of controlling the new energies that America adored."

It was bad enough for Powell that he was pulling against such a social tide. He also had to deal with the likes of William Gilpin, who had traded his soapbox for the governor's mansion in Denver; he had to fight with the provincial newspapers, the railroads, and all the others who were already there and had a proprietary interest in banishing the Great American Desert; he had to deal with western members of Congress who could not abide anyone calling their states arid (although a hundred years later, when the Bureau of Reclamation had become their prime benefactor, members of Congress from these same states would argue at length over whose state was the *more* arid and hostile).

Powell seemed at first to have everything going in his favor. The West was coming hard up against reality, as more hundreds of thousands of settlers ventured each year into the land of little rain. His exploits on the Colorado River had made him a national hero, the most celebrated adventurer since Lewis and Clark. He was on friendly if not intimate terms with a wide cross-section of the nation's elite—everyone from Henry Adams to Othniel C. Marsh, the great paleontologist, to Carl Schurz, the Interior Secretary, to Clarence King, the country's foremost geologist, to numerous strategically placed members of Congress. By 1881, he was head of both the Bureau of Ethnology and the Geologic Survey, two prestigious appointments that made him probably the most powerful, if not the most influential, scientist in America. But none of this prestige and power, none of these connections, was a match for ignorance, nonsense, and the nineteenth century's fulsome, quixotic optimism. When he testified before Congress about his report and his irrigation plan, the reception from the West—the region with which he was passionately involved, the region he wanted to *help*—was icily hostile. In his biography of Powell, Wallace Stegner nicely characterized the frame of mind of the typical western booster-politician when he surveyed Powell's austere, uncompromising monument of facts:

> What, they asked, did he know about the West? What did he know about South Dakota? Had he ever been there? When? Where? For how long? Did he know the average rainfall of the

James River Valley? Or the Black Hills? . . . [Did he] really know anything about the irrigable lands in the Three Forks country in Montana? They refused to understand his distinction between arid and subhumid, they clamored to know how their states had got labelled "arid" and thus been closed to settlement. . . . [W]hat about the artesian basin in the Dakotas? What about irrigation from that source? So he gave it to them: artesian wells were and always would be a minor source of water as compared to the rivers and the storm-water reservoirs. He had had his men studying artesian wells since 1882. . . . If all the wells in the Dakotas could be gathered into one county they would not irrigate that county.

Senator Moody [of South Dakota] thereupon remarked that he did not favor putting money into Major Powell's hands when Powell would clearly not spend it as Moody and his constituents wanted it spent. We ask you, he said in effect, your opinion of artesian wells. You think they're unimportant. All right, the hell with you. We'll ask somebody else who will give us the answer we want. Nothing personal.

The result, in the end, was that Powell got some money to conduct his Irrigation Survey for a couple of years—far less than he wanted, and needed—and then found himself frozen permanently out of the appropriations bills. The excuse was that he was moving too slowly, too deliberately; the truth was that he was forming opinions the West couldn't bear to hear. There was inexhaustible land but far too little water, and what little water there was might, in many cases, be too expensive to move. Having said this, held to it, and suffered for it, Powell spent his last years in a kind of ignominy. Unable to participate in the settlement of the West, he retreated into the Bureau of Ethnology, where his efforts, ironically, helped prevent the culture of the West's original inhabitants from being utterly trampled and eradicated by that same settlement. On September 23, 1902, he died at the family compound near Haven, Maine, about as far from the arid West as he could get.

Powell had felt that the western farmers would stand behind him, if not the politicians themselves; there he made one of the major miscalculations of his life. "Apparently he underestimated the capacity of the plains dirt farmer to continue to believe in myths even while his nose was being rubbed in unpleasant fact," Stegner wrote. "The press and a good part of the public in the West was against him more than he knew. . . . The American yeoman might clamor for government assistance in his trouble, but he didn't want any that would make him change his thinking."

What is remarkable, a hundred years later, is how little has changed. The disaster that Powell predicted—a catastrophic return to a cycle of drought—did indeed occur, not once but twice: in the late 1800s and again in the 1930s. When that happened, Powell's ideas—at least his insistence that a federal irrigation program was the only salvation of the arid West—were embraced, tentatively at first, then more passionately, then with a kind of desperate insistence. The result was a half-century rampage of dam-building and irrigation development which, in all probability, went far beyond anything Powell would have liked. But even as the myth of the welcoming, bountiful West was shattered, the myth of the independent yeoman farmer remained intact. With huge dams built for him at public expense, and irrigation canals, and the water sold for a quarter of a cent per ton— a price which guaranteed that little of the public's investment would ever be paid back—the West's yeoman farmer became the embodiment of the welfare state, though he was the last to recognize it. And the same Congress which had once insisted he didn't need federal help was now insisting that such help be continued, at any cost. Released from a need for justification, released from logic itself, the irrigation program Powell had wanted became a monster, redoubling its efforts and increasing its wreckage, both natural and economic, as it lost sight of its goal. Powell's ideal was a future in which the rivers of the American West would help create a limited bounty on that tiny fraction of the land which it made sense to irrigate. It is hard to imagine that the first explorer of the Colorado River would have welcomed a future in which there might be no rivers left at all.

CHAPTER TWO

The Red Queen

W hile Los Angeles moldered, San Francisco grew and grew. The city owned a superb natural harbor—the best on the Pacific Coast, one of the best in the world. When gold was struck in the Sierra Nevada foothills, 150 miles across the Central Valley, San Francisco became the principal destination of the fortune seekers of the world. The names of the camps suggested the potency of the lure: New York-of-the-Pacific, Bunker Hill, Chinese Camp, German Bar, Georgia Slide, Nigger Hill, Dutch Corral, Irish Creek, Malay Camp, French Bar, Italian Bar. Those who found their fortunes were inclined to part with them in the nearest haven of pleasure, which was San Francisco. Those who did not discovered that they could do just as well providing the opportunities. With oranges going for $2 apiece at the mines, and a plate of fresh oysters for $20 or more, it was a bonanza for all concerned.

In 1848, the population of San Francisco was eight hundred; three years later, thirty-five thousand people lived there. In 1853 the population went past fifty thousand and San Francisco became one of the twenty largest cities in the United States. By 1869, San Francisco possessed one of the busiest ports in the world, a huge fishing fleet, and the western terminus of the transcontinental railroad. It teemed with mansions, restaurants, hotels, theaters, and whorehouses. In finance it was the rival of New York, in culture the rival of Boston; in spirit it had no competitor.

Los Angeles, meanwhile, remained a torpid, suppurating, stunted little slum. It was too far from the gold fields to receive many fortune seekers on their way in or to detach them from their fortunes on the way out. It sat forlornly in the middle of an arid coastal basin, lacking both a port and a railroad. During most of the year, its water source, the Los Angeles River, was a smallish creek in a large bed; during the few winter weeks when it was not—when supersaturated tropical weather fronts crashed into the mountains ringing the basin—the bed could not begin to contain it, and the river floated neighborhoods out to sea. (For many years, Santa Anita Canyon, near Pasadena, held the United States record for the greatest rainfall in a twenty-four-hour period, but it may be more significant to state that the twenty-six inches that fell in a day were nearly twice the amount of precipitation that Los Angeles normally receives in a year.) Had humans never settled in Los Angeles, evolution, left to its own devices, might have created in a million more years the ideal creature for the habitat: a camel with gills.

The Spanish had actually settled Los Angeles long before they ever saw the Golden Gate. It was more convenient to Mexico and, from an irrigation farmer's point of view, it was a more promising place to live. By 1848, the town had a population of sixteen hundred, half Spanish and half Indian, with a small sprinkling of Yankees, and was twice the size of San Francisco. A decade later, however, San Francisco had grown ten times as large as Los Angeles. By the end of the Civil War, when San Francisco was the Babylon of the American frontier, Los Angeles was a filthy pueblo of thirteen thousand, a beach for human flotsam washed across the continent on the blood tide of the war. One of the town's early pioneers, a farm boy whose family had emigrated from Iowa, described it as a "vile little dump . . . debauched . . . degenerate . . . vicious."

If anything could be said to have saved Los Angeles it was its reputation as a haven from persecution, a place where one could lose oneself. Since the ranks of the persecuted include those who are too virtuous for their fellow citizens, as well as those who are not virtuous enough, sooner or later the city was bound to attract the victims of mobocracy. And the most persecuted among the virtuous in nineteenth-century America were, besides peaceful Indians and runaway slaves and Mennonites and Quakers, the members of the Mormon faith.

After fleeing Illinois for Utah, the Mormons had always been obsessed with finding escape routes to the sea. The first irrigation canals were still being dug beside the Wasatch Range when Brigham Young dispatched a party of his most loyal disciples, in 1851, to follow Jedediah Smith's old route to the coast. When they crossed the San Bernardino Mountains, they found themselves in a huge arid basin that reminded them of home and was only a day or two from the sea. The streams were less reliable than those in Utah—the southern mountains received a scantier snowpack that never lasted halfway through the summer—but the San Bernardinos got decent winter rain, and artesian wells below them flowed like geysers. With money earned by selling food and supplies at usurious prices to adventurers bound through Utah for the gold fields, the Mormons purchased a huge chunk of land from an old Spanish rancho. The soil was good, the climate was ideal, and no one was better at irrigation farming than Mormons. Before long they were supplying much of the basin with food. In 1857, the U.S. Cavalry marched on Utah and Brigham Young ordered all distant settlements abandoned, but the Mormons' achievement had left its mark. A Presbyterian colony was soon established nearby, then a Quaker colony, then an ethnic colony of Germans. In this freakish climate—semitropical but dry, ocean-cooled but lavishly sunny—you could grow almost anything. Corn and cabbages sprouted next to oranges, avocados, artichokes, and dates. The capitalists of San Francisco did not remain oblivious; the Southern Pacific ran a spur line to Los Angeles in 1867, finally linking it to the rest of the world. On this same line, huge San Bernardino Valencias found their way to the 1884 World's Fair in New Orleans, where they attracted crowds. No one could imagine *oranges* grown in the western United States. It was then and there, more or less, that the phenomenon of modern Los Angeles began.

They came by ship, they came by wagon, they came by horse. They came on foot, dragging everything they could in a handcart, but the real hordes came by train. In 1885, the Atchison, Topeka, and Santa Fe Railroad linked Los Angeles directly with Kansas City, precipitating a fare war with the Southern Pacific. Within a year, the cost of passage from Chicago had dropped from $100 to $25. During brief periods of mad competition, you could cross two-thirds of the continent for a dollar. If you were asthmatic, tubercular, arthritic, restless, ambitious, or lazy—categories that pretty well accounted for Los Angeles' first flood of arrivals—the fares were too cheap to pass up. Out came Dakota farmers who despaired at the meager profits they made growing wheat. *You could grow oranges.* Out came Civil War veterans looking for an easy life, failures looking for another chance, and the usual boom-town complement of the slick, the sharp, and the ruthless.

The first boom began in the early 1880s and culminated in 1889, when the town transacted $100 million worth of real estate—in today's economy, a $2 billion year in Idaho Falls. Fraud was epic. Hundreds of unseen, paid-for lots were situated in the bed of the Los Angeles River, or up the nine-thousand-foot summits of the San Gabriel Range. The boom was, predictably, short-lived. In 1889, a bank president, a newspaper publisher, and the town's most popular minister all fled to Mexico to spare themselves jail terms, and a dozen or more victims took their own lives. By 1892, the population had dropped by almost one-half, but the bust was followed quickly by an oil boom, and enough fortunes were being made (the original Beverly Hillbillies were *from* Beverly Hills, then a patch of jackrabbit scrub overlying an oil basin) to pack the arriving trains again. Los Angeles soon drew close to San Francisco in population and was crowing with glee. "The 'busting of the boom' became but a little eddy in the great stream," enthused the Los Angeles *Times*, "the intermission of one heartbeat in the life of . . . the most charming land on the footstool of the Most High . . . the most beautiful city inhabited by the human family." Only one thing stood in the way of what looked as if it might become the most startling rise to prominence of any city in history—the scarcity of water.

The motives that brought Harrison Gray Otis, Harry Chandler, and William Mulholland to Los Angeles were the same that would eventually bring millions there. Otis came because he had been an incontrovertible, if not quite an ignominious, failure. He was born in Marietta, Ohio, and as a young man held a series of unspectacular jobs—a clerk for the Ohio legislature, a foreman at a printing plant, an editor of a veterans' magazine. His one early taste of glory came during the Civil War, in which he fought on the Union side, acquired several wounds and decorations, and ultimately rose to the rank of captain. *Captain* Harrison Gray Otis. He liked the title well enough to think himself deserving of a sinecure, and after the war he drifted out to California in search of one. What he got was an appointment as government agent on the Seal Islands, some frigid, treeless, wind-blasted humps of rock in the Bering Sea. His chief duty there was to prevent the poaching of walrus and seals, an assignment that suited Otis better than he knew, since he bore an odd resemblance to the former and had a disposition to match. He was a large blubbery man with an intransigent scowl, an Otto von Bismarck mustache and a goatee, and a chronic inability to communicate in tones quieter than a yell, whether he was debating the American role in the Pacific or telling someone to pass the salt. "He is a damned cuss who doesn't seem to feel well unless he is in a row with someone," one among his legion of enemies would later remark.

The Seal Islands post was a humiliation that Otis, who was more ambitious than he was clever, couldn't afford to pass up. But after three years he had had enough, and he returned, bilious and frustrated, to California, where he got a job as editor of a local newspaper in Santa Barbara. Otis hated Santa Barbara. It was a hangout of the privileged classes, smug, snobbish, and perfectly content to remain small. Otis despised inherited wealth and class, but he despised a town that was disdainful of growth even more. He believed in it, perfervidly, just as he believed in those who started with nothing and dynamite their way to success. "Hustlers . . . men of brain, brawn, and guts" were the people he admired most, even if he had less in common with them than he thought. Otis would pursue a sinecure as a greyhound chases a rabbit, and it was his rotten luck at it, more than anything else, that finally caused his success. Trying to get himself appointed marshal of California, he was offered the job of consul in Tientsin, an insult that was more than he could bear. In 1881, Otis quit the paper in Santa Barbara and moved his family to Los Angeles.

The city was still small when Otis arrived, but it was already served by several newspapers, one of which, the *Times and Mirror*, was owned by a small-time eastern financier named H. H. Boyce. Boyce was looking for a new editor, and, though the pay was a miserable $15 a week, Otis took the job. Perhaps because he was fuming about the pay, or perhaps because he knew that time was running out, Captain Otis then made one of the bolder decisions of his life. He took all of his savings and, to help offset the low pay, convinced Boyce to let him purchase a share in the newspaper. Privately he was thinking that someday, perhaps, he could force H. H. Boyce out.

Harry Chandler came to Los Angeles for his health. He grew up in New Hampshire, a cherubic child with cheeks like Freestone peaches. His falsely benign appearance, which stayed with him all his life, made him a popular boy model among advertisers and photographers. But cherubic Harry was a rugged individualist and a ferocious competitor, and if there was money involved he would rarely pass up an opportunity or a dare. While at Dartmouth College, he accepted someone's challenge and dove into a vat of starch—a display that nearly ruined his lungs. Advised by doctors to recuperate in a warm and dry climate, he bought a ticket to Los Angeles. Arriving there, he moved from flophouse to flophouse because none of his fellow tenants could endure his hacking cough. When he was thoroughly friendless and nearly destitute, Harry met a sympathetic doctor who suffered from tuberculosis and owned an irrigated orchard near Cahuenga Pass, at the head of the San Fernando Valley. Would Harry like a job picking fruit?

The work was hard but invigorating. Before long, Harry felt almost cured. The work was also surprisingly lucrative. The doctor was as uninterested in money as Harry was interested, and let him sell a large share of what he picked. In his first year, Harry made $3,000. It was a small fortune, and inspired in Harry an awed faith in the potential of irrigated agriculture and, most particularly, agriculture in the San Fernando Valley. With the proceeds, Harry began to acquire newspaper circulation routes, which, at the time, were owned independently of the newspapers and bought and sold like chattel. Before long, he was a child monopolist, owning virtually all the routes in the city.

By 1886, Harrison Gray Otis had finally managed to hound H. H. Boyce out of the Los Angeles *Times and Mirror*. It was a pyrrhic victory, however, because Boyce had immediately established a rival paper, the *Tribune*, and engaged Otis in an all-out circulation war. With the allegiance of whoever dominated the circulation routes, one or the other was certain to win. It was Otis's luck that he got to Harry Chandler first. Within days, the *Tribune* began to disappear mysteriously from people's doorsteps, and its delivery boys simultaneously contracted a contagion. Meanwhile, new subscribers began to flock, like moths scenting pheromones, to the *Times*. Boyce was broken within months. Before Otis had much chance to gloat, however, he learned that the defunct *Tribune*'s printing plant had secretly acquired a new owner, whose name was Harry Chandler, and that the tactics that they had used together against Boyce could just as easily be turned against the *Times*. Otis, who bore lifelong grudges over provocations infinitely smaller than this, was realistic enough to know when he was had. Besides, this mild-appearing young man was the embodiment of every quality he admired. As a result, the *Times* acquired a new circulation manager and guiding light, whose name was Harry Chandler, and in 1894 Harry Chandler acquired a new father-in-law, whose name was Harrison Gray Otis.

William Mulholland came to Los Angeles more or less for the hell of it. He was born in 1855 in Dublin, Ireland, where his father was a postal clerk. At fifteen, he signed on as an apprentice seaman aboard a merchant ship that carried him back and forth along the Atlantic trade routes. By 1874 he had had enough, and spent a couple of years hacking about the lumber camps in Michigan and the dry-goods business in Pittsburgh, where his uncle owned a store. It was in Pittsburgh that Mulholland first read about California. He had just enough money to get to Panama by ship, and after landing in Colón, he traversed the isthmus on foot and worked his way north aboard another ship, arriving in San Francisco in the summer of 1877. Being back on a ship had renewed Mulholland's taste for the sea, and, after a brief failure at prospecting in Arizona—where he also fought Apaches for pay—he decided to ship out at San Pedro, the port nearest Los Angeles. He had ten dollars to his name. Anxious to make a little extra money, he joined a well-drilling crew. "We were down about six hundred feet when we struck a tree. A little further we got fossil remains. These things fired my curiosity. I wanted to know how they got there, so I got hold of Joseph Le Conte's book on the geology of the country. Right there I decided to become an engineer."

In his official photograph for the Los Angeles Department of Water and Power, which was taken when he was nearly fifty, Mulholland still looks young. He is wearing a short-brimmed dark fedora and a dark pinstripe suit; a luxuriant silk cravat circumnavigates a shirt collar that appears to be made of titanium; from a thick, bushy mustache sprouts a lit cigar. The face is supremely Irish: belligerence in repose, a seductive churlish charm. Once, in court, Mulholland was asked what his qualifications were to run the most far-flung urban water system in the world, and he replied, "Well, I went to school in Ireland when I was a boy, learned the Three R's and the Ten Commandments—most of them—made a pilgrimage to the Blarney Stone, received my father's blessing, and here I am." He began his engineering career in 1878 as a ditch-tender for the city's private water company, clearing weeds, stones, and brush out of a canal that ran by his house. One day Mulholland was approached by a man in a carriage who demanded to know his name and what he was doing. Mulholland stepped out of his ditch and told the man that he was doing his goddamned job and that his name was immaterial to the quality of his goddamned work. The man, it turned out, was the president of the water company. Learning this, Mulholland went to the company office to collect his pay before being fired. Instead, he was promoted.

The Sierra Nevada blocks most of the weather fronts moving across California from the Pacific, so that a place on the western slope of the range may receive eighty inches of precipitation in a year, while a place on the east slope, fifty miles away, may receive ten inches or less. The rivers draining into the Pacific from the West Slope are many and substantial, while those emptying into the Great Basin from the

East Slope are few and generally small. The Owens River is an exception. It rises southeast of Yosemite, near a gunsight pass that allows some of the weather to come barreling through, heads westward for a while, then turns abruptly south and flows through a long valley, ten to twenty miles wide, flanked on either side by the Sierra Nevada and the White Mountains, which rise ten thousand feet from the valley floor. The valley is called the Owens Valley, and the lake into which the river empties—used to empty—was called Owens Lake. Huge, turquoise, and improbable in a desert landscape, it was the shrunken remnant of a much larger lake that formed during the Ice Ages. Due to a high evaporation rate and, for its size, a modest rate of inflow, the lake was more saline than the sea, but it supported two species of life in the quadrillions: a salt-loving fly and a tiny brine shrimp. The soup of shrimp and the smog of flies attracted millions of migratory waterfowl, a food source whose startling numbers were partially responsible for inducing some of the valley's first visitors to remain. "The lake was alive with wild fowl," wrote Beveridge R. Spear, an Owens Valley pioneer. "Ducks were by the square mile, millions of them. When they rose in flight, the roar of their wings . . . could be heard . . . ten miles away. . . . Occasionally, when shot down, a duck would burst open from fatness which was butter yellow."

The greater attraction, however, was the river. When whites arrived in the 1860s, Paiute Indians who had learned irrigation from the Spanish were already diverting some of the water to raise crops. In traditional pioneer fashion, the whites trumped up some cattle-rustling charges against the Indians, which appear to have led to the murder of a white woman and a child. The pious Owens Valley citizens then murdered at least 150 Paiutes in retaliation, driving the last hundred into Owens Lake to drown. They then took over the Indians' land, borrowed their irrigation methods, and began raising alfalfa and pasture and fruit. By 1899, they had established several ditch companies and had put some forty thousand acres under cultivation.

The huge new silver camp at Tonopah, Nevada, consumed most of what the valley grew. With prosperity, several thriving towns sprang up: Bishop, Big Pine, Lone Pine, Independence. The irrigated valley was postcard-pretty, a narrow swath of green in the middle of the high desert, with 14,495-foot Mount Whitney, the highest peak between Canada and Mexico, looming over Lone Pine and the river running through. Mark Twain came to visit, and Mary Austin, who was to become a well-known writer, came to live. But the entrance that most excited the valley people was that of the United States Reclamation Service (later renamed the Bureau of Reclamation). The Service was an unparalleled experiment in federal intervention in the nation's economy, and was being watched so closely by skeptics in Congress that it could not afford to have any of its first projects fail. To Frederick Newell, the first Reclamation Commissioner, the Owens Valley looked like a place where he could almost be guaranteed success. The people were proven irrigation farmers—a rarity in the non-Mormon West; the soil could grow anything the climate would permit; the river was underused; and there was a good site for a reservoir. Sixty thousand additional acres were irrigable, and all of them could be gravity-fed. In early 1903, just a few months after the Service was created, a team of Reclamation engineers was already trooping around the valley, gauging streamflows and making soil surveys. Sixty thousand new acres would even make it worthwhile to run a railroad spur to Los Angeles. Los Angeles, everyone thought, was going to make the Owens Valley rich.

Fred Eaton thought differently. Eaton had been born in Los Angeles in 1856; his family had founded Pasadena. Most of the Eaton men were engineers, and when they looked around them it seemed that half of what they saw they had built themselves; it gave them an overpowering sense of pride-in-place. Fred had gone into hydrologic engineering, which is to say that he pretty much taught it to himself, and by the time he was twenty-seven, he was superintendent of the Los Angeles City Water Company. As San Francisco had bloomed into pseudo-Parisian splendor, Fred Eaton had chafed. When Los Angeles finally began to take on the appearance of a place with a future, he had been intensely proud. But he was one of the few people who understood that this whole promising future was an illusion. With artesian pressure still lifting fountainheads of water eight feet into the air, no one believed that someday the basin would run out of water. Few understood that the occasional big floods in the Los Angeles River were testimony to the *absence* of rain: that the basin was normally so dry there wasn't enough ground cover to hold the rain when it fell. The annual flow of the Los Angeles River (that which ran aboveground) represented only a fifth of 1 percent of the runoff of the state, and because of the pumping the flow was dropping fast, from a hundred cubic feet per second in the 1880s to forty-five cfs in 1902. If growth continued, the population and the water would fall hopelessly out of balance. Everyone was living off tens of thousands of years of accumulated groundwater, like a spendthrift heir squandering his wealth. No one knew how much groundwater lay beneath the basin or how long it could be expected to last, but it would be insane to build the region's future on it.

There was no other source of water nearby. Deserts lay on three sides of the basin, an ocean on the fourth. The nearest large rivers were the Colorado and the Kern, but to divert them out of their canyons to Los Angeles would require pumping lifts of thousands of feet—an impossibility at the time. It would also require a Herculean amount of energy.

But there was, 250 miles away, the Owens River. It might not be quite sufficient for the huge metropolis forming in Eaton's imagination, but it was large enough; there was water for at least a million people. Indeed, Eaton was one of the few Los Angeleans who knew the river even existed. Its distance from Los Angeles was staggering, but its remoteness was overshadowed by one majestically significant fact: Owens Lake, the terminus of the river, sat at an elevation of about four thousand feet. Los Angeles was a few feet above sea level. The water, carried in pressure aqueducts and siphons, could arrive under its own power. Not one watt of pumping energy would be required. The only drawback was that the city might have to take the water by theft.

During their years together at the Los Angeles City Water Company, Fred Eaton and Bill Mulholland became good friends, thriving on each other's differences. Eaton was a western patrician, smooth and diffident; Mulholland an Irish immigrant with a musician's repertoire of ribald stories and a temperament like a bear's. Eaton thought so much of Mulholland that he groomed him to be his successor, and when Eaton left the company in 1886 to pursue a career in politics and seek his fortune, Mulholland was named superintendent. In the years that followed, Fred Eaton would become messianic about the water shortage he saw approaching. The only answer, he told Mulholland, was to get the Owens River. At first, Mulholland found the idea preposterous: going 250 miles for water was out of the question, and Mulholland didn't much believe in surface-water development anyway. Damming rivers meant forming reservoirs, and in the heat and dryness of California, reservoirs would evaporate huge quantities of water. It made more sense to slow down the rainfall as it returned to the ocean and force more of it into the aquifer. Mulholland preached soil and forest conservation thirty years before its time. He wanted to seed the whole basin, and when he said that the deforestation of the mountainsides would reduce the basin's water supply, everyone thought he was slightly nuts. He had his men filling gullies and installing infiltration galleries and checkdams all over the place. Everything he did, however, was nullified by the basin's growth.

By 1900, Los Angeles' population had gone over 100,000; it doubled again within four years. During the same period, the city experienced its first severe drought. Even with lawn watering prohibited and park ponds left unfilled, the artesian pressure, as Eaton had predicted, began to drop. Gushes became gurgles, then dried up. Pumps were frantically installed. By 1904, the pressure was low enough to prompt Mulholland to begin shutting irrigation wells in the San Fernando Valley, which lay across the Hollywood Hills and fed both the aquifer and the river. The farmers were furious, and Mulholland began spending a lot of time in court. The Los Angeles City Water Company was eventually taken over by the city, and Mulholland was retained in command. (The city didn't have much choice in the matter. Mulholland was such a seat-of-the-pants engineer that the plan of the entire water system resided mainly in his head; the most elemental schematics and blueprints did not exist.) In late 1904, the newly created Los Angeles Department of Water and Power issued its first public report. "The time has come," it said, "when we shall have to supplement the supply from some other source." With that simple statement William Mulholland was about to become a modern Moses. But instead of leading his people through the waters to the promised land, he would cleave the desert and lead the promised waters to them.

There is a widely held view that Los Angeles simply went out to the Owens Valley and stole its water. In a technical sense, that isn't quite true. Everything the city did was legal (though its chief collaborator, the U.S. Forest Service, did indeed violate the law). Whether one can justify what the city did, however, is another story. Los Angeles employed chicanery, subterfuge, spies, bribery, a campaign of divide-and-conquer, and a strategy of lies to get the water it needed. In the end, it milked the valley bone-dry, impoverishing it, while the water made a number of prominent Los Angeleans very, very rich. There are those who would argue that if all of this was legal, then something is the matter with the law.

It could never have happened, perhaps, had the ingenuous citizens of the Owens Valley paid more attention to a small news item that appeared in the Inyo *Register*, the valley's largest newspaper, on September 29, 1904. The item began: "Fred Eaton, ex-mayor of Los Angeles, and Fred [sic] Mulholland, who is connected with the water system of that city, arrived a few days ago and went up to the site of the proposed government dam on the [Owens] River." The person who took them around, the story continued, was Joseph Lippincott, the regional engineer for the Reclamation Service. It wasn't so much this small piece of news that should have aroused the valley's suspicions. It was the fact that Lippincott had already taken Eaton around the valley twice before.

The valley had no particular reason to distrust J. B. Lippincott, although a search into his background would have dredged up a revelation or two. As a young man out of engineering school, he had joined John Wesley Powell's Irrigation Survey, the first abortive attempt to launch a federal reclamation program in the West, but had lost his job soon thereafter when Congress denied Powell funding. Embittered by the experience, Lippincott migrated to Los Angeles, where, by the mid-1890s, he had built up a lucrative practice as a consulting engineer. In 1902, when the Reclamation Service was finally created, its first commissioner, Frederick Newell, immediately thought of Lippincott as the person to launch its California program. He had a good reputation, and he understood irrigation—a science few engineers were familiar with. The post, however, meant a substantial cut in salary, and Lippincott insisted on being allowed to maintain a part-time engineering practice on the side. Newell and his deputy, Arthur Powell Davis (who was John Wesley Powell's nephew), were a little wary; in a fast-growing region with little water, a district engineer with divided loyalties could lead the Service into a thicket of conflict-of-interest entanglements. The centerpiece of the Service's program in California was to be the Owens Valley Project, and there were already rumors that Los Angeles coveted the valley's water. One of the Service's engineers, in fact, had raised this issue with Davis; with Lippincott, a son of Los Angeles, in charge, a collision between the city and the Service over the Owens River might leave the city with the water and the Service absent its reputation. But the Service's early leadership, unlike those who succeeded them, suffered from a certain lack of imagination. "On the face of it," Davis scoffed, "such a project is as likely as the city of Washington tapping the Ohio River."

The only person who seemed suspicious when Lippincott began showing Eaton and Mulholland around the Owens Valley again and again was one of his own employees, a young Berkeley-educated engineer named Jacob Clausen. His apprehensions had been aroused during Eaton's second visit, when Lippincott and Eaton had ridden up to the valley from Los Angeles by way of Tioga Pass and Clausen, at Lippincott's request, had met them at Mono Lake. On the way down the valley, Lippincott insisted that they stop at the ranch of Thomas Rickey, one of the biggest landowners in the valley. Rickey's ranch was in Long Valley, an occluded shallow gorge of the Owens River, hard up against the giant Sierra massif, which contained the reservoir site the Reclamation Service would have to acquire in order for its project to be feasible. Eaton had told Clausen that he wanted to become a cattle rancher and was interested in buying Rickey's property if he was willing to sell. As they visited the ranch, however, he seemed much more interested in water than in cattle. Clausen understood the dynamics of the Owens Valley Project—the streamflows, the water rights, the interaction of ground and surface water—better than anyone, and Lippincott asked him to explain to Eaton how the project would work. Eaton hung on his every word, and that, Clausen was to testify later, "was exactly what Lippincott wanted." The two Los Angeleans were good friends, and Eaton had been the first to dream of Los Angeles going to the Owens Valley for water. Was it so farfetched, Clausen would remember thinking to himself, to believe that Lippincott was out to help Los Angeles steal the valley's water?

If Clausen's suspicions were aroused, those of his high superiors remained utterly dormant, even though they would soon have equal reason to suspect Lippincott of being a double agent for Los Angeles. In early March of 1905, Lippincott had sent his entire engineering staff to Yuma, Arizona, on the Colorado River, to move the Yuma Irrigation Project forward at a faster pace. Work on the Owens Valley Project had been held up by winter and by the delayed arrival of a piece of drilling equipment which was on order. During the hiatus, the Reclamation Service received a couple of applications for rights-of-way across federal lands from two newly formed power companies in the Owens Valley. Each was interested in building a hydroelectric project, and Lippincott had to decide which, if any, of the plans could coexist with the Reclamation project. Unable or unwilling to look into the matter himself, Lippincott might have waited for one of his engineers to return later in the spring, but he wanted to dispose of the issue, so he decided to appoint a consulting engineer to look into the matter for him. And though there were dozens of engineers in Los Angeles and San Francisco among whom he could have chosen, he decided to turn to his old friend and professional associate Fred Eaton.

The news that Lippincott had hired Fred Eaton to decide on a matter that could affect the whole Owens Valley Project left his superiors stunned, but their response, typically, was one of bafflement rather than anger. "I fail to understand in what capacity he is acting" was the only response Arthur Davis managed to give.

Eaton himself had no questions about the capacity in which he was acting, though the public face he presented was very different.

With his letter of introduction from Lippincott and an armload of freshly minted Reclamation maps, he strode into the government land office in Independence, claiming to represent the Service on a matter of vital importance to the Owens Valley Project. For the first three days, however, his investigations had nothing to do with the hydroelectric plans. Poring over land deeds in the office's files—deeds to which he might have had no access as a private citizen—Eaton jotted down a wealth of information on ownership, water rights, stream flows—things Los Angeles had to know if and when it decided to move on the Owens Valley's water. Handsome and charming, Eaton even managed to get the land office employees to help him, unaware that the information they were digging out had nothing to do with the matter that had allegedly brought Eaton there. When he finally had what he felt he needed, he turned to the official matter at hand.

The problem of the conflicting power-license applications was straightforward; there could only be one resolution. One of the two power companies, the Owens River Water and Power Company, held water rights senior to those of its competitor, the Nevada Power Mining and Milling Company. Its rights even predated those of the Reclamation Service, and if it was refused its application it might cause the Service some real legal embarrassment. In addition, its plan of development was far more compatible with the Reclamation project than the Nevada company's; Jacob Clausen had taken a cursory look at both and decided that the Nevada company's project could reduce the Long Valley reservoir to a glorified mudflat during the peak summer irrigation season, when water was needed most. To Clausen, the applications were hardly worth a second look, and he couldn't understand why Lippincott had even bothered to hire someone to review them so carefully. The Owens River company deserved a conditional go-ahead, the Nevada company decidedly did not. But Clausen was far too naive to understand the complexity of such matters: One of the founders of, and partners in, the Nevada Power Mining and Milling Company was a rancher named Thomas B. Rickey.

Eaton's baffling recommendation in favor of the Nevada Power Mining and Milling Company threw Clausen into a state of apoplexy. When Lippincott formally endorsed his judgment a few weeks later, Clausen finally understood that something was terribly wrong, but how wrong even he could not fathom. On the 6th of March, exactly three days after Lippincott had hired Eaton as his personal representative in the matter of the power company applications, the city of Los Angeles had quietly hired its own consultant to prepare a report on the options it had in its search for water. The report had taken only a couple of weeks to prepare—most of the information was in Mulholland's office, and the conclusion was foregone anyway—and the consultant had received an absurdly grandiose commission of $2,500, more than half his annual salary. It was not so much a commission as a bribe. The money, however, was well spent: the name of the consultant was Joseph B. Lippincott.

One other person besides Jacob Clausen had begun to follow the comings and goings of Eaton, Lippincott, and Mulholland with more than detached interest—Wilfred Watterson, the president of the Inyo County Bank. Wilfred and his brother, Mark, were the most popular citizens in the Owens Valley. Their family had founded the bank, and Wilfred and Mark, when still in their twenties, became president and treasurer. Both were attractive young men, but Wilfred in particular was strikingly handsome. He had clean-cut, perfect features, an absolutely even gaze, and the erect, confident air of a nineteenth-century optimist. In his elegant clothes, Wilfred could have passed easily for Bat Masterson instead of a small-town banker. The lending policies of the Inyo County Bank were as much of an aberration as its owners. The Wattersons rarely refused a loan and often stretched out debts; they displayed a strong interest in the valley's survival and a casual, almost careless attitude toward money.

Wilfred's suspicions that Los Angeles was engineering a water grab had begun to simmer when word got around that Fred Eaton, the would-be cattle rancher, was offering some astonishingly generous sums for land with good water rights. There were stories that Eaton would make an offer that already seemed generous, and, if a landowner gambled and tried to raise him, Eaton would readily meet his terms. It was hard for Wilfred to nail any of this down, because no one wanted to let the Wattersons know that he was thinking of selling out—not after they had loaned money with such abandon up and down the valley—but the stories were enough to make Wilfred skeptical about Eaton's true intentions. Was he rich enough to pay those prices? Where did he get the money?

Watterson's suspicions became intensely aroused one day in the early summer of 1905 when an unidentified young man arrived in the valley, went directly to the Inyo Bank, and displayed a written order from Fred Eaton to pick up a parcel in a safe deposit box. As soon as he had it in his hand, the young man left with unseemly haste and stalked down the street in the direction of the post office. Watterson sprang up from his desk and asked the teller who the man was. He was Harry Lelande, the Los Angeles city clerk—the official legally charged with handling any transactions for the city that involved transfers of water or land.

Watterson burst out the door and ran down the street in the direction in which Lelande had disappeared. He found him across the street from the post office.

Watterson ambled up to Lelande, accosted him in his disarming manner, and said, "I'm sorry, Mr. Lelande, but there's a small formality we forgot to carry out at the bank."

Lelande looked perplexed. Watterson asked him to follow him back to the bank.

Once they were safely inside the president's office, Watterson offered the clerk a seat and some coffee, then walked casually to the door and locked it. "We want the deed back," he said.

Lelande looked stricken. "What deed do you mean?" he asked.

"The deed by which your city is going to try to rape this valley," Watterson answered.

"I haven't any idea what you're talking about."

"Maybe this will help," said Watterson. He opened his desk drawer, removed a revolver, and put it on top of the desk.

Lelande's mouth opened. "I can't give you something I don't have," he begged.

Watterson stood up and hovered menacingly over the clerk. "Take off your coat and trousers," he said.

Lelande, badly frightened, obliged.

Watterson turned all of his pockets inside out and found nothing. He ordered Lelande to get dressed and take him to his room at the Hotel Bishop.

"Eaton's been buying land in an underhanded way to secure water for the city of Los Angeles, hasn't he?" Watterson said to Lelande on the way over. He was inventing the theory as he walked, but Lelande's agonized expression told him he was right. "You've paid high prices not because you're dumb but because you're smart. You're masquerading as investors and all you're going to invest in is our ruin."

Lelande kept insisting that he didn't know what Watterson was talking about. At the hotel, Watterson nearly tore apart his room, but found none of the documents Lelande had extracted from Eaton's box. It was obvious that Lelande had been so fearful of being discovered that he had immediately run to the post office and mailed the deed. Without the document, Watterson had nothing to go on but his hunches, and he was forced to let Lelande go. But, his temper notwithstanding, he knew he would have had to let him go anyway; the clerk had done nothing against the law. Neither, from what he knew, had Eaton. Was it possible, Watterson asked himself, that a distant city could destroy the valley he and his family had worked so hard and gambled so recklessly to build up, and never step outside the law?

Meanwhile, the $2,500 contract accepted by Joseph Lippincott from Los Angeles was, if not exactly illegal, an apparent violation of the most basic ethical standards for government officials. Newell had let Lippincott off with another fatherly lecture, but everyone in the Reclamation Service had heard about it, and since the Service had been created as an answer to the epic graft and fraud associated with the General Land Office, some of Lippincott's associates were furious with him. By July of 1905, Newell realized that the whole thing might blow up in his face; he had to do something to contain the damage. As a result, he decided to appoint a panel of engineers to review the conflict between the Reclamation project and the water needs of Los Angeles and decide whether the Owens Valley Project should move forward, be put on hold, or be abandoned. Newell felt that Lippincott, as the senior engineer most familiar with the project, should sit on the panel. To his and Lippincott's astonishment, several Reclamation engineers said they would refuse to sit next to him. Lippincott now realized that he, too, would have to mount a damage-control operation in a hurry. On July 26, the night before the panel was scheduled to convene, he dashed off a telegram to Eaton that read, "Reported to me and publicly accepted that you had represented yourself as connected with Reclamation and acting as my agent in Owens Valley. As this is entirely erroneous and very embarrassing to me, please deny publicly or the Service will be forced to do so." The truth of Lippincott's denial can best be judged by Fred Eaton's reaction, which was incendiary. He received the telegram in the federal land office in Independence, where he was still trying to masquerade as Lippincott's agent. After reading it he felt compelled to vent his spleen on the nearest person available, agent Richard Fysh. "Eaton said he had a telegram from Mr. Lippincott and it was a damned hot one," said Fysh later in a deposition, "and he, Eaton, did not like it a little bit, as it put him in a wrong light."

Newell's panel of engineers was convened in San Francisco on July 27. After two days of hearing divided opinions (Clausen testified in favor of continuing, Lippincott in favor of abandonment), the panel reached a unanimous verdict. The Owens Valley Project should not be sedulously pursued, they recommended; the needs of Los Angeles had become too great an issue. But neither should it be formally abandoned until a more persuasive case could be made for doing so. Los Angeles would have to demonstrate that it had absolutely no choice but to go to the valley for water, and it would have to prove that it had the resources to carry out such a gigantic undertaking on its own. Such a recommendation, the panel added, was of course based on the assumption that the Reclamation project was still feasible.

Which, unbeknownst to anyone but Eaton and a select handful of Los Angeles officials, it was not. Four months earlier, after completing his consultant duties for Lippincott, Eaton had gone back to see the stubborn Thomas Rickey, who held the key piece of land in the valley—the land the city had to have in order to block the federal project—but who had refused to sell. In Eaton's hand was his recommendation that Rickey's hydroelectric company be allowed to usurp its competitor's claim on the main power sites on the river. That, Eaton thought, was the sweetener that would surely make Rickey sell. After hours of pleading and cajoling, however, the rancher still held out. In disgust, Eaton finally stood up, roughly shook Rickey's hand, and stomped out the door. As he was standing at the railroad depot, waiting for the train that would take him back to Los Angeles, Rickey raced up in his carriage. He had had a sudden change of heart; for $450,000, he told Eaton, he would sell him an option clear on the ranch, including the Long Valley reservoir site.

Eaton's jubilation was so great he couldn't restrain himself. He ran to the telegraph office and shot off a cryptic message to Mulholland. "The deal is made," he wired. All it had required was "a week of Italian work."

Los Angeles now had most of what it needed, but Mulholland still wanted some additional water rights in order to kill the Reclamation project once and for all. Within hours of receiving Eaton's telegram, he was frantically organizing an expedition of prominent Los Angeleans to the Owens Valley, using the pretext that they were investors interested in developing a resort. The group included Mayor Owen McAleer and two prominent members of the water commission. For them to see the river firsthand was crucial, Mulholland reasoned, because he and Eaton would need more money to buy the last water rights they wanted, and the city could not legally appropriate money toward a project that hadn't even been described, let alone authorized. A group such as this could easily free up some money in the Los Angeles business community if they fathomed how much water there was.

It went exactly as planned. The group arrived in the valley on the cusp of spring, when even small tributaries of the Owens River were overflowing; days after they returned, Eaton and Mulholland had all the money they needed. They requisitioned an automobile and raced off to the valley by the shortest route, across the Mojave Desert—probably the first time anyone crossed it by car. After a week of frantic, furtive buying, the two men returned. "The last spike has been driven," Mulholland announced to the assembled water commissioners. "The options are all secured."

Like all the other newspaper publishers in the city, Harrison Gray Otis had been operating under a self-imposed gag rule. Although the publishers knew what was going on, not a word of Mulholland and Eaton's stealthy grab of water options had appeared in the papers. However, on July 29, the same day the Reclamation panel reached its verdict, Otis could no longer contain himself. Under a headline that read, "Titanic Project to Give the City a River," the whole unauthorized story spilled out in the Los Angeles *Times*.

Otis seemed to take particular satisfaction in the way Fred Eaton had hoodwinked the greedy but guileless rubes in the Owens Valley. "A number of the unsuspecting ranchers have regarded the appearance of Mr. Eaton in the valley as a visitation of Providence," the *Times* chortled. "In the eyes of the ranchers he was land mad. When they advanced the price of their holdings a few hundred dollars and he stood the raise, their cup of joy fairly overflowed. . . . The farmer folk in the Owens River Valley think that he has gone daffy on stock raising. To them he is a millionaire with a fad." The paper even admitted that the town of Independence, whose neighboring ranchers had been made offers they couldn't refuse, was faced with financial ruin, but it refused to let such a fact spoil its enjoyment of a good joke. The paper also recalled in excruciating detail Joseph Lippincott's career as a double agent, apparently thinking it was doing him a favor. "In the consummation of the great project that is to supply Los Angeles with sufficient water for all time, great credit is given to J. B. Lippincott," it said. "Without Mr. Lippincott's interest and cooperation, it is declared that the plan never would have gone through. . . . *Guided by the spirit of the Reclamation Act* . . . he recognized the fact that the Owens River water would fulfill a greater mission in Los Angeles than if it were to be spread over acres of desert land. . . . Any other government engineer, a nonresident of Los Angeles and not familiar with the needs of this section, undoubtedly would have gone ahead with nothing more than the mere reclamation of the arid lands in view" (emphasis added). It was praise that was to damn Lippincott for the rest of his life.

There was nothing quite as revealing in the *Times*'s story, however, as its very lead sentence: "The cable that has held the San Fernando Valley vassal for ten centuries to the arid demon," it gushed in a spasm of metaphorical excess, "is about to be severed by the magic scimitar of modern engineering skill."

There was something very strange about that sentence. All along, the Owens River had been portrayed as a matter of life or death to the city of Los Angeles. No one had ever said a word about the San Fernando Valley.

Sesquipedalian tergiversation was the strong suit of Harrison Gray Otis, along with slander, meanness, biliousness, and the implacable pursuit of a good old-fashioned grudge. Under his ownership, the *Times* was less a newspaper than a kind of mace used to bludgeon and destroy his enemies, who, and which, were many. (Otis often said that he considered objectivity a form of weakness.) The Democratic Party was "a shameless old harlot"; labor leaders were "corpse defacers," labor unions "anarchic scum"; California's preeminent reformer, Governor (later Senator) Hiram Johnson, was "a born mob leader—a whooper— a howler—a roarer." The newspaper owned by Otis's former partner, H. H. Boyce, was the "Daily Morning Metropolitan Bellyache," while Boyce himself was "a coarse vulgar criminal." William Randolph Hearst and his *Examiner*, more serious rivals than Boyce, were, interchangeably, "Yellow Yawp." Even innocent bystanders were vaporized by the General's ire. One morning Otis was greeted by a new neighbor who happened to mispronounce his name. "Good morning, General Ah-tis," said the man cheerily. "It's *O*-tis, you goddamn fool," the General bellowed back.

General Harrison Gray Otis. Otis's military coronation had come through the offices of President William McKinley as a reward for volunteering to send young men into the Philippine jungles during the Spanish-American War. By the time he returned to the States, the twentieth century had dawned, and Otis was utterly unprepared for it. Unions were organizing, the open shop was threatened, and even in Los Angeles the Socialists—the *Socialists*—were getting ready to run a candidate for mayor. Anti-unionism became breakfast fare for *Times* readers, as predictable as sunrise, and Otis was soon ordained public enemy number one by organized labor in the United States— no mean feat for a newspaper publisher in a remote western city. It was a notoriety he loved. To celebrate it, Otis commissioned a new headquarters that resembled a medieval fortress—it even had a parapet with turrets and cannon slots—and had a custom touring car built with a cannon mounted on the hood. The effect of all this on his enemies was inspirational. Hiram Johnson was addressing a crowd in a Los Angeles auditorium when someone in the audience, who knew that Johnson's talent for invective surpassed even the General's, yelled out, "What about Otis?" Johnson, all prognathous scowl and murderous intent, took two steps forward and began extemporaneously. "In the city of San Francisco we have drunk to the very dregs of infamy," he said in a low rumble. "We have had vile officials, we have had rotten newspapers. But we have had nothing so vile, nothing so low, nothing so debased, nothing so infamous in San Francisco as Harrison Gray Otis. He sits there in senile dementia with gangrene heart and rotting brain, grimacing at every reform, chattering impotently at all the things that are decent, frothing, fuming, violently gibbering, going down to his grave in snarling infamy. This man Otis is the one blot on the banner of southern California; he is the bar sinister on your escutcheon. My friends, he is the one thing that all Californians look at when, in looking at southern California, they see anything that is disgraceful, depraved, corrupt, crooked, and putrescent—*that*," concluded Johnson in a majestic bawl, "that is Harrison Gray Otis!"

The vitriol that Otis and his rivals hurled at one another, however, could be turned off instantly if some more important matter was at hand. In the avaricious social climate of southern California, that usually meant an opportunity to make money; and in the dry climate of southern California, money meant water.

The first sign something was afoot came in the weeks following the *Times*'s disclosure of Mulholland and Eaton's daring scheme, when Otis's newspaper took time out from its usual broadsides to laud the future of the San Fernando Valley, an encircled plain of dry, mostly worthless land on the other side of the Hollywood Hills. "Go to the whole length and breadth of the San Fernando Valley these dry August days," the paper editorialized on August 1. "Shut your eyes and picture this same scene after a big river of water has been spread over every acre, after the whole expanse has been cut up into five-acre, and in some cases one-acre, plots—plots with a pretty cottage on each and with luxuriant fruit trees, shrubs and flowers in all the glory of their perfect growth. . . ." Again on October 10, a so-called news story began, "Premonitory pains and twitches: The San Fernando Valley has caught the boom. It appears just about ready to break. . . ."

What was odd about this was that there was as yet no guarantee—at least none publicly offered by Mulholland—that the San Fernando Valley was going to receive any of the Owens Valley water. In the first place, the route of the aqueduct had not yet been disclosed; it might go through the valley, but then again it might not. Secondly, the voters had not even approved the aqueduct, let alone voted for a bond issue to finance it. Mulholland had been saying that the city had surplus water sufficient for only ten thousand new arrivals. If that was so, and if the city was expected to grow by hundreds of thousands during the next decade, where was this great surplus for the San Fernando Valley to come from? In those days, the valley was isolated from Los Angeles proper; it sat by itself far outside the city limits. In theory, the valley couldn't even *have* the city's surplus water, assuming there was any— it would be against the law.

The truth, which only a handful of people knew, was that William Mulholland's private figures were grossly at odds with his public pronouncements; it was the same with his intentions. Despite his talk of water for only ten thousand more people, there was still a big surplus at hand. (During the eight years it would take to complete the aqueduct, in fact, the population of Los Angeles rose from 200,000 to 500,000 people, yet no water crisis occurred.) The crisis was, in large part, a manufactured one, created to instill the public with a sense of panic and help Eaton acquire a maximum number of water rights in the Owens Valley. Mulholland and Eaton had managed to secure water rights along forty miles of the Owens River, which would be enough to give the city a huge surplus for years to come. But Mulholland was not saying that he would *use* any of the surplus; in fact, he seemed to be going out of his way to assure the Owens Valley that he would not. For example, the proposed intake for the aqueduct had been carefully located downstream from most of the Owens valley ranches and farms, so that they could continue to irrigate; Mulholland would later tell the valley people that his objective was simply to divert their unused and return flows.

In truth, Mulholland planned to divert every drop to which the city held rights as soon as he could. Like all water-conscious westerners, he lived in fear of the use-it-or-lose-it principle in the doctrine of appropriative rights. If the city held water rights that went unused for years, the Owens Valley people might successfully claim them back. But where would he allow the surplus to be used?

Privately, Mulholland planned to lead the aqueduct through the San Fernando Valley on its way to the city. In his hydrologic scheme of things, the valley was the best possible receiving basin; any water dumped on the earth there would automatically drain into the Los Angeles River and its broad aquifer, creating a large, convenient, non-evaporative pool for the city to tap. It provided, in a word, free storage. That it was free was critically important, because Mulholland, intentionally or not, had underestimated the cost of building the aqueduct, and to build a large storage reservoir in addition to the aqueduct would be out of the question financially. Even had it been feasible, Mulholland was deeply offended by the evaporative waste of reservoirs; he was much more inclined to store water underground.

Mulholland had an even more important reason for wanting to include the San Fernando Valley in his scheme. Under the city charter, Los Angeles was prohibited from incurring a debt greater than 15 percent of its assessed valuation. In 1905, that put its debt limit at exactly $23 million, which was what he expected the aqueduct to cost. But the city already had $7 million in outstanding debt, which left him with a debt ceiling too low to complete the project. After coming this far—securing the water rights, organizing civic support—he wouldn't have the money to build it!

Mulholland, however, was clever enough to have thought of a way out of this dilemma. If the assessed valuation of Los Angeles could be rapidly increased, its debt ceiling would be that much higher. And what better way was there to accomplish this than to *add to the city?* Instead of bringing more people to Los Angeles—which was happening anyway—*the city would go to them.* It would just loosen its borders as Mulholland loosened his silk cravat and wrap itself around the San Fernando Valley. Then it would have a new tax base, a natural underground storage reservoir, and a legitimate use of its surplus water in one fell swoop.

Anyone who knew this, and bought land in the San Fernando Valley while it was still dirt-cheap, stood to become very, very rich.

The person who finally began to figure it all out was Henry Loewenthal, the editor of Otis's despised rival newspaper, William Randolph Hearst's *Examiner.* The *Examiner* had been skeptical of the aqueduct plan from the beginning, though it did not oppose it outright; Loewenthal's editorials merely made a point of questioning Mulholland's sense of urgency and, on occasion, his figures. But even such mild skepticism was more than enough to enrage Otis, who attributed Loewenthal's doubts to the fact that the *Times* had scooped the *Examiner* about the aqueduct story. "Anyone but a simpleton or a poor old has-been in his dotage would sing very low over a failure like that," snarled Otis in an editorial, "but the impossible Loewenthal insists on emphasizing his own incompetency."

Such invective simply instilled in Loewenthal a passionate urge to outscoop Otis, and, in the process, catch him with his hand in the till. There *must* be some hanky-panky, Loewenthal surmised. Otherwise why Otis's sudden interest in a desolate valley? And why did

Otis's number-one enemy, E. T. Earl, rival publisher of the *Express*, seem as enthusiastic as Otis? In the past, Earl had opposed nearly anything Otis endorsed, and vice versa, as a simple matter of dignity. But now Otis, Earl, and virtually all the rival newspapers, except his own, were united on perhaps the most controversial issue Los Angeles had ever faced. Why? Loewenthal decided to send a couple of his top reporters to the courthouse in San Fernando to find out.

The co-conspirators hadn't even bothered to cover their tracks. They could have invented blind trusts, paper corporations, or some other ruse to conceal their identity; but there they were, caught in the open on an exposed plain.

On November 28, 1904—just six days after Joseph Lippincott was paid $2,500 to help steer his loyalties in the direction of Los Angeles— a syndicate of private investors had purchased a $50,000 option on the Porter Land and Water Company, which owned the greater part of the San Fernando Valley—sixteen thousand acres all told. Innocent enough. But the investors had then waited to consummate their $500,000 purchase until March 23, 1905—*the same day* that Fred Eaton had telegraphed the water commission that the option on the Rickey ranch in Long Valley was secured. On that day, as anyone who had access to Mulholland's thinking knew, Los Angeles was all but guaranteed 250,000 acre-feet of new water—an amount that would leave the city with a water surplus for at least another twenty years. And the only sensible place to use the surplus water was in the San Fernando Valley.

Was the timing mere coincidence? The names of the investors who made up the secret land syndicate strongly suggested that it was not. In fact, their identity had given Loewenthal the scoop of his dreams. The only way he could improve its impact was to wait for exactly the right moment to go to press.

Loewenthal knew that the San Francisco *Chronicle* was, in a vague way, on to the same story. He also knew the *Chronicle* was not nearly as methodical in its investigations as his paper, and would probably publish rumors without supporting facts. On August 22, just as Loewenthal supposed, the *Chronicle* ran a story, unsupported by evidence, to the effect that the Owens Valley aqueduct was somehow linked to a land-development scheme in the San Fernando Valley. Two days later, the *Times* derisively dismissed the allegations in an editorial which, to Loewenthal's delight, ran under the heading "Baseless Rumors." On that same morning, the *Examiner*'s story went to press.

The San Fernando land syndicate, the *Examiner* revealed, was composed of some of the most influential and wealthy men in Los Angeles.

There was Moses Sherman, a balding school administrator from Arizona who had moved to Los Angeles and become a trolley magnate—one of the most ruthless capitalists in a city that was legendary for same. (By coincidence, Moses Sherman also sat on the board of water commissioners of Los Angeles; the syndicate could not have prayed for a better set of eyes and ears.) Then there was Henry Huntington, Sherman's implacable rival in the rush to monopolize the region's transportation system. There was Edward Harriman, the chairman of the Union Pacific Railroad and a rival of both Sherman and Huntington. There was Joseph Sartori of the Security Trust and Savings Bank, and *his* rival, L. C. Brand of the Title Guarantee and Trust Company. There was Edwin T. Earl, the publisher of the *Express*; William Kerckhoff, a local power company magnate; and Harry Chandler, Otis's son-in-law, the tubercular young man with the minister's face, the gambler's heart, and the executioner's soul. But Loewenthal reserved the best for last. The person who had signed the check securing the $50,000 option on the immense San Fernando property was the same person who, that very morning, had dismissed talk of such a nest of land speculators as lies—Brigadier General Harrison Gray Otis.

"This is the prize for which the newspaper persons . . . are working and the size of it accounts for their tremendous zeal," wrote Loewenthal, almost squealing with delight. "The mystery of the enterprise is how it happens that Messrs. Huntington and Harriman, who let no one into their [previous] land purchasing schemes, but who bought up everything for themselves, consented to let the other in." Loewenthal was, of course, enough of a cynic to know exactly why they had. The participants, taken together, represented the power establishment of southern California with an exquisite sense of proportion. Railroads, banking, newspapers, utilities, land development—it was a monopolists' version of affirmative action. Besides, William Kerckhoff was a prominent conservationist and friend of Gifford Pinchot, the chief of the U.S. Forest Service, whose influence with President Theodore Roosevelt could prove invaluable. Harriman's railroad owned a hundred miles of right-of-way along the aqueduct path that the city would need permission to cross, and Huntington owned the building that housed the regional headquarters of the Reclamation Service! Including Earl and Otis, the two feuding neighbors and publishers, was the master stroke. Like a couple of convicts bound together by a ball and chain, neither could betray the other without exposing himself.

The *Examiner*'s exposé had Harrison Gray Otis venting steam from both nostrils and ears, but he didn't dare look the accusations directly in the eye, so in the ensuing weeks he tried to hide behind "Mr. Huntington's" skirts, as if Huntington had been solely responsible for the syndicate and he—Otis—had been an innocent seduced into joining, as a fresh young wayward girl is seduced into sex. Where Otis couldn't weasel out, he blazed away. "Had Hearst's . . . yellow atrocity been the first to announce the plans of the Water Board, it would have claimed the project as its own conception and inauguration," he raved.

Its front page would have shrieked in poster type about "The *Examiner's* solution to the water problem," and the public would have been deafened with yawp about how the *Examiner* "discovered Owens River," laid out plans to bring the water to Los Angeles and showed the engineers how to build the aqueduct. The line would have been dubbed, "The Great Hearst Aqueduct," or "The *Examiner* Pipe-line," and Loewenthal the Impossible would have been the Moses of Los Angeles, who smote the rock of Mount Whitney with the rod of his egotism and caused the water to flow abundantly. Deprived of the opportunity for mendacious self-glorification . . . the foolish freak vents its impotent rage in snarling under its breath. . . . The insane desire of the *Examiner* to discredit certain citizens of Los Angeles has at last led it into the open as a vicious enemy of the city's welfare, its mask of hypocrisy dropped and its convulsed features revealed.

In the end, though, the broadsides between the rival papers were all sound and fury, signifying not much. Ever since their foremost minister had fled prosecution for land fraud, the citizens of Los Angeles had grown accustomed to scandal, and the city's temperament was quite comfortable with graft. Henry Loewenthal would later speak of a "spirit of lawlessness that prevails here, that I have never seen anywhere else." Nature was also smiling on the Owens Valley scheme. On August 30, a week before the scheduled referendum on the aqueduct, the temperature climbed to 101 degrees. The city had gone its usual four months without rain, and there would likely be two rainless months to come. On September 2, Hearst himself rode down from San Francisco in his private railroad car for a quiet palaver with the city's oligarchs. As men of commerce, they understood each other, and Hearst had recently been bitten by the presidential bug; if he was truly serious about the White House, he could use their help. When the meeting was over, the publisher strode into the *Examiner's* offices, barked Loewenthal into acquiescence, and personally wrote an editorial recommending a "yes" vote. Samuel T. Clover's *Daily News*, the only paper on record opposing the aqueduct, lobbed a potential bombshell when it reported that the city's workers, under cover of darkness, were dumping water out of the reservoirs into the Pacific to make them go dry, thus assuring a "yes" vote. But Mulholland's lame explanation that they had merely been "flushing the system" was widely believed.

On September 7, 1905, the bond issue passed, fourteen to one.

To the Los Angeles *Times*, it was a "Titanic Project to Give the City a River." To the Inyo *Register*, it was a ruthless scheme in which "Los Angeles Plots Destruction, Would Take Owens River, Lay Lands Waste, Ruin People, Homes, and Communities." That sensational headline actually belied the feeling in the valley somewhat. Few people thought, at first, that things would be so bad. A number of the ranchers had made out well selling their water rights, and they would be able to keep their water for years, until the aqueduct was built. The city had bought up nearly forty bank miles of the river and would probably dry up the lower valley, but the upper valley, except for Fred Eaton's purchase of the Rickey estate, had been left mostly intact. When Eaton moved up from Los Angeles as promised and began his new life as a cattle rancher, the valley people were reassured. After a while, they even began to fraternize with him.

Mulholland, meanwhile, had begun his own campaign to mollify the people of the valley, a campaign in which he was joined, somewhat more bellicosely, by the Los Angeles *Times*, which featured headlines such as "Ill-feeling Ridiculous" and "Owens Valley People Going Off as Half-Cock." Inyo County's Congressman, Sylvester Smith, was an influential member of the House Public Lands Committee, and since the city would have to cross a lot of public land it would have to deal with him. Meanwhile, Theodore Roosevelt, the bugaboo of monopolists, had just been elected to a second term. He would never let the Owens Valley die for the sake of Henry Huntington, Harrison Gray Otis, and their cronies in the San Fernando Valley syndicate. On top of all this, the Owens was a generous desert river, with a flow sufficient for two million people. It was laughable to think of Los Angeles growing that big, so even under the worst of circumstances there would be water enough for all. The reasoning was very sensible, the logic very sound, and it was fatefully wrong.

There was one person who knew that it was. She was Mary Austin, the valley's literary light, who had published a remarkable collection of impressionistic essays entitled *Land of Little Rain* that won her recognition around the world. In the course of her writing she had spent long hours with the last of the Paiutes, the Indians who had lived in the valley for centuries until they were instantly displaced by the whites. The Paiutes showed her what no one else saw—that order and stability are the most transient of states, that there is rarely such a thing as a partial defeat. In a subsequent book, a novella about the Owens Valley water struggle called *The Ford*, she wrote about what happens when "that incurable desire of men to be played upon, to be handled," runs up against "that Cult of Locality, by which so much is forgiven as long as it is done in the name of the Good of the Town." Mary Austin was convinced that the valley had died when it sold its first water right to Los Angeles—that the city would never stop until it owned the whole river and all of the land. One day, in Los Angeles for an interview with Mulholland, she told him so. After she had left, a subordinate came into his office and found him staring at the wall. "By God," Mulholland reportedly said, "that woman is the only one who has brains enough to see where this is going."

No sooner had the city gotten the aqueduct past the voters than it faced the more difficult task of getting it past Congress. Most of the lands it would traverse belonged to the government, so the city would have to appeal for rights-of-way. The Reclamation project, though moribund, was still not officially deauthorized, which was, at the very least, a nuisance to the city. But deauthorization could prove to be even worse, because tens of thousands of acres that the Service had withdrawn would return to the public domain and be available for homesteading. Homesteading in California was another name for graft; half of the great private empires were amassed by hiring "homesteaders" to con the government out of its land. If the withdrawn lands went back to the public domain, every available water right would be coveted by speculators for future resale to the city. Mulholland seemed to believe that the city would never require more water, but others, notably Joseph Lippincott, thought him wrong. The withdrawn lands had to be kept off-limits at all costs.

The instrument for achieving this wishful goal was a bill introduced at the behest of Mulholland's chief lawyer, William B. Matthews, by Senator Frank Flint of California, a strong partisan of Los Angeles and urban water development in general. The bill would give the city whatever rights-of-way it needed across federal lands and hold the withdrawn lands in quarantine for another three years, which would presumably give the city enough time to purchase whatever additional water or land it might need. Flint's bill reached the Senate floor in June of 1906, and flew through easily. Its next stop, however, was the House Public Lands Committee, where it crashed into Congressman

Sylvester Smith. Smith was an energetic and charming politician, a former newspaper publisher from Bakersfield with a sense of public duty and enough money to maintain an ironclad set of principles. The idea of Harrison Gray Otis and Henry Huntington becoming vastly richer than they already were on water abducted from his district inflamed his well-developed sense of outrage. Smith knew what he was up against, however, and realized that his best defense was to appear utterly reasonable. As a result, he said that he was willing to acknowledge the city's need for more water, that he was willing to let it have a substantial share of the Owens River, and that he was willing to grant the aqueduct its necessary rights-of-way. He was not willing, however, to do any of this in the way the city wanted. He suggested a compromise. Let the Reclamation Service build its project, including the big dam in Long Valley—a dam that could store most of the river's flow. The water could then be used first for irrigation, and because of the valley's long and narrow slope, the return flows would go back to the lower river, where they could be freely diverted by Los Angeles. The city would sacrifice some of the water it wanted, the valley would sacrifice some irrigable land. It was, Smith argued, an enlightened plan: sensible, efficient, conceived in harmony. It was the only plan under which no one would suffer. He would add only two stipulations: the Owens Valley would have a nonnegotiable first right to the water, and any surplus water could not be used for irrigation in the San Fernando Valley.

Smith's proposal was obviously anathema to the San Fernando land syndicate, and to the city as well. The chief of the Geologic Survey doubted that it would work, and even if it did, for the West's largest city to settle for leftover water from a backwater oasis of fruit and cattle ranchers was, to say the least, humiliating. The city might have to beg for extra water in times of drought or go to court to try to condemn it. If the Owens Valley held on to its first rights and expanded its irrigated acreage, Los Angeles might soon have to look for water again, and the only river in sight was the Colorado, a feckless brown torrent in a bottomless canyon which the city could never afford to dam and divert on its own. Smith's proposal led directly to one unthinkable conclusion: at some point in the relatively near future, Los Angeles would have to cease to grow.

What was William Mulholland's response? He took a train to Washington, held a summit meeting with Smith and Senator Flint, and decided to do what any sensible person would have done: he accepted the compromise.

If it was a smokescreen, as it appears to have been, it was a brilliant move. (Mulholland seems to have been a far better political schemer than he was a hydrologist and civil engineer.) For one thing, it put Sylvester Smith off guard, making him believe that the reconciliation he wanted to effect was a success. For another, it gave Los Angeles some critical extra time to plead its case before the two people who might help the city get everything it wanted: the President of the United States, Theodore Roosevelt, and the man on whom he leaned most heavily for advice—Gifford Pinchot.

Pinchot was the first director of Roosevelt's pet creation, the Forest Service, but that was only one of his roles. He was also the Cardinal Richelieu of TR's White House. Temperamentally and ideologically, the two men fit hand in glove. Both were wealthy patricians (Pinchot came from Pittsburgh, where his family had made a fortune in the dry-goods business); both were hunters and outdoorsmen. Though their speeches and writings rang of Thomas Jefferson, at heart Pinchot and Roosevelt seemed more comfortable with Hamiltonian ideals. Roosevelt liked the Reclamation program because he saw it as an agrarian path to industrial strength, not because he believed—as Jefferson did—that a nation of small farmers is a nation with a purer soul. Pinchot espoused forest conservation not because he worshiped nature like John Muir (whom he privately despised) but because the timber industry was plowing through the nation's forests with such abandon it threatened to destroy them for all time. Roosevelt was a trust-buster, but only because he feared that unfettered capitalism could breed socialism. (For evidence he only had to look as far as Los Angeles, where Harrison Gray Otis was whipping labor radicals into such a blind, vengeful froth that two of them blew up his printing plant in 1910 and killed twenty of their own.) The conservation of Roosevelt and Pinchot was utilitarian; their progressivism—they spoke of "the greater good for the greatest number"—had a nice ring to it, but it also happens to be the progressivism of cancer cells.

On the evening of June 23, Senator Frank Flint left his offices on Capitol Hill for a late meeting with the President. It was a hot and muggy night, and Roosevelt seemed in an irritable mood. Behind him, however, stood a man who seemed a model of coolness and decorum, Gifford Pinchot. Flint, who had just received an intensive coaching from Matthews and Mulholland, began a passionate appeal.

Smith's so-called compromise, he said, was nothing less than capitulation. Los Angeles had agreed only in despair; it was going to run out of water any day and it couldn't afford to be filibustered to death in Congress. Smith's prohibition on using surplus water in the San Fernando Valley left the city no choice but to leave any surplus in the

Owens Valley or dump it in the ocean. In the first case, water rights the city had purchased at great expense might revert to the valley under the doctrine of appropriative rights; in the second case, the city would violate the California constitution, which forbade "inefficient use" of water. The real estate bust of 1889 had depopulated the city by one-half. Imagine what a water famine would do! All of the city's actions in the Owens Valley had been legitimate. It had paid for its water, fair and square, and it wanted to let the valley survive. But there was only so much water, and it was a hundredfold—a thousandfold, said Smith—more valuable to the state and the nation if it built up a great, strong, progressive city on America's weakly defended western flank instead of maintaining a little agrarian utopia in the high desert.

It was a rousing speech—the kind of speech that Roosevelt liked to hear. It was, in fact, just the kind of speech *he* would have made.

Roosevelt turned to his other visitor. "What do you think about this, Giff?"

"As far as I am concerned," Pinchot answered coolly, "there is no objection to permitting Los Angeles to use the water for irrigation purposes."

It was as simple as that. Roosevelt did not even bother to call in the Interior Department's lawyers or the Geologic Survey's hydrologists to ask whether Flint's argument was sense or nonsense. He never invited Sylvester Smith to give his side of the argument. He didn't even tell Smith or his own Interior Secretary, Ethan Hitchcock, about his decision; they found out about it secondhand a day and a half later. Hitchcock, a wealthy, principled man in the style of Sylvester Smith, had been profoundly embarrassed by the two-faced behavior of his employee J. B. Lippincott, and had been looking for a way to make amends to the Owens Valley. Flabbergasted and infuriated by the President's decision, Hitchcock raced over to the White House, where Roosevelt refused to hear him. Instead, he forced him to suffer the humiliation of helping him draft a letter explaining "*our* attitude in the Los Angeles water supply question." As Hitchcock stood by, impotent and enraged, Roosevelt wrote, "It is a hundred or a thousandfold more important to state that this water is more valuable to the people of Los Angeles than to the Owens Valley." The words could have come right out of William Mulholland's mouth.

The Otis-Sherman-Huntington-Chandler land syndicate was, potentially, enough of an embarrassment to Roosevelt's antimonopolist image that he felt compelled to add an amendment to Flint's bill prohibiting the city from reselling municipal water for irrigation use.

In the opinion of the House Public Lands Committee, however, the stipulation was "meaningless." "This water will belong absolutely to Los Angeles," said the bill's sponsor, echoing the sense of the committee, "and the city can do as it pleases. . . ." Which it would.

Roosevelt's support for Flint's bill was only the beginning of the aid and comfort he was to give to the most powerful city on the Pacific Coast. When the Reclamation Service officially annulled the Owens Valley Project in July of 1907, the hundreds of thousands of acres it had withdrawn were not returned to the public domain for home-steading, on Roosevelt's orders—just as Mulholland wished. It was a decision without precedent, and its result was that the handful of rich members of the San Fernando syndicate could continue using the surplus water in the Owens River that thousands of homesteaders might have claimed instead. Ethan Hitchcock had promised that such a decision, which he already foresaw when Roosevelt closed ranks behind Los Angeles, would be made over his dead body, but Roosevelt spared his life by firing him first. And when the city, immensely satisfied with the result, asked Pinchot whether he couldn't go a step further, the chief of the Forest Service decided to include virtually all of the Owens Valley in the Inyo National Forest.

The Inyo National Forest! With six inches of annual rainfall, the Owens Valley is too dry for trees; the only ones there were fruit trees planted and irrigated by man, some of which were already dying for lack of water. This didn't seem to bother Pinchot, nor did the fact that his action appears to have been patently illegal. The Organic Act that created the Forest Service says, "No public forest reservation shall be established except to improve and protect the forest . . . or for the purpose of creating favorable conditions of water flow, and to provide a continuous supply of timber for the use and necessities of the United States; but it is not the purpose of these provisions . . . to authorize the inclusion . . . of lands more valuable for the mineral therein, *or for agricultural purposes*, than for forest purposes" (emphasis added). The valley's irrigated orchards were infinitely more valuable than the barren flats and scattered sagebrush that characterized the new national forest, so Pinchot's action was incontrovertibly a violation of the legislation that put him in business. He lamely countered that he was simply acting to protect the quality of Los Angeles' water; but since much of the treeless acreage he included in the Inyo National Forest lay *below* the intake of the aqueduct, it was a flimsy excuse. As a formality, Pinchot was obliged to send an investigator to the Owens Valley to recommend that he do what he had already made up his mind to do. He sent three before he found one who was willing to go along. "This is not a government by legislation," lamented Sylvester Smith on the Senate floor, "it is a government by strangulation."

In July of 1907, with the reclamation project in its grave and the Owens Valley imprisoned inside a national forest without trees, Joseph Lippincott resigned from the Reclamation Service and immediately went to work, at nearly double his government salary, as William Mulholland's deputy. He remained utterly unchastised. "I would do everything over again, just exactly as I did," he said as he departed.

The one thing that no one seems to have thought about in all this was that the people of Owens Valley were only human, and there was just so much they could take.

The aqueduct took six years to build. The Great Wall of China and the Panama Canal were bigger jobs, and New York's Catskill aqueduct, which was soon to be completed, would carry more water, but no one had ever built anything so large across such merciless terrain, and no one had ever done it on such a minuscule budget. It was as if the city of Pendleton, Oregon, had gone out, by itself, and built Grand Coulee Dam.

The aqueduct would traverse some of the most scissile, fractionated, fault-splintered topography in North America. It would cover 223 miles, 53 of them in tunnels; where tunneling was too risky, there would be siphons whose acclivities and declivities exceeded fifty-grade. The city would have to build 120 miles of railroad track, 500 miles of roads and trails, 240 miles of telephone line, and 170 miles of power transmission line. The entire concrete-making capacity of Los Angeles was not adequate for this one project, so a huge concrete plant would have to be built near the limestone deposits in the grimly arid Tehachapi Mountains. Since there was virtually no water along the entire route, steampower was out of the question and the whole job would be done with electricity; therefore, two hydroelectric plants would be needed on the Owens River to run electric machinery that a few months earlier had not even been invented. The city would have to maintain, house, and feed a work force fluctuating between two thousand and six thousand men for six full years. And it would have to do all this for a sum equivalent, more or less, to the cost of one modern jet fighter.

The workers would have to supply their own hard-shelled derby hats, since hard hats did not yet exist, and even if they had the city couldn't afford them. They would live in tents in the desert without liquor or women—although both were available nearby and ended up consuming most of the aqueduct payroll. They would eat meat that spoiled during the daytime and froze at night, since the daily temperature range in the Mojave Desert can span eighty degrees. Nonetheless, the men would labor on the aqueduct as the pious raised the cathedral at Chartres, and they would finish under budget and ahead of schedule. If you asked any of them why they did it, they would probably say they did it for the chief.

The loyalty and heroics that Mulholland inspired in his workers were a perpetual source of wonder. For six years he all but lived in the desert, patrolling the aqueduct route like a nervous father-to-be pacing a hospital waiting room—giving advice, offering encouragement, sketching improvised solutions in the sand. In sandstorms, windstorms, snowstorms, and terrifying heat, his spirits remained contagiously high. Pilfering, which can add millions to the cost of a modern project, was almost unknown. Although the pay was terrible— Mulholland simply couldn't afford anything more—he initiated a bonus system that shattered records for hard-rock tunneling. (The men were in a race with the world's most illustrious tunnelers, the Swiss, who were digging the Loetchberg Tunnel at the same time.)

Throughout the entire time, Mulholland showed the better side of a complex and sometimes heartless character. If he wandered through a tent city and discovered that a worker's wife had just had a baby, he would stop long enough to show her the proper way to change a diaper. He would sit down and eat with the men and complain louder than anyone about the food. In lieu of newspapers, his wit was breakfast conversation. Once, when a landslide sealed off a tunnel with a man still inside, Mulholland arrived to check on the rescue effort.

"He's been in there three days, so I don't suppose he's doing so well," said the supervisor, a mirthless Scandinavian named Hansen.

"Then he must be starving to death," said Mulholland.

"Oh, no, sir," said the supervisor. "He's getting something to eat. We've been rolling him hard-boiled eggs through a pipe."

"Have you?" said Mulholland archly. "Well, then, I hope you've been charging him board."

"No, sir," said the flustered Hansen. "But I suppose I should, eh?"

And Los Angeles loved Mulholland even more than the men, because its reward would be infinitely greater than theirs—to the thirsty city, he was Moses. And he was that greater rarity, a Moses without political ambition. When a move was afoot a few years later to run him for mayor, Mulholland dismissed it with a typical bon mot: "I would rather give birth to a porcupine backwards than become the mayor of Los Angeles." But nothing that William Mulholland ever said or did quite matched the speech he gave when, on November 5, 1913, the first water cascaded down the aqueduct's final sluiceway into the San Fernando Valley. It had been a day of long speeches and waiting, and the crowd of forty thousand people was restless. Mulholland himself was exhausted; his wife was very ill, and he had slept only a few hours in several nights. When the white crest of water finally appeared at the top of the sluiceway and cascaded toward the valley, an apparition in a Syrian landscape, Mulholland simply unfurled an American flag, turned toward the mayor, H. H. Rose, and said, "There it is. Take it."

It was the high point of Mulholland's life and career.

Very little of the water that was, according to Theodore Roosevelt, a hundred or a thousandfold more important to Los Angeles than to the Owens Valley would go to the city for another twenty years. All through the teens and early twenties, the San Fernando Valley used three times as much aqueduct water as the city itself, the vast part of it for irrigation. During one particularly wet year, every drop of the copious flow of the aqueduct went to irrigate San Fernando Valley crops; the city took nothing at all. Understandably, this news enraged the people of the Owens Valley. For Los Angeles to take their water to fill their washtubs and water glasses was one thing. For it to turn their valley back to desert so that another desert valley, owned by rich monopolists, could bloom in its place was quite another.

The teens and early twenties, however, were extraordinarily wet years—the same wet years that caused the Reclamation Service to overestimate dramatically the flow of the Colorado River—and there was water enough for everyone. The irrigated acreage in the San Fernando Valley rose from three thousand acres in 1913—the year both the completion of the aqueduct and the annexation of the valley occurred—to seventy-five thousand acres in 1918. Even so, the Owens Valley lost few of its orchards and irrigated pasturelands, and the new railroad to Los Angeles and the silver mine at Tonopah fed in enough wealth to allow the town of Bishop to build a grand American Legion Hall and Masonic Temple, those cathedrals of the rural nineteenth century.

The same uncharacteristically engorged desert river that was keeping the Owens Valley green was responsible, in Los Angeles, for the most transfixing change. Santa Monica Boulevard, once a dry dusty strip, became an elegant corridor of palms; in Hollywood, where the motion picture industry had risen up overnight, outdoor sets resembled New Guinea; and since most Los Angeleans were immigrants from the Middle West, every bungalow had a green lawn. The glorious anomaly of a fake tropical city with a mild desert climate brought people from everywhere. Dirt farmers came from Arkansas; Aldous Huxley moved from England. The Chamber of Commerce, an Otis creation, kept them coming. They arrived on the Union Pacific, a Harriman railroad, and once they were there, the *Times*, an Otis and Chandler newspaper, urged everyone to settle in the San Fernando Valley, an Otis and Chandler property. Few could afford automobiles, so they got around on Sherman and Huntington trolley cars between Sherman-and-Huntington-built homes and Sherman and Huntington resorts in the San Gabriel and San Bernardino Mountains.

As Otis never tired of saying, this was the promised land. All things were possible; anyone could get rich; the cardinal sin was doubt. During the nadir of the Depression, when the city was invaded by homeless Okies so destitute they sat hollow-eyed in the parks and gnawed on the crusts thrown out for the pigeons, the *Times* sent them this holiday greeting: "Merry Christmas! Look pleasant! Chin up! A gloomy face never gets a good picture. The great battles are fought by Caesars and their fortunes, by Napoleons and their stars. Faith still does the impossible! Merry Christmas! Catch the tempo of the times. You have your life before you, and, if you are growing old, the greatest adventure of all is just around the corner. Earth may have little left in reserve, but heaven is ahead! Merry Christmas!" The only greater fraud than such blather from Otis and Chandler's newspaper was the overflowing desert river on which all depended.

In the West, drought tends to come in cycles of about twenty years, and the next drought arrived on schedule. The years 1919 and 1920 were a premonition; rainfall was slightly below average. It rose back to average—a measly fourteen inches—in 1921 and went slightly over that in 1922. Then it crashed. Ten inches in 1923; six inches in 1924; seven inches in 1925. In Florida, a seven-inch rainstorm may occur two or three times a year, but Los Angeles was trying to look like Florida, and grow even faster, on a fifth of its precipitation, and when the drought struck it kept going on a tenth. Mulholland had expected 350,000 people by 1925, but had 1.2 million on his hands instead. The city was growing fifteen times faster than Denver, eleven times faster than New York. And though the city at its core had become a metropolis, Los Angeles County led the nation in the value of its agricultural output. All of this agriculture depended on irrigation, which, together with the phenomenal urban growth, depended on a river draining Mount Whitney two hundred miles away.

As the drought intensified, the Owens River moved perilously close to overappropriation. The problem was not only that the river was small, but also that no carryover storage existed—nothing but some small receiving reservoirs around the basin and the snowfields in the Sierra. The Los Angeles Aqueduct was essentially a run-of-the-river project. If the river didn't run, the city collapsed.

It the city and the Owens Valley were to continue sharing the river, carryover storage would have to be built; otherwise, one place or the other would lose its water during a drought. Mulholland, of course, knew this, but still refused to build the dam at Long Valley. He blamed it on the city's fragile finances, but that was a poor excuse; the real reason was that he and his old friend Eaton had had a nasty falling-out.

Fred Eaton had not even bothered to attend the dedication of the aqueduct in 1913, though its existence was owed mainly to him. He had bought the initial water rights the city needed with his own money, taking a considerable risk; had the voters failed to approve the bond referendum, he would have been drowning in both unusable water and debt. The city had paid him quite adequately for the right, but it had not made him a multimillionaire. Originally, Eaton had hoped to operate the Owens Valley end of the aqueduct as a private concession, which could have made him incredibly rich, but Frederick Newell and Roosevelt had dashed that dream, insisting that the project be municipally owned from end to end. Eaton had also had some bad luck in the cattle business, and had to switch ignominiously to chickens. He was sixty-five years old; it was time things finally went right. The one item of real value Eaton owned was the reservoir site on the ranch he had purchased from Thomas Rickey. Ideally, a dam built at the site ought to be 140 feet high, the approximate depth of the gorge; that would create a reservoir large enough to provide for both the city and the valley during all but the worst droughts. A damsite of such importance to the city—a site which, if developed, would drown a good portion of his ranch—was worth a lot of money, as far as Eaton was concerned. When Mulholland asked him what his price was, Eaton said $1 million. Mulholland, who seemed personally indifferent to money (though he was reputedly the highest-paid civil servant in California), laughed him off. Time and time again he asked Eaton to accept a reasonable offer—$500,000, perhaps, or a little more—and each time his offer was more angrily refused. By 1917, the two old friends were no longer on speaking terms.

As the drought intensified, Mulholland begged the city fathers to end their abject deification of growth. The only way to solve the city's water problem, he grumbled aloud, was to kill the members of the Chamber of Commerce. When he was ignored, he began to regulate irrigation practices in the San Fernando Valley. First he forbade the irrigation of alfalfa, a low-value, water-demanding crop; then he pro- hibited winter planting. When these measures proved inadequate, he swallowed his disdain for surface storage and began building reser- voirs in the basin—first the Hollywood Reservoir, then a much larger dam in San Francisquito Canyon, a deep fissure in the shaky, shaly topography of the Santa Paula hills.

With the tens of thousands of people pouring in each year, every- thing was a stopgap measure. By the early 1920s, Mulholland was already lobbying for an aqueduct from the Colorado River. This, how- ever, put him on a collision course with Harry Chandler, who owned 860,000 acres in Mexico that relied on the Colorado, and who was so greedy that, despite his enormous wealth, he put the interests of his Mexican holdings above the welfare of the city he had created out of whole cloth. Chandler's opposition, together with fierce feuding among the Colorado River Basin states, kept the Boulder Canyon Project Act, which would create the storage reservoir that any Colorado River aqueduct would need, bottled up for years. Frustrated at every turn, Mulholland reached the end of his tether sometime in 1923. The only answer, he decided, was to do what Mary Austin had predicted the city would ultimately do—dry the Owens Valley up.

The trouble began where troubles usually begin, in the heart. Wilfred and Mark Watterson, the brothers who symbolized the Owens Valley's mortmain and its success, had a young uncle named George, only ten years older than Wilfred. George's attitude toward his nephews was less avuncular than competitive. Somehow, in competition, George always lost. When Wilfred and George had filed rival claims on a mining right, Wilfred won. George had always wanted to own the first automobile in the valley, but one day he looked down the street and saw Wilfred drive up in a yellow Stanley Steamer, mobbed by adoring crowds. Wilfred and Mark were treasurer of this, president of that; George Watterson was not even a has-been—he was a never-was.

George Watterson had an ally in bitterness. Some years earlier, a lawyer with an adventuresome bent named Leicester Hall had wan- dered into the Owens Valley from Alaska, taken one look at Wilfred and Mark's sister Elizabeth, and fallen helplessly in love. Elizabeth, however, had spurned him and married Jacob Clausen, Lippincott's former assistant—a symbol, like the Watterson brothers, of resistance to Los Angeles. The Owens Valley was a gossipy place, and the hatred

that George Watterson felt for his nephews and the bitterness that
Hall felt toward Jacob Clausen were well known. The city had its
agents in the valley, and they had ears. When William Mulholland
invited George Watterson, Hall, and their friend William Symons
down for dinner at his club one evening, they were happy to come.

The tactic was the old reliable one: the lightning strike. Symons
was the president of the McNally Ditch, which held the oldest and
largest water right among all the irrigation cooperatives in the valley.
Hall and George Watterson were officers in the Bishop Creek Ditch
and the Owens River Canal Company. On March 15, 1923, the three
men returned to the valley and went immediately to work. "Leave
none of the ranchers out," Mulholland had told them. "We want them
all." Within twenty-four hours, Watterson, Symons, and Hall owned
options on more than two-thirds of the McNally Ditch's water rights.
They had paid as much as $7,500 per cubic second-foot of water, and
the total cost to the city was more than $1 million—the price Fred
Eaton had wanted for access to his damsite.

The size and length of an irrigation ditch depend critically on the
number of people who use it. Since all the irrigators must spend a
substantial amount of time maintaining it—clearing out weeds, desilting it, repairing earthslides—losing just a few farmers can put a
terrible burden of responsibility on those who remain. So many farmers who belonged to the McNally Ditch had sold out that the cooperative was quickly put out of its misery; those who remained couldn't
possibly maintain it by themselves, so ultimately they would have to
sell out, too. By the time the three men moved on the other ditch
companies, however, pockets of resistance had formed, and they had
to seek out the more avaricious or vulnerable souls. Hall had managed
to raid the confidential files of the collective ditch companies, making
off with critical information about who was in financial trouble, who
was a poor farmer, who was inclined to move on. He and his collab-
orators, therefore, didn't waste much time on people who were unlikely
to yield to temptation; they knew who would. But their strategy—a
strategy of division and attrition—was especially cruel, not only because it placed an even larger burden of responsibility on the farmers
and ranchers who held out, but because it pitted neighbor against
neighbor, wife against husband, brother against brother.

Meanwhile, the master strategist, off in Los Angeles, was sixty-nine
years old and a changed man. Thirty years earlier, Mulholland had
spent his idle hours in a cabin at one of the city's outlying reservoirs,
reading the classics and planting poplars. When the city had first
talked about tapping the Owens River, his concern about the valley's

welfare led him to suggest that the city plant millions of trees which the residents could sell for firewood to the barren mining camps in Nevada—until someone informed him that so many trees would suck up enough groundwater to bleed the river dry. In his later years, however, the William Mulholland who had read Shakespeare and quoted Alexander Pope was hardly recognizable. No person ever put his imprint on an agency as strongly as Mulholland left his on the Los Angeles Department of Water and Power, and that agency was now using secret agents, breaking into private records, and turning neighbors into mortal foes. And, worst of all, Mulholland was ignoring a solution that would have satisfied everyone—a dam at Long Valley—out of petty niggardliness and almost fanatical pride.

In 1980, there were few people still alive who remembered Mulholland, but one who did was Horace Albright, the director of the National Park Service under Herbert Hoover. Albright could no longer remember the year—he was eighty-two—but it was probably 1925 or 1926, and he was a young park superintendent invited to attend a testimonial dinner for Senator Frank Flint, the man who had engineered the dubious federal decisions that allowed the Owens Valley aqueduct to be built. Albright was seated at Mulholland's table, a couple of chairs away, and midway through dinner he felt a rough tap on his shoulder.

"You're from the Park Service, aren't you?" Mulholland demanded more than asked.

"Yes, I am," said Albright. "Why do you ask?"

"Why?" Mulholland said archly. "Why? I'll tell you why. You have a beautiful park up north. A majestic park. Yosemite Park, it's called. You've been there, have you?"

Albright said he had. He was the park's superintendent.

"Well, I'm going to tell you what I'd do with your park. Do you want to know what I would do?"

Albright said he did.

"Well, I'll tell you. You know this new photographic process they've invented? It's called Pathé. It makes everything seem lifelike. The hues and coloration are magnificent. Well, then, what I would do, if I were custodian of your park, is I'd hire a dozen of the best photographers in the world. I'd build them cabins in Yosemite Valley and pay them something and give them all the film they wanted. I'd say, 'This park is yours. It's yours for one year. I want you to take photographs in every season. I want you to capture all the colors, all the waterfalls, all the snow, and all the majesty. I especially want you to photograph the rivers. In the early summer, when the Merced River roars, I want to see that.' And then I'd leave them be. And in a year I'd come back, and take their film, and send it out and have it developed and treated by Pathé. And then I would print the pictures in thousands of books and send them to every library. I would urge every magazine in the country to print them and tell every gallery and museum to hang them. I would make certain that every American saw them. And then," Mulholland said slowly, with what Albright remembered as a vulpine grin, "and then do you know what I would do? I'd go in there and build a dam from one side of that valley to the other and *stop the goddamned waste!*"

"It was the tone of his voice that surprised me," Albright said. "The laughingly arrogant tone. I don't think he was joking, you see. He was absolutely convinced that building a dam in Yosemite Valley was the proper thing to do. We had few big dams in California then. There were hundreds of other sites, and there were bigger rivers than the Merced. But he seemed to want to shake things up, to outrage me. He almost *wanted* to destroy."

It was the same tone, the same bitter and unreasoning quarrelsomeness, that Mulholland displayed when a reporter from the *Times* asked him why there was so much dissatisfaction in the Owens Valley. "Dissatisfaction in the valley?" said Mulholland mockingly. "Yes, a lot of it. Dissatisfaction is a sort of condition that prevails there, like foot and mouth disease." It was the same unreasoning rage that made him say, when his war of attrition against the Owens Valley had finally caused events to take a drastic turn for the worse, that he half regretted the demise of so many of the valley's orchard trees, because now there were no longer enough live trees to hang all the troublemakers who lived there.

Trees or no trees, that George Watterson, Leicester Hall, and William Symons had not yet been lynched themselves said something about the valley's self-restraint. Symons and Watterson had prudently taken to carrying sidearms, but, aside from an occasional curse or jeer, they were left alone. The valley thought it had a better means of taking revenge on the city than assassinating its agents. Soon after the McNally Ditch coup was engineered, the ditch companies that still had control of their water began opening their headgates and letting water flood uselessly over their fields. Before long, only a trickle was reaching the intake of the aqueduct. Mulholland demanded that the diversions stop, but the farmers refused. In exasperation, he tried a bit of double psychology: he sent more purchasing agents to reinforce Watterson, Symons, and Hall, and at the same time sent his attorney,

William Matthews, to meet with the ranchers to see if the matter could still come to an amicable settlement. Just hours before Matthews was scheduled to sit down with the ditch companies, however, Mulholland went into one of his sudden fits of anger and telephoned his maintenance crews to demolish the intake of the largest diverter, the Big Pine Canal.

The reaction was instantaneous. The leaders of the Big Pine Company were the worst people Mulholland could have chosen to antagonize: the Watterson brothers, a resort operator and speculator named Karl Keough, and Harry Glasscock, the incendiary editor of the *Owens Valley Herald*. As soon as news arrived of what was happening, a posse of twenty men, bristling with guns, roared out to the canal intake. As guns were trained on Mulholland's crew, the rest of the men dumped their equipment into the Owens River. The valley mood veered suddenly from bitterness to wild exuberance. "Los Angeles, it's your move now," exulted the Big Pine *Citizen*. And yet the Big Pine farmers were soon to prove as indifferent to the valley's fate as the members of the McNally Ditch. When Mulholland shrewdly responded with ever higher offers for the cooperative's water rights, a majority (not including the Watterson brothers) finally agreed to sell out for a price of $15,000 per second-foot, twice what the city had paid for the McNally Ditch rights. Mulholland was jubilant, but victory carried a heavy price. To satisfy his vendetta against his oldest friend, he had now spent twice what the Long Valley damsite would have cost, and made himself evil incarnate throughout an entire valley as well.

As the farmers who held out felt increasingly alone, their methods grew more and more violent. On May 21, 1924, a group of men "broke" into the Watterson brothers' warehouse, "stole" three cases of dynamite, and blew a large section of the aqueduct to smithereens. From that moment on, William Mulholland refused to refer to anyone in the Owens Valley by any other name than "dynamiter." Then, in August, Leicester Hall, who had been warned to stay away forever, returned to the valley and was abducted from a restaurant as he ate. He was driven blindfolded to a road's end, where he found himself facing a grim-looking group of men and a noose strung over a tree. Hall saved himself by uttering the Freemason's distress call; there were so many Masons among the valley population that one was in the gang of would-be lynchers, and he managed to talk the others out of murder. But the dynamitings continued. When the Department of Water and Power released a report that recommended "destroying all irrigation"—those were the exact words—in the valley, and it turned out that the main author was Joseph P. Lippincott, the response was a fresh series of blasts. Glasscock's paper was now openly counseling sabotage. The Ku Klux Klan, sensing a perfect battle stage between "Hollywood"—which was to say, cities, big business, liberalism, and Jews—and the small-town, revanchist values it cherished, was sending recruiters into the valley and getting good results. Even Fred Eaton, after holding himself aloof, finally entered the fray against the city of which he had been mayor. "Wherever the hand of Los Angeles has touched Owens Valley," he wrote in a letter to the editor, "it has turned back into desert."

Joseph Lippincott, whose one admirable quality may have been prescience, had said twenty years earlier that the Owens Valley was doomed as soon as Los Angeles obtained its first water right. Mulholland, however, kept insisting blindly that the valley could live on—he didn't say how—even as he turned life there into a kind of hell.

No one knew when his neighbor would be approached and persuaded to sell out; no one knew when the city would move to condemn; no one knew when the armed guards who patrolled the aqueduct would receive orders to shoot to kill. "Suspicions are mutual and widespread," a visitor from Los Angeles observed. "The valley people are suspicious of each other, suspicious of newcomers, suspicious of city men, suspicious, in short, of almost everybody and everything. . . . Owens Valley is full of whisperings, mutterings, recriminations. . . ." It seemed only a matter of time before the onset of real war.

On November 16, 1924, as the drought continued to hold Los Angeles in a deadly grip, a caravan of automobiles rumbled slowly southward through the town of Independence. In the first car, behind drawn blinds, sat the grim figure of Mark Watterson. The cars turned toward the Alabama Hills, a small range of barren rises at the foot of the Sierra escarpment. Weaving through the hills was the Owens River aqueduct, and somewhere along its course were the Alabama Gates. In wetter times, the gates had turned floodwaters in the aqueduct onto the desert to keep them from straining the capacity of the siphons below. They hadn't been used in years, but they still worked. When the caravan arrived at the gatehouse, a hundred men got out of the cars, walked up to the spillway, and turned the five huge wheels that moved the weirs. For the first time in many years, the Owens River flowed back across the desert into Owens Lake.

The effect of the seizure was electrifying. Mulholland was in a murderous rage. He dispatched two carloads of armed city detectives to take back the gates, but news of their imminent arrival prompted the local sheriff to go down to meet them. "If you go up there and start trouble," he told the detectives, "I don't believe you will live to tell the tale." They never went. Mulholland, in the meantime, secured a court injunction against the seizure, but when the papers were served to the men at the gates they threw them into the water.

And then, to everyone's surprise, what could easily have produced bloodshed turned into a picnic. Wives, children, grandmothers, and dogs joined the lawbreakers. Tom Mix was filming a movie nearby, and when he heard what was happening he sent over his salutations and his orchestra. By evening a huge cloud of smoke began to rise from the scene, but it came from a barbecue pit. After dinner, the sheriff arrived and joined in. The crowd was now seven hundred strong, and the strains of "Onward Christian Soldiers" filled the desert night.

Events were finally swinging to the Owens Valley's side. To Mulholland's disgust, even the Los Angeles *Times*, now that Otis was dead, was sympathizing with the lawbreakers. "These farmers are not anarchists or bomb throwers," it said in an editorial, "but in the main honest, hardworking American citizens. They have put themselves hopelessly in the wrong by taking the law into their own hands, but that is not to say that there has not been a measure of justice on their side." Meanwhile, as Mark Watterson led the seizure of the Alabama Gates, Wilfred had wisely gone to Los Angeles to closet himself with the Joint Clearinghouse Association, a roundtable of the city's bankers. After several hours, he emerged and sent Mark a telegram. "If the object of the crowd at the spillway is to bring their wrongs to the attention of the citizens of Los Angeles, they have done so one hundred percent," he wired. "I feel sure that the wrongs done will be remedied."

But such a simple happy ending could occur only on a Hollywood movie lot. As soon as the Alabama Gates were released and Wilfred Watterson had returned home, the bankers with whom he had met rejected his price for the consolidated valley water rights, to which he swore they had agreed. Meanwhile, Mulholland's public relations department was flooding the state with a booklet "explaining" the Owens Valley crisis. "Never in its history has the Owens Valley prospered and increased in wealth as it has in the past twenty years," it said. And it was true, as long as you looked at only the first nineteen of those years; in the twelve subsequent months, the city had almost brought the valley to its knees. Shops and stores were closing for lack of business—thousands of people had already moved out—but Mulholland dismissed pleas for reparations out of hand. If business was down, he said, the shopkeepers could move, too.

The first order to shoot to kill came on May 28, 1927, a day after the No Name Siphon, a huge pipe across a Mojave hill, lay in shards, demolished by a tremendous blast of dynamite. As city crews hauled in 450 feet of new twelve-foot pipe, another blast destroyed sixty feet of the aqueduct near Big Pine Creek. On June 4, another 150 feet went sky high. In response, a special train loaded with city detectives armed with high-velocity Winchester carbines and machine guns rolled out of Union Station for the Owens Valley. Roadblocks were erected on the highways; all cars with male occupants were searched; floodlights beamed across the valley as if it were a giant penitentiary. Miraculously, though the Owens Valley water war had gone on for more than twenty years, though it had turned violent during the past three, there were still no corpses. Harry Glasscock, however, was predicting in his editorial columns that the aqueduct would "run red with human blood," and no one was prepared to argue with him. But before it could happen fate cast a plague on both houses. First came the collapse of the Watterson banks, and the revelation that the Owens Valley's leading citizens were felons. Then, a few months later, came the collapse of the Saint Francis Dam.

The relationship between George Watterson and his two nephews had gone from one of competitiveness to one of bitterness to one of rancid hatred. In the early months of 1927, George saw his opportunity to invest in their final ruin. Four years of drought and rapidly declining business had left all five branches of the Inyo County Bank severely weakened. At the same time, the election of a new governor, Clement Young, on a huge infusion of campaign cash from A. P. Giannini and his Bank of Italy had resulted in the liberalization of the state banking laws, mainly to Giannini's advantage. It was no surprise, then, when George Watterson filed, in the name of the Bank of Italy, an application to launch a competitive bank in Inyo County. But it was no surprise either when the state banking commissioner voided the application on the strength of Wilfred Watterson's testimony that the bank was a front which Los Angeles would use to drive the valley into submission. Nor was it a surprise when, in response, an infuriated George Watterson, with considerable help from the Los Angeles Department of Water and Power, began a dirt-gathering investigation into his nephews' bank. The surprise was what they ultimately found.

To say that the Wattersons had played fast and loose with their investors' capital was an understatement. For at least the past two years, they had been using the amalgamated capital of the Owens Valley to shore up their failing financial empire—their resort, the mineral water company, their tungsten mine. They had recorded deposits in other banks that were never made, recorded debits that were already paid, entered balances that never existed on ledger sheets.

They had loaned the entire life savings of their friends and neighbors to enterprises which were, at best, unlikely to succeed. When it was all tallied, there was a $2.3 million discrepancy between the bank books and reality. The brothers had always been the valley's best and last hope. Now they were going to go to jail for embezzlement and fraud.

They had done it, they said, for the good of the valley, and as outrageous as it sounded, it was probably true. None of the money had ever left Inyo County. With the irrigation economy dying at the hands of Los Angeles, the valley's only chance of surviving at all was to develop its minerals, its mining, its potential for tourism. During the trial, people who had lost everything nodded and agreed. Even as the Wattersons were being charged with thirty-six counts of embezzlement and grand theft, the citizens of Owens Valley were pledging $1 million to keep them in business.

It was too late. On August 4, 1927, all five banks were permanently closed. People wandered over to gawk at their final sign of defeat, a bitter message posted on the door: "This result has been brought about by the past four years of destructive work carried on by the city of Los Angeles."

The prosecuting attorney was a lifelong friend of both Wilfred and Mark. If he had not been the prosecutor, he said, he would have agreed to be a character witness. He cried openly as he made his final argument, and the judge and jury wept along with him. On November 14, the Wattersons were sent to San Quentin for ten years, later reduced to six. As the train taking them to San Francisco passed outside Bishop, someone was putting up a sign. It read, "Los Angeles City Limits."

William Mulholland had only four months to savor his triumph.

By refusing to pay Fred Eaton the $1 million he wanted for his reservoir site, Mulholland had left himself short of water storage capacity. It was a serious situation to begin with, and it was compounded by the drought, the dynamitings, and the phenomenal continuing influx of people. His power dams were also running day and night, spilling water into the ocean before it could be reused. The water he had obtained at such expense and grief was being wasted. As a result, he turned to the dam he had under construction in San Francisquito Canyon, and, ignoring the advice of his own engineers, decided to make it larger.

The reservoir behind the enlarged Saint Francis Dam reached its capacity of 11.4 billion gallons in early March of 1928, and immediately began to leak. Few dams fail to leak when they are new, but if they are sound they leak clear water. The water seeping around the abutment of the Saint Francis Dam was brown. It was a telltale sign that water was seeping through the canyon walls, softening the mica shale and conglomerate abutment.

It was also a sign that William Mulholland chose, if not exactly to ignore, then to disbelieve. After all, it was *his* dam. Would the greatest engineering deparment in the entire world build an unsafe dam? To reassure the public, Mulholland and his chief engineer rode out to the site on March 12 for an inspection. The last of the season's rains was falling, and muddy water was running from a nearby construction site. After a perfunctory look, Mulholland decided that the site was the source of the mud, and pronounced the dam safe. On the same night, at a few minutes before midnight, its abutment turned to Jell-O, and the reservoir awoke from its deceptive slumber and tore the dam apart.

There are few earthly phenomena more awesome than a flood, and there is no flood more awesome than several years' accumulation of rainfall released over the course of an hour or two. The initial surge of water was two hundred feet high, and could have toppled nearly anything in its path—thousand-ton blocks of concrete rode the crest like rafts. Seventy-five families were living in San Francisquito Canyon immediately below the dam. Only one of their members, who managed to claw his way up the canyon wall just before the first wave hit, survived. Ten miles below, the village of Castaic Junction stood where the narrow canyon opened into the broader and flatter Santa Clara Valley. When the surge engulfed the town, it was still seventy-eight feet high. Days later, bodies and bits of Castaic Junction showed up on the beaches near San Diego.

The flood exploded into the Santa Clara River, turned right, and swept through the valley toward the ocean. It tore across a construction camp where 170 men were sleeping, and carried off all but six. A few miles below, Southern California Edison was building a project and had erected a tent city for 140 men. At first, the night watchman thought it was an avalanche. As it dawned on him that the nearest snow was fifty miles away, the flood crest hit, forty feet high. The men who survived were those who didn't have time to unzip their canvas tents, which were tight enough to float downstream like rafts. Eighty-four others died.

When the flood went through Piru, Fillmore, and Santa Paula it was semisolid, a battering ram congealed by homes, wagons, telephone poles, cars, and mud. Wooden bridges and buildings were instantaneously smashed to bits. A woman and her three children clung to a floating mattress until it snagged in the upper branches of a tree. They survived. A rancher who heard the deluge coming loaded his family in his truck and began to dash to safety. As he stopped by his neighbors' house and ran to the door to warn them, the flood arrived and swept his family out to sea. A four-room house was dislodged and floated a mile downstream without a piece of furniture rearranged; when the dazed owners came to inspect it, they found their lamps still upright on their living-room tables. A brave driver trying to outrace the flood could not bring himself to pass the people waving desperately along the way; his car held fourteen corpses when it was hauled out of the mud. The flood went on, barely missing Saticoy and Montalvo, and, at five o'clock in the morning, went by Ventura and spent itself at sea.

Hundreds of people were dead, twelve hundred homes were demolished, and the topsoil from eight thousand acres of farmland was gone. William Mulholland, whose career lay amid the ruins, was still alive, but as he addressed the coroner's inquest he bent his head and murmured, "I envy the dead." After a feeble effort to put the blame on "dynamiters," he took full responsibility for the disaster.

But the great city his aqueduct had created was, for the moment at least, willing to forgive him. "Chief Engineer Mulholland was a pitiable figure as he appeared before the Water and Power Commission yesterday," the Los Angeles *Times* reported on March 16. "His figure was bowed, his face lined with worry and suffering. . . . Every commissioner had the deepest sympathy for the man who has spent his life for the service of the people of Los Angeles . . . his Irish heart is kind, tender, and sympathetic."

Nine separate investigations eventually probed the collapse of the Saint Francis Dam. No one is even sure how many lives were lost, but a likely total is around 450: it would become one of the dozen worst peacetime disasters in American history. The precise cause of the collapse was never officially determined, but when an investigator dropped a piece of the rock abutment into a glass of water, it dissolved in a few minutes. It was also learned that Mulholland had ordered the reservoir filled fast—a violation of a cardinal engineering rule—because he didn't want Owens River water to go to waste.

The city took full responsibility for all losses and paid most of the claims without contest, which cost it close to $15 million. For much less than that, Mulholland not only could have bought the Long Valley site, but built the dam, too.

In the ensuing months, in hearing after hearing, Mulholland was dragged through an agonizing reappraisal of his career. It was learned that two other dams in whose design and construction he participated as a consultant eventually collapsed, and a third had to be abandoned when partially built. He was a bold engineer, an innovative engineer; he was also a reckless, arrogant, and inexcusably careless engineer. His fall from grace was slow, awful, and complete. By the time he wearily resigned, in November of 1928, at the age of seventy-three, his reputation was sullied beyond redemption. His wit and his combativeness vanished in retirement, and even in the company of his per-fervidly loyal children he often lacked the energy to speak. He told them, "The zest for living is gone."

The city finally settled with Fred Eaton, who lost almost everything in the collapse of the Watterson banks, for $650,000. A few weeks later, the two old and broken men moved to heal their twenty-year rift. Lost in despondency at home, Mulholland received a message that Eaton, who had since returned to Los Angeles, would like to see him. Without a word, he got his hat and strode out the door. Eaton had suffered a stroke; he needed a cane to walk, and he looked ancient. "Hello, Fred," said Mulholland as he approached Eaton's bedside. Then both of them broke down and wept.

The dam in Long Valley was ultimately built, and the reservoir that formed behind it, which was named Lake Crowley in honor of a priest who devoted the latter part of his life to healing the rift between city and valley, was, in its day, one of the largest in the country. By then, however, all hope of fruitful coexistence had died. On a map, the Owens Valley was still there, but it had ceased to exist as a place with its own aspirations, its own destiny. By the mid-1930s, Los Angeles was landlord of 95 percent of the farmland and 85 percent of the property in the towns. In the town of Independence, the Eastern California Museum, which tells the story of the battle largely from the valley's side, sits on land leased from the city.

Los Angeles leased some of the land back to farmers for a while, but the unpredictability of the water supply discouraged most of those who tried to carry on. There might be enough for twenty or thirty thousand acres in wetter years; then there might be enough for only three or four thousand. As the city grew, the river became utterly appropriated; when that happened, the Department of Water and Power sank wells and began depauperating the aquifer, as would hap-pen—as is happening—in so many places in the West. The last of the ranchers quit in the 1950s and the economy shifted to tourism; most of those who remain now pump gas, rent rooms, or serve lunch to the skiers and tourists driving through on Highway 395. By the 1970s, even that tenuous existence was threatened; the aquifer was so drawn-down that desert plants which can normally survive on the meagerest capillary action of groundwater began to die, and the valley went beyond desert and took on the appearance of the Bonneville Salt Flats. When the winds of convection blow, huge clouds of alkaline dust boil off the valley floor; people now live in the Owens Valley at some risk to their health. The city has refused every request that it limit its groundwater pumping, just as it has refused to stop diverting the creeks that feed Mono Lake to the north—another casualty of its unquenchable thirst. Some sporadic dynamitings began to occur again in the 1970s, and reporters arrived eager to cover the "second Owens Valley War," but the war was long since over—there was nothing left to win.

As for Otis, Chandler, Sherman, and the rest of the syndicate that called itself the San Fernando Mission Land Company, they became rich—phenomenally rich. While presiding over the San Fernando Valley's metamorphosis from desert to agricultural cornucopia, they used the profits to constantly acquire more land. In 1911, Chandler, Otis, and Sherman purchased another 47,500 acres nearby and began to develop them—the biggest subdivision in the world. Within a year, they were assembling the third-largest land empire in the history of the state, the 300,000-acre Tejon Ranch, straddling Los Angeles and Kern counties. (Besides the Los Angeles *Times*, the Tejon Ranch, undiminished in size, remains the principal local asset of the Chandler family.) In a speech given in 1912, Theodore Roosevelt singled out Otis as "a curious instance of the anarchy of soul which comes to a man who in conscienceless fashion deifies property at the expense of human rights." But Roosevelt, as much as anyone, was responsible for setting this anarchic soul loose. No one knows how great a profit the syndicate realized from the initial seventeen thousand San Fernando acres, but one writer, William Kahrl, estimates that Chandler was worth as much as $500 million when he died, and the San Fernando Valley was the soil from which this incredible fortune grew. It may not have been the most lucrative land scam in United States history, but it ranked somewhere near the top.

Between the arrival of William Mulholland and his death, Los Angeles grew from a town of fifteen thousand into the then most populous desert city on earth. Today it is the second-largest, barely surpassed by Cairo. Its obsessive search for more water, however, was never to end. While Lake Crowley was filling, the city was already completing its aqueduct to the Colorado River, whose construction almost precipitated a shooting war with Arizona, a rival as formidable as the Owens Valley was weak. And though the first Colorado River aqueduct was supposed to end its water famines forever—as was the Owens River aqueduct—the city was soon planning a second Colorado River Aqueduct and plotting to seize half of the Feather River, six hundred miles away, at the same time. No sooner had it managed to do all of that than the city fathers were secretly meeting with the Bureau of Reclamation, mapping diversions from rivers a thousand miles distant in Oregon and Washington. Like the Red Queen, Los Angeles runs faster and faster to stay in place.

No one says or remembers much about the Reclamation Service's involvement in the Owens Valley story, which is ironic, because nothing in its history may have affected the interests of the nation-at-large quite as much. Almost as soon as it was created—well before it metamorphosed into the mighty Bureau of Reclamation—the agency found itself working on behalf of the wealthy and powerful and against the interests of the constituency it was created to protect, the small western irrigation farmer. In California, to a surprising degree, it has done so ever since. Small farmers do not matter much in the worldly scheme of things; if they did, their numbers would not be declining by the tens of thousands every year. But large farmers do, and explosively growing desert cities do, too, and the Bureau of Reclamation, after learning this lesson in the Owens Valley, would remember it well. Its largest dam is San Luis in central California; its most magnificent dam is Hoover. Above all, the Bureau loves to build great dams, and were it not for Los Angeles, the odds are low that either Hoover or San Luis would exist.

The Owens River created Los Angeles, letting a great city grow where common sense dictated that one should never be, but one could just as well say that it ruined Los Angeles, too. The annexation of the San Fernando Valley, a direct result of the aqueduct, instantly made it the largest city in the world in geographic size. From that moment, it was doomed to become a huge, sprawling, one-story conurbation, hopelessly dependent on the automobile. The Owens River made Los Angeles large enough and wealthy enough to go out and capture any river within six hundred miles, and that made it larger, wealthier, and a good deal more awful. It is the only megalopolis in North America which is mentioned in the same breath as Mexico City or Djakarta— a place whose insoluble excesses raise the specter of some majestic, stately kind of collapse. In *The Water Seekers*, Remi Nadeau, a city historian, says, "They brought in so much water for so many people that few cared any more whether Los Angeles grew at all. . . . Indeed, one might say that . . . they have brought in too much water. For if California now has enough water to more than double in population, then much of California is doomed to be insufferable."

That, in any event, is the way it appears some days from atop Mulholland Drive.

CHAPTER THREE

First Causes

When archaeologists from some other planet sift through the bleached bones of our civilization, they may well conclude that our temples were dams. Imponderably massive, constructed with exquisite care, our dams will outlast anything else we have built—skyscrapers, cathedrals, bridges, even nuclear power plants. When forests push through the rotting streets of New York and the Empire State Building is a crumbling hulk, Hoover Dam will sit astride the Colorado River much as it does today—intact, formidable, serene.

The permanence of our dams will merely impress the archaeologists; their numbers will leave them in awe. In this century, something like a quarter of a million have been built in the United States alone. If you ignore the earthen plugs thrown across freshets and small creeks to water stock or raise bass, then fifty thousand or so remain. These, in the lexicon of the civil engineer, are "major works." Even most of the major works are less than awesome, damming rivers like the Shepaug, the Verdigris, Pilarcitos Creek, Mossman's Brook, and the North Fork of the Jump. Forget about them, and you are left with a couple of thousand really big dams, the thought of whose construction staggers the imagination. They hold back rivers our ancestors thought could never be tamed—the Columbia, the Tennessee, the Sacramento, the Snake, the Savannah, the Red, the Colorado. They are sixty stories high or four miles long; they contain enough concrete to pave an interstate highway from end to end.

These are the dams that will make the archaeologists blink—and wonder. Did we overreach ourselves trying to build them? Did our civilization fall apart when they silted up? Why did we feel compelled to build so many? Why five dozen on the Missouri and its major tributaries? Why twenty-five on the Tennessee? Why fourteen on the Stanislaus River's short run from the Sierra Nevada to the sea?

We know surprisingly little about vanished civilizations whose majesty and whose ultimate demise were closely linked to liberties they took with water. Unlike ourselves, future archaeologists will have the benefit of written records, of time capsules and so forth. But such things are as apt to confuse as to enlighten. What, for example, will archaeologists make of Congressional debates over Tellico Dam, where the vast majority ridiculed the dam, excoriated it, flagellated it—and then allowed it to be built? What will they think of Congressmen voting for water projects like Central Arizona and Tennessee-Tombigbee—projects costing three or four billion dollars in an age of astronomical deficits—when Congress's own fact-finding committees asserted or implied that they made little sense?

Such debates and documents may shed light on reasons—rational or otherwise—but they will be of little help in explaining the psychological imperative that drove us to build dam after dam after dam. If there is a Braudel or a Gibbon in the future, however, he may deduce that the historical foundations of dams as monumental as Grand Coulee, of projects as nonsensical as Tennessee-Tombigbee, are sunk in the 1880s, a decade which brought, in quick succession, a terrible blizzard, a terrible drought, and a terrible flood.

The great white winter of 1886 came first. The jet stream drove northward, grazed the Arctic Circle, then dipped sharply southward, a parabolic curve rushing frigid air into the plains. Through December of 1886, the temperature in South Dakota barely struggled above zero. A brief thaw intervened in January, followed by a succession of monstrous Arctic storms. Week after week, the temperature fell to bottomless depths; in the Dakotas, the windchill factor approached a hundred below. Trapped for weeks, even for months, in a warp of frozen treeless prairie, thousands of pioneers literally lost their minds. As the last of the chairs were being chopped and burned, settlers contemplated a desperate hike to the nearest town, unable to decide whether it was crazier to stay or to leave. No one knows how many lost their lives, but when the spring thaw finally came, whole families were discovered clutching their last potatoes or each other, ice encrusted on their staring, vacant eyes.

But the settlers' suffering was merciful compared to that of their cows. On the woodless plains, barns were rare. Cattle were turned out into blizzards to survive by their wits, which they don't have, and which wouldn't have done them much good anyway. They were found piled by the hundreds at the corners of fenced quarter sections, all facing southeast; even when a storm abated, the survivors were too traumatized to turn around, and they died a night or two later under a listless winter's moon. It was a winter not just of horrendous cold but of gigantic snows, horizontal broadsides that reduced visibility to zero and stung the cattle like showers of needles. Twenty-foot drifts filled the valleys and swales, covering whatever frozen grass was left to eat. At night families would lie awake listening to their cows' dreadful bawls, afraid to go out and have the wind steal their last resources of warmth. Anyway, there was nothing they could do.

The toll was never officially recorded. Most estimates put the loss of cattle at around 35 percent, but in some regions it may have been nearer 75 percent. In sheer numbers, enough cows died to feed the nation for a couple of years. Much of the plains' cattle industry was in financial ruin. The bankrupt cattle barons dismissed thousands of hired hands, who were forced to find new careers. When the snows of 1886 melted, Robert Leroy Parker, a young drover, cattle rustler, and part-time bank robber with a reputation, had more recruits on his hands than he knew what to do with. He organized them into a gang known as the Wild Bunch and called himself Butch Cassidy. The Wild Bunch and the scores of outlaw bands like them worked the banks, the railroads, and the Pinkerton agents into a murderous froth. To others, however, they were a moral weight on the mind. Many of the outlaws had been "good boys," former ranch hands and farmers, occupations that everyone hoped would domesticate the West and cure it of its cyclical agonies of boom and bust. But weather was the ultimate arbiter in the American West. Unless there was some way to control it, or at least minimize its effects, a good third of the nation might remain uninhabitable forever.

As if to confirm such a prophecy, the decade following the great white winter was a decade when the western half of the continent decided to dry up. Like most droughts, this one came gradually, building up force, nibbling away at the settlers' fortunes as inexorably as their cattle nibbled away the dying grass. The sun, to which the settlers had so recently offered prayerful thanks, turned into a despotic orb; as Hamlin Garland wrote, "The sky began to scare us with its light." In July of 1888, at Bennett, Colorado, the temperature rose to 118, a record that has never since been equaled in the state. It was the same throughout the West, as an immense high-pressure zone sat immobile across the plains. Orographic clouds promising rain formed over the Rockies, were boiled off in midair, and disappeared. The atmosphere, it seemed, had been permanently sucked dry.

By 1890, the third year of the drought, it was obvious that the theory that rain follows the plow was a preposterous fraud. The people of the plains states, still shell-shocked by the great white winter, began to turn back east. The populations of Kansas and Nebraska declined by between one-quarter and one-half. Tens of thousands went to the wetter Oklahoma territory, which the federal government usurped from the five Indian tribes to whom it had been promised in perpetuity and offered to anyone who got there first. Meanwhile, the windmills of the farmers who remained north were pumping up sand instead of water, and the huge dark clouds on the horizon were not rain but dust. The great cattle freeze of the white winter had been, in retrospect, a blessing in disguise. Had several million more cows been around to graze the dying prairie grasses to their roots, the Dust Bowl of the 1930s could have arrived half a century early.

When statistics were collected a few years later, only 400,000 home-steading families had managed to persevere on the plains, of more than a million who tried. The Homestead Acts had been a relative success in the East; west of the hundredth meridian, however, they were for the most part a failure, even a catastrophic failure. Much of the blame rested on flaws in the acts themselves, and on the imper-fections of human nature, but a lot of it was the fault of the weather. How could you settle a region where you nearly froze to death one year and expired from heat and lack of water during the next eight or nine?

The drought that struck the West in the late 1880s did not occlude the entire continent. In the spring of 1889, the jet stream that had bypassed the West was feeding a thoroughfare of ocean moisture into the eastern states. In the mountains of Pennsylvania, it rained more or less continuously for weeks. The Allegheny and Susquehanna rivers became swollen surges of molten mud. Above Johnstown, Pennsyl-vania, on the South Fork of the Conemaugh River, a tributary of the Allegheny, sat a big earthfill dam built thirty-seven years earlier by the Pennsylvania Canal Company; it was, for a while, the largest dam in the world. Pounded by the rains, infiltrated by the waters of the rising reservoir, the dam was quietly turning into Cream of Wheat. On May 31, with a sudden flatulent shudder, it dissolved. Sixteen billion gallons of water dropped like a bomb on the town below. Before anyone had time to flee, Johnstown was swallowed by a thirty-foot wave. When the reservoir was finally in the Allegheny River, sending it far over its banks, the town had disappeared. Four hundred corpses were never positively identified. The number of dead was eventually put at twenty-two hundred—twice as many casualties as in the burning of the *General Slocum* on the East River in 1904; many more than in the San Francisco earthquake and fire; nine times as many as in the Chicago fire. The only single disaster in American history that took more lives was the hurricane that struck Galveston, Texas, eleven years later. The Johnstown flood was significant if only for this sheer loss of life; but it was also an indictment of privately built dams.

The rapid rise of the federal irrigation movement in the early 1890s was due in part to this succession of overawing catastrophes. But it had just as much to do with the fact that by the late 1880s, private irrigation efforts had come to an inglorious end. The good sites were simply gone. Most of the pioneers who had settled successfully across the hundredth meridian had gone to Washington and California and Oregon, where there was rain, or had chosen homesteads along streams whose water they could easily divert. Such opportunities, however, were quick to disappear. Groundwater wasn't much help either. A windmill could lift enough drinking water for a family and few cattle; but it would require thirty or forty windmills, and reliable wind, to lift enough water to irrigate a quarter section of land—a disheartening prospect to a farmer with no money in a region with no wood.

Even if their land abutted a stream with some surplus water rights, few farmers had the confidence, cooperative spirit, and money to build a dam and lead the stored water to their lands through a long canal. It was one thing to throw a ten-foot-high earthen plug across a freshet in order to create a two-acre stock pond—though even that taxed the resources of most farmers in the West, who had invested all their savings simply to get there from Kentucky or Maine. It was quite another thing to build a dam on a stream large enough to supply a year-round flow, and to dig a canal—by horse and by hand—that was long enough, and deep enough, and wide enough, to irrigate hundreds or thousands of acres of land. The work involved was simply stupefying; clearing a field, by comparison, seemed like the simplest, most effortless job.

The farmers' predicament, on the other hand, was an opportunity for the legions of financial swashbucklers who had gone west in pursuit of quick wealth. In the 1870s and 1880s, hundreds of irrigation companies, formed with eastern capital, set themselves to the task of reclaiming the arid lands. Almost none survived beyond ten years. At the eighth National Irrigation Congress in 1898, a Colorado legislator likened the American West to a graveyard, littered with the "crushed and mangled skeletons of defunct [irrigation] corporations . . . [which] suddenly disappeared at the end of brief careers, leaving only a few defaulted obligations to indicate the route by which they departed."

There was, indeed, a kind of cruel irony in the collapse of the irrigation companies. Most of them operated in the emphatically arid regions—the Central Valley of California, Nevada, Arizona, southeastern Colorado, New Mexico—where agriculture without irrigation is daunting or hopeless, but otherwise the climate is well suited for growing crops. The drought, on the other hand, struck hardest in the region just east of the hundredth meridian, where, in most years, a nonirrigating farmer had been able to make a go of it. Kansas was emptied by the drought and the white winter, Nevada by irrigation companies gone defunct. In the early 1890s, the exodus from Nevada, as a percentage of those who hung on, was unlike anything in the country's history. Even California, in the midst of a big population boom, saw the growth of its *agricultural* population come to a standstill in 1895.

California, the perennial trend-setting state, was the first to attempt to rescue its hapless farmers, but the result, the Wright Act, was another in the long series of doomed efforts to apply eastern solutions to western topography and climate. The act, which took its inspiration from the township governments of New England, established self-governing mini-states, called irrigation districts, whose sole function was to deliver water onto barren land. Like the western homestead laws, it was a good idea that foundered in practice. The districts soon buckled under their responsibilities—issuing bonds that wouldn't sell, building reservoirs that wouldn't fill, allocating water unfairly, distributing it unevenly, then throwing up their hands when anarchy prevailed. Elwood C. Mead, then the state engineer of Wyoming and probably the country's leading authority on irrigation, called the Wright Act "a disgrace to any self-governing people." George Maxwell, a Californian and founder of the National Irrigation Association, said "the extravagance or stupidity or incompetence of local [irrigation] directors" had left little beyond a legacy of "waste and disaster." Though the Wright Act was in most ways a failure, Colorado, thinking it had learned something from California's mistakes, adopted its own version, which added a modest subsidy for private irrigation developers in order to improve their odds of success. By 1894, under Colorado's new program, five substantial storage reservoirs had been built. Three were so poorly designed and situated that they stored no water at all; the fourth was declared unsafe and was never even filled;

and the fifth was so far from the land it was supposed to irrigate that most of the meager quantity of water it could deliver disappeared into the ground before it got there.

In that same year—1894—Senator Joseph Carey of Wyoming, thinking he had learned something from California's and Colorado's mistakes, introduced a bill that offered another approach: the federal government would cede up to a million acres of land to any state that promised to irrigate it. But, by some elusive reasoning, the states were forbidden to use land as the collateral they would need to raise the money to build the irrigation works—and land, at the time, was the only thing of value most of them had. Sixteen years later, using a generous estimate, the Carey Act had caused 288,553 acres to come under irrigation throughout the entire seventeen-state West—about as much developed farmland as there was in a couple of counties in Illinois.

As the private and state-fostered experiments with irrigation lay in shambles, many of the western reclamation advocates heaped blame on the East and "Washington" for not doing more to help, just as their descendants, four generations later, would vilify Jimmy Carter, an easterner and southerner, for not "understanding" their "needs" when he tried to eliminate some water projects that would have subsidized a few hundred of them to the tune of hundreds of thousands of dollars apiece. In each case, the West was displaying its peculiarly stubborn brand of hypocrisy and blindness. Midwestern members of Congress were understandably uneager to subsidize competition for their own farmer constitutents, but they had little to do with making reclamation fail; the West was up to the task itself. Its faith in private enterprise was nearly as absolute as its earlier faith that settlement would make the climate wetter. John Wesley Powell, a midwesterner, knew that all the private initiative in the world would never make it bloom. Theodore Roosevelt, an easterner, had returned from the West convinced that there were "vast areas of public land which can be made available for . . . settlement," but only, he added, "by building reservoirs and main-line canals impractical for private enterprise." But the West wasn't listening. For the first time in their history, Americans had come up against a problem they could not begin to master with traditional American solutions—private capital, individual initiative, hard work—and yet the region confronting the problem happened to believe most fervently in such solutions. Through the 1890s, western Senators and Congressmen resisted all suggestions that reclamation was a task for government alone—not even for the states, which had failed as badly as the private companies, but for the national government. To believe such a thing was to imply that their constituents did not measure up to the myth that enshrouded them—that of the indomitable individualist. When they finally saw the light, however, their attitude miraculously changed—though the myth didn't—and the American West quietly became the first and most durable example of the modern welfare state.

The passage of the Reclamation Act of 1902 was such a sharp left turn in the course of American politics that historians still gather and argue over why it was passed. To some, it was America's first flirtation with socialism, an outgrowth of the Populist and Progressive movements of the time. To others, it was a disguised reactionary measure, an effort to relieve the mobbed and riotous conditions of the eastern industrial cities—an act to save heartless capitalism from itself. To some, its roots were in Manifest Destiny, whose incantations still held people in their sway; to others, it was a military ploy to protect and populate America's western flank against the ascendant Orient.

What seems beyond question is that the Reclamation Act, or some variation of it, was, by the end of the nineteenth century, inevitable. To resist a federal reclamation program was to block all further migration to the West and to ensure disaster for those who were already there—or for those who were on their way. Even as the victims of the great white winter and the drought of the 1880s and 1890s were evacuating the arid regions, the trains departing Chicago and St. Louis for points west were full. The pull of the West reached deep into the squalid slums of the eastern cities; it reached back to the ravined, rock-strewn farms of New England and down into the boggy, overwet farmlands of the Deep South. No matter what the government did, short of erecting a wall at the hundredth meridian, the settlement of the West was going to continue. The only way to prevent more cycles of disaster was to build a civilization based on irrigated farming. Fifty years of effort by countless numbers of people had resulted in 3,631,000 acres under irrigation by 1889. There were counties in California that contained more acreage than that, and the figure included much of the easily irrigable land. Not only that, but at least half the land had been irrigated by Mormons. Each additional acre, therefore, would be won at greater pain. Everything had been tried—cheap land, free land, private initiative, local initiative, state subsidy—and everything, with a few notable exceptions, had failed. One alternative remained.

There seemed to be only one politician in the arid West who fathomed his region's predicament well enough to end it. He had emigrated to San Francisco from the East, made a fortune through a busy law

practice and the inheritance of his father-in-law's silver mine, moved
to Nevada, and in 1888 launched the Truckee Irrigation Project. It was
one of the most ambitious reclamation efforts of its day, and it failed—
not because it was poorly conceived or executed (hydrologically and
economically, it was a good project) but because squabbles among its
beneficiaries and the pettiness of the Nevada legislature ruined its
hopes. In the process Francis Griffith Newlands lost half a million
dollars and whatever faith he had in the ability of private enterprise
to mount a successful reclamation program. "Nevada," he said bitterly
as his project went bust in 1891, "is a dying state."

Newlands, who succeeded at everything else he tried, gave up on
irrigation, ran for Congress, and won. For the remainder of the decade,
he kept out of the reclamation battles, if only to give everyone else's
solutions an opportunity to fail. All the while, however, he was waiting
for his moment. It came on September 14, 1901, when a bullet fired
by an anarchist ended the life of President William McKinley.

Theodore Roosevelt, the man who succeeded McKinley as President, was, like Francis Newlands, a student and admirer of John Wesley Powell. Infatuated with the West, he had traveled extensively there
and been struck by the prescience and accuracy of Powell's observations. Roosevelt was first of all a politician, and had no interest in
sharing Powell's ignominious fate; nonetheless, he knew that Powell's
solutions were the only ones that would work, and he wanted a federal
reclamation effort badly. A military thinker, he was concerned about
Japan, bristling with expansionism and dirt-poor in resources, and
knew that America was vulnerable on its underpopulated western
flank. A bug for efficiency, he felt that the waste of money and effort
on doomed irrigation ventures was a scandal. Roosevelt was also a
conservationist, in the utilitarian sense, and the failure to conserve—
that is, use—the water in western rivers irritated him. "The western
half of the United States would sustain a population greater than that
of our whole country today if the waters that now run to waste were
saved and used for irrigation," he said in a speech in December of
1901. For all his enthusiasm, however, Roosevelt knew that his biggest
problem would be not the eastern states in Congress but the myth-
bound western bloc, whose region he was trying to help. His second-
greatest problem, ironically, would be his chief ally, Francis Newlands.

As soon as Roosevelt was in the White House, Newlands introduced
a bill creating a federal program along the lines suggested by Powell.
But the bitterness he felt over his huge financial loss was so strong
that he described his bill in language almost calculated to infuriate
his western colleagues, who were clinging to the myth that the hostile

natural forces of the West could be overcome by individual initiative. In a long speech on the floor of Congress, Newlands said outright that the legislation he was introducing would "nationalize the works of irrigation"—which was like saying today that one intended to nationalize the automobile industry. Then he launched into a long harangue about the failures of state reclamation programs, blaming them on "the ignorance, the improvidence, and the dishonesty of local legislatures"—even though many of his listeners had recently graduated from such legislatures themselves. He even suggested that *Congress* should have no oversight powers, implying that he distrusted that body as much as he did the thieves, opportunists, and incompetents whom he saw controlling the state legislatures.

Newlands' bill, as expected, ran into immediate opposition. When it came up for a vote in March, it was soundly defeated. Western members then began to support a rival bill, proposed by Senator Francis E. Warren of Wyoming, that contained none of the features Newlands wanted. By February of 1902, Warren's bill was finally passed by the Senate and seemed destined to become law. At that point, however, fate and Theodore Roosevelt intervened. Mrs. Warren became gravely ill, necessitating the Senator's return to Wyoming. In Warren's absence, Roosevelt leaned on Newlands to tone down his language, and before long the Congressman was describing his defeated measure, which he had already reintroduced, as a "conservative" and "safe" bill. Roosevelt still wouldn't risk supporting it, but he came up with a brilliant ploy. Announcing his "sympathy with the spirit" of Warren's bill, he said he would support it with "a few minor changes." The person whom he wanted to make the changes and lead the bill through Congress was Wyoming's young Congressman-at-large, Frank Mondell, the future Republican leader of the House. Mondell had a weakness for flattery and a less than athletic mind, and Roosevelt was a master at exploiting both. Before long, he had persuaded Mondell to incorporate as "minor changes" in Warren's bill almost all of Newlands' language. Roosevelt then softened up his eastern opposition with some implied threats that their river and harbor projects might be in jeopardy if they did not go along—a strategy that has seen long useful service. By the time Warren returned from Wyoming, Newlands' bill, disguised as his own, had cleared both houses. On June 17, 1902, the Reclamation Act became law.

The newly created Reclamation Service exerted a magnetic pull on the best engineering graduates in the country. The prospect of reclaiming a desert seemed infinitely more satisfying than designing a steel mill in Gary, Indiana, or a power dam in Massachusetts, and the graduates headed west in a fog of idealism, ready to take on the most intractable foe of mankind: the desert. But the desert suffers improvement at a steep price, and the early Reclamation program was as much a disaster as its dams were engineering marvels.

The underlying problems were politics and money. Under the terms of the Reclamation Act, projects were to be financed by a Reclamation Fund, which would be filled initially by revenues from sales of federal land in the western states, then paid back gradually through sales of water to farmers. (It should be mentioned right away that the farmers, under the law, were exempted from paying interest on virtually all of their repayment obligations—a subsidy which was substantial to begin with, and which was to become breathtaking in later decades, as interest rates topped 10 percent. In some cases, the interest exemption alone—which is, of course, an indirect burden on the general taxpayer—has amounted to a subsidy of ninety cents on the dollar.) Section 9 of the Reclamation Act implied, if it didn't require, that all money accruing to the Reclamation Fund from sales of land in any given state should be spent in that state as well. Frederick Newell, the Service's first director, was particularly anxious to locate a few projects in each state anyway, because that might dispel some of the antipathy that had attended the Service's creation. By 1924, twenty-seven projects were completed or under construction. Of those, twenty-one had been initiated before the Service was even half a decade old.

The engineers who staffed the Reclamation Service tended to view themselves as a godlike class performing hydrologic miracles for grateful simpletons who were content to sit in the desert and raise fruit. About soil science, agricultural economics, or drainage they sometimes knew less than the farmers whom they regarded with indulgent contempt. As a result, some of the early projects were to become painful embarrassments, and expensive ones. The soil turned out to be demineralized, alkaline, boron-poisoned; drainage was so poor the irrigation water turned fields into saline swamps; markets for the crops didn't exist; expensive projects with heavy repayment obligations were built in regions where only low-value crops could be grown. In the Bureau of Reclamation's quasi-official history, *Water for the West*, Michael Robinson (the son-in-law of a Commissioner of Reclamation) discreetly admits all of this: "Initially, little consideration was given to the hard realities of irrigated agriculture. Neither aid nor direction was given to settlers in carrying out the difficult and costly work of clearing and leveling the land, digging irrigation ditches, building roads and houses, and transporting crops to remote markets. . . ."

Robinson also acknowledges the political pressures that have bedeviled the Reclamation program ever since it was born. The attitude of most western members of Congress was quaintly hypocritical: after resisting this experiment in pseudosocialism, or even voting against it, they decided, after it became law, that they might as well make the best of it. "The government was immediately flooded with requests for project investigations," Robinson writes. "Local chambers of commerce, real estate interests, and congressmen were convinced their areas were ideal for reclamation development. State legislators and officials joined the chorus of promoters seeking Reclamation projects. . . . Legislative requirements and political pressures sometimes precluded careful, exhaustive surveys of proposed projects. . . . Projects were frequently undertaken with only a sketchy understanding of the area's climate, growing season, soil productivity, and market conditions."

Congress's decision, in passing the act, to ignore much of John Wesley Powell's advice made things worse. Powell had proposed that in those inhospitable regions where only livestock could be raised, settlers should be allowed to homestead 2,560 acres of the public domain—but allocated enough water to irrigate only twenty. The Reclamation Act gave everyone up to 160 acres (a man and wife could jointly farm 320 acres), whether they settled in Mediterranean California or in the frigid interior steppes of Wyoming, where the extremes of climate rival those in Mongolia. You could grow wealthy on 160 acres of lemons in California and starve on 160 acres of irrigated pasture in Wyoming or Montana, but the act was blind to such nuances. And by building so many projects in a rush, the Reclamation Service was repeating its mistakes before it had a chance to learn from them.

All of these problems were compounded by the fact that few settlers had any experience with irrigation farming—nor were they required to. They overwatered and mismanaged their crops; they let their irrigation systems silt up. Many had optimistically filed on more acreage than they had resources to irrigate, and they ended up with repayment obligations on land they were forced to leave fallow. From there, it was a short, swift fall into bankruptcy. Fifty years earlier, the ancestors of the first Reclamation farmers had endured adversity by putting their faith in God and feeding themselves on game. But this was the twentieth century; the game was vanishing, and government was replacing God as the rescuer of last resort. As Michael Robinson wrote, "Western economic and social determinants were changing rapidly. Nineteenth-century irrigation pioneers were better suited to endure hardships than settlers who struggled to survive on Federal Reclamation projects after 1902. In the nineteenth century, wild game was plentiful, livestock could graze on the public domain outside irrigated areas, and the settlers were inured to privation." And so, after a few years of trial and a lot of error, the Reclamation Act began to undergo a long and remarkable series of "reforms."

The first reform was humble—a $20 million loan from the Treasury to the bankrupt Reclamation Fund to keep the program from falling on its face. It was approved in 1910, the same year that Section 9— the ill-advised clause promoting the construction of projects where they couldn't work—was repealed. New projects were also required to have the explicit consent of the President before they were launched. A paper reform, however, is not necessarily a reform in real life. Every Senator still wanted a project in his state; every Congressman wanted one in his district; they didn't care whether they made economic sense or not. The Commissioner of Reclamation and the President were only human. If Congress authorized a bad project and voted funding for it, a President might have good reasons not to veto the bill—especially if it also authorized a lot of things the President *did* want. Congress caught on quickly, and was soon writing "omnibus" authorization bills, in which bad projects were thrown in, willy-nilly, with good ones. (Later, Congress would learn a new trick: attaching sneaky little amendments authorizing particularly wretched projects to legislation dealing with issues such as education and hurricane relief.) As a result, instead of weeding out or discouraging bad projects, the "reforms" began to concentrate on making bad projects work—or, to put it more bluntly, on bailing them out.

The first of these adjustments came in 1914, when the repayment period, which had been set in the act at a rather unrealistic ten years, was extended to twenty. It was quite a liberal adjustment, but failed to produce any measurable results. By 1922, twenty years after the Reclamation Fund began, only 10 percent of the money loaned from the Reclamation Fund had been repaid. Sixty percent of the irrigators—an astounding number—were defaulting on their repayment obligations, even though they paid no interest on irrigation features.

In 1924, Congress commissioned a Fact Finder's report on the Reclamation program, which recommended an even more drastic adjustment—raising the repayment period from twenty years to forty. No sooner was that done, however, than the most chronic and intractable problem of twentieth-century American agriculture began to appear: huge crop surpluses. Production and prices reached record levels during the First World War; when the war ended, production remained high, but crop prices did not. The value of all crops grown on Reclamation land fell from $152 million in 1919 to $83.6 million in 1922—as morose a statistic as the number of farmers in default. With their profits shriveling, the beleaguered farmers were reluctant to pay for water they were beginning to regard as rightful recompense for attempting to civilize the desert, especially when the Reclamation Service, in most cases, didn't dare shut it off when they refused to pay. So Congress took further steps to bail the Reclamation program out, rerouting royalties from oil drilling and potassium mining to the Reclamation Fund on the theory that the West, while being stripped of its mineral resources, ought to get something in return. But even after all these measures had been adopted a number of projects continued to operate at a hopeless loss.

Nonetheless, the psychic value of the Reclamation farms remained high. The only relief in a pitiless desert landscape, their worth was computed in almost ethereal terms, as if they were art. And their investment value to speculators remained high, too. An acre which in pre-project years was worth $5 or $10—if that—was suddenly worth fifty times as much. At such prices, many farmers found the temptation to sell out irresistible; by 1927, at least a third of the Reclamation farmers had. The buyers were usually wealthy speculators who figured they could absorb some minor losses for a while—especially if they could convince Congress to give them tax breaks—as long as they could make money when agricultural prices went back up. The Salt River Project in Arizona was notable for having been all but taken over by speculators. Elwood Mead, who succeeded Newell and Arthur Powell Davis as Commissioner of Reclamation, called speculation "a vampire which has done much to destroy the desirable social and economic purposes of the Reclamation Act." But the big, distant new owners were often better at paying their water bills than the stone-broke small farmers, so the Reclamation Service, in a number of instances, turned a blind eye toward what was going on. It was a case of lawlessness becoming de facto policy, and it was to become more and more commonplace.

Part of the reason the Reclamation Service (which metamorphosed, fittingly, into the *Bureau* of Reclamation in 1923) seemed so hapless at enforcing its social mandate had to do with the Omnibus Adjustment Act of 1926, one of those well-meaning pieces of legislation that make everything worse. Intended to clamp down on speculation, the act demanded that landowners owning excess amounts of land sign recordable contracts in which they promised to sell such lands within a designated period, at prices reflecting the lands' pre-project worth. But the contracts were to be signed with the local irrigation district acting as wholesaler of the Bureau's water—not with the Bureau itself. It was an ideal opportunity to camouflage acreage violations, since the same people who were in violation of the Reclamation Act often sat on the local irrigation district's board of directors.

A more important and insidious reason, however, had to do with the nature of the Bureau itself. "There was a tendency for some engineers to view public works as ends in themselves," admits Michael Robinson. "Despite official declarations from more sensitive administrators that 'Reclamation is measured not in engineering units but in homes and agricultural values' . . . the Service regarded itself as an 'engineering outfit.' "

That may have been the understatement of the year. To build a great dam on a tempestuous river like the Snake was terrifically exhilarating work; enforcing a hodgepodge of social ideals was hardly that. Stopping a wild river was a straightforward job, subjugable to logic, and the result was concrete, heroic, real: a dam. Enforcing repayment obligations and worrying about speculators and excess landowners was a cumbersome, troublesome, time-consuming nuisance— a nuisance without reward. Was the Bureau to abandon the most spellbinding effort of modern times—transforming the desert into a garden—just because a few big landowners were taking advantage of the program, just because some farmers couldn't pay as much as Congress hoped?

There were to be still more "reforms" tacked onto the Reclamation Act: reforms extending the repayment period to fifty years, setting water prices according to the farmers' "ability to pay," using hydroelectric revenues to subsidize irrigation costs. It wasn't until the 1930s, however, that the Reclamation program went into high gear. In the 1920s and early 1930s, the nation's nexus of political power still lay east of the Mississippi River; the West simply didn't have the votes to authorize a dozen big water projects each year. Western politicians who were to exercise near-despotic rule over the Bureau's authorizing committees in later years, men like Wayne Aspinall and Bernie Sisk and Carl Hayden, were still working their way up the political ranks. (In 1902, the year the Reclamation program began, Arizona was still ten years away from becoming a state.) Presidents Harding and Coolidge were ideological conservatives from the East who sternly resisted governmental involvement in economic affairs, unless it was an opportunity for their friends to earn a little graft. And even Herbert Hoover, though a Californian and an engineer, was not regarded by the western water lobby and the Bureau as a particularly loyal friend.

All of this was to change more abruptly than the Bureau of Reclamation and its growing dependency could have hoped. The most auspicious event in its entire history was the election to the presidency in 1932 of a free-wheeling, free-spending patrician. The second most auspicious event was the passage, during the five-term Roosevelt-Truman interregnum, of several omnibus river-basin bills that authorized not one, not five, not even ten, but dozens of dams and irrigation projects at a single stroke. Economics mattered little, if at all; if the irrigation ventures slid into an ocean of debt, the huge hydroelectric dams authorized within the same river basin could generate the necessary revenues to bail them out (or so it was thought). It was a breathtakingly audacious solution to an intractable problem, and the results were to be breathtaking as well. Between Franklin Roosevelt and the river-basin approach—which, in an instant, could authorize dams and canals and irrigation projects from headwaters to river mouth, across a thousand miles of terrain—the natural landscape of the American West, the rivers and deserts and wetlands and canyons, was to undergo a man-made transformation the likes of which no desert civilization has ever seen. The first, and perhaps the most fateful, such transformation was wrought in the most arid and hostile quarter of the American West, a huge desert basin transected by one comparatively miniature river: the Colorado.

CHAPTER FOUR

An American Nile (I)

*Ours was the first and will doubtless be the last party of whites
to visit this profitless locale.*

—Lieutenant Joseph Christmas Ives,
on sailing up the Colorado River
to a point near the present location
of Las Vegas, in 1857

The Colorado is neither the biggest nor the longest river in the American West, nor, except for certain sections described in nineteenth-century journals as "awful" or "appalling," is it the most scenic. Its impressiveness and importance have to do with other things. It is one of the siltiest rivers in the world—the virgin Colorado could carry sediment loads close to those of the much larger Mississippi—and one of the wildest. Its drop of nearly thirteen thousand feet is unequaled in North America, and its constipation-relieving rapids, before dams tamed its flash floods, could have flipped a small freighter. The Colorado's modern notoriety, however, stems not from its wild rapids and plunging canyons but from the fact that it is the most legislated, most debated, and most litigated river in the entire world. It also has more people, more industry, and a more significant economy dependent on it than any comparable river in the world. If the Colorado River suddenly stopped flowing, you would have four years of carryover capacity in the reservoirs before you had to evacuate most of southern California and Arizona and a good portion of Colorado, New Mexico, Utah, and Wyoming. The river system provides over half the water of greater Los Angeles, San Diego, and Phoenix; it grows much of America's domestic production of fresh winter vegetables; it illuminates the neon city of Las Vegas, whose annual income is one-fourth the entire gross national product of Egypt—the only other place on earth where so many people are so helplessly dependent on one river's flow. The greater portion of the Nile, however, still manages, despite many diversions, to reach its delta at the Mediterranean Sea. The Colorado is so used up on its way to the sea that only a burbling trickle reaches its dried-up delta at the head of the Gulf of California, and then only in wet years. To some conservationists, the Colorado River is the preeminent symbol of everything mankind has done wrong—a harbinger of a squalid and deserved fate. To its preeminent impounder, the U.S. Bureau of Reclamation, it is the perfection of an ideal.

The Colorado has a significance that goes beyond mere prominence. It was on this river that the first of the world's truly great dams was built—a dam which gave engineers the confidence to dam the Columbia, the Volga, the Paraná, the Niger, the Nile, the Zambezi, and most of the world's great rivers. The dam rose up at the depths of the Depression and carried America's spirits with it. Its electricity helped produce the ships and planes that won the Second World War, and its water helped grow the food. From such illustrious and hopeful beginnings, however, the tale of human intervention in the Colorado River degenerates into a chronicle of hubris and obtuseness. Today, even though the Colorado still resembles a river only in its upper reaches and its Grand Canyon stretch—even as hydrologists amuse themselves by speculating about how many times each molecule of water has passed through pairs of kidneys—it is still unable to satisfy all the demands on it, so it is referred to as a "deficit" river, as if the river were somehow at fault for its overuse. And though there are plans to relieve the "deficit"—plans to import water from as far away as Alaska—the twenty million people in the Colorado Basin will probably find themselves facing chronic shortages, if not some kind of catastrophe, before any of these grandiose schemes is built—if, indeed, one is ever built.

One could almost say, then, that the history of the Colorado River contains a metaphor for our time. One could say that the age of great expectations was inaugurated at Hoover Dam—a fifty-year flowering of hopes when all things appeared possible. And one could say that, amid the salt-encrusted sands of the river's dried-up delta, we began to founder on the Era of Limits.

In terms of annual flow, the Colorado isn't a big river—in the United States it does not even rank among the top twenty-five—but, like a forty-pound wolverine that can drive a bear off its dinner, it is unrivaled for sheer orneriness. The virgin Colorado was tempestuous, willful, headstrong. Its flow varied psychotically between a few thousand cubic feet per second and a couple of hundred thousand, sometimes within a few days. Draining a vast, barren watershed whose rains usually come in deluges, its sediment volume was phenomenal. If the river, running high, were diverted through an ocean liner with a cheesecloth strainer at one end, it would have filled the ship with mud in an afternoon. The silt would begin to settle about two hundred miles above the Gulf of California, below the last of the Grand Canyon's rapids, where the river's gradient finally moderated for good. There was so much silt that it raised the entire riverbed, foot by foot, year by year, until the Colorado slipped out of its loose confinement of low sandy bluffs and tore off in some other direction, instantly digging a new course. It developed an affection for several such channels, re-turning to them again and again—Bee River, New River, Alamo River, big braided washes that sat dry and expectant in the desert, waiting for the river to return. The New and Alamo channels drove into Mexico, then veered back north into the United States, a hundred-mile semi-loop, and ended at the foot of the Chocolate Mountains, where the delinquent river would form a huge evanescent body of water called the Salton Sea. After a while, the New and Alamo channels would themselves silt up and the Colorado would throw itself back into its old bed and return to the Gulf of California, much to the relief of the great schools of shrimp, the clouds of waterfowl, and the thousands of cougars, jaguars, and bobcats that prowled its delta. The Salton Sea would slowly evaporate and life would return to normal, for a while. The river went on such errant flings every few dozen years—a vanishing moment in geologic time, but long enough so that the first people who tried to tame it had no idea what they were in for.

The first of these tamers was an eastern developer with a grandiose imagination, a bulldog chin, a shock of steel-wool hair, and a name suggestive of his temperament. In 1892, Charles Rockwood saw the Colorado River for the first time and became obsessed. Sitting north of it, an appendage of the vast Sonoran Desert of southern California and Arizona, were hundreds of thousands of absolutely flat acres built by its ancient delta, fertile land where you could grow crops twelve months of the year. All that stood in the way of cultivation was an annual rainfall of 2.4 inches, about the lowest in the United States. Despite the imposing nature of the task, the temptation to play God with the river and turn this brutal desert green was too much for Rockwood to resist. After traveling halfway around the world for fi-nancial support, he seduced the most famous private irrigationist of his day, George Chaffey, into joining forces with him. By 1901, Rockwood and Chaffey had cut a diversion channel, and a good portion of the river was pouring over fields in what had once been called the Valley of the Dead (in grand nineteenth-century fustian tradition, Rockwood renamed it Imperial Valley). Within eight months, there were two towns, two thousand settlers, and a hundred thousand acres ready for harvesting.

By 1904, however, the artificial channel had already silted up, and a bypass had to be cut. It silted up. Another bypass was cut; it too silted up. Finally, after much negotiation, the developers persuaded the Mexican government to let them cut still another channel below the border. Because it was meant as a temporary expedient while the original channel was cleaned out in advance of the spring floods, the Mexican channel had the flimsiest of control gates. As luck would have it, the spring floods arrived two months early. In February, a great surge of snowmelt and warm rain spilled out of the Gila River, just above the Mexican channel, and made off with the control gate. For the first time in centuries, the river was back in its phantom channel, the Alamo River, heading for its old haunt, the Salton Sink. As the surge advanced across the Imperial Valley, it cut into the loamy soil at a foot-per-second rate, forming a waterfall that marched backward toward the main channel. Even as their fields were being eaten and as their homes swam away, the valley people came out by the hundreds to see this apparition, a twenty-foot falls moving backward at a slow walk. By summer, virtually all of the Colorado River was out of its main channel, and the Salton Sink had once again become the Salton Sea.

Chaffey had had some differences with Rockwood and got out of the California Development Company a short while earlier with his reputation intact, leaving his erstwhile partner ruined. But the Southern Pacific Railroad had already invested too much money in a spur line to the valley to watch it abandoned to fate, so it took Rockwood's company into receivership and set about trying to tame the river. For the next two years, Edward H. Harriman, the railroad magnate, and the Colorado River fought nose to nose. Southern Pacific trains crawled back and forth across the valley like caterpillars, carrying rock and gravel to plug the half-mile breach. But 1905, 1906, and 1907 were some of the wettest years in the Colorado Basin's history. In 1907, the river sent a record twenty-five million acre-feet—eight quadrillion gallons—to the gulf. The floods, one following another, casually ripped Harriman's brush weirs to shreds; his miles of driven piles were uprooted and washed away. Finally, in February of 1907, after laughing away the railroad's best efforts, the river decided to lull. With mad energy, the SP crews finally secured the breach. When the next surge came down, the weirs held, and the river, dumping silt ten times faster than the trains, began rebuilding its own confinement.

Victory or no, the Colorado River was a rampant horse in a balsa corral. The only way to control it effectively, and to give the farmers some insurance against its countervailing tendency to dry up, was to build a dam—a huge dam—to lop the peaks off the floods and provide storage during droughts. The problem with such a dam, from the point of view of the basin at large, was that California was then the only state in a position to use the water. Wyoming, Arizona, Nevada, and New Mexico were still mostly uninhabited. Colorado and Utah had a few hundred thousand people each, but they had scarcely begun to tap the Colorado River and its tributaries; most of Utah's irrigation had been developed in another basin. California, on the other hand, was gaining people like no place on earth, and most of the growth was occurring in the south. The Imperial Valley could have immediately used three or four million acre-feet of the river, the consumption of all the upper-basin states and then some. The Coachella Valley, farther north, and the Palo Verde and Yuma projects could swallow another million acre-feet. Los Angeles, growing like a gourd in the night, would soon overrun its Owens Valley supply; the next logical source of water—the *only* logical source—was the Colorado River. Under simple appropriative-rights doctrine, the water would belong to California as soon as it began to use it. If California perfected its rights in court, it would, in effect, monopolize a huge portion of the river for itself. And the *real* injustice in all of this was that California contributed nothing to the river's flow. Nearly half the runoff came from Colorado and another third from Wyoming and Utah. Arizona and New Mexico contributed very little; Nevada and California, nothing at all. California's efforts to get the dam authorized by Congress were soon beaten back. Finally, it realized that if it wanted the dam and a reliable share of the river, it would have to sit down with its neighbor states and divide it up.

The negotiation of the Colorado River Compact took place in 1922 under the guidance of Commerce Secretary Herbert Hoover at Bishop's Lodge, a swank resort outside Santa Fe, New Mexico. For the time spent debating and drafting it—about eleven months—and its reputation as a western equivalent of the Constitution, the compact didn't settle much. Using the Reclamation Service's estimated average flow of 17.5 million annual acre-feet, the delegates from the seven states divided the river arbitrarily at Lee's Ferry, Arizona—a point just below the Utah border—into two artificial basins. California, Arizona, and

Nevada were the lower basin; the other four states were the upper
basin; pieces of New Mexico and Arizona were in both. Each basin
was allotted 7.5 million acre-feet. How they were to divide that among
themselves was their problem. Of the remainder, 1.5 million acre-feet
were reserved for Mexico, and the final million acre-feet were apportioned, with extreme reluctance on the part of some, as a bonus to the
lower basin, whose delegates had threatened to walk out of the negotiations if they didn't get a better deal.

The compact was signed by the delegates in November of 1922;
they then took it home for ratification by the voters or legislatures of
their respective states, which quickly tore it to shreds. California
wouldn't ratify without a conjugal authorization of Boulder Canyon
Dam and a new canal running exclusively through American territory
to Imperial Valley, a demand that gave the upper basin fits. Arizona
wanted to divide the lower basin's apportionment before it ratified
anything. Harry Chandler, probably the most influential human being
in the Southwest—he talked through his vast wealth and his newspaper—was delighted by the compact and the authorization of the
dam, but he was too greedy to tolerate an All-American Canal, which
would divert the river right above his 860,000 acres in Mexico, so he
ended up opposing everything. George Maxwell, the head of the National Reclamation Association, should have been in favor of Boulder
Dam, but out of principle he opposed anything Harry Chandler liked.

In 1928, after six years of paralysis, Congress took matters into its
own hands. It authorized Boulder Dam and the All-American Canal
on the condition that at least six of the seven states ratify the compact,
and that California limit its annual diversion to 4.4 million acre-feet
per year. That implied only 2.8 million for Arizona (Nevada got 300,000
acre-feet), which was less than it wanted. Arizona, as a result, became
the one state that refused to ratify, an act of defiance that would
muddle things for another thirty-five years. At the time, however, its
vote wasn't needed, and the other states' ratification led forthwith to
the California Limitation Act and, subsequently, to passage of the
Boulder Canyon Project Act. All of this appeared to settle matters:
the basin could now embark on an orgy of growth the likes of which
the West had never seen. And it did settle things, temporarily at least,
except for one small matter: the average annual flow of the Colorado
River was nowhere near 17.5 million acre-feet.

In 1930, the American West had a population of eleven million people,
about the population of New York State. Half of the people were in
California, by far the most populous and modern of the western states.

When Californians traveled, however, they went mainly on dirt roads. The drive from San Francisco to Lake Tahoe, which is now done in three or four hours, was a two-day adventure or ordeal, depending on one's point of view. The city's great bridges had not yet been built. San Jose was not yet a city of thirty thousand, Silicon Valley a stronghold of orchards and roaming mountain lions. In some of the other states, the usual means of locomotion was still a horse and wagon. Electricity and telephones were unknown in most rural communities, and didn't reach the more remote ones until the 1950s. In the midst of this same depopulated, untrammeled region, however, the engineering wonder of all time was about to rise.

In Oakland, California, an egomaniacal small-time construction tycoon named Henry J. Kaiser had followed the passage of the Boulder Canyon Project Act with consuming interest. Obsessed with his niche in history, Kaiser was still enough of a realist to know that he could not begin to build such a dam alone. So he called up his friend W. A. Bechtel to ask if he was interested in making a joint bid. Dad Bechtel was a horse-drawn Fresno-scraper kind of contractor; most of his business was road paving, his most noteworthy innovation a folding toothbrush which he carried on trips. Outside of northern California, and even there, the Bechtel Corporation was all but unknown. "I don't know, Henry" was Bechtel's response when Kaiser, flushed with excitement, got him on the phone. "It sounds a little ambitious to me."

A thousand miles away in Utah, two sheep-ranching Mormon brothers named W. H. and E. O. Wattis were as captivated by the Boulder Canyon Project as Kaiser, and just as unable to undertake it themselves. The Wattises' other business, the Utah Construction Company, specialized in something as mundane as Bechtel's paving contracts: laying railroad bed. Lately, however, they had taken on a new partner, a maverick Mormon banker with Keynesian leanings who talked about deficit financing while candidate Franklin Roosevelt was still promising a balanced budget. His name was Marriner Eccles, and the reward he was about to receive for his ideological flexibility was an influential position on the Federal Reserve Board. The Wattises had also been in contact with Harry Morrison and Morris Knudsen, two engineers formerly with the Bureau of Reclamation who had gone into business together in Boise, Idaho. And they had spoken with Frank Crowe, another former Bureau engineer whose enthusiasm for Boulder Dam was as obsessive as Kaiser's. Morrison had just returned from a trip east, where he had tried to influence the financial community to back a bid on the dam. He was told by the western bankers that there wasn't a company west of the Mississippi they would trust to take on something like this. But one thing would lead to another. Before long, the Wattises were talking with Bechtel and Kaiser, and Henry and Dad were in touch with some other firms—J. F. Shea Construction of Los Angeles, McDonald and Kahn of San Francisco, General Construction of Seattle. In February of 1931, during a meeting at the Engineers Club in San Francisco, the first of the West's supercompanies was born. There were eight firms altogether, but Kaiser couldn't resist borrowing a name from the tribunal before which the tongs, the Chinese equivalent of the Mafia families, took their grievances. At his insistence, the executives agreed to call their joint venture Six Companies, Inc. Hocking everything but their shirts, they could barely scrape together the few million dollars they would need to buy enough equipment to begin the job. When the Bureau auctioned off the job, however, it was Six Companies' amazingly low bid, in the amount of $48,890,995.50, that won. Once again, sang the Los Angeles *Times*, the West had "laughed at logic and driven [its] destiny over obstacles that rational minds deemed insuperable."

The first eighteen months of work on Boulder Canyon Dam involved the construction of a new Colorado River. Four diversion tunnels were blasted through the rock of the box canyon, two on the Nevada side and two on the Arizona side, each of them three-quarters of a mile long. Their diameter was spacious enough to accommodate a jumbo jet shorn of its wings—a capacity that was needed mainly as insurance against an errant flood of 200,000 cubic feet per second, or more. The task required the excavation of three and half million tons of rock with enough dynamite to level Toledo. On November 13, 1932, four tremendous explosions blew out the entrances and exits of the two Arizona tunnels. The dust had not yet settled when a caravan of trucks lumbered onto a trestle bridge built downstream from the tunnel entrances and began dumping rocks and earth in the river's path. Finding itself blockaded, the Colorado slowly roiled and rose in frustration; sensing an escape route, it rode off into the tunnels. In a matter of hours, the river had been lured out of a bed it had occupied since the Grand Canyon was formed.

No sooner was the Colorado flowing through the canyon walls than the crews began replacing the flimsy trestle dam with a far more substantial cofferdam; then, for good measure, they built another below. Made of earth and rock and faced with concrete, the upper cofferdam measured 450 by 750 by 96 feet. Half a century earlier, it would have been the largest dam in the world, but its usefulness was to be measured in months.

When the cofferdams were finished, the engineers turned to the next task—stripping the canyon abutments to expose fresh clean solid rock. Because the dam would rise more than seven hundred feet, there was no crane big enough to do the job; it would have to be done by hand. The four hundred men whose job it was to clean the walls were known as high-scalers. Those who persevered—seven were killed on the job—spent months hanging four or five hundred feet in the air, drilling holes in the rock, inserting dynamite, and praying they would be hauled to safety before it exploded. Because the canyon was so tight, they also had to blast out space for portions of the huge powerhouse, the intake towers, and the penstock headers. Some of the rock amphitheaters they created could have held an orchestra.

Besides the hazards of the construction work (the falling rock, the explosives, electrocution, behemoth machines); besides the hazards of off-hours (fist fights, drunken binges, social diseases from the whores who camped about); besides all this, there was the heat. The low-lying parts of the Colorado and Sonora deserts are the hottest corner of North America, and we are speaking of temperatures in open, ambient air. The Colorado's box canyon held heat like an oven with the door open about an inch. Workers sometimes sacrificed eggs to see if they would actually fry on a sun-fired rock. The first death from heat prostration occurred a few days after construction began, and so many men collapsed that some of the crews finally forced a shutdown, demanded a pay raise, and ultimately staged a strike. The strike, however, did no good. Next to Boulder City was an encampment of tents and shanties known as Ragtown, where the unemployed waited by the hundreds for someone to give up, be fired, or die. "One of the myths about the Depression," Arthur Miller, the playwright, once said, "is that it brought everyone closer together. Actually, it just made everyone more voracious." "They will work under our conditions, or they will not work at all," proclaimed W. H. Wattis. And they did, at a base pay of $4 per day.

It was in 1933 when the explosive din suddenly stopped and an eerie silence descended on Boulder Canyon. The canyon walls were finally clean, the abutments sculpted, the cofferdams in place. Nearly three years after work had begun, the dam was still a figment of the imagination. Now it was time to dig down to bedrock.

The bed of the Mississippi River is hundreds, even thousands, of feet deep in silt. The Columbia and the Missouri flow over alluvial wash as thick as Arctic glaciers. On the Colorado, however, to everyone's amazement, bedrock was struck at forty feet. A milled piece of sawtimber was found resting at the bottom of the muck, obviously of very recent origin. Since white men had begun to settle the region, perhaps eighty years before, a huge flood had evidently washed the entire channel clean. No one seemed bothered by the certainty that all of the silt constantly being relocated along the entire 1,450-mile length of the river would be forever imprisoned behind Hoover and the other dams soon to be built.

In June of 1933, the foundation was finally ready, and the first of the wooden forms that would be used to lay concrete was being built. The concrete—sixty-six million tons of it—created one of the most vexing problems the engineers had faced, a problem peculiar to large dams. The dam's size and weight would generate superpressures and insulating mass that would both generate and retain heat. Though the dam would appear solid, it would be, in reality, a pyramid of warm pudding. Left to its own devices, Boulder Dam would require 100 years to cool down. Moreover, the cooling would be uneven, and the resultant shrinkage and warping would leave the structure fissured and cracked. After weeks of wondering what to do, the engineers finally agreed on a solution. As each form was poured, one-inch pipe would be laid through it at five-foot intervals; frigid water from a cooling plant would then be run through the pipes until convection cooling had lowered the temperature of the concrete to forty-three degrees near the base and seventy-two degrees near the crest. Since the amount of pipe required, if it had been laid out in a straight line, would have reached to Big Sur on the central California coast, this was no mean refrigeration plant. Converted to ice-making, it could have chilled a couple of million cocktails a day. Instead, it reduced a century of cooling time to something like twenty months.

When visitors were led to the canyon rim to watch Boulder Dam on the rise, there was usually a long moment of silence, a moment when the visitors groped for something appropriate to say, something that expressed proper awe and reverence for the dazzling, half-formed monstrosity they saw. The dam defied description; it defied belief. Standing on the upstream side of it, two on each flank, were the intake towers, marvelous fluted concrete columns rising 395 feet from platforms that had been blasted halfway up the canyon walls. The towers were as high as forty-story buildings, and someone who had never been to New York or Chicago or Philadelphia would never have seen a man-made stucture that high. But the crest of the dam rose nearly to the tops of the towers, and its foundation was *hundreds of feet* below their base. Its seamless curve swept across the canyon and imbedded itself in each side, a gigantic but somehow graceful intrusion. The men working on top were not even ants; they hardly qualified as fleas.

Stretching overhead, from canyon rim to canyon rim, was a thick cable on which hung suspended a sixteen-ton bucket that lowered fresh concrete into the forms. Although it was big enough to accommodate a Buick, the bucket seemed incapable of ever filling the dimensions of Hoover Dam—the name it was ultimately to acquire. But twenty-four hours per day, 220 cubic yards an hour, it did. After two years of pouring, the dam was finally topped out. On March 23, 1935, it stood 726 feet and 5 inches tall.

When the engineers surveyed what they had built, it seemed impossible to believe that anything so immense could fail to hold back the Colorado River under every conceivable circumstance. Between 1907 and 1917, however, the wettest period on record, the river had discharged nearly enough water to fill the reservoir during several years: twenty-four million acre-feet; twelve million; twenty-five and a half million; fourteen million; twenty million; nineteen million; twenty million. Hidden within the figures were big floods, periods when the river flowed at 100,000 or 200,000 cubic feet per second for weeks in a row. If such a flood happened to hit when the reservoir was full, the full force of it would have to be spilled; the penstocks leading to the power plant would never be able to handle it. But 200,000 cfs sent over the top of the dam could erode it like a seawall in a storm. The dam, therefore, required spillways on either side, and to allow for the unforeseen and the incredible they were to be built to handle 400,000 cubic feet per second—nearly twice the Columbia River's flow. The spillway troughs were excavated on the canyon sides of the intake towers and led into the vast diversion tunnels hollowed through the walls. Like everything else about the dam, they were designed curvilinear and graceful, with immense brass drum gates shaped like diamond heads. Set down in a spillway channel, the *Bismarck* would have floated clear. Some of the project engineers wistfully suggested that turbines be installed at the spillway outlets, even if they operated only during floods. With the penstocks and the outlet works both generating power, the dam, during brief periods, could have electrified the state of California.

Nothing, however, was more astonishing than the speed with which all of it was built. As the nation languished in the Depression, as plant after plant remained idle and company after company went bankrupt, Hoover Dam was being built at a breathtaking pace. The eyes of the country were fixed on it in awe. A landmark event—the completion of a spillway, the installation of the last generator—was front-page news. The initial excavations for the diversion tunnels had begun on May 16, 1931. The river was not detoured from its channel until November, and the cofferdams were not completed until April of 1933. But two years later, all the blocks in the dam were raised to crest elevation, and a year later everything was finished: spillways, powerplant, penstocks, generators, galleries, even the commemorative plaque in the frieze alongside U.S. Highway 93, which ran across the top. The first electrical power, from what was then the largest power plant in the world, was produced in the fall of 1936. The greatest structure on earth, perhaps the most significant structure that has ever been built in the United States, had gone up in under three years.

The difference in climate between the eastern and western United States—the fact that the East generally gets enough rainfall to support agriculture, while the West generally does not—is easily the most significant distinction between those two regions. It is also obvious that there are significant distinctions within each region as well. For example, oranges grow well in central Florida; they do not in South Carolina, a few hundred miles north. The climate in Duluth, Minnesota, is quite different from that in Chicago, a mere day's drive away.

In the West, however, climatic differences far more striking than these may occur within the same state, even within the same county. In the Willamette Valley of Oregon, a farmer can raise a number of different crops without irrigation; there is usually a summer drought, but it is short, and even if he decides not to depend entirely on rainfall, a few inches of irrigation water—instead of the hundred inches used by some farmers in California and Arizona—will usually do. Two hours away, on the east side of the Cascades, rainfall drops to a third of what the Willamette Valley ordinarily receives; not only that, but the whole of eastern Oregon is much higher than the section west of the Cascades, and lacks a marine influence, so the climate is far colder as well. It can be forty above zero in Eugene and ten below zero in Bend, a two-hour drive to the east. In eastern Oregon, not only must a farmer irrigate but he is extraordinarily limited, compared to his Willamette Valley counterpart, in the types of crops he can grow.

Around Bakersfield, California, an irrigation farmer can raise the same crops that one sees growing in Libya, southern Italy, Hawaii, and Iraq: pistachios, kiwis, almonds, grapes, olives, melons, crops whose value per cultivated acre is astonishingly high. An hour's drive away, across the Tehachapi Mountains, lies the Antelope Valley, a high-desert region with a cold interior climate that can bring frost in May, and where little but alfalfa and grass can be grown. Both Bakersfield and the Antelope Valley are within Kern County, whose climatic extremes are rather typical of California, and, for that matter, of many counties throughout the West. Air conditioners and furnaces in two relatively nearby towns—Phoenix and Flagstaff—may be running at the same time; one end of a county may be plagued by floods while another is plagued by drought.

The reason for all this is mainly topographic: the mountains that block weather fronts and seal off the interior from the ocean's summer cooling and winter warmth (the prevailing westerly winds of the northern hemisphere give the ocean a much wider influence in the West than in the East, reaching as far away as Idaho); the tectonic upheavals that pushed much of the interior West, even the flat mountainless sections, to elevations higher than a mile. The significance of it, from the standpoint of water development, is that it makes infinitely greater economic sense to build dams and irrigate in warmer regions than in colder ones—even if it makes infinitely greater *political* sense to do otherwise.

When John Wesley Powell explored the American West, he duly noted these bewildering extremes of climate. Powell knew that irrigation was an expensive proposition, and that a few inches of extra rainfall or a couple of thousand feet of elevation difference would mean a project that was worth developing or, on the other hand, a project that would require heavy subsidization. A farmer raising fruit or two annual crops of tomatoes in the Imperial Valley might earn ten times more per irrigated acre than a farmer raising alfalfa at six thousand feet in Colorado; yet it might cost far more to deliver water to the Colorado farmer because his water might have to be pumped uphill, out of deep river canyons, while the Imperial Valley lay near sea level below Hoover Dam. The Imperial Valley farmer could pay enough for water to allow the government to recoup its enormous investment in dams, canals, and other irrigation works; the Colorado farmer might be able to repay, at best, a dime on every dollar.

What Powell did not foresee, however, was the Colorado River Basin arbitrarily divided, with each half given an equal amount of water. To him, such a false partitioning might have seemed absurd, for it made far better sense to irrigate in the lower basin than in the upper. But he could not imagine that the blind ambition of the Bureau and the political power of the upper basin would join forces to try to pretend that a mile of elevation difference, and the staggering climatic difference such a disparity implies, did not exist.

Simply stated, the problem with most of the upper basin was that it was too high, too dry, and too cold. Land that was well suited to

irrigation in a topographic sense—meaning that a river flowed through
a wide valley with good soil which lay below a natural damsite somewhere in the mountains above—often sat at altitudes above five thousand feet. Virtually the whole state of Wyoming, for example, lies at
an altitude of six thousand feet or higher. Much of Colorado is over a
mile high; most of Utah is over four thousand feet. In Cheyenne, Wyoming, the frost-free season is barely four months. In such a climate,
one can grow only low-value crops—alfalfa, irrigated pasture, wheat—
which require much acreage to produce a meager income. Not only
that, but some such crops—irrigated pasture in particular—require a
lot of water, up to three times more than some high-value crops: oranges, tomatoes, nuts, even lettuce.

In 1915, it made sense to build a few economically ill-advised
projects in the interior West anyway, in order to reduce its abject
reliance on imported food and offer some economic stability to the
region. And, in fact, dozens of marginal projects were built in the Rocky
Mountain and northern plains states during the first thirty years of
Reclamation's reign. But it began to make less and less sense by 1945,
after tens of billions of dollars had been invested in an efficient transportation system that forever ended the isolation of places like Cheyenne and helped bring them into the nation's economic mainstream.
And it made even less sense by 1955, when the nation was burying
itself under mountains of surplus crops—often the same crops (wheat,
barley, corn) that had to be grown in the high, cold intermountain
West.

What all of this meant—to the taxpayers, anyway—was that the
overwhelming share of the cost of any so-called self-financing project
in the upper Colorado Basin would end up being subsidized by them.
The cost of the projects would be so great, the value of the crops so
low, and the irrigators' ability to pay for water so pitiful that to demand that they repay the taxpayers' investment in forty years, even
allowing for the exemption from interest payments, would be to lead
them into certain bankruptcy. Some of the older, *better* projects had
already had some of their repayment contracts sneakily extended by
several decades, and there was absolutely no evidence that they could
be repaid even then. But, on the other hand, to imagine Congress
booting farmers off Reclamation projects because they couldn't meet
their payment obligations was unthinkable. The taxpayers would have
to bail them out, even if bailing them out meant a long-term bill of
billions and billions of dollars.

How well the Bureau's leadership understood this is a good question—although the secret correspondence in the Bureau's files reveals

that they knew a lot more than they let on in public. (In the 1920s, Federick Newell, the former Reclamation commissioner, was already decrying the "sentimentality" of the federal irrigation program, through which, he said, money was "deftly taken from the pockets" of the taxpayers.) What is *true*, of course, does not necessarily *matter* in a political sense, and that was particularly the case in the American West, and even more so in the upper basin. By the 1950s, California was already using its full 4.4 million acre-foot entitlement to the Colorado River and planning batteries of new pumps that would allow it to suck up 700,000 acre-feet of additional flows. The Bureau, having built Hoover Dam mainly for California's benefit, was now embarking on the Central Valley Project, a project of absolutely breathtaking scope that was exclusively for California. As far as the upper basin was concerned, it was time for some equity. And equity was only the half of it. If there was surplus water in the river—water which the upper basin owned but wasn't yet able to use—and California began "borrowing" it, would that imperial-minded state deign to give it back? The imperative for the upper basin was to develop its share of the Colorado River as fast as possible, whether the projects that could be built there made sense or not. And it was the basin's unbelievably good fortune that in the 1940s, Congress would give it a money-making machine that would allow it to do so—a machine that became known as the cash register dam.

A cash register dam was to be a dam with an overriding, if not a single, purpose: to generate electricity for commercial sales. If electricity would bring in many millions of dollars in annual revenues which could be used to subsidize irrigation projects that hadn't a prayer of paying back the taxpayers' investment. The dams were an invention spawned by something the Bureau of Reclamation called river-basin "accounting," which was itself spawned by something it called river-basin "planning."

River-basin planning, at least, made a certain amount of sense. A river like the Arkansas, which rises in the Colorado Rockies and empties into the Mississippi in an utterly different time zone and topography and climate, invites competing and potentially incompatible uses. Upstream, it is valuable for irrigation; downstream, it is valuable for inland navigation. If the Bureau diverts too much water for upstream irrigation, there won't be enough water available downstream to justify the Army Corps of Engineers' efforts to turn the lower river into a freeway for barges—an obsession it has been pursuing on virtually every large river in the country. The dilemma could also work in reverse; if the Corps got a head start on the lower sections of a river, the Bureau could find itself unable to get any upriver projects authorized. The creation of the Tennessee Valley Authority marked the first time a major river system was "viewed whole," even if the natural river virtually disappeared as a result. The TVA was regarded as such a success by the administration of Franklin Roosevelt that it began to demand, if not more quasi-dictatorial authorities like the TVA, then at least a coordinated plan of development between the Bureau and the Corps. This was river-basin "planning," and, except for the fact that no one ever spent more than a minute or two thinking about the value of a river in its natural state, it made some degree of sense.

River-basin "accounting" was a horse of a different color, though the Bureau developed a propensity to use "planning" and "accounting" interchangeably. With river-basin accounting, one could take all the revenues generated by projects in any river basin—dams, irrigation projects, navigation and recreation features—and toss them into a common "fund." The hydroelectric dams might contribute ninety-five cents of every dollar accruing to the fund, while the irrigation features might contribute only a nickel (and cost three times as much to build and operate as the dams), but it wouldn't matter; as long as revenues came in at a pace that would permit the Reclamation Act's forty-year repayment schedule to be met, the whole package could be considered economically sound. It was as if a conglomerate purchased a dozen money-losing subsidiaries while operating a highly profitable silver mine—a case of horribly bad management which, nonetheless, still leaves the company barely in the black.

Michael Robinson, the Bureau's semiofficial historian, exhibits no compunction about admitting any of this in the Bureau's authorized history, *Water for the West*:

> By the late 1930s, the high cost of projects made it increasingly difficult for Reclamation engineers to meet economic feasibility requirements. In the early 1940s, the Bureau devised the plan of considering an entire river basin as an integrated project. It enabled the agency to derive income from various revenue-producing subfeatures (notably power facilities) to fund other works not economically feasible under Reclamation law.

> Thus, by offsetting construction and development costs against pooled revenues the Bureau was able to demonstrate the economic feasibility for the entire, pooled program. In 1942 this method was used for the first time in planning a basinwide development program for the Bighorn River in Wyoming. All

benefits and income from producing units were lumped together to establish overall feasibility. In 1944, the Bureau's
"Sloan Plan" for the development of the Missouri River followed the same formula . . . [and] encouraged the Bureau to
*enthusiastically prepare basinwide plans for several western riv-
ers.* . . . [Emphasis added]

"Enthusiastically" is a bit of an understatement. The beauty of
river-basin "accounting," from the Bureau's point of view, was that
it would be literally *forced* to build dams. The engineering mentality
which, Robinson himself admits, came to dominate the Bureau's thinking in the 1930s and 1940s created an institutional distaste for irrigation projects. They were a necessary nuisance that provided the
rationale for what Bureau men really loved to do: build majestic dams.
In the past, however, the infeasibility of many projects put a damper
on their ambitions, because if a project didn't make economic sense,
they lost the rationale they needed to build a dam to store water. With
river-basin accounting, the equation was stood on its head: a lot of
bad projects—economically infeasible ones—created a rationale for
building *more*, not fewer, dams. The dams—all with hydroelectric
features, of course—would be required to compensate for the financial
losses of the irrigation projects; the losses would miraculously vanish
in the common pool of revenues.
      River-basin "accounting," then, was a perversion of a sensible
idea—that idea being to plan the "orderly" (a favorite Bureau word)
development of a river basin from headwaters to mouth. But even if
it subverted logic, economics, and simple common sense, it was essential to the Bureau's survival as an institution and to the continued
expansion of irrigation in the high, arid West. On the other hand, it
was something akin to a blanket death sentence for the free-flowing
rivers in sixteen states.
      What the upper basin of the Colorado lacked, because of its ele-
vation, in feasible irrigation projects it more than made up—for the
same reason—in sites for cash register dams. High and mountainous,
geologically young, the basin had deep valleys and tight plunging
gorges ideal for dams—gorges in which ran rivers that fed the main
Colorado and could be included, under the bizarre new logic of river-
basin accounting, in any grand basinwide scheme. The rivers, draining
arid and semiarid regions, may not have held much runoff, but a very
high dam on a small river can yield as much hydroelectricity as a low
dam on a much larger one; that is the beauty of what dam engineers
call hydrologic head: velocity of falling water does the work of volume,

of mass. There was Glen Canyon on the main-stem Colorado, Powell's favorite riverine haunt, an ideal site for a six-hundred-foot dam. There was Flaming Gorge on the Green, and Red Canyon—each a perfect site for a gigantic curved-arch, thin-wall dam approaching Hoover in size. There was the Black Canyon of the Gunnison, an almost sheer thousand-foot gorge with several sites for high dams. The Dolores, the Yampa, the White, even smaller streams like the Animas and San Miguel and Little Snake—each had at least one site for a cash register dam. Since the dams would have to be large compared to the meager river flows, they would be expensive to build. But that wouldn't matter; the Bureau had the Treasury at its disposal.

All the upper basin needed, then, was Congressional clout—that, and a Reclamation Commissioner who believed in dams for dams' sake. And it was the upper basin's further good fortune that, near the end of his third term, Franklin Roosevelt would appoint such a man as his Commissioner of Reclamation. His name was Michael Straus.

Mike Straus was the unlikeliest commissioner the Bureau ever had. For one thing, he was an easterner; for another, he was a newspaperman. On top of that, he was rich. By temperament, Straus was an exact opposite of the slide-rule engineers who had guided the Bureau during its forty-odd years. He was an anomaly down to his very genes. Straus had married into the Dodge family, and his brother-in-law was Eliot Porter; he had wealth and social connections, too. While typical Bureau of Reclamation families spent their vacations on houseboats cruising the reservoirs that Daddy built, Straus went to the family retreat at Spruce Head Island, on Maine's Penobscot Bay. It was their island—all of it. Straus could have spent his life clipping coupons, safari hunting, or writing the hyperventilating prose that was his second love. But there was nothing on earth that gave Mike Straus quite as much boyish, exuberant satisfaction as erecting dams. In eight years as Commissioner of Reclamation, he would become responsible for as many water projects as any person who ever lived.

Straus had been selected at the close of the war by Harold Ickes, himself a newspaperman, after the Roosevelt administration had endured twelve years of relatively plodding Bureau leadership under Elwood Mead, Harry Bashore, and John Page. Straus was Ickes's alter ego—a newspaperman, a liberal, a fighter, a curmudgeon. Franklin Roosevelt, who equated wealth with energy and idealism, heartily endorsed the appointment. It was a brilliant stroke. For all his man-of-the-people reputation, FDR felt paranoia about the common man. His secret fear was that the Depression would begin anew after the war, and the returning veterans would be unable to find jobs; ultimately, they might revolt. In the Bureau of Reclamation, FDR had a vast job-creating engine, an agency that remade the western landscape into a place where the dispossessed could go. In Mike Straus, he had a commissioner who would stoke the engine until the rivets began to pop.

Like a lot of people who inherit or marry wealth, Straus viewed money abstractly. A million was a number, budgets were a nuisance, feasibility reports were a waste of time. And, having abandoned a career that asked for a constant objective adherence to facts, he soon acquired an easygoing way with the truth. "Facts," said one of his successors as commissioner, Floyd Dominy, "didn't mean a god-damned thing to him."

Straus was a spectacle. He was shambling, big as a bear, a terrible dresser, and a slob. "The characteristic Mike Straus pose," remembered Dominy, "was for him to plant his feet on his desk, almost in your face, and lean back in his swivel chair flipping cigarette ashes all over his shirt. At the end of the day, there was a little mound of ash behind his seat. He was an uncouth bastard! He carried one white shirt with him on trips. I remember one night when Reclamation was throwing a party, and a cub reporter came by and asked me where to find Mike Straus. I just said, 'Go upstairs and look for the guy who reminds you of an unmade bed.' "

There was something else about Mike Straus: his arrogance. Once, in the very early 1950s, he got on a plane without reconfirming his reservation, which one was required to do in those days. The plane turned out to be overbooked, and since Straus had not reconfirmed, he was the one who was supposed to be bumped. The flight attendants invited him off the plane, but Straus refused to budge; he pretended not to hear. As a whole plane full of passengers cursed him under their breaths, Mike Straus sat there like a pig in goo. Finally, the captain had to ask for volunteers to bump themselves so that the plane could take off. There weren't a lot of flights in the early 1950s, and the passengers would have to wait a long time for another one. But Straus appeared unmoved; he wasn't even embarrassed. "It didn't faze Mike a bit," said a Reclamation man who was with him. "He thought he was performing the greatest work in the country, and he felt like the holiest bureaucrat in the land."

Cavalier, arrogant, mendacious, and whatever else he was, Mike Straus was also an idealist. A good stalwart liberal in the New Deal tradition, he believed in bringing the fruits of technology to the common man. He bore a ferocious grudge against the private utilities of the West, who denied reasonably priced power (or power at all) to rural areas struggling against adversity on every side, and who bought space in magazines and (he was convinced) bribed reporters to rail against the Bureau's public-power dams. Straus also made some tentative efforts to crack down on the big California growers who were setting up dummy corporations and trusts in order to farm tens of thousands of acres illegally with subsidized Bureau of Reclamation water. In so doing, he infuriated the growers' and the utilities' friends in Congress, and a group of them finally decided to get rid of him. Since Straus served at the President's whim and had Harry Truman's blessing, it was useless to demand he be fired, so the politicians tried another tack. In 1949, they pinned an obscure rider onto the public-works appropriations bill that specifically withheld the salaries of Michael Straus and his regional director in California, Richard Boke. The independently wealthy Straus remained as commissioner—without pay. His enemies were upset, and that is putting it mildly. "Straus made them so mad I thought they might put out a contract on his life," says Floyd Dominy. "I have done what no good Republican has been able to do," Straus wrote to his friend Bill Warne, a former assistant commissioner then in Iran, "and that is to unite the Republican party on at least one platform and provide them with one program—to wit, who can fire Straus first."

However, as the big growers in California and the private western utilities were trying to get rid of Mike Straus, the upper basin was cultivating him just as assiduously. The population of the basin had grown substantially since the Colorado River Compact was signed, but the growth of irrigated agriculture had remained well behind. Most irrigation was by simple diversion, without benefit of reservoir storage. During droughts, the farmers were flirting with disaster; during floods, they watched millions of acre-feet escape to the lower basin unused. The farmers on the other side of the Front Range, on the perfectly flat expanse of the plains, had topography working for them; they could easily lead a diversion channel out of a river such as the Platte, fill a small offstream basin, and have a ready-made storage reservoir for a fraction of the cost of an on-stream dam. The West Slope farmers— those sitting in the Colorado River drainage—were at a terrific natural disadvantage, having no way to store their water and (in the case of some) being at a higher elevation besides. Meanwhile, California was now using up its entire entitlement and still growing by leaps and bounds. If the upper basin didn't hurry and begin using its own entitlement, California seemed certain to try to "borrow" it; if it succeeded, and millions of people then depended on that water, how would the upper basin ever get it back? But how, on the other hand, were Colorado, Utah, and Wyoming ever to use their share of the river if they couldn't afford to build dams themselves and if high-altitude Reclamation projects could never pay themselves back?

The answer, frantically conceived by Mike Straus's Bureau during the last days of his reign—much of it was laid out in the weeks after Eisenhower, who was certain to fire Straus, was already President-elect—was the Colorado River Storage Project. Behind the innocuous name was something as big as the universe itself. In a press release that accompanied the legislation's transmittal to Congress in early 1953—days before Ike's inauguration—Straus described it rather modestly as "a series of ten dams having a storage capacity of 48.5 million acre-feet." What he failed to mention was that 48.5 million acre-feet was more than all the existing reservoirs on the main-stem Colorado and all the tributaries could hold—more than the combined capacity of Lake Havasu, Theodore Roosevelt Lake, Apache Lake, Bartlett Reservoir, San Carlos Reservoir, Painted Rock Reservoir, plus the then largest reservoir on earth, Lake Mead. The ten dams would, according to Straus, capture "several times the total annual flow of the river." In fact, with the lower basin reservoirs already holding close to forty million acre-feet, between *five* and *eight* times the long-term annual flow of the river would be captured, depending on whose estimate you believed—a storage-to-yield ratio that was not approached by any other river in the world, no matter how used. The annual evaporation from all these huge, exposed bodies of water, languishing under the desert sun, would itself exceed the storage capacity of all but a few reservoirs in the nation.

It wasn't, however, the mere magnitude of the project that set it apart. What set it apart was the way irrigation and power production were linked. The earliest projects were designed exclusively as irrigation projects; if any power was incidentally generated, it was sold to project farmers at bargain rates. With Hoover Dam, the Bureau took a big plunge into public power; nearly two-thirds of its hydroelectricity went to light Los Angeles. However, when Angelenos paid their power bills, they weren't subsidizing the farmers in the Imperial and Coachella valleys who were irrigating with Lake Mead water; they were merely paying back the cost of the dam.

The Colorado River Storage Project would be utterly and fatefully different. Anyone who bought electricity at market rates from the dams—and 1,622,000 kilowatts, an enormous amount at that time, was planned—would be subsidizing irrigation in the upper basin. *Eighty-five cents* of every dollar spent on irrigation features would be subsidized by power revenues. Every time they flicked a switch, electricity consumers in the region would be helping a farmer plant alfalfa at six thousand feet to feed a national surplus of beef.

The Bureau was strikingly candid about the dismal economics of irrigation in the upper basin. "The [upper basin] farmers can't pay a dime, not one dime," lamented the Bureau's chief of hydrology, C. B. Jacobsen, to a Congressional committee. And as if to demonstrate how far Congress had come in accepting the subsidization of an entire region, Jacobsen's words fell on sympathetic ears. Western members, even those whose districts were well outside the basin, lined up to support the bill—perhaps because they expected their own uneconomical projects to be supported in return. For the first time, a majority of eastern members seemed indifferent, neutral, or even sympathetic— perhaps because *they* had Corps of Engineers projects they wanted built which might require the western members' support. Even the Eisenhower administration decided to give the Colorado River Storage Project lukewarm support, though it violated every conservative principle Ike had ever espoused.

The most effective opposition, by far, came from Paul Douglas, the urbane Senator from Illinois, who, ironically, had played a pivotal role in the creation of the New Deal. When World War II broke out, Douglas was fifty years old, a former economics professor at the University of Chicago who had become a reform-minded Chicago alderman. He promptly enlisted in the Marines, talked himself out of a desk job, and got to the front lines of the Pacific theater. He was gravely wounded at Peleliu and again at Okinawa, and was lucky to return alive. Elected to the Senate after the war, Douglas brought all of his determination and iconoclastic, brilliant thinking to Washington with him. He was—perhaps because of his economics background—the first architect of the New Deal who seemed to sense that something had gone drastically wrong. And the worst perversion of the New Deal ideas that *he*, at least, had in mind was the Reclamation program, subsidizing high-altitude desert farmers so they could grow the same crops some of Douglas's farmer constituents were being paid *not* to grow—so serious had America's crop-surplus problem become now that Europe was back in production again.

In a series of memorable debates on the Senate floor, Douglas, tall, athletic, and white-haired, went after the Colorado River Storage Project hammer and tongs. At Glen Canyon Dam, he told his colleagues, the cost of hydroelectricity per kilowatt would be $463; at Echo Park Dam, it was over $600; at Central Utah, it was $765; at Flaming Gorge, it was more than $700. "Let us compare that cost with the average cost in the Tennessee Valley of $166 per kilowatt of capacity. At Bonneville, the average cost was only $115. At Hoover, the cost was only $112. At Grand Coulee, the cost was only $90. . . . [I]t is extraordinary that an administration which has declared public power to be creeping socialism, which has put the lid on additional dams on the Columbia, should go up into the mountains of Colorado and there locate public power projects where the cost will be three, four, or five times what they would be at these other locations. . . . I am not saying that the administration wishes to have this project fail. But I will say that if the administration had wished to discredit the public power system, it could not have proceeded in any better fashion than it has done in this instance." And he couldn't help noticing, said Douglas sarcastically, that certain Senators who opposed public power in the Tennessee Valley and the Columbia Basin had suddenly emerged as great champions of public power when it was to come from cash register dams in the mountains of Colorado.

The power features, however, were, as Douglas knew, not the worst aspect of the storage project, but the best. The worst, by far, was the irrigation. "The original projects," he lectured his colleagues, "tended to be at low altitudes and in fertile soil, and to involve low costs. . . . Now we are being asked to irrigate land in the uplands, at altitudes between five thousand and seven thousand feet, where the growing season is short and the chief products will be hay, corn, livestock, and alfalfa. . . . There exists an interesting tendency for Senators in those States to congregate on the Committee on Interior and Insular Affairs and the Committee on Appropriations, which consider irrigation and reclamation bills. There is a sort of affinity, just as sugar draws flies." For the benefit of his colleagues and the Bureau, whose economists had labored mightily to put the CRSP in the best possible light, Douglas had sat down and figured out the per-acre costs of the various projects himself. The Silt River Project in Colorado, for example, would cost $674 per acre; the Paonia project, $873 per acre; the Central Utah Project, the most expensive of the lot, $1,757. If one calculated interest, Paonia would go up to $2,135 per acre, Central Utah to $3,953 per acre. These were the mid-1950s, when land prices in the West were still dirt-cheap. Most of the land whose conversion to irrigation would cost thousands of dollars an acre was not worth more than $50 per acre, and that, in many cases, was being generous. "In my state of Illinois," Douglas pointed out, "the price of the most fertile natural land in the world is now between $600 and $700 per acre. In the largest project of all, the Central Utah Project, the cost would be nearly $4,000 an acre—six times the cost of the most fertile land in the world."

If an investment of $2,000 an acre could create reclaimed land worth $2,000 an acre, that would be one thing. But even after being supplied with irrigation water, the upper-basin lands would be worth nowhere near that. "What is to be grown on the land?" asked Douglas. "Of the sixteen projects reported, eight of them were stated as being suitable for livestock only, through the raising of alfalfa and pasture. Seven were stated as being primarily for livestock, but with some fruit and vegetable production . . . 95 percent of the projects contemplate the production of alfalfa or grain or are directly or indirectly for the feeding of cattle. As a consequence, this land, *after irrigation*, will not be worth very much, probably not more than from $100 to $150 per acre—$150 per acre at the outside. Yet we are being asked to make an average expenditure of $2,000 an acre on land which, when the projects are finished, will sell for only $150 per acre."

Douglas's western colleagues, of course, had no answer to this; his math was correct, his reasoning impeccable. All they could do was stand the rhetoric of their nineteenth-century predecessors on its head; instead of praising the fertile soil and glorious climate of the West, they talked about how miserable and uninhabitable their home states were. "The Senator from Illinois has correctly stated that we have little rain," said Joseph O'Mahoney of Wyoming. "I say to him, 'Pity us. Let us store the rainwater which for thousands of years has been rolling down the Colorado River without use. Please have some pity on the area, which is the arid land area of the country. It wants to conserve the great natural supply of water which the Almighty placed there, for man to use, if he has the intelligence and the courage to use it.' "

All of Paul Douglas's eloquence and logic, as it turned out, were a poor match for appeals such as O'Mahoney's and the growing Congressional power of the arid West. O'Mahoney and Clinton Anderson of New Mexico, representing Colorado Basin states, were powerhouses on the Senate Interior Committee; Carl Hayden of Arizona ruled Appropriations; Wayne Aspinall of western Colorado was the ascendant power at the House Interior Committee. The Colorado River Storage Project also enjoyed overwhelming public support, not just among the western farmers, but among their city brethren, too; conservatives, liberals, Democrats, Republicans—ideology meant nothing where water was concerned. The only serious public opposition came from southern California (which was expected) and from conservationists, who were horrified at the prospect of watching three of the most magnificent river canyons in the West filled by giant, drawn-down reservoirs: Glen Canyon on the main Colorado and Flaming Gorge and Echo

Park on the Green. Each of these reservoirs would be as long as smaller eastern states; Glen Canyon would stretch back for nearly two hundred miles behind the dam, not even counting tentacles of water that would reach up side canyons and tributary streams. But in those days conservationists didn't count for much. The Sierra Club had just one full-time person, whose name was David Brower, on its paid staff.

The outcome was foreordained. California had gotten Hoover Dam, Parker Dam, Davis Dam, the Imperial and Coachella projects, and water and power for Los Angeles. Now the upper basin would get its share. After minimal debate on the floor, the CRSP bill passed both Houses and was signed into law by Eisenhower in April of 1956. The estimated cost of everything was around $1.6 billion, but it would, of course, be substantially more. Never in U.S. history had so little economic development been proposed at such an exorbitant public cost, for all the billions were buying, besides extremely expensive public power, were a few patches of new irrigated lands whose composite size was smaller than Rhode Island. The subsidies, it turned out later, would be worth as much as $2 million per farm, perhaps five times as much as the farms themselves were worth. But even if the Colorado River Storage Project seemed like utter folly, the Bureau of Reclamation and its sometime collaborator and arch-rival, the Army Corps of Engineers, were on a tear.

CHAPTER FIVE

The Go-Go Years

The U.S. economy had fallen flat on its face several times before. In the years after the Great Crash, however, it could not pick itself back up. Things were worse in 1930 than in 1929, worse in 1931 than in 1930. By 1932, millions of people had lost all faith and hope—in the nation, in the capitalist system, in themselves.

The person whom Americans elected to pull the country out of the abyss came across as a genial aristocrat; in some ways, though, he was as close to being a benevolent despot as a democracy can allow. Franklin Roosevelt's own Treasury Secretary, Henry Morgenthau, said that the President never saw himself as "anything else but a ruler." Carl Jung met him and came away saying, "Make no mistake, he is a force—a man of superior but impenetrable mind, but perfectly ruthless, a highly versatile mind which you cannot see." But the President, a man of greater charm and persuasiveness than ruthlessness, was adored by most of the country no matter what he did. Had Gerald Ford or Lyndon Johnson tried to pack the Supreme Court, they probably would have been impeached; when Roosevelt tried it, nearly half the country thought it was a good idea. After seeing Roosevelt in action, Republicans who had voted for Hoover prayed to God to forgive them. Even God must have felt humbled by the new President; in a popularity contest conducted among New York City schoolchildren, Roosevelt outpolled Him.

Franklin Roosevelt said that he wanted to be remembered as the greatest conservationist and the greatest developer of all time. In a country with a population barely greater than Germany's and with fifteen times the landmass, it seemed possible to be both. FDR's conservation was not scientific, as his cousin Teddy's was to a great degree, but instinctive. At Hyde Park, he had spent afternoons planting thousands of little trees. Why not plant millions of them on the high plains to break the wind and conserve the soil? A lot of scientists laughed and said it would never work, but it did. FDR thought up the Civilian Conservation Corps, too, and it became the most popular of all his programs.

What TR and FDR did have in common was an acute awareness of the limits of capitalism. The former Roosevelt saw the seeds of capitalism's self-destruction in monopoly and rapacious business practice, the latter saw them in chronic depression and unemployment. In 1933, when he assumed the Presidency, nearly a quarter of the U.S. population was without visible means of support. Declaring a bank holiday was one way to arrest the widespread financial panic that was costing millions of workers their jobs, but the only thing that would make a real dent in the horrifying unemployment figures was to build public works: bridges, highways, tunnels, parks—dams.

The person whom Roosevelt put in charge of much of the apparatus of recovery was Harold Ickes, a stolid, round, owlish, combative ex-newspaperman who grew to love his nickname, "the old curmudgeon." (Because of Ickes's high-pitched squawk of a voice, Roosevelt, in private, called him Donald Duck.) Ickes ran not only the Interior Department—in which were the Bureau of Reclamation, the Civilian Conservation Corps, the National Park Service, and the Fish and Wildlife Service—but the Public Works Administration as well. The PWA was a catch basin of programs with a chameleon identity (it was also known as the Civil Works Administration and the Works Progress Administration) and interchangeable leaders (first Harry Hopkins, then Ickes, then Hopkins again). In a few years, it had overseen the building of the Lincoln Tunnel, the Washington Zoo, the Triborough Bridge, Fort Knox, Denver's water-supply system, a deepwater port at Brownsville, Texas, the huge Camarillo Hospital in southern California, and the causeway to Key West. It built a dozen fantasyland bridges along Oregon's coast highway. Above all, it built dams.

Under Roosevelt and Ickes, the Bureau of Reclamation underwent some fundamental changes, the most obvious of which was in size. From two or three thousand employees under Herbert Hoover—a very large federal agency in its day—the Bureau mushroomed into an elephantine bureaucracy with a staff of nearly twenty thousand by the time Roosevelt died. Headquarters was the top floor of the gigantic new Interior building in Washington—the Bureau's offices were above those of the Interior Secretary himself—but the real work was done out of the Bureau's sprawling engineering complex in west Denver, where it designed its mighty dams. Then there were regional offices, field offices, project offices. When Jim Casey, who was to become deputy chief of planning in the 1960s, first went to work for the Bureau in Nebraska, he found himself amid nine hundred fellow employees. "This wasn't even a regional office," remembers Casey. "This was just a field office. I never had the faintest idea what everyone did and neither did they." And very few of the Bureau's people had anything to do with the actual physical construction of the dams; that work was contracted out to the engineering firms, the Bechtels and Morrison-Knudsens, that had become instant giants after cutting their teeth on Hoover Dam. The Bureau's nineteen thousand-odd employees merely planned projects, supervised projects, and looked for new projects to build.

There were also some fundamental changes in the Bureau's approach, in its character. In the beginning, FDR was content to let it be run, as it had been in the past, by engineers. Elwood Mead, who, after John Wesley Powell, was the most illustrious reclamationist in America, headed the agency until his death in 1936. He was succeeded by John C. Page and Harry Bashore, engineers who had come up through the ranks. As commissioners, Mead, Page, and Bashore, remembering Congress's exasperation over the Bureau's early failures and cognizant that the nexus of power still lay east of the Mississippi River, tended to be somewhat modest in their ambitions. And if they lapsed from time to time, Ickes, at least in the beginning, was prepared to restrain them himself. "Commissioner Mead, of course, is always in favor of any new reclamation project," he wrote in a sarcastic memo to Roosevelt bemoaning a Bureau proposal. "That is his job." As the economics of reclamation played themselves out, however, and the salvation of the program lay in the construction of big public-power dams—which most of the electric utilities and much of the Republican Party regarded as anathema—the role of the commissioner abruptly changed. In the new reclamation era, a commissioner needed to be someone very much like Ickes; a fighter, a public-power ideologue, and, above all, a salesman. There was no better candidate then Ickes's close friend, fellow newspaperman, and faithful subordinate, Mike Straus.

Public relations and salesmanship, skills few engineers possess, were second nature to Mike Straus. "Born with a gold-plated irrigation shovel ready to be placed in her hands," reads a Straus press release dated June 5, 1952, "Reclamation's Golden Jubilee baby arrived at

Washington's Yakima Memorial Hospital at 12:45 today, the daughter of Mr. and Mrs. Donald T. Dunn of Moses Lake, Washington. . . . The baby was born on the eve of the fiftieth anniversary of federal Reclamation, and the child has been adopted by the National Reclamation Association. . . .

"Michael W. Straus, Commissioner of the Bureau of Reclamation, declared in a congratulatory message that 'the Reclamation program must be pushed forward with utmost speed so the Dunn child and all the other kiddies born this year will have a happier and more secure life on the land through Reclamation development. . . . We should be starting today on [new] development so that there will be a Reclamation farm ready for baby Dunn.' "

Whether the cost of supplying water to baby Dunn's ex-desert was utterly beyond reason; whether she even wanted to spend her life on an irrigation farm; whether the country, already suffocating under mountainous farm surpluses by 1952, really needed her production— these were the kinds of questions which the Bureau, after eight years of Mike Straus, would rarely ask again.

To Mike Straus, millionaire dam builder, economic feasibility mattered little, if at all. Once, on a visit to the Bureau's regional office in Billings, Montana, Straus rented the town's only theater and demanded that all the employees show up in the evening for a "pep talk." The Billings office was in charge of the upper Missouri Basin, where the greatest concentration of physically possible but economically unfeasible projects happened to be located. When the employees had filed in and taken their seats, Straus slouched against a lectern on the stage and launched into a tirade against them for doing their jobs. "I don't give a damn whether a project is feasible or not," he thundered at his astonished staff. "I'm getting the money out of Congress, and you'd damn well better spend it. And you'd better be here early tomorrow morning ready to spend it, or you may find someone else at your desk!"

The Great Depression and the Roosevelt administration, together with the pyramid-scheme economics of the river-basin accounts, were more than enough to launch the federal dam-building program on a forty-year binge. It probably wouldn't have needed the Dust Bowl— but it helped.

Since the blizzards and drought of the 1880s and 1890s, the farmers of the western plains had been playing a game of "Mother May I?" with nature. When the isohyet of twenty inches of rainfall maundered westward, they advanced. When it moved eastward, they retreated—

some of them, anyway. Through most of the first three decades of the twentieth century, the line stayed close to the lee side of the Rockies. The teens and 1920s, in particular, were years of extraordinary and consistent rainfall. Millions and millions of acres of shortgrass prairie west of the hundredth meridian, land already depauperated by livestock overgrazing during the last century, were converted to the production of wheat, whose price had reached record levels during the war. "Everything in the country was going full blast," wrote Paul Sears in his book *Deserts on the March.* "It was the most natural thing in the world for the plains farmers, whose cattle business had prospered during the war and who had been encouraged to try dry farming, to attempt the growing of wheat on a huge scale. The soil was loose and friable; the land was theirs to use as they saw fit." Even in the wettest years of the 1920s, the high-plains wheat rarely grew taller than someone's knee; sometimes it was ankle-high, and during a dry year it wouldn't come up at all. Everyone knew the wet years wouldn't last, and everyone knew that the loose soil, with the wheat stubble disked under, had nothing to hold it if drought and wind should coincide. But everyone was making money.

The first of the storms blew through South Dakota on Armistice Day, November 11, 1933. By nightfall, some farms had lost nearly all their topsoil. "Nightfall" was a relative term, because at ten o'clock the next morning the sky was still pitch-black. People were vomiting dirt. Machinery, fences, roads, shrubs, sheds—everything was covered by great hanging drifts of silt. "Wives packed every windowsill, door frame, and keyhole with oiled cloth and gummed paper," William Manchester wrote, "yet the fine silt found its way in and lay in beach-like ripples on their floors." As a gallon jug of desert floodwater, after settling, contains a quart and a half of solid mud, the sky seemed to be one part dust to three parts air. A naked human tethered outside would have been rendered skinless—such was the scouring power of the dirt-laden gales. Huge numbers of jackrabbits, unable to close their eyes, went blind. That was a blessing. It gave the human victims something to eat.

The storms, dozens of them, continued through the spring and summer of 1934. An old physician in southwestern Nebraska wrote in his diary, "Wind forty miles an hour and hot as hell. Two Kansas farms go by every minute." With the temperature up to 105 degrees and the horizon lined with roiling clouds that seemed to promise ten inches of rain but delivered three feet of dirt, the plains took on a phantasmagorical dreadfulness. The ravenous storms would blow for days at a time, eating the land in their path, lifting dust and dirt high enough to catch the jet stream, which carried it to Europe. In 1934, members of Congress took time out from debating the Taylor Grazing Bill—designed to control overgrazing on the public lands—to crowd the Capitol balcony and watch the sky darken at noon. From the look of the western horizon, half the continent could have been on fire. The Taylor Act was passed in that year, despite efforts by some western members to weaken it even as their states were sailing over their heads. Between storms, when visibility sometimes increased to five or six miles, people in the Oklahoma and Texas panhandles, in Kiowa and Crowley counties in Colorado, in Texas's Gaines County on the New Mexico border, in 756 counties in nineteen states that were ultimately affected, watched their world turn into the Sahara.

The Dust Bowl was triggered by the same fatal congelation of hope and drought that caused the plains to empty half a century earlier. The longest severe drought in the nation's history—the one that Bureau of Reclamation planners, ever optimistic, now use as their "worst-case scenario"—began to descend over the West in 1928. For seven years in a row, precipitation remained below normal. The snow that fell on the plowed-up fields of the Dakotas was so light that the ground, bereft of insulation, froze many feet down; the snow evaporated without penetrating and the spring rains, those that came, slid off the frozen ground into the rivers, leaving the land bare. The virgin prairie, grazed well within its carrying capacity by thirty million buffalo, could probably have withstood the wind and drought; ravaged by too many cattle and plowed up to make way for wheat, it could not. If not the worst man-made catastrophe in history, it was, at least, the quickest.

By 1934, the National Resources Board reported that thirty-five million acres—Virginia and then some—had been essentially destroyed; 125 million acres—an area equivalent to Virginia plus Ohio plus Pennsylvania plus Michigan plus Maryland—were severely debilitated, and another hundred million acres were in marginal shape. "We're through," wrote a wheat baron from the shortgrass territory. "It's worse than the papers say. Our fences are buried, the house hidden to the eaves, and our pasture, which was kept from blowing by the grass, had been buried and is worthless now. We see what a mistake it was to plow up all that land, but it's too late to do anything about it." In the wake of the Dust Bowl, the short-term prospect was bankruptcy; the long-term prospect was the migration of three-quarters of a million itinerant paupers to California, Washington, and Oregon.

As the grizzled Okies advanced on California in their ancient LaSalles, Dodges, and Model T's, mattresses and washbasins strapped to the rooftops, they seemed to represent, as Arthur Schlesinger, Jr., wrote, "the threat of social revolution by a rabble of crazed bankrupts and paupers—a horrid upheaval from below . . . which could only end in driving all wealth and respectability from the state." Since the population of Hall County, Texas, to cite one example, had dropped from forty thousand to one thousand, and the states of North and South Dakota lost at least 146,000 people, the laws of probability demanded that there had to be at least a grain of respectability in the human tide—mayors and preachers were migrating along with toothless dirt farmers and petty thieves—but to those who were being invaded, the Okies were an appalling mob. They had to be settled somewhere; anywhere but here.

One of the more promising places to settle them was the Central Valley of California—more specifically, in the arid and morosely bleak southern two-thirds known as the San Joaquin Valley. The irrigation of the San Joaquin Valley—60 percent of all the prime farmland in a state made up mostly of mountains and high desert—had been an unrequited obsession with California for half a century. By the late 1800s, a few parts of it had been privately reclaimed by farmers and irrigation districts rich enough to build small dams, but most of the valley was a vista of wild blond grassland and wheat. Then came cheap oil, electricity, and the motorized centrifugal pump. Finally freed from all constraints but nature's (irrigation would last only as long as the finite aquifer held out), the farmers began pumping in the finest California tradition—which is to say, as if tomorrow would never come.

By 1930, a million and a half acres were under irrigation in the San Joaquin Valley, and a subterranean thicket of 23,500 well pipes had sucked up so much groundwater that the prognosis for irrigation was terminal within thirty to forty years. In some places, the water table dropped nearly three hundred feet. It was a predicament of their own making, but the farmers were not about to blame themselves; guilt-free life-styles took root in the San Joaquin Valley long before Marin County became a trendsetter. Having exhausted a hundred centuries' worth of groundwater in a generation and a half, they did what any pressure group usually does: run to the politicians they ordinarily despise and beg relief.

Thanks to the stunning wealth irrigation farming had produced, California came rolling out of the 1920s like Jay Gatsby in his alabaster phaeton. Agriculture *was* California; there were no sprawling defense and aerospace industries, there was no Silicon Valley. To give all of this up was unthinkable, even if it was the middle of the Depression. The rescue project which the legislature approved in 1933 not only was bold, it was almost unimaginable. If built, it would be by far the biggest water project in history. It would capture the flows not just of the San Joaquin River, which drained the southern half of the Sierra Nevada, but of the Sacramento, which drained the northern half and some of the Coast Range. It was planned to capture two-thirds of the runoff of the nation's second-largest state, and would move water through thousands of miles of canals and relocate rivers, quite literally, from one end of the state to the other. In normal times, California might even have had the means to begin building it. But this was the Depression, and California, rich as it was, still had to go to the New York bond markets for cash. The voters had no sooner approved a $170 million bond issue (a colossal sum considering the time and circumstances) than the bottom fell out of the market. No sooner had that happened, however, than Franklin Roosevelt landed in the White House.

On FDR's orders, the Bureau of Reclamation officially took over the Central Valley Project in December of 1935. By then the Great Plains had dissolved into the Dust Bowl and the first hundreds of thousands of Okies were rattling into California. In the face of such destitution and calamity, the dams going up were a thrilling sight. The grandest of them was rising in a wild madrone and digger pine canyon on the upper Sacramento River; 602 feet high, it would top out 124 feet lower than Hoover, but it would be half again as wide, an immense, curvilinear, gravity-arch curtain of concrete whose name would first be Kennett, then Shasta. On the San Joaquin River, a big squat dam called Friant was being built at the same time; a third huge earth-and-rock structure would be erected later on the Trinity River, shoveling water from the Klamath drainage to the Sacramento. All together, the dams would give the project an annual yield of more than seven million acre-feet of water, enough to irrigate a million and a half, two million, perhaps three million acres—depending on how much was supplemental irrigation for existing farms and how much was new land. But all of this effort would create, at most, jobs and farms for 100,000 displaced people. (Most of the refugees would actually become migrant workers—wetbacks with Oklahoma accents and white skin.) The biggest public-works project in the world, in other words, was not nearly big enough to soak up the huge tide of the dispossessed. FDR knew that, and that was why he had announced, before the Central Valley Project was even officially underway, "in definite and certain terms, that the next great . . . development to be undertaken by the Federal Government must be that on the Columbia River."

D aughter of ice, orphan of fire, the Columbia River emerged sometime within the relatively recent past, say twenty million years
ago, and for most of its ancestral existence followed a course straight
westward toward Seattle and Puget Sound. Seattle, of course, was not
there when the Columbia first rose. Neither was Puget Sound. Neither
was Washington. Most of what we call the Pacific Northwest is accreted
terrain—a landmass of exotic origin that migrated up from somewhere
around the Equator, riding the Pacific Plate, and glommed on. When
the Pacific and North American plates began to collide millions of
years ago, the Pacific Plate was at first subducted into North America's
basement. Down there, it encountered the large fraction of the planet
that is still molten, and began to crowd it. The lava had nowhere to
go but up.

To geologists, the age of the Columbia River Basalts was a particularly exciting time. The vulcanism lasted ten million years or so, and
covered a broad area. You can see the evidence in the cindered lava
beds of Idaho and western Oregon, in the columnar basalts of Devil's
Postpile in the Sierra Nevada, in the smoking cones of the Cascades.
The Cascade volcanoes, which formed recently—Mount St. Helens is
probably no more than fifteen thousand years old—are the last embers
of a giant bonfire which began to end, according to the available evidence, about seven million years ago. By then, the Pacific and North
American plates had begun to equalize, grinding against each other
like teeth and causing a chaos of earthquakes and volcanos beyond
anything imaginable in our time. The Columbia River flowed during
the whole period of eruption; constantly smothered by lava dams, it
must have changed course hundreds of times. As the vulcanism subsided, the river began to enjoy the first quietude of its long existence,
which lasted several million years—until the ice came.

The continents of snow that slid down from the North Pole during
the Ice Ages stopped somewhere along a latitudinal line defined by
Seattle, Spokane, and Great Falls, Montana. Where the topographic
conditions were right, however, some of them went farther, huge pen-
insulas of ice that protruded a hundred or two hundred miles south.
Near the present location of Lake Coeur d'Alene in western Idaho, an
ice lobe laid itself across the path of the voluminous melt pouring from
the mile-high glacial walls and blocked it, forming what may well
have been history's most prodigious dam. Confronted by a wall of ice

thousands of feet high, the runoff pooled and backed into a reservoir referred to by geologists as Glacial Lake Missoula. Frigid, ephemeral, hundreds of feet deep, the lake covered an area roughly the size of Lake Michigan and contained half as much water. At some point, as the lake deepened behind the ice dam, the dam must have begun to float—ice being lighter than a corresponding volume of water. The flood probably came in a sudden instantaneous release, like the collapse of Teton Dam, and emptied Lake Missoula within a couple of weeks. The volume of the flood is anyone's guess; Larry Meinert, a geologist at Washington State University at Pullman, says a reasonable estimate is ten times the combined flow of all the rivers in the world. The modern topography of the Northwest was pretty well formed by then; most of Lake Missoula searched out the main stream of the Columbia as its route to the sea. Inundated by a flood surge of 230 million cubic feet per second, the Columbia's spacious canyon was a thimble holding a dinosaur egg. In the upper stages the flood was probably twenty miles wide, confined by steeper valleys, but as it poured across the old lava plains of central Washington it spread into a flowing tumult as wide as Indiana. In places, the water excavated canyons overnight, extensive channels scoured through bedrock that remain such a dominant feature of the landscape that central Washington is more often referred to by geologists as "the channeled scablands." The big channels are known as coulees—Rocky Coulee, Lind Coulee, Esquazal Coulee. The biggest of all—seven hundred feet deep, five miles across, more than fifty miles long—is called the Grand Coulee.

Lake Missoula—greater and lesser incarnations of it—formed and reformed at least six times. The last time was about seventeen thousand years ago; by then there may have been humans living in the region. All of the land swept by the floods was stripped absolutely to bedrock. The glaciers, however, had left behind mountains of fine silt—the ground-up surface of Canada—and the winds distributed it around the region with a generous universality. The silt, known as loess, makes for extremely good farmland, and in some parts of Washington, such as the Palouse region below the Blue Mountains, it accumulated to depths of nearly two hundred feet. Rainfall is sparse behind the Cascades—ten to twenty inches is the norm—but loess has outstanding water-retentive qualities. Through this fortuitous coincidence, the soil neither washed away nor blew away—it grew a cover of blond grass and stayed put, waiting for the white man to arrive. That, in any case, is what white men thought. One spot in particular, around the Grand Coulee, was astonishingly suited for irrigation farming. There were more than a million acres of fine soil on the benchlands, a natural storage reservoir in the coulee itself, and, in the river canyon, a favorable site for a dam. A very, very large dam.

In 1933, the Columbia was by far the biggest river anyone had ever dreamed about damming. Bigger than the Colorado, bigger than the Snake, bigger than the Klamath, bigger than the Rio Grande—about twice as big, in fact, as all of those put together—it was the fourth biggest river in North America. Swelling out of the Purcell Range in Canada, it took off for the ocean like an express train on a route mapped by the Olympic Torch Committee: for three hundred miles it went straight for Alaska, until it picked up the melt from Columbia Glacier, an icefield the size of Chicago; then it turned south; then west; then south again; then east; then south; then west again to the sea. By the time it crossed the U.S. border, it was already so large that the Pend Oreille, a tributary larger than the Colorado, could be swallowed without appreciable effect. At the Dalles, the virgin Columbia had an average flow in excess of 200,000 cubic feet per second, one of the largest rivers anywhere with enough of a drop to contain rapids. Such a volume and such a drop—all of it in a confined canyon—made the river ideal for hydroelectricity; it had a power potential out of proportion even to its vast size. In 1933, it could, if fully developed, have generated enough electricity for everyone living west of the Mississippi River.

For all its power potential, the idea of building a large hydroelectric dam at Grand Coulee was regarded by many people as insane. The Northwest had plenty of smaller rivers, much more easily dammed. The region, in 1930, had only three million inhabitants, and 70 percent of the rural people had no electricity. Even a tenth of its power potential could not be used—especially with Bonneville Dam going up downriver. The Bureau of Reclamation had surveyed the soils of the Grand Coulee benchlands in 1903 and found them excellent, but it had said nothing about building a dam. Major General George Goethals, with the Panama Canal under his belt, came to size up the task and backed off; he recommended a run-of-the-river irrigation diversion instead. Herbert Hoover, himself an engineer and an enthusiast about the dam that was to bear his name, said that construction of a dam at Grand Coulee was "inevitable," that it should be built "at the earliest possible date," but from the zeal with which he pursued the goal he might have been talking about the Second Coming. Even the Columbia's propensity to drown low-lying Portland and Vancouver—it could raise a flood of a million cubic feet per second without too much effort—left the Corps of Engineers unmoved. Only three institutions in the entire country seemed interested in Grand Coulee Dam: the

Wenatchee (Washington) *Daily World,* the Bureau of Reclamation, and the new President of the United States.

Franklin Roosevelt first heard about Grand Coulee from Nat Washington, a descendant of George Washington's brother, who approached him about it at the Democratic National Convention in 1920, when FDR was James M. Cox's running mate. The future President was intrigued, but in a mild way; it would cost a fortune, and FDR, in those days, was still promising to balance the budget. By 1933, however, the Grand Coulee project would have been invented by Roosevelt if someone else hadn't thought of it first. It was colossal and magnificent—a purgative of national despair. It would employ tens of thousands. It could settle tens of thousands more on irrigated lands in a region whose inhabitants, in the late 1920s, consisted of a ferryman and a couple of hay farmers. It was loathed by the Republican conservatives and the private-power interests. Perhaps best of all, it was regarded by none other than the president of the American Society of Civil Engineers as "a grandiose project of no more usefulness than the pyramids of Egypt." To Roosevelt, that remark was as good a reason as any to build it.

And it was built on a foundation of deception.

In 1931, the Corps of Engineers finally pronounced the construction of a concrete dam at Grand Coulee feasible. What the Corps had in mind, however, was a low dam, rising two or three hundred feet from bedrock—a dam similar to its own Bonneville Dam downstream, useful only for regulating navigation flows and for hydroelectricity. The Bureau, however, was not interested in a low dam. The pump lift from the reservoir surface to the canyon rim would be at least five hundred feet; such a lift was beyond the capacity of any pumps in existence at the time, and even if they *had* existed their enormous appetite for power would make any irrigation project infeasible in an economic sense. A high dam was absolutely necessary for an irrigation project, not only because it would knock twenty stories off the pump lift, but because it would produce a vast amount of surplus hydroelectricity to handle the still impressive pump lift and generate enough revenue to subsidize the cost of water so that the farmers could afford it.

The problem with a high dam, however, was Congress. Confronted on all sides by calamity and cries for relief, Congress was not about to appropriate $270 million (about twelve times more in today's money) to build a white elephant of a dam in a remote corner of the country where hardly anyone lived. As it happened, however, Congress had undermined its own intention by giving FDR blanket authority,

under the Public Works Administration and the National Industrial
Recovery Act, to select and fund "emergency" projects that would
assist the relief effort. Why not use some of that money to get started
with a low dam—and then switch horses in midstream?

Nowhere is there absolute proof that this is the strategy FDR had
in mind. The circumstantial evidence is merely overwhelming. In 1933,
he designated $63 million, the greatest sum ever for any single purpose,
from the Public Works Administration under Section 202 of the National Industrial Recovery Act to begin construction on a low dam at
Grand Coulee. At that point, there was no question of intent; a low
dam was specifically mentioned in the appropriation. A few months
later, the construction contract for the dam was let to a consortium
of engineering firms that went by the acronym MWAK. The contract
also specified a low dam. The $63 million was spent in a hurry; by
1935, cofferdams were already in place and the permanent dam's foundation was rising in the riverbed. It was not, however, a foundation
for a low dam—*it was the foundation of a high dam.*

In interviews, no engineer who worked on Grand Coulee Dam
would admit that the Bureau and FDR had a high dam in mind all
along and quietly decided to hoodwink a Congress which they knew
would never authorize it. Nonetheless, no other explanation seems
plausible. Charles Weil, the Bureau engineer charged with concrete
inspection, said that a "substantial" amount of the high dam foundation's concrete had already been poured before the Roosevelt administration went to Congress in 1935 with a request to change the
authorization from a low dam to a high dam. Still, he insisted that
the Bureau never tried to deceive anyone. "I wouldn't say that the
Bureau tried to mislead Congress," Weil offered. "But it had to keep
in mind what Congress was willing to fund." That, of course, is another
way of saying that the Bureau chose to mislead Congress. In the beginning, before construction began, a high dam was out of the question.
After $63 million had been spent building a foundation for it, however,
a *low* dam was out of the question; at the very least, it wouldn't have
made much sense. The Bureau had presented Congress with a *fait
accompli* in the form of a gigantic foundation designed to support a
gravity dam 550 feet tall. To build a two-hundred-foot dam on it would
have been like mounting a Honda body on the chassis of a truck.

Phil Nalder, who rose from draftsman to manager of the entire
Columbia Basin Project, was as circumspect as Weil about the Bureau's motives and strategy. According to Nalder, "The Bureau determined belatedly that a low dam would have been impractical at the
site." But that, of course, is something the Bureau must have recog-

nized all along. There was nothing "impractical" about building a low dam for power and navigation, but building a low dam for an irrigation project was hopelessly impractical. Nalder, at least, was a bit more candid about whether the evidence didn't suggest that Roosevelt and the Bureau had pulled a fast one on the Congress. "Well, if you look at the evidence superficially," he said, "it would certainly appear that way."

The issue of a high dam versus a low dam involved much more than power production and the fate of the irrigation project. It also involved the fate of the greatest spawning run of salmon in the world. During the Depression, salmon was the one high-protein food most people could afford; it was still so abundant that it cost about ten cents for a one-pound can. America's Atlantic salmon were almost wiped out by then; virtually all domestic salmon came from Alaska and the West Coast, and the greatest run—equal to or greater than all the streams and rivers in Oregon and California combined—went up the Columbia River. Some of the fish branched off into the lower tributaries to spawn, but the majority went far up the river into the higher tributaries, beyond Grand Coulee. Many salmon could probably have gotten past a low dam; today, tens of thousands manage to circumnavigate the Dalles, John Day, and Bonneville dams through fish ladders every year. A high Grand Coulee Dam, however, would block their passage forever. A fifty-story wall rising straight out of the river would form an ultimate obstruction—hopeless and forbidding. A fish ladder, built at a proper gradient, would have to run for many miles, cut into sheer canyon walls. No one was even talking about building it; the cost might approach the price of the dam. (Fish facilities at Bonneville Dam's second powerplant, built many years later, would end up costing $65 million, almost one-fourth the cost of the powerplant itself.) If the high dam spelled doom for most of the salmon in the Columbia River, however, it did perform a miraculous service which, at the time, was utterly unforeseen. It probably won the Second World War.

It is hard to imagine today, when big public-works projects such as New York's Westway are held up for fifteen years in the courts, what the go-go years were like. In 1936, the four largest concrete dams ever built—Hoover, Shasta, Bonneville, and Grand Coulee—were being erected at breakneck speed, all at the same time. In Montana, Fort Peck Dam, the largest structure anywhere except for the Great Wall—which took a third of the Chinese male population a thousand years to build—was going up, too. The age of dams reached its apogee in the 1950s and 1960s, when hundreds upon hundreds of them were thrown up, forever altering the face of the continent—but most of those dams were middle-sized, squat, utilitarian, banal. The 1930s were the glory days. No dam after Hoover has ever quite matched its grace and glorious detail. Shasta Dam looks rundown now—the Grecian pavilions are rotting, the face is water-stained—but it was nearly as majestic as Hoover when it was built, and quite a bit bigger.

Symbolic achievements mattered terribly in the thirties, and the federal dams going up on the western rivers were the reigning symbols of the era. A few years earlier it had been the great skyscrapers that served as the landmarks of American achievement. In the late 1920s, they were rising simultaneously, too—the Empire State Building, the Chrysler Building, the Bank of Manhattan, 70 Pine Street, the Lincoln-Leveque Tower in Columbus, and the Carew Tower in Cincinnati—but just as they were being finished, the capitalist engine that had built them fell into ruin. In a slip of time, the mantle of achievement passed from private enterprise to public works. The dams announced that America could still do remarkable things; they also said that the country would never be the same. The centralized welfare state that everyone decries, and nearly everyone depends on to some degree, is said to have emerged from the war, the Depression, and the Great Society. It might be more accurate to say that it was born in the rivers of the American West.

Hoover was big; Shasta was half again as big; Grand Coulee was bigger than both together. Many of the workers who came up to build it were those who had just finished Hoover. When they imagined it filling this huge U-shaped canyon, they were speechless. "When they worked on Hoover they thought it made everything else look like nothing," says Phil Nalder. "When they saw what we were going to build here they said it made Hoover look like nothing."

After a while, visitors being taken around the damsite became tired of the phrase "largest in the world." The mass (10.5 million cubic yards) and crest length (four-fifths of a mile) were, for a concrete dam, the largest and longest in the world. The concrete-mixing plant, the spillway, the generators, the powerhouse, the pumps, the penstocks, and the pump lift from the reservoir to the irrigated benchlands would all be the largest in the world, and as the dam went up the engineers were still scratching their heads about how to lift such an immense volume of water thirty stories high. The turbines, the scroll casings, the conveyor belts, the forms, the cofferdams, and the concentration of brothels and bars within a five-mile radius were also the largest in the world. The dam's dimensions—height and length—were roughly those of the Golden Gate Bridge—it was not quite as high or long—but it was *solid*, and, at the base, five times as wide. Grand Coulee would use more lumber—130 million board feet—than any edifice ever built, but it was a tiny fraction of the dam's total mass, and none of it was even visible. Like Hoover, the dam was so massive it would ordinarily have required hundreds of years to cool down, and cooling pipe had to be laid through it at close intervals. Laid out in a straight line, the pipe would have connected Seattle to Chicago.

The astonishing thing about Grand Coulee—about the whole era—was that people just went out and built it, built anything, without knowing exactly how to do it or whether it could even be done. There were no task forces, no special commissions, no proposed possible preliminary outlines of conceivable tentative recommendations. Tremendous environmental impacts, but no environmental impact statements. When Chuck Weil applied for a job on Grand Coulee, he didn't know the first thing about concrete; before long, he was inspecting more concrete than anyone in history. Phil Nalder was trained as an electrical engineer; he started as a tracer (one rung below draftsman) and, later on, was put in charge of the whole project. Once, well into construction, a mudslide the size of a small mountain came off one side of the canyon and threatened to cover the foundation of the dam. To stabilize it, the Bureau ran around the Northwest looking for the biggest refrigeration units it could find; then it ran supercooled brine through the slide and froze it while construction continued. No one had ever tried it before, but it worked. When one of the cofferdams sprang a huge leak, it was plugged with old mattresses. The dam was finished and in service by September of 1941, an unbelievable sight. The three largest ocean liners in the world could have sat atop its crest like bathtub toys.

Much of the country thought Grand Coulee was marvelous, but it was so gigantic a project that it had to invite some kind of attack. Private utilities, not quite brave enough to lambast so popular a creation, were suspected of bribing journalists to write diatribes against it. One writer, Walter Davenport, went out to see the dam for *Collier's* magazine; it was, he reported, in the middle of a "dead land, bitter with alkali," shunned "even by snakes and lizards," where "the air you breathe is full of the dust of dead men's bones." But Ickes and Mike Straus cooked up the idea of hiring Woody Guthrie as a "research assistant" to write some songs in praise of the dams. Guthrie, an itinerant Okie guitar picker, toured the Northwest like a prince in a chauffeured car, composing paeans to water and power like "Talking Columbia":

You jus' watch this river 'n pretty soon
E-everybody's gonna be changin' their tune. . . .
That big Grand Coulee 'n Bonneville Dam'll
Build a thousand factories f'r Uncle Sam. . . .
    'N ev'rybody else in the world
    Makin' ev'rything from sewin' machines
    To a-tomic bedrooms, 'n plastic . . .
    E-everything's gonna be made outa plastic.

Uncle Sam needs wool, Uncle Sam need wheat
Uncle Sam needs houses 'n stuff to eat
Uncle Sam needs water 'n power dams,
Uncle Sam needs people 'n the people need land.
    Don't like dictators none much myself,
    What I think is the whole world oughta be run by
    E-electricity. . . .

What Guthrie sensed, and what Franklin Roosevelt knew by 1939, was that America stood an excellent chance of going to war. It would be a war won or lost not so much through strategy as through production. Germany had the greatest industrial capacity in Europe; Japan's was the greatest in the Orient. In the balance stood the United States. And since this would be a war of, more than anything, air power, the critical material was going to be aluminum. It would be, at least, until the critical material became plutonium.

In the nineteenth century, aluminum had a street value close to gold's—a function of the amount of energy needed to produce it and the type of energy required. It takes twelve times as much energy to produce raw aluminum as it does to make iron, and since the process is electrolytic, it has to be done with electricity. Until another process is invented, nothing else will do. The one-thousand-ounce aluminum Pope's cap installed in the pinnacle of the Washington Monument when it was completed in the mid-nineteenth century was the largest ingot of its day. After the First World War, aluminum became cheaper, though still not common. The raw material, the production flow, the manufacturing patent, and the end uses were pretty much controlled by the Aluminum Company of America, which was to vertical integration what William Randolph Hearst was to yellow journalism. Hearst, at least, had competition; Alcoa didn't—except from Adolf

Hitler, who made Germany the world leader in aluminum production soon after seizing power, for reasons the Allies did not immediately discern. When the first electricity began to flow out of Bonneville Dam, the Corps of Engineers' big power and navigation dam three hundred miles downriver, the government tried to induce Alcoa's potential competitors to build plants in the Northwest by offering them bargain rates, but nobody was particularly interested. By the time the Japanese bombed Pearl Harbor, however, the luxury of persuasion could no longer be afforded. The government simply went out and built the plants itself.

No one knows exactly how many planes and ships were manufactured with Bonneville and Grand Coulee electricity, but it is safe to say that the war would have been seriously prolonged at the least without the dams. Germany's military buildup during the 1930s gave it a huge start on Britain and France. When Hitler invaded Poland and war broke out in Europe, the United States was, militarily speaking, of no consequence; we had fewer soldiers than Henry Ford had auto workers, and not enough modern M-1 Garand rifles to equip a single regiment. By 1942, however, we possessed something no other country did: a huge surplus of hydroelectric power. By June of that year, 92 percent of the 900,000 kilowatts of power available from Grand Coulee and Bonneville Dams—an almost incomprehensible amount at the time—was going to war production, most of it to building planes. One writer, Albert Williams, estimates that "more than half the planes in the American Air Forces were built with Coulee power alone." After France capitulated, England was left hanging by a thread. It was rescued by a European sky suddenly full of American planes. The Columbia River was a traffic jam of barges carrying bauxite to the smelters in Longview, Washington. By the middle of the war, almost half of the aluminum production in the country was located in the Northwest—nearly all of it going toward the war effort. American planes were being downed almost as fast as they could be produced. German planes, however, were being downed faster than they could be produced. The Nazis had neither the raw materials nor the electricity to produce what they needed fast enough.

In late 1940, when Grand Coulee Dam was being completed, people had been saying that its power would go begging until the twenty-first century. Twenty-two months later, all of its available power was being used and the defense industries were screaming for more. As the first six generators were being installed, the next two units were still being manufactured and wouldn't be ready for power production for some weeks. The war was at such a critical juncture that some weeks was too long. The Bureau collected every outsize piece of transportation equipment it could find, took the two generators waiting to be installed at Shasta Dam, and laboriously moved them to Grand Coulee instead. Shasta's generators were thirty thousand kilowatts smaller than Grand Coulee's, and the turbines revolved in the wrong direction: Grand Coulee's went clockwise, Shasta's went counterclockwise. The Bureau solved the problem by installing the Shasta units in the wrong pits and excavating tunnels to the proper ones next door, so the water could surge in from the right side. After the war, the engineers had to invent some mammoth excavation devices to shoehorn them out.

The Westinghouse generators built for Grand Coulee were rated for a maximum output of 105,000 kilowatts each, which was the capacity of a good-sized oil power plant that could run, say, Duluth. For the entire duration of the war, they ran at 125,000 kilowatts, twenty-four hours a day, without a glitch. "We would shut one down only when it was absolutely necessary," says Phil Nalder. "You'd stand there in the powerhouse and feel that low vibration, that low but incredibly powerful vibration, and you'd feel certain that they were going to burn themselves up. And you'd think that maybe the course of history depended on these damned things. But they never overheated, so we just ran them and ran them. God knows, they were beautifully made. By the end of the war, at Grand Coulee, we were generating 2,138,000 kilowatts of electricity. We were the biggest single source of electricity in the world. The Germans and the Japanese didn't have anything nearly that big. Imagine what it would have been like without Grand Coulee, Hoover, Shasta, and Bonneville. At the time, they were ranked first, second, third, and fourth in the world. We had so much power at Grand Coulee that we could afford to use two generators just to run Hanford."

Although few of the people who lived there knew it at the time, the strange squat structures going up in 1943 at the Hanford Reservation, an ultrasecret military installation along the Columbia River near Richland, Washington, were intimately connected to the Manhattan Project. A lot of the history is well-known now: how Niels Bohr was smuggled out of Nazi-occupied Denmark in the wheel well of a British balsa-wood aircraft; how pacifistic Albert Einstein urged Franklin Roosevelt to build the bomb before the Nazis did; how thousands of technicians and scientists descended on the tiny mountain hamlet of Los Alamos, New Mexico, to figure out how to build their catastrophically explosive device. The key material was plutonium-239, an element virtually unknown in nature which has just the right fissile

characteristics for an atomic bomb. The problem with plutonium—
aside from its being fiendishly toxic—is that its production is energy-
consumptive in the extreme. The amount of electricity used by the
eight plutonium-production reactors at Hanford is still classified information, but a good guess is fifteen or twenty megawatts each—
perhaps 160 megawatts in all. Nowhere else in a country involved in
a gigantic war effort could one have found that kind of power to spare.

In the end, the Axis powers were no match for two things: the
Russian winters, and an American hydroelectric capacity that could
turn out sixty thousand aircraft in four years. We didn't so much
outmaneuver, outman, or outfight the Axis as simply outproduce it.

The main stem of the Columbia River didn't have a single dam on it
until 1933, when the Puget Sound Power and Light Company went
out on its own and built a run-of-the-river dam called Rock Island,
which produced 212,000 kilowatts of power—a mind-boggling amount
in its day. Five years later, Bonneville Dam was finished and generated
almost three times as much power. In 1941 came Grand Coulee; in
1953, McNary Dam; in 1955, Chief Joseph Dam; in 1957, The Dalles,
contributing 1,807,000 kilowatts to the seven million or so that had
already been wrung out of the river. In that same year, the Grant
County Public Utility District finished Priest Rapids Dam, which
added another 788,500 kilowatts. In 1961, the Chelan County PUD
came back and built Rocky Reach Dam, with a capacity one million
kilowatts greater than the dam by which it had gotten things off to a
start twenty-eight years before. And it still wasn't over. In 1963, the
Grant County PUD added Wanapum Dam and another 831,250 kilowatts. In 1967, the Douglas County PUD completed Wells Dam. The
Corps of Engineers, which had built Bonneville and Chief Joseph and
McNary and The Dalles, got back into the picture in 1968 with John
Day Dam, whose 2,160,000 kilowatts were second only to Grand Coulee. In that year, the Canadians finally joined in, building Keenleyside
Dam, whose sole purpose was to equalize the upper river's flow
throughout the year for the benefit of navigation and power production. In 1973, they added Mica Dam, which formed the largest reservoir
on the river in a remote wilderness not far from the Columbia's headwaters. Thirteen tremendous dams in forty years.

And these were just the *main-stem* dams. As they were going up,
the Columbia tributaries were also being chinked full of dams. Libby
Dam on the Kootenai River. Albeni Falls and Boudary dams on the
Pend Oreille. Cabinet Gorge and Noxon Rapids dams on the Clark
Fork. Kerr and Hungry Horse on the Flathead. Chandler and Roza

dams on the Yakima. Ice Harbor Dam, Lower Monumental Dam, Little Goose Dam, Lower Granite Dam, Oxbow Dam, Hells Canyon Dam, Brownlee Dam, and Palisades Dam on the Snake. Dworshak Dam on the North Fork of the Clearwater. Anderson Ranch Dam on the South Fork of the Boise. Pelton and Round Butte dams on the Deschutes. Big Cliff, Foster, Green Peter, and Detroit dams on the three forks of the Santiam River. Cougar Dam on the South Fork of the McKenzie. Dexter, Lookout Point, and Hills Creek dams on the Willamette. Merwin Dam, Yale Dam, and Swift Dam on the Lewis River. Layfield and Mossyrock dams on the Cowlitz. Thirty-six great dams on one river and its tributaries—a dam a year. The Age of Dams.

The Corps of Engineers and the region's public utilities played a big role in the damming of the Pacific Northwest because it had in abundance what the rest of the region lacked—water—so many of the dams were built for flood control, nagivation, or power. Everywhere else in the West, however, where deserts were the rule and irrigation was the be-all and end-all of existence, the Bureau reigned supreme. Within its first thirty years, it had built about three dozen projects. During the next thirty years, it built nineteen dozen more. The Burnt River Project, the Cachuma Project, the Mancos Project, the Ogden River Project, the Collbran Project, the Gila Project, the Pine River Project, the Palisades Project, the Weber Basin Project, the Columbia Basin Project, and the Central Valley Project. Shasta Dam, Parker Dam, Friant Dam, Davis Dam, Laguna Dam, Canyon Ferry Dam, Cascade Dam, Flaming Gorge Dam. Cedar Bluff Lake, Paonia Reservoir, Kirwin Reservoir, Webster Reservoir, Pathfinder Reservoir, Waconda Lake, Clair Engle Lake, Lake Berryessa, Lake C. W. McConaughy, Enders Reservoir, Box Butte Reservoir. The Tucumcari Project, the Palo Verde Project, the San Angelo Project, the Canadian River Project, the Crooked River Project, the Kendrick Project, the Hubbard Project, the Hyrun Project, the Eden Project, the W. C. Austin Project, the Colorado–Big Thompson Project, the Pecos River Basin Water Conservation Project ("conservation" meaning, in this case, the virtual drying-up of the Pecos River), the Mercedes Division, the Middle Rio Grande Project. Trinity Dam, Keswick Dam, Folsom Dam, Morrow Point Dam, Blue Mesa Dam. The Oroville-Tonasket Unit of the Okanogan Similkameen Division of the Columbia Basin Project. Glen Canyon Dam. Lake Powell, Jewel of the Colorado.

By 1956, the Congress had voted 110 separate authorizations for the Bureau of Reclamation, some encompassing a dozen or more irrigation projects and dams. Of these, seventy-seven—nearly three-quarters—were authorized between 1928 and 1956, along with hundreds of projects built by the Corps of Engineers in the East and West. In that astonishingly brief twenty-eight-year period between the first preparations for Hoover Dam and the passage of the Colorado River Storage Project Act, the most fateful transformation that has ever been visited on any landscape, anywhere, was wrought.

It was profound change—profound and permanent. You can levee a river, dredge it, riprap it, channelize it, straighten it, do almost anything to it except build a dam on it, and unless you maintain your works diligently, nature will soon take the river back. Simple diversion works of great ancient civilizations collapsed not long after the civilizations themselves did; for the most part, a remnant here and there is all that remains. Had the Assyrians built Grand Coulee Dam, however, it would sit exactly where it does today, looking exactly as it did when it was built. The only thing different is that the dam would no longer function as a dam. It would be a waterfall. The reservoir behind it would have long since silted up.

And the effects would go far beyond the natural world. In the Northwest, the dams produced so much cheap hydroelectricity that hundreds of thousands of people who flocked to the region during and after the war did not bother to insulate their homes. Insulation was expensive; electricity was dirt-cheap. In 1974, $196.01 worth of power from Con Edison in New York would have cost $24 if purchased from Seattle City Light. (For decades, the Northwest and British Columbia have had the highest rates of electricity consumption in the world.) The result was that by the 1970s, to everyone's amazement, the seemingly limitless hydroelectric bonanza was coming to an end; brownouts were being predicted for the 1980s. Since the good damsites were gone, the region's utilities and their federal power broker, the Bonneville Power Administration—another product of the go-go years—launched a program of coal and nuclear powerplant construction which, viewed in retrospect, seems more like dementia than the rational, orderly planning it was purported to be. Of the twenty-four thousand-megawatt plants that were to be built under the Washington Public Power Supply System—one a year—five were begun, only to be scrapped or mothballed, half-completed, a few years later, threatening to cause the biggest municipal bond default in history. The cost of their construction, driven by inflation and hyperactive interest rates, drove electricity rates up, which immediately drove demand down, which drove rates further up, which drove demand further down—a self-perpetuating vortex known among municipal bond traders and their hapless victims and the region's hollow-eyed utilities as "death spiral." No one knows where this fiasco—now referred to simply by the power consortium's onomatopoetic acronym WPPSS—may end, but more than $6 billion has been invested in nuclear plants that may never produce a watt of power. The blame for it—if it is worth laying blame at all—has to fall on the region's forty-year love affair with dams.

It was, of course, a love affair not limited to the Northwest or even the West. The whole country wanted more dams. In Appalachia, the Tennessee Valley Authority had an answer to poverty: dams. No river in the entire world has as much of its course under reservoirs as the Tennessee; by the late 1960s, it was hard to find a ten-mile free-flowing stretch between dams. The Missouri is a close second; about seven hundred miles in its middle reaches became a series of gigantic stair-step reservoirs. In Texas and Oklahoma, between 1940 and 1975, something like eight million acres of land were submerged by artificial lakes. Much of this land was in the eastern part of those states; it was exceptionally fertile (as were the bottomlands along the Tennessee) and visited by adequate rainfall, making it some of the best farmland in the nation. No one seemed bothered by the spectacle of a government creating expensive farmland out of deserts in the West while drowning millions of acres of perfect farmland in the East. If there was a stretch of free-flowing river anywhere in the country, our reflex action was to erect a dam in its path.

There were legitimate reasons, of course, to build a fair number of those thousands of dams. Hydropower obviously was one; the Columbia dams helped prevent the horror of Nazism from blackening the entire world. Some new irrigation projects made economic sense, as late as the 1940s and 1950s (though virtually none did after then). The Tennessee and Red rivers were prone to destructive floods, as was the Columbia—as were many rivers throughout the country. A better solution, in many cases, would have been to discourage development in floodplains, but the country—least of all the Congress—wasn't interested in that. For a dam, whether or not it made particularly good sense, whether or not it decimated a salmon fishery or drowned a gorgeous stretch of wild river, was a bonanza to the constituents of the Congressman in whose district it was located—especially the engineering and construction firms that became largely dependent on the government for work. The whole business was like a pyramid scheme—the many (the taxpayers) were paying to enrich the few— but most members of Congress figured that if they voted for everyone else's dams, someday *they* would get a dam, too.

And this, as much as the economic folly and the environmental damage, was the legacy of the go-go years: the corruption of national politics. Water projects came to epitomize the pork barrel; they were the oil can that lubricated the nation's legislative machinery. Important legislation—an education bill, a foreign aid bill, a conservation bill—was imprisoned until the President agreed to let a powerful committee chairman tack on a rider authorizing his pet dam. Franklin Roosevelt had rammed a lot of his public-works programs through a Congress that was, if not resistant, then at least recumbent. A generation or two later, however, it was Congress that was writing omnibus public-works bills authorizing as much as $20 billion worth of water projects at a stroke and defying threats of presidential vetoes. Most members who voted for such bills had not the faintest idea what was in them; they didn't care; they didn't dare look. All that mattered was that there was something in it for them. What had begun as an emergency program to put the country back to work, to restore its sense of self-worth, to settle the refugees of the Dust Bowl, grew into a nature-wrecking, money-eating monster that our leaders lacked the courage or ability to stop.

CHAPTER SIX

Rivals in Crime

On the 16th of August, 1962, Major General William F. Cassidy, the director of civil works for the United States Army Corps of Engineers, gave a speech titled "The Future of Water Development" before a gathering of his peers in Davis, California. Considering what Cassidy had to say, the speech attracted surprisingly little attention in the press.

"Before white men came to North America," the General began, "it is estimated that about one million Indians inhabited the region between the Canadian border and the Gulf of Mexico. The streams were unpolluted, the forests still stood, and the plow had not broken the plains. They had the resources of a continent at their disposal, and about four hundred acres of arable land for every man, woman, and child. Yet they often starved—because they lacked the capacity to develop their resources.

"In the 1890s," Cassidy continued, "the United States had about seventy-seven acres of cultivated land per person. Before World War I, about four acres. Today, we have only about two acres of cultivated land per person. Yet the United States maintains the highest level of living known to history, it exports food to other nations, and it has even accumulated substantial surpluses of a few crops. This is the result of increasingly intensive resource development."

But all of this resource development had been a mere warm-up exercise compared with what was still to come. "During the next twenty years," Cassidy went on, "we estimate that we will have to provide some 320 million acre-feet of reservoir storage at a cost of about $15 billion; about thirteen thousand miles of new or improved inland waterways; about sixty new or improved commercial harbors; thirty million kilowatts of hydroelectric power-generating capacity; some eleven thousand miles of levees, floodwalls, and channel improvements; and recreational facilities for perhaps 300 million visitors at our reservoirs. . . ." If all of that seemed "unduly large or visionary," Cassidy admonished, "let us remember the responsibilities our nation is facing."

It is worth taking a moment to put some of these figures in perspective. In 1962, the total amount of federally built reservoir storage in the nation was somewhere around 300 million acre-feet. In twenty years, Cassidy wanted to more than double that. Every year, the Mississippi River carries about 355 million acre-feet of water out to sea, the runoff of most of the United States from Pennsylvania to Montana. In twenty years, according to the Corps of Engineers, we were going to put the equivalent of 90 percent of that water behind dams. In 1962, there were 37,342 megawatts of installed hydroelectric generating capacity in the United States; by 1982, that figure was nearly to double. By 1962, nearly all the major rivers in the United States—long reaches of the Mississippi, the Snake, the Columbia, the Illinois, the Missouri, the Sacramento, the Susquehanna, the Red, the Delaware, the Tennessee, the Apalachicola, the Savannah—had been dredged, realigned, straitjacketed, riprapped, diked, leveed, stabilized, and otherwise made over in order to accommodate barge and freighter traffic. In twenty years, we were going to add or "improve" thirteen thousand more miles.

And this Promethean agenda was going to be possible, according to the director of civil works, because we were "about to enter an era of unprecedented cooperation in planning water resource development to meet future needs. . . . The walls which formerly separated various spheres of interest are crumbling under the pressure of manifold needs."

Even allowing for the temper of the times, Cassidy's prophecy, in retrospect, seems one of derangement more than vision. Nineteen years later, the $15 billion which was to construct 320 million acre-feet of reservoir storage would barely suffice to build ten million acre-feet of new storage in California—had it been politically possible to do it. It was hard to imagine thirteen thousand miles of new or "improved" navigable waterways without envisioning barges bumping against the Rocky Mountains or poking into bulrushes at the headwaters of southern streams. Even had there been money to build all those reservoirs, there wasn't any room for them—as Cassidy was almost willing to admit. "In many intensively occupied river basins," he said, using the military jargon of which the Corps is inordinately fond, "we . . . face a very difficult task in finding sites for the reservoirs needed to support future growth"—thus raising the prospect of a nation requiring so many new dams to feed water and electricity into its hyperventilating economy that it would flood itself right off the land and find itself forced to go about its business aboard houseboats.

Actually, the General's vision was to mutate into irony as fabulous as the prophecy itself. He was right in one sense—you did not build such incredible works to carry water from areas of "surplus" to areas of "deficit" without intricate political compromises among the states involved and unprecedented collaboration between the agencies that would presumably do the job, the Bureau of Reclamation and the Corps. But he was dead wrong in predicting that such harmonious relations would ever be. And if a single entity could be blamed for this—because it schemed constantly against its would-be confederate, because it seized every opportunity to build any senseless project it could, because it worked diligently, if unwittingly, to give water development a bad name—it was none other than his own agency, the U.S. Army Corps of Engineers.

In California, where Cassidy gave his speech—at the very moment, in fact, when he was giving his speech—the Corps of Engineers was shamelessly trying to steal from the Bureau of Reclamation at least one major project the Bureau had intended to build for years. It had already done it several times before, in California and elsewhere. Across the entire West, the Corps, as opportunistic and ruthless an agency as American government has ever seen, was trying to seduce away the Bureau's irrigation constituency; it was toadying up to big corporate farmers who wanted to monopolize whole rivers for themselves; it was even prepared to defy the President of the United States. As a result, the business of water development was to become a game of chess between two ferociously competitive bureaucracies, on a board that was half a continent plus Alaska, where rivers were the pawns and dams the knights and queens used to checkmate the other's ambition. But the Corps and the Bureau played a little too well and a little too long for their own good. While they were fighting over a Lake Ontario-size reservoir in the middle of Alaska, and over countless squalid little projects desired by local interest groups, an unprecedented water crisis was gathering on the southern high plains—a crisis tailor-made for their own limitless ambition which, in the end, they would do nothing about. The Corps and the Bureau wasted so much money on frivolous projects which didn't so much solve the nation's water situation as satisfy the greed of powerful interests and their own petty ambitions that in the 1980s, despite dozens of new dams and reservoirs built during the intervening years, a water crisis loomed larger than in 1962. Within the next half century, as much irrigated land is likely to go out of production—land that grows nearly 40 percent of our agricultural exports—as the Bureau of Reclamation managed to put *into* production during its entire career. And though projects to rescue those regions remain on the drawing boards, the age when they might have been built seems to have passed.

The Corps of Engineers, the construction arm of the United States Army, was baptized during the Revolutionary War, when a group of engineers in the Continental Army built a breastwork on Bunker Hill. In 1794, the Corps was officially christened with its current name and divided into a civilian and a military works branch. The civil works branch, which was to become by far the larger of the two, began modestly enough, clearing driftwood and sunken ships out of rivers and harbors and occasionally doing a bit of dredging. It also played a role in the early exploration and surveying of the nation. The Corps' great work—and its transmutation into one of history's most successful bureaucracies—began late in the nineteenth century, when it took upon itself the task of restyling America's largest rivers to accommodate barge traffic and, occasionally, deep-draft ships. At the same time, it found a role for itself in flood control, which it first accomplished by building levees and dikes, and then, after denying for years that reservoirs could control floods, by building flood-control reservoirs. And it built them at a pace that would have left the most ambitious pharaoh dazzled—something like six hundred in sixty years.

The Army Engineers have so many hands in so many different types of work that their various activities sometimes cancel each other out. The Corps drains and channels wetlands—it has ruined more wetlands than anyone in history, except perhaps its counterpart in the Soviet Union—yet sometimes prohibits the draining and dredging of wetlands by private developers and other interests. (This was a role forced on the Corps by the Congress, not one it undertook voluntarily.) Its dams control flooding, while its stream-channelization and wetlands-drainage programs cause it. Its subsidization of intensive agriculture—which it does by turning wetlands into dry land, so they may then become soybean fields—increases soil erosion, which pours into the nation's rivers, which the Corps then has to dredge more frequently.

Cynics say this is all done by design, because the Corps of Engineers' motto, "Building Tomorrow Today," really ought to be "Keep Busy." Its range of activities is breathtaking: the Corps dams rivers, deepens rivers, straightens rivers, ripraps rivers, builds bridges across rivers, builds huge navigation locks and dams, builds groins on rivers and beaches, builds hatcheries, builds breakwaters, builds piers, and repairs beach erosion (finally fulfilling the first stage of a destiny conservationists have long wished on it: carrying sandpiles from one end of the country to the other and back again). The works for which the Corps is most famous—or notorious, depending on one's point of view—are the monumental inland navigation projects such as Red River, Tennessee-Tombigbee, and Arkansas River. However, though each of these may cost billions to begin with, and hundreds of millions to maintain, the opportunities for such work are pretty thin. Opportunities for serious work come most frequently in the form of flood-control and water-supply dams.

The Corps confined its activities mainly to the East and Middle West until the Great Depression—it is widely, and falsely, regarded as the "eastern counterpart" of the Bureau of Reclamation—but the temptations of the West ultimately proved too much to resist. Throughout much of the East, it is hard to find a decent spot for a dam. There are few tight gorges and valleys, or there are few natural basins behind them, or there are too many people along rivers who would have to be moved. (Not that uprooting and relocating people particularly bothers the Army Engineers; it is more a matter of expense.) The West, however, is a dam builder's nirvana, full of deep, narrow canyons and gunsight gaps opening into expansive basins. The West is also more sparsely populated, and has floods—enormous floods—because its precipitation tends to be both erratic and highly seasonal, and because of this, the groundcover, compared with the East, is spare. With little in the way of grass or forests or wetlands to hold it back, runoff during the storms is extreme. Small streams, even tiny creeks, have flowed at rates approaching the country's largest rivers. They rarely flow like this for long, but a few minutes is all it takes to float away a town. Bijou Creek in Colorado, nearly always dry, has gone over 400,000 cubic feet per second after an eight-inch rainstorm. California's Eel River peaked at 765,000 cfs—the flow of the Mississippi and the Columbia combined—during the Christmas flood of 1964.

On many rivers in the West a dam built for irrigation will incidentally control floods. But the equation also works in reverse: a flood-control dam, by evening out a river's flow year-round, makes it useful for irrigation. And if the Corps of Engineers builds the dam, and calls it a flood-control dam, the water is free.

The Kings, the Kaweah, the Tule, and the Kern are the southernmost rivers flowing out of the Sierra Nevada into the Central Valley of California. They are the only rivers that do not ultimately end up in either the Sacramento or the San Joaquin drainage, because a low rise of land in the upper San Joaquin Valley, south of Fresno, effectively divides the valley into two hydrologic basins. The southernmost one, which receives the runoff of the Kings, Kaweah, Tule, and Kern, is known as Tulare Basin. Historically, the four rivers of Tulare Basin went into two terminal lakes, Tulare and Buena Vista, which appeared and disappeared every year like phantoms. During the wet winters, the lakes would begin to fill; they would reach their largest size in May, after twenty feet of Sierra snow had melted into them in a matter of weeks; and, in all but the wettest years, they would evaporate so quickly under the glaring summer sun that they were dry again, or mostly dry, around September. Tulare, the more impressive of the two lakes, often grew larger than Lake Tahoe, though it was not more than a few feet deep. From year to year, its shoreline would shrink or grow by miles. It was a wonderful sight to see all of that water glimmering amid the merciless dryness of the San Joaquin Valley in summer, and the lakes were a stopover for millions of migrating ducks, geese, and sandhill cranes.

Before World War II, most of the agricultural lands around Tulare and Buena Vista lakes—and the lakes themselves—were owned by four private landholders. They were, in a sense peculiar to California, "family" farms. Buena Vista Lake and the land around it was the largest remnant of the million-acre domain amassed by Henry Miller, and later squandered by a succession of dissolute heirs. The property encompassed about eighty thousand acres, seven times the area of Manhattan Island. The adjacent Kern County Land Company, the estate originally put together by Miller's archenemies James Ben Ali Haggin and Lloyd Tevis, was even larger. According to testimony by Senator Paul Douglas before the Senate Interior Committee in 1958, the company controlled some 1.1 million acres in 1939, of which 413,300 acres were in California—most of it in Kern County. (The Kern County Land Company later became the main agricultural holding of the Tenneco Corporaton, one of the nation's largest conglomerates.) The Salyer and Boswell farming empires were in and around Tulare Lake, each of them comprising tens of thousands of acres. Since most large California growers also lease land, the total acreage under their control could only be guessed at; they may not have known themselves. Without a doubt, however, Salyer, Boswell, Kern County Land, and Miller and Lux were among the very largest and richest farmers in the entire world.

To the four companies, Tulare and Buena Vista Lake were both a convenience and a nuisance. Usually, as the lakes shrank, their exposed beds would be quickly planted with grains or row crops, which were irrigated by pumping back the remaining water. After particularly wet winters, however—and there had been a string of them in the 1940s—the Sierra snowmelt kept filling them into July and August, by which time it was too late to plant. Both water and available land were therefore unpredictable, and, though farmers around the world have learned to live with unpredictableness, it is something that California's big growers, accustomed as they are to perfect summer weather and unfailing man-made rain through irrigation, intensely dislike.

Although Tulare and Buena Vista lakes were privately owned, for the most part, the rivers that fed them were in the public domain. The four big farming companies held rights to a substantial amount of their water, but there were still big surpluses in all but the driest years—especially in the larger rivers, the Kings and the Kern. Had those surpluses been directed elsewhere in the valley, they could have created a great many small irrigated farms. If the rivers were going to be developed—if any agency of government was to develop them— it was a job for the Bureau of Reclamation. The only problem with that rationale was that the big growers wanted all of the water for themselves, they wanted the government to develop it for them, and they didn't want to have to pay for it.

Someday, if anyone has the inclination or the ability to penetrate the wall of secrecy behind which the Corps of Engineers has always managed to carry on its affairs, we may hear from its own mouth— from incriminating letters, memoranda, or confessions of its officials— why it was so eager to develop the Kings and the Kern—to ally itself unabashedly with a handful of huge land monopolies and, in the process, shove the Bureau off two made-to-order small-farm irrigation projects. The only obvious explanation (which is probably the correct one) is that it sensed the growing unpopularity of the acreage limitations of the Reclamation Act. Here was an unparalleled opportunity to establish a beachhead in a region where the natural topography and demand for water could give it new work for decades to come. No stranger to power politics, the Corps knew that its best hope of long-range success was a quick, dramatic demonstration of its abilities. The best way to ensure that was to pick a group of beneficiaries who were nearly as potent a political force as the Corps itself. If this was indeed its reasoning, then it reasoned well.

In 1937, the Bureau of Reclamation was just beginning its detailed feasibility investigations of the Kings and Kern River projects; it had, in fact, already been authorized to build the Kings River Project on the basis of cruder reconnaissance studies alone. In the very same year that the Bureau began its investigations, however, the Corps went to the House Flood Control and Appropriations Committees and extracted an authorization and some money to perform investigations of its own on these same two rivers—rivers which, in effect, had already been promised to the Bureau. It was a brazen act. The Bureau was incensed, and Harold Ickes, the Interior Secretary, was apoplectic. Nonetheless, neither the Bureau nor Ickes could do anything to stop the Corps; they were, in effect, in a race. The National Resources Planning Board, one of FDR's superagencies, pleaded with the agencies to plan a unified project, then practically ordered them to do so. But they refused. As a result, in 1940, Congress received two separate reports on developing the Kings and the Kern: one on a traditional Reclamation project, the other on a project that purported to be for flood control, but which, by controlling the rivers' runoff and drying up Tulare Lake, would irrigate a roughly equal amount of land.

It was a bureaucratic battle that was to drag on for more than five years. Sympathies in California, where the Bureau had a lot of support from smaller farmers, were divided—as they were in Congress. The Roosevelt administration, however, was emphatically on the side of the Bureau of Reclamation. FDR felt so strongly about the matter that on the 5th of May, 1941, he wrote a personal letter to the chairman of the House Flood Control Committee, saying, "A good rule for Congress to apply in considering these water projects, in my opinion, would be that the dominant interest should determine which agency should build and operate the project." Obviously, Roosevelt said, the dominant interest was irrigation. "Not only that, but Kings River had already been authorized for construction by the Bureau of Reclamation; to [reauthorize] would only lead to needless confusion."

But the Flood Control Committee was practically married to the Corps of Engineers, and ignored Roosevelt's recommendation; the committee quickly authorized Kings River for construction by the Corps. With Ickes lobbying furiously on behalf of the Bureau, however, the full Congress refused to go along.

At that point, FDR made what would, in retrospect, look like a fateful mistake. The United States had by then entered the Second

World War; to squander precious funds on a water project when there was still no demonstrable need for it seemed foolish. Even a few hundred thousand dollars would have given the Bureau enough of a head start, at least on the already-authorized Kings River project, to thwart the Corps' ambition. But Roosevelt refused to recommend any money in his budget.

To the Corps of Engineers, the Bureau's inability to move repre- sented a last chance. In 1942, without any clear authorization from Congress, it began to construct an "emergency" flood diversion struc- ture on the lower Kings. Although its action outraged those members who sided with the Bureau, and who saw what the Corps was trying to do, they could not bring the full Congress, now utterly preoccupied with the war, to waste its time debating such a trivial issue. Besides, the Corps' works didn't seem like much, a mere diversion gate. But it wasn't the size of the works so much as the fact that the Corps had established its beachhead in the Tulare Basin before the Bureau ever got to turn a shovelful of earth. The Corps also made sure the flood- waters were diverted where they could do some economic good— toward the lands of the big growers.

Nothing much happened with the Kings River and Kern River projects during the middle war years. By 1944, however, Europe's farmlands and economy were in ruins; overnight, the United States had become the breadbasket of the world. Now, at last, the two projects seemed to make some sense. In his budget request for fiscal year 1945, FDR included a request of $1 million to permit the Bureau to begin work on the Kings River. The House, dominated by the Flood Control Committee, immediately took the appropriation out; the Senate threw it back in. Finally, hearings had to be scheduled to try to resolve the matter.

It was at those hearings that the Corps of Engineers demonstrated where its true loyalty lay. Although the White House had left absolutely no doubt that it was strongly behind a Reclamation project, and ex- pected the rest of the administration to support its position, the Corps of Engineers chose not to; instead, Chief of Engineers Raymond A. Wheeler displayed outright defiance of his commander in chief. Tes- tifying at the hearings, Wheeler gave no support at all to the Roosevelt position, a breach of loyalty that made Harold Ickes, the ultimate Roosevelt loyalist, absolutely livid. Meanwhile, the deputy chief was busy undermining the administration's position back in California. In a speech to a group of business leaders in Sacramento, Major General Thomas Robins said that Californians were being denied "necessary flood control" by "a lot of arguments that are neither here nor there."

If the state would only "wake up and get the water first and then decide what to do with it," he said, "she would be a lot better off." Otherwise, by the time the dams are built "we may all be dead." What Robins didn't say is that most Californians *wouldn't be able* to use the water in the Kings and the Kern if the Corps built the dams. It had already announced that it would build dams, but not aqueducts; therefore, the water couldn't go anywhere but down the river channels, and the big growers owned nearly all the land on both sides. The Corps had also announced that if its projects offered incidental irrigation benefits, it would not apply the Reclamation Act and its acreage laws. What all of this meant was that if the Corps built the Kings and Kern dams, nearly all of the water could be used by four agribusiness giants and a handful of oil companies owning land nearby—which were to become agribusiness giants themselves.

In the end, the hearings resolved nothing. Congress was still deadlocked. Sensing this, it came up with an inimitable solution to a paralysis of its own making: it authorized the Kings River Project for construction by *both* the Bureau and the Corps. Whichever could convince the appropriations committees to give it money first would end up building it.

The fight now began to get serious. In its budget request for fiscal year 1947, the Truman administration said that the War Department's earlier requests to begin construction on both the Kings and Kern river were to be considered "officially eliminated." There were to be no further requests from the Corps of Engineers pending "a decision by the President as to the course to be followed on these works." In his personal testimony during the appropriations hearings, however, the Chief of Engineers calmly announced that "we are ready to make a definite recommendation to undertake the construction"—a remark that could only be interpreted as smug defiance once again of his commander in chief.

Had Roosevelt not died, the Corps might well have lost the battle. But Harry Truman lacked the romantic feeling about the Reclamation program that Roosevelt had, and he was from a state where the Corps was generally loved. Ickes, the old curmudgeon, was gone, too, replaced by the more conciliatory Cap Krug. In the end, the Corps simply played a waiting game, confident that the growers' friends in Congress would extract money with which it could begin work on both the Kings and the Kern—which they soon did. Truman was so angry that he impounded the first funds, but he gradually lost interest in the whole affair. By 1948, he and Krug had given up. The Kings and Kern rivers belonged to the Corps.

The Army Engineers did accede to Truman's request that they collect a one-time user fee from the growers. The figure settled on was $14,250,000, which covered just a third of the $42,072,000 cost of Pine Flat Dam. Considering the tens of thousands of new acres that would be opened to double-crop production when the floodwaters were stored in the Pine Flat and Isabella reservoirs, the "user fee" was more tokenism than anything else.

The covert liaison between the Corps of Engineers and the world's largest irrigation farmers was to live on. A few years later, the Corps added insult to injury by damming the Kaweah and the Tule rivers, which, by rights, should have been Reclamation rivers, too. But as an example of government subsidizing the wrong people, for the wrong reasons, nothing would quite equal its performance thirty-five years later in the Tulare Lake floods of 1983.

During the El Niño winter of 1983, when the eastern Pacific's resident bulge of high pressure migrated to Australia and the storm door was left open for months, much of California got double or triple its normal precipitation. The previous year hadn't been much different. By the early spring of 1983, all four Corps of Engineers dams were dumping hundreds of thousands of acre-feet over their spillways as the largest snowpack in the annals of official California weather records melted. Because the farmlands in what used to be Tulare Lake were now protected by dikes, most of the water couldn't enter its old basin and had to go elsewhere. When the floodwaters began encroaching on nearby towns, the Corps of Engineers spent $2.7 million in emergency funds to erect levees around them. There was nothing inherently wrong with that, except that 80,000 acres of old lake bottom—land that could have absorbed the floods—remained dry; one need only have breached one of the levees that had since been built around the ex-lake. But the Tulare Lake Irrigation District, dominated by Salyer and Boswell, wouldn't have that, so the growers convinced the Corps to spend taxpayers' money on levees in order that their land, the natural catch basin for the floods, could remain in subsidized production.

However, El Niño was soon to prove too much even for the big growers and the Army Engineers. By March of 1983, the flooding rivers were out of control and one of the lake levees was breached, inundating thirty thousand acres of farmland. The Tulare Lake Irrigation District immediately applied to the Corps for a permit to pump out the water and send it over the Tulare Basin divide into the San Joaquin River, which feeds San Francisco Bay. There was nothing inherently wrong with that idea, either—the bay and the Delta normally can use all the fresh water they can get—except that at least one of the reservoirs upstream had been illegally planted with a species of fish called white bass, which got flushed down by the floodwaters and were already flourishing in the reincarnated Tulare Lake. White bass are a voracious, opportunistic, highly adaptable type of rough fish and love to eat young salmon and striped bass. (Salmon and white bass have never managed to coexist, anywhere.) Unless a fish screen below the pumps could guarantee that 100 percent of the white bass would be removed before entering the San Joaquin, the bay and Delta's two most valuable commercial and sports fish would be threatened with extinction. Just a handful of escaped white bass of opposite sexes could be enough to seal their doom.

Even though no fish screen has *ever* operated 100 percent effectively, the Corps of Engineers, ignoring a cacophony of protest from sportsmen in several states, issued another "emergency" permit on Friday, October 7, 1983, to allow the pumping to begin. The growers hadn't even waited for the permit; the pumps were all in place and ready to operate, and television reporters who arrived to take a look at things were scared away by armed guards. The pumps howled to life minutes after the permit was issued. The California Department of Fish and Game had strung a gill net across the river below the fish screen, just in case. On Saturday morning, not twenty-four hours after the pumping began, the net yielded four white bass. The pumps were shut off, and Fish and Game—as if to underscore the catastrophic consequences of releasing white bass—poured a thousand gallons of rotenone, a virulent pesticide, into six miles of river around the fish screens. Everything in that stretch of river—crappies, black bass, white bass, catfish, crayfish, ducks—died a ghastly death. A week later, Fish and Game performed a second mass poisoning. Then, satisfied that there was no danger to humans, it allowed the pumps to start up again. Every legal effort to stop them failed. Virtually all of the water was pumped out of the lake, and although there is no evidence yet that white bass got into the San Joaquin River and migrated down to the Delta and bay, they could just as well be there; no one knows. If they are—and some sportsmen think it is inevitable that white bass will reach the Delta—then the last remnant of central California's once prolific salmon fishery may soon be a thing of the past.

It would have been one thing, this whole game of Russian roulette with the most important anadromous fishery in the state, if the drowned lands in Tulare Lake were pumped out so they could grow valuable food. Most of them, however, have been planted in cotton for years. And as the lake was being pumped out, they were not even growing cotton. In March of 1983, just four days after the levee was breached and the floodwaters began to fill Tulare Lake, several of the big corporate farmers applied to the Department of Agriculture for enlistment in the Payment-in-Kind (PIK) program, which had recently been created to relieve the nation's chronic problem of surplus crop production. Thanks to PIK, they would receive free grain from bulging silos in exchange for not planting crops. The Boswell Company alone got $3.7 million worth of wheat in exchange for keeping fourteen thousand acres idle. (Boswell has consistently received more money from agricultural price support programs than any other farmer in the entire nation.) No one knows how much the other farmers got, but most of the eighty thousand acres of the old lake bed were registered in PIK—even as they were underwater.

In his personal epitaph on the Kings and Kern saga, written in 1951, Harold Ickes lambasted the Corps as "spoilsmen in spirit . . . working hand in glove with land monopolies." He called it a "willful and expensive . . . self-serving clique . . . in contempt of the public welfare" which had the distinction of having "wantonly wasted money on worthless projects" to a degree "surpassing any federal agency in the history of this country. . . . [N]o more lawless or irresponsible group than the Corps of Army Engineers," Ickes concluded, "has ever attempted to operate in the United States either outside of or within the law. . . . It is truly beyond imagination."

The Corps' success in bouncing the Bureau of Reclamation off a project it had already been authorized to build, and three other projects where it should have been the one to build, had the effect Ickes foresaw. An effort was immediately launched by the state's growers to repeal all the constraining features of the Reclamation Act—the acreage limitation, the prohibition on leasing, the requirement that farmers must live within fifty miles of their land—as it applied to the Central Valley Project. (Naturally, all the subsidies were to be retained.) Even though the campaign failed, the Corps' record in California made the irrigation lobby throughout the entire West sit up and take notice. The Bureau of Reclamation was a good thing, but the *Corps*—the Corps of Engineers was a dream come true.

At the same time the Corps and the Bureau of Reclamation were fighting over the rivers of the southern Sierra Nevada, they were engaged in a battle of more epic proportions over the Missouri River. The historical significance of that battle would be greater, too—not only because the Missouri is a much bigger and more important river than the Kings or the Kern, but because, in defiance of common sense, economics, and even simple hydrology, the Missouri was an instance where *both* agencies managed to win.

The Missouri River is, after the Columbia, the biggest river in the American West, though it takes it a long time to grow to size. The Columbia, rising prodigiously out of the rain forests of the Purcell Mountains in Canada, is like a Clydesdale horse, big and powerful at birth. The Missouri, still small after going a distance in which the Columbia becomes huge, is a scavenger of a river, struggling to attain size. It isn't until the North Dakota border, nearly a thousand miles from its source, where the Yellowstone River adds a surge out of the Absaroka and Big Horn Mountains, that the Missouri begins to look impressive. The river turns south, capturing the Platte and the Niobrara and the Kansas and the James, and then east again. By the time it has gone two and a half thousand miles and joined the Mississippi, it is the twelfth-longest river in the world; however, because of the aridity of the basin it drains, the Missouri is only the seventh-ranking river in the country in terms of annual flow.

Meager for its huge watershed and length, the virgin Missouri also flowed erratic in the extreme. At Hermann, Missouri, the discharge to the Mississippi has been measured as low as forty-two hundred cubic feet per second and, in June of 1944, as high as 892,000 cubic feet per second, enough water in a day to satisfy New York City for a year. Its course was as unpredictable as its volume. Flowing across the glacial outwash of the plains, the Missouri is unconfined by a true canyon; it is held in check, more or less, by low bluffs as far apart as ten miles. Even these bluffs, in the river's days of freedom, existed pretty much at the Missouri's whim. Within its wide and crumbly confinement, the virgin Missouri writhed like a captive snake. Seemingly permanent islands and bottomlands covered by meadows and trees would seduce farmers down to the river; then they would disappear, never to return, when the river made a lateral migration of a half mile in a single day. Boats often marooned on what had been the main channel the day before; whole neighborhoods on the river bluff sometimes dropped in when the Missouri chewed its banks.

Until 1940, when the Corps of Engineers finished Fort Peck Dam and created, for reasons that were and still are less than obvious, a 140-mile-long flood-control reservoir in the arid heart of Montana, the Missouri River was almost completely uncontrolled. There were two reasons for this. One was that the river didn't show promise of carrying

Three godfathers of the newly reclaimed West. AT LEFT: John Wesley Powell, who got things moving. BELOW, LEFT: Michael Straus, the millionaire commissioner of reclamation, who under FDR and Truman threw up hundreds of dams. BELOW, RIGHT: Floyd Dominy, the two-fisted commissioner who rode reclamation's falling star.

Mules lugging sections of the Los Angeles Aqueduct into place. At the time, no motorized vehicle existed that could haul anything so heavy.
(*Photo Department of the Los Angeles Department of Water and Power*)

The Owens Valley before the Los Angeles Aqueduct was completed.
(*Photo Department of the Los Angeles Department of Water and Power*)

The three main actors, from Los Angeles' standpoint, in the Owens Valley episode. AT RIGHT: Fred Eaton, the ex-mayor who ultimately felt betrayed by the city he helped create. BELOW, LEFT: J. B. Lippincott, who acted as a double agent in behalf of the city. BELOW, RIGHT: William Mulholland, the man who brought the water. (*Photo Department of the Los Angeles Department of Water and Power*)

ABOVE AND BELOW: Two views of Los Angeles—the squalid pueblo in 1869, and the megalopolis, at once tawdry and glitzy, that water built, in the late 1950s. (*Photo Department of the Los Angeles Department of Water and Power*)

OPPOSITE: Rare photos of the Saint Francis Dam, before and after its collapse. After the disaster, the Los Angeles Department of Water and Power attempted to acquire and hoard as many photos as it could find; it didn't release them until many years later. A virtually identical dam, which creates the Hollywood Reservoir, was faced with earth and seeded with grass and trees so people living below it would be less inclined to think about the Saint Francis catastrophe, which, according to official records, killed more people than the San Francisco earthquake. (*Photo Department of the Los Angeles Department of Water and Power*)

A section of the just-completed Los Angeles Aqueduct crosses the Mojave Desert.
(*Photo Department of the Los Angeles Department of Water and Power*)

Looking like a masterwork left by the Romans, Theodore Roosevelt Dam stands athwart the Salt River in Arizona. The Bureau of Reclamation's first great structure—and the prototype of all high, curved-arch dams—Roosevelt Dam was constructed entirely of huge stone blocks hewn from cliffs in the Salt River Canyon.　　　　　　　　　　　　　　　　　　　　　　　　(*Bureau of Reclamation*)

Still the architectural masterpiece among all the world's dams, Hoover rises seventy stories from the bed of the Colorado River. Though Hoover appears minuscule compared to Lake Mead, whose length is greater than a hundred miles—it widens considerably a few miles upriver—the dam may outlast the reservoir. (*Bureau of Reclamation*)

Grand Coulee Dam under construction in June of 1938. Appearances are deceptive: the width of the dam is four fifths of a mile. (*Bureau of Reclamation*)

much barge traffic—at least compared to other big rivers like the
Mississippi and the Illinois—so the Corps of Engineers didn't have a
good reason to improve it for navigation. Even if it had wanted to, the
task of making such an erratic, muddy, unconfined river suitable for
navigation was overwhelming. The Missouri habitually flooded Kansas City and other towns along its course, but until a major federal
flood-control act was passed in 1937—and until the Corps abandoned
the doctrine, which it had held to with Ptolomeic rigidity, that reservoirs don't control floods—the Army Engineers had little interest in
doing much about it.

The Bureau hadn't built much in the upper Missouri Basin, either,
for the same reason that it hadn't built much along the upper Colorado
and its tributaries: irrigation farming in cold, high-altitude terrain
was usually a losing proposition. It had investigated the basin thoroughly, and by 1907 it had nine projects underway there, mainly for
political reasons: the Missouri Basin states contributed a lot of money
to the Reclamation Fund. But of the nine projects, not a single one
was going to pay for itself within the forty-year term required by the
amended Reclamation Act. The nine projects together owed the Treasury and the Reclamation Fund $55,755,000, but had repaid only
$17,518,000, even though they were exempted from paying interest.
At the rate that revenues—which depended more than anything else
on the irrigators' meager ability to pay—were dribbling in, the projects wouldn't be repaid within two hundred years, if ever.

The only way to steer reclamation away from utter financial disaster in the Missouri Basin was to subsidize it with hydropower revenues. Hydroelectric output being a function of two variables—volume
of water and height of drop—it made good sense, from the Bureau's
point of view, to build high dams along the upper tributaries to generate as much power as possible. The stored water could then be used
to irrigate adjacent agricultural land, and hydropower revenues would
cover the inevitable losses. Glenn Sloan, an assistant engineer in the
Billings office, had begun to draw the outlines of such a basinwide
project in the late 1930s, and was reasonably close to finishing his
report in 1943, when the Missouri decided to go on a rampage. It
produced three big floods—in March, May, and June—and during the
last one Omaha and Kansas City were navigable by boat. The Corps'
regional office happened to be in Omaha, and its petulant director,
Lewis Pick, who would later become the Chief of Engineers, was nearly
chased by the river to higher ground. To a military man like Pick, it
was an unforgivable insult. "I want control of the Missouri River!" he

is said to have barked at his subordinates. Before the end of the year, Pick had dispatched to Washington a twelve-page report on harnessing the Missouri, which was to become known as the Pick Plan.

The trouble with the Pick Plan and the Sloan Plan—which was frantically completed after the Bureau learned about the Pick Plan—was that you could logically build one or the other, but not both. The Corps wanted to build a few dams on upriver tributaries, although, in locating them, it paid no attention at all to irrigation. It also wanted to erect fifteen hundred miles of new levees. All of that was dwarfed, however, by what the Corps planned to do to the river between Fort Peck Reservoir and Yankton, South Dakota. The plan called for five dams and reservoirs, all of them of monstrous size. Garrison Dam, in western North Dakota, was the largest, and would, as the Corps took pains to point out, contain twenty-five times as much material as the Great Pyramid of Cheops. Two and a half miles long, 210 feet high, the dam would be the second-biggest structure on earth (Fort Peck Dam was larger). The Washington Monument would stick out of it like a spike in a railroad tie. The other dams—Oahe, Gavins Point, Big Bend, Fort Randall—would be smaller, but large enough to dwarf almost anything else around. Eight hundred miles of the Missouri would be transformed into a chain of huge, turbid reservoirs. The six main-stem dams would back up almost ninety million acre-feet of water, sufficient to turn Pennsylvania into a shallow lake. The whole scheme—if one believed the Corps' figures, which have always been notoriously low—would cost $660 million, in 1944 dollars.

There was almost nothing about the Corps' plan that the Bureau liked. The dams were all too low or poorly situated to draw the power potential out of the river. (The Corps usually installed about as much public power as it felt the private power companies would tolerate, and it was no surprise to anyone that the Western Power Company became a champion of the Pick Plan, not the Sloan Plan.) The storage was, with a few exceptions, far downriver from the lands the Bureau wanted to irrigate, and a lot of it was in the middle of unirrigable wastelands, which made the Bureau furious. The Missouri's potential as a navigable waterway—that was one of the main justifications of the Pick Plan—was, as far as the Bureau was concerned, shamelessly overstated; to spend more than half a billion dollars on a river channel that would never carry more than a few hundred barges a year was a criminal waste of scarce money and water. It was wasteful in other ways as well. One of the reservoirs, Garrison, would drown the best winter cattle range in North Dakota. Although the Bureau had flooded its share of productive river bottomlands, this was an instance where it was troubled by the idea. As for flood control, Glenn Sloan, who understood the hydrodynamics of the Missouri River as well as anyone alive, said in Congressional testimony that "the 1943 flood could have been regulated to a safe capacity . . . at Sioux City, Omaha, and Kansas City with only two million acre-feet in storage." But the Corps was talking about creating *sixty million* acre-feet of new reservoir storage.

The Corps of Engineers' obsession with humbling the wild Missouri River seemed to derive mainly from the fact that Colonel Pick was mad at it. (Although, needless to say, in the wake of the war his agency, its staff swollen by the thousands, was eager for new work.) According to Henry Hart, a journalist and historian who covered the Pick-Sloan controversy in the 1940s and later wrote a book about the Missouri, the Corps "relied for justification entirely on the public sense of shock at the disruption caused by floods." Nonetheless, the Pick Plan went through the House Rivers and Harbors Committee without a hitch, and passed the full House in the spring of 1944, while still under consideration in the Senate. It seemed only a matter of weeks before it became law.

The Bureau of Reclamation, meanwhile, felt so threatened by the Pick Plan that it had quickly produced a plan of its own that was equally ambitious, and only slightly more susceptible to logic. Reconnaissance studies of reservoir and irrigation sites were conducted with such haste that, even within the Bureau, they were referred to as "windshield reconnaissance"—an allusion to $30 million reservoirs being plotted from behind the windshields of moving cars. The Bureau spewed out project recommendations like popcorn. The final Sloan Plan was a catch basin of ninety dams and several hundred individual irrigation projects; among other things, it called for fifteen reservoirs on three meager tributaries in the Dakotas. The Sloan Plan, however, soon acquired some powerful supporters, too. By the end of 1943, the Congress had two irreconcilable plans before it. The lobbies behind them were about equally matched. Under the circumstances, there was only one thing to do: adopt them both.

The impetus came from FDR himself, though the result was not exactly what he intended. With the Bureau and the Corps stalemated, Roosevelt decided to break the impasse by sending Congress a strongly worded letter saying that the solution to developing the Missouri Basin was to create a regional authority, similar to the TVA, and take development out of both agencies' hands. That was more than the Corps and the Bureau had bargained on. On October 15, 1944, Glenn Sloan and a representative of Colonel Pick (who had since gone off to build the Ledo Road in Burma) sat down in a meeting which is probably historic for what it accomplished in a given amount of time. On October 17, two days later, they emerged to announce that the Pick Plan and the Sloan Plan had been "reconciled." Had anyone taken a closer look—hardly anyone did—he would have seen that the reconciliation amounted to the adoption, virtually intact, of both agencies' plans. With the single exception of a dam at Oak Creek, South Dakota, originally proposed by the Corps of Engineers, the Pick-Sloan "compromise" included every dam and project in the original and separate plans, plus some additions which the agencies had somehow managed to overlook. Critics such as James Patton, the president of the National Farmers' Union, called it a "shameless shotgun wedding," and calculated that instead of saving the taxpayers money, it would cost them at least $250 million in redundant features. Henry Hart acidly observed that "reconciliation meant chiefly that each agency became reconciled to the works of the other."

The most significant aspect of the reconciliation was that the two agencies had agreed to spend $1.9 billion of the taxpayers' money (an estimate which would, as usual, turn out to be much too low) on a whole whose parts, according to their earlier testimony, would cancel out each other's usefulness. The second most significant aspect was that the Bureau agreed to let the Corps go ahead and build its huge main-stem reservoirs first. "The Corps got the here and now," says David Weiman, a lobbyist who would later be hired to fight several of the Bureau's projects by the same farmers who were supposed to benefit from their existence. "The Bureau got the then and there."

One of the least-known consequences of water development in America is its impact on the Indians who hadn't already succumbed to the U.S. Cavalry, smallpox, and social rot. Although many of the tribes had been sequestered on reservations that were far from the riverbottoms where they used to live, some tribes had been granted good riverbottom reservation land—either because the lands were prone to flooding, or because the government was occasionally in a generous mood.

The three tribes whom Lewis and Clark encountered along the Missouri River in North Dakota were the Mandan, the Hidatsa, and the Arikara. Perhaps because they were generally peaceful and had helped the explorers (Lewis and Clark spent their first winter with the Mandan, and their adopted Shoshone-Mandan interpreter, Sacajawea, probably saved their lives), the associated Three Tribes were later rewarded with some of the better reservation land in the West: miles of fertile bottoms along the serpentine Missouri, which they used mainly for raising cattle. These were the same lands that the Bureau of Reclamation considered the best winter cattle range in the state, and which it said ought never to be drowned by a reservoir. Under the Corps of Engineers plan, however, the Three Tribes' reservation would sit directly under the reservoir behind Garrison Dam.

The Corps had, of course, taken extraordinary care not to inundate any of the white towns that were situated along the river. The reservoir behind Oahe Dam, which would be more than 150 miles long, would stop just shy of Bismarck, North Dakota. Pierre, the capital of South Dakota, would sit safely inside a small reservoir-free zone between the tail end of Lake Francis Case and the upper end of Lake Oahe; were it not for the town, the two reservoirs would have virtually touched, nose to tail. Chamberlain, South Dakota, nestled between the reservoirs formed by Big Bend and Fort Randall Dams, was similarly spared. The height of Garrison Dam was reduced by twenty feet so that the surface level of the reservoir would be 1,830 feet above sea level, not 1,850 feet as originally planned. It was a loss of several million acre-feet of storage exclusively for the benefit of Williston, North Dakota, a small part of which could have been subject to inundation during wet years.

For the sake of the Fort Berthold Indian Reservation, where the Mandan and Arikara and Hidatsa lived, no such intricate gerrymandering of reservoir outlines was even tried. Garrison Dam, which the Corps justified largely because of its flood-control benefits downstream, was going to cause horrific local flood damage the moment its reservoir began to fill. Virtually every productive acre of bottomland the tribes owned would go under.

Colonel Lewis Pick, the architect of the tribes' inundation, was the embodiment of a no-nonsense military man. Pick liked to punctuate his conversation with Cagney-style "See? See?"s; these were not questions—they were commands. When first assigned to the Missouri River Division during the early part of the Second World War, he ordered all of his staff to work a series of continuous seven-day weeks. On the first Sunday after the order was given, Pick spied on all his top officers and summarily dismissed those who were not at their desks. Later, when he was in Burma, he fired a whole team of surveyors for laying out a technically perfect road which, in his opinion, would take too long to build. Instead, he designed a treacherous road that could be finished slightly sooner.

Since what Pick proposed to do to the Indians was the most calamitous thing that had happened to them in their history, he might have had the good grace to leave the proceedings through which the tribe would be compensated to someone else. But Pick was a take-charge type. He not only insisted on participating; he insisted on running them himself.

Initially, the Three Tribes pleaded with the government not to build Garrison Dam at all. "All of the bottom lands and all of the bench lands on this reservation will be flooded," wrote the business council of the Three Tribes in an anguished resolution condemning the plan.

> Most of it will be underwater to a depth of 100 feet or more. The homes and lands of 349 families, comprising 1,544 individuals, will be covered with deep water. The lands which will be flooded are practically all the lands which are of any use or value to produce feed for stock or winter shelter. We are stockmen and our living depends on our production of cattle. . . . All of our people have lived where we now are for more than 100 years. Our people have lived on and cultivated the bottom lands along the Missouri River for many hundreds of years. We were here before the first white men stepped foot on this land. We have always kept the peace. We have kept our side of all treaties. We have been, and now are, as nearly self-supporting as the average white community. We recognize the value to our white neighbors, and to the people down stream, of the plan to control the River and to make use of the great surplus of flood waters; but we cannot agree that we should be destroyed, drowned out, removed, and divided for the public benefit while all other white communities are protected and safe-guarded by the same River development plan which now threatens us with destruction. . . .

However, when the Interior Department, the parent agency of the Bureau of Indian Affairs, threw itself behind the plan, the Three Tribes saw the futility of abject resistance. What they asked for as compensation, considering the agony they were about to be put through, was pitiful enough. First, they wanted at least an equivalent amount of compensatory land. Since it would inevitably be poorer land, they also wanted twenty thousand kilowatt-hours per year of electricity, mainly to run the pumps they would need to bring water, once freely available from the river, up from depths of three hundred feet or more on the arid plains. They asked for permission to graze and water their cattle along the margins of the reservoir, and for first rights to the timber which the reservoir would flood. They wanted a bridge built across a narrow reach of the reservoir so their people could maintain contact with one another (the reservoir would effectively split the reservation in half). Otherwise, they would have to spend hours driving around its endless shore or brave violent winds and waves trying to cross its surface by boat.

One small faction of the Three Tribes, led by a flamboyant young radical named Crow Flies High, remained opposed to any compromise at all. As negotiations were already underway between the Interior Department, the Corps, and the Tribal Business Council, a delegation from the dissident faction burst into the room in ceremonial dress and began disrupting the proceedings. The leader of the group, who was probably Crow Flies High, went up to Colonel Pick and made an obscene gesture. Pick turned the color of uncooked liver. It was an insult, he said lividly, that he would remember as long as he lived.

On the basis of that petty insult, Pick stormed out of the negotiations, never to return. As far as he was concerned, all of the points of agreement that had already been reached were null and void. When Arthur Morgan, the first director of the Tennessee Valley Authority— and the one person who kept the memory of the Indians' tragedy alive—visited the Three Tribes some time later, however, he discovered a different sentiment as to why Pick had walked out. There was, he wrote, "a nearly unanimous opinion that the Corps welcomed the attack of the Crow Flies High group because it provided a semblance of justification for ignoring the clear terms of the law. . . ."

Before the negotiations were interrupted, the Corps had offered the Indians some scattered property on the Missouri benchlands to replace the bottomlands they would lose. ("I want to show you where we are going to place you people," a local Congregationalist minister quoted Pick as saying.) Under the law, all compensatory lands were to be "comparable in quality and sufficient in area to compensate the said tribes for the land on the Fort Berthold Reservation." It was up to the Secretary of the Interior, Cap Krug, to decide whether the criteria had been met. As Krug well knew, there was no land in North Dakota that could adequately compensate the tribes for prime winter cattle range in a river valley. He had decided, therefore, to accede to the Indians' other demands for water, at-cost hydroelectric power, and first timber and mineral rights. Since even this appeared to be too little, he also agreed to pay them $5,105,625 for the 155,000 acres they would lose. It was only $33 an acre, but it was better than nothing.

Colonel Pick, however, was still smoldering over the indignity he had suffered, and he had his good friends in Congress. A few months after Krug announced that he was prepared to meet most or all of the Indians' terms, the disposition of their case was removed by Congress from Interior's hands and given to the Committee on Interior and Insular Affairs. The committee soon tore up Interior's version of the bill and wrote its own version exactly along the lines suggested by Pick. The Fort Berthold tribes would not even be permitted to fish in the reservoir. Their cattle would not be allowed to drink from it, or graze by it. The right to purchase hydroelectricity at cost was abrogated. The tribes were forbidden to use any compensatory money they received to hire attorneys. They were not even allowed to cut the trees that would be drowned by the reservoir, except in one case, and there, according to the new terms, *they were not permitted to haul them away*.

On May 20, 1948, Secretary Krug ceremoniously signed the bill disposing of the Fort Berthold matter in his office in Washington. Despite some intervention by the Interior Department, most of the Corps' vengeful provisions were still intact. Standing behind Krug, alongside a slouching Mike Straus of the Bureau of Reclamation and a scowling General Pick, was handsome George Gillette, the leader of the tribal business council, in a pinstripe suit. "The members of the tribal council sign this contract with heavy hearts," Gillette managed to say. "Right now the future does not look good to us." Then, as Krug reached for a bundle of commemorative pens to sign the bill, and as the assembled politicians and bureaucrats looked on embarrassed or stony-faced, George Gillette cradled his face in one hand and began to cry.

To eliminate any possibility that Congress or the President might succumb to a tender conscience and eliminate Garrison Dam from the Pick-Sloan Plan, the Corps had already begun work on it in 1945, three years before the agreement with the Indians was signed. In fact, it would spend $60 million on ambiguously authorized "preliminary" work on the dam between 1945 and 1948. A number of members of Congress protested that such work was, if not outright illegal, then certainly a moral wrong. But the one party that might have gone to court for a ruling—the Fort Berthold tribes—had been forbidden to spend any of their compensatory money on attorneys.

The Fort Berthold Indians have never recovered from the trauma they underwent. Their whole sense of cohesiveness was lost, and they adjusted badly to life on the arid plains and in the white towns. But no humiliation could have been greater than for them to see the signs that were erected around the reservoir as it slowly filled, submerging the dying cottonwoods and drowning the land they had occupied for at least four hundred years. In what looked to the Indians like a stroke of malevolent inspiration, the Corps of Engineers had decided to call the giant, turbid pool of water Lake Sacajawea.

As is the case with most schemes that involve a dazzling transmogrification of nature, this is a story without an end, and a later chapter will say something about the likely consequences of trapping most of the Missouri's silt behind six great dams. For now, it is worth looking briefly at what the Pick-Sloan plan has wrought.

The Corps' six Missouri River reservoirs, which cost $1.2 billion to build even then, have undoubtedly lowered the flood crests all the way down to New Orleans—though they did not prevent a disastrous flood in the early 1970s, when the Mississippi widened by several miles and caused tens of millions of dollars in damage. Barge traffic hasn't come close to the Corps' projections; in 1984, traffic on the entire navigable stretch of the Missouri amounted to only 2.9 million tons, an infinitesimal percentage of the 590 million tons carried by the Mississippi system. The small port of Lorain, Ohio, handled nearly five times as much. The worst natural damage was the flooding of some of the best riparian waterfowl habitat in the world. A former director of the U.S. Fish and Wildlife Service, John Gottschalk, remembers walking along the undammed middle Missouri for five miles and flushing countless flocks of pheasants and migrating ducks; today, one would be lucky to see anything at all. The birds thrived in the spacious, secluded bottomlands and oxbow pools and marshes, and those are almost entirely gone.

Had the Missouri been left to the Bureau of Reclamation exclusively, things wouldn't necessarily have turned out much better. However, because the projects would, for the most part, have been well upriver, the Fort Berthold Reservation wouldn't have been drowned, a lot of riparian waterfowl habitat in the heart of the Central Flyway wouldn't have been inundated, and the dams, being high rather than wide, would likely have produced a lot more hydroelectricity for their size. The irrigation projects the Bureau planned might have been losers in an economic sense, but the Missouri, if it had to be intensively developed, might have been more useful irrigating crops than providing free transit—at enormous public expense—for a handful of barges.

The Bureau, of course, was not to be denied, either, if it could help it. Ever since the 1950s, it has been trying, without too much success, to build the irrigation projects authorized by the Pick-Sloan Plan—the "then-and-later" dams over which the Corps' reservoirs took precedence. The O'Neill Project on the Niobrara River in Nebraska, the Narrows Dam on the South Platte in Colorado, the Garrison and Oahe projects in the Dakotas—projects that have become some of the most controversial in the nation—were all authorized by that same misbegotten act. The Bureau, of course, knew well enough that few, if any, of those projects made economic sense, and at least one of its officials, in private, was willing to admit it. In 1955, future commissioner Floyd Dominy, then chief of the Irrigation Division, received an angry letter from two old farmer friends from Nebraska, Claire and Donald Hanna. The Hannas were dryland farmers, and they were incensed that the Bureau's Ainsworth Project—one of the Pick-Sloan bunch—might literally force them into irrigation farming. "I am really not happy about the Ainsworth Project," Dominy confessed in his letter of reply of April 15, 1955. ". . . My views about the impropriety and damn foolishness involved in the construction of irrigation projects in relatively good dry land areas at the present have been repeatedly expressed. . . . As dear and honored friends I am troubled as to how to advise you," Dominy went on. "The local towns and businessmen wanted it [the Ainsworth Project]. They could see themselves growing fat on large-scale construction payrolls. They could see something to be gained by increasing the number of farm families in their service area. Like the usual selfish citizen they were willing to accept this increase to their personal larder without thought as to the burden to be placed on the Federal tax payer."

Predictably, Dominy managed to overcome such scruples after he was appointed commissioner. In the 1960s and 1970s, the Bureau launched a mighty effort to push forward the Garrison and Oahe projects, enormous diversions from Lake Sacajawea and Oahe Reservoir to compensate North and South Dakota (if not the dispossessed tribes) for the land drowned by the Corps. But the irrigation canals and local storage reservoirs would have consumed nearly as much productive farmland as the irrigation water would have created—in the case of the Garrison Diversion Project, 220,000 acres for canals and reservoirs versus 250,000 new acres irrigated. In addition, Garrison, in its original version, would have converted some 73,000 more acres of superb waterfowl habitat—prairie marshes and potholes used by hundreds of thousands of migrating ducks—into farm fields. Not only that, but it could easily have introduced parasites and competitive trash fish from the Missouri into streams emptying into Lakes Winnipeg and Manitoba, threatening very productive pike and lake trout fisheries. The Canadians, in fact, had been screaming objections into the deaf ear of the Bureau for years, and even sent a series of stern diplomatic protests to the State Department.

In the late 1970s, the Oahe Diversion Project, after reigning for years as the biggest political issue in South Dakota, was defeated by the very same farmers for whose alleged benefit it was to be built, and one of its principal champions, former Senator George McGovern, saw his political career buried with it. McGovern's surprising loss, according to local political insiders, had as much to do with his unwavering support for a suddenly unpopular Oahe Project as it did with the campaign mounted against him by the ultraconservative Right. Garrison, in 1985, was partly completed and still alive, and a bobtailed version of the project seems likely to be built, irrigating perhaps 130,000 acres (devoted mainly to surplus crops) at a cost of $1,650,000 per farm. Energy requirements for pumping, which totaled 288,000 kilowatts under the original plan, would also be reduced, but how the dams can pump water to 130,000 acres *and* sell power at market rates to subsidize the water costs is a question that no one, least of all the Bureau, can answer.

As for the other dozens of projects assigned to the Bureau, few have been built, but the Corps, despite the antipathy of the local citizenry, has tried to steal away even these. On May 9, 1963, Dominy's regional chief in Billings, Bruce Johnson, reported that the Corps "is not dismayed by the opposition" to two newly proposed dams on the Missouri, Fort Benton and High Cow Creek—nor, Johnson said, would "the highly preliminary stage of the basic investigations . . . deter them. . . . They will, I think, seek authority to build both dams." As a result, Johnson advocated "that we grit our collective teeth and decide to use the reconnaissance data that we have so we can go to the 'hill' just as quickly as the Corps does and ask for authority [to build] without additional investigation." If such "additional investigation" (reconnaissance data are usually based on a mere desultory look) disclosed that either dam would be a waste of money, that was the taxpayers' problem.

And, by now, it is, even if Fort Benton and High Cow Creek dams have not yet been built. Between the money-losing irrigation ventures in the Missouri Basin and the river's mediocre power potential, the Missouri Basin "Fund" appears to be in unhealthy financial shape. The problem is, no one knows exactly how bad things are. According to a Carter administration audit, Missouri Basin power is *already* vastly oversubscribed, and the Corps and the Bureau, employing some complex economic chicanery even the auditors couldn't quite decipher,

may be borrowing on "anticipated" revenues from as far away as the
next century, just as New York City did in the early 1970s before some
of its elected officials almost went to jail. "Our conclusion is that the
financial posture of Pick-Sloan is, at best, based on an uncertainty,"
the auditors wrote. "At worst, it is based on an unreality."

The "reconciliation" of the Pick Plan and the Sloan Plan had taken
a mere two days; the political fallout, the environmental damage, and
the drain on the Treasury that have resulted seem likely to go on
forever.

It was back in California, meanwhile, that the bitterest rivalry and
the most vicious infighting between the Bureau and the Corps continued to occur. Awkward, expensive, and redundant as it was, Congress
had at least come up with some kind of division of responsibilities in
the Missouri Basin. The same applied to the Columbia Basin. In California, however, the two giant bureaucracies were left pretty much
to fight it out among themselves. Their rivalry was a wonderful opportunity for the state's irrigation lobby; the growers could sit back
and smile coyly as they were madly pursued by rival suitors in hard
hats. But it was an equally wonderful opportunity in the 1960s for
Governor Pat Brown, under whose leadership the state was trying, all
by itself, to complete the most expensive water project ever built.

On Wednesday, January 27, 1965, a highly secret meeting was held
in the office of California's resources secretary, Hugo Fisher. In attendance were most of the oligarchs of water development in California: Pat Brown's water resources director, Bill Warne; Robert Pafford,
the Bureau of Reclamation's regional director; Brigadier General Arthur Frye of the Corps of Engineers; Ralph Brody, the chairman of the
California Water Commission; state senator James Cobey; and assem-
blyman Carley Porter, the chief author of the bill that authorized the
State Water Project in 1959. The purpose of the meeting was to discuss
the future of Marysville Dam.

With all the big Sierra rivers developed close to their limits, precious few good damsites were left anywhere in the state except on the
North Coast. There were, however, still three rather marginal sites in
the Sierra foothills which the Bureau wanted to develop in order to
augment the water supply of the Central Valley Project. One was New
Melones on the Stanislaus River. Another was Auburn Dam on the
American River. The third was Marysville Dam on the Yuba River. Of
the three rivers, the Yuba, at the time, had the blackest reputation. It
had devastated Yuba City and forced the evacuation of twenty thousand people during an awesome flood in 1955, and just a few weeks

earlier it had flooded menacingly again during the great Christmas storm of 1964. The dam, therefore, was of interest not only to the Bureau but to the Corps of Engineers. No one, however, was more interested in it than the governor of California and his director of water resources, Bill Warne.

Brown and Warne had an unenviable dilemma on their hands, even if they brought it on themselves. They had misrepresented, either unintentionally or by design, the cost of the State Water Project, and were left without the funds to finish it. The state had signed binding contracts to deliver 4,230,000 acre-feet of water; the $1.75 billion bond issue that the voters had approved, however, would not even suffice to build Oroville Dam, San Luis Dam, and the 444-mile aqueduct down the San Joaquin Valley and over the mountains to Los Angeles. All of those works could deliver a safe yield of only 2.5 million acre-feet of water. Somehow, the state had to come up with nearly two million additional acre-feet—quite an imposing agenda. The water was there, on any number of northern California rivers. But since the voters had just shouldered the most expensive state bond issue in history, the money to develop it was emphatically not.

Marysville Dam, therefore, was exactly the opportunity Brown and Warne were looking for, provided the State Project could gain rights to some or all of the water and contribute little or nothing to the cost of the dam. No federal dam could be built without the consent of the governor of the state. The question, then, was which of the potential builders would give the state what it wanted: the Bureau or the Corps.

The report on the meeting which Bob Pafford of the Bureau sent to his superior in Washington, Commissioner Floyd Dominy, began on a gloomy note. The state was very much inclined to let the Corps build Marysville, for obvious reaons. There would be no federal claim on the water in a purported flood-control reservoir and no strings attached to its use, as there would be if the Bureau built the dam. Pafford, however, held out one hopeful prospect to Dominy: "California might be willing to recommend changing their position from one of strong support for immediate construction and operation of Marysville Reservoir by the Corps of Engineers, with the State taking the conservation water via Title III, to one of support for immediate authorization of Marysville for early construction by the Corps, but with the project to be integrated fully with the Central Valley Project."

The Title III of which Pafford spoke referred to a section of the federal Water Supply Act of 1958, which allowed water from a federal dam to be sold to another political entity, such as a city or state, provided the water was used only for municipal or industrial purposes—that is, not for irrigation. The provision, in fact, owed its existence to the earlier battle on the Kings and the Kern rivers; there was so much resentment over the fact that the state's biggest growers had gotten an enormous supply of water virtually free from the Corps that a number of Congressmen vowed never to let it happen again, and the result was Title III. But Brown and Bill Warne's predicament was that the State Water Project was first and foremost an *irrigation* project. The specter of water famine in southern California gave the project its moral justification, and Los Angeles offered the assessed property wealth needed to guarantee the bonds, but the first deliveries of water would go to the big corporate farmers in the San Joaquin Valley. Los Angeles wasn't scheduled to receive its full entitlement for many years—and, in fact, would take only a fraction of each year's entitlement all through the 1970s and early 1980s, permitting most of the water to be sold, at bargain prices, as "surplus" water to the same big growers in the San Joaquin Valley.

Pafford told Dominy that he had cautioned Warne about the inherent legal risks in trying to use the water from a federally built reservoir to augment a state project whose main purpose, at least for now, was irrigation. "I pointed out that authorization for the sale of water under Title III . . . might severely limit the use of this water, since the Act referred to the use of water only for municipal and industrial purposes." In other words, if the Corps built the dam, the whole arrangement would be quite naked—the Kings and the Kern all over again—and probably enjoinable in a court of law. If the Bureau built the dam, on the other hand, the water, in theory, would have to go into the Central Valley Project and the State Water Project's main beneficiaries, the big San Joaquin growers, could not touch a drop.

There had to be a way out of the predicament, and Bill Warne was canny and cynical enough to come up with it. What if the Corps built the dam on the promise that the water would "someday" flow into the Central Valley Project? And what if, since there was still plenty of surplus water sloshing around in the CVP, the Bureau let the State Water Project "borrow" it for a while?

What Warne wanted to do, Pafford confided to Dominy (and he apparently thought it wasn't a bad idea), was create exactly such an arrangement as "a test of this Act"—by which he meant Title III of the Flood Control Act of 1958. If the Corps built the reservoir and the state took the water with no strings attached, they were risking a head-on collision with existing law. But promising the water to the Bureau—eventually—might provide a legal out; Congress should have no objection to California's "borrowing" it for twenty or thirty years. That would give the state time enough to climb out from under the staggering pile of debt that the construction of its huge water project had dumped on it, and to build the necessary works to develop the full 4,230,000 acre-feet of water it was legally obligated to provide.

If the Bureau acceded to Bill Warne's plan, it wouldn't get to do what it loved best: build a dam. But at least it would get *something* out of it instead of being boxed out entirely, as it was on the Kings and the Kern and the Kaweah and the Tule.

That it was willing to let itself be so used is an indication of how desperate the Bureau had become since the Corps of Engineers began trying to muscle its way into its domain. If anything, Pafford ought to have been incensed by Warne's idea. After all, the Bureau, more than any other single entity, had made California into the wealthy and populous state it was. It built Hoover Dam for Los Angeles and the Imperial Valley, and Parker and Davis and Imperial dams as well. It built the Central Valley Project to rescue the growers from economic suicide by groundwater overdraft. It was paying nearly half the cost of the world's fourth-largest dam, San Luis, which would store water jointly for both the CVP and the State Project; without such assistance, the California project might have fallen on its face. And on top of this, the state's big farmers, when they began to receive cheap "surplus" water from the State Project, would be in a position to engage in cutthroat competition with the Bureau's constituency, the smaller farmers. From the point of view of many small farmers—and this would be borne out later in actual fact—the expansion of the State Water Project was one of the worst things that could happen.

Not only did outrage fail Bob Pafford, but he was willing to go the Corps one better—so badly did he want an opportunity to construct a new dam. Should the Bureau, instead of the Corps, be allowed to build Marysville Dam, he wrote Dominy, "I restated our offer to make water from Marysville available to the state on an interim basis *at a price no greater* than under a Title III arrangement with the Corps" (emphasis added). Then he added a cryptic, furtive remark: "It was concluded that it would not be necessary to include this possibility in the State's comments."

And no wonder! What Pafford was proposing was, if not illegal, then at the ragged margin of the law. Where did the Reclamation Act permit the Bureau to sell deeply subsidized water to a state on an "interim" basis, when the state would turn right around and resell it to some of the largest corporate farmers in the world? Where did it allow the Bureau to promise to match any price offered by the Corps *before it even knew what the cost of developing the water would be*?

The answer was, nowhere. But the Bureau was willing to sell sub-sidized water from one of its dams to California, which would turn around and resell it, at bargain "surplus" rates, to thirty- and forty-thousand-acre farmers who had the economic muscle to drive the Bureau's chief clients and dependents—the state's smaller farmers—out of business. And it was willing to do this, to play with federal law and forsake its small-farmer constituency, simply because its archrival might snatch away a damsite it wanted for itself.

Nineteen years later, Robert Pafford, then retired, never saw a bit of irony in this position. During an interview in 1983, he down-played the rivalry between the Bureau and the Corps—"we had our points of contention, but it was nothing serious"—and said that he "didn't blame Bill Warne for playing both ends against the middle." After all, he said, "you might have done the same thing yourself." Warne, whom he referred to as a "great guy," had "a legal obligation to deliver water to his own constituency—those contracts with the San Joaquin farmers were valid contracts and he couldn't just ignore them." But why would the Bureau go out of its way to help a group of giant corporate farmers who might put the Bureau's little farmers out of business? "Our farmers had water at $3.50 an acre-foot. No way anyone is going to compete with that." Involuntarily, Pafford admitted what the Bureau has always tried to deny: that its cheap water gives its client farmers an unfair advantage over all the other farmers of the nation.

Two years after the secret 1965 meeting, however, no agreement had been reached on Marysville Dam; evidently, neither federal party would yield (Floyd Dominy, a proud man, privately loathed the Corps of Engineers, and probably refused to go along with Pafford's recommendation), and each had the power to hang the project up in Congress, which appropriated no money for construction by anyone. By 1966, in fact, Pat Brown and Bill Warne evidently realized that their strategy of pitting the Bureau against the Corps could backfire on them. If the agencies became *too* competitive, then nothing might get built, just as a pair of rutting elk can lock horns so hopelessly that both of them starve. As a result, that year saw the formation of a new suprabureaucratic entity called the California State-Federal Interagency Group—William E. Warne, chairman—which immediately issued a call for a Herculean amount of water development, most of it on the undammed rivers of the isolated, rainy North Coast. On a big wall map depicting what they wanted to build, one saw, traced in red, a Dos Rios Reservoir and an English Ridge Reservoir and a Sequoia Reservoir and an Etsel Reservoir and a Panther Reservoir and a Frost

Reservoir and a Sebow Reservoir and a Mina Reservoir on the Eel—eight reservoirs on three forks of a middle-size river that was nearly dry from June to October. And that wasn't all. There was a Baseline and a Dinsmore Dam on the Van Duzen. There was a Butler Valley and an Anderson Ford and an enlarged Ruth Reservoir (a smaller one already existed) on the Mad River. Sounding like the Creator himself, Bill Warne, in his introduction to the report, described how the waters had been divided up: "The upper main Eel above the Middle Fork, as shown on attached Chart One, was assigned to the Bureau of Reclamation. . . . The main Eel between the Middle Fork and the South Fork was assigned to the Corps of Engineers . . . the Van Duzen River Basin to the Bureau, the Lower Eel to the Corps. . . ." Nothing was to be built by California itself, though it, of course, would reap the rewards, especially its big growers, who would receive millions of acre-feet of water via long tunnels drilled through the Coast Range. The whole scheme, said Warne in his introductory remarks, represented "a new chapter in California's illustrious history of water planning."

But it wouldn't quite work out that way. In fact, the feuding agencies were about to lock horns and starve over the first two dams on their priority lists.

The North Coast dam that the state and the Corps passionately wanted to build was Dos Rios, a seventy-three-story earthfill wall across the Middle Fork of the Eel that would capture twice as much water as Shasta Lake; it would create one of the biggest reservoirs in the West. The dam the Bureau wanted to build first was English Ridge, which would create a smaller reservoir on the main Eel a short distance away. Since each project required a hard-rock tunnel, about twenty miles long, to shunt the water to where it was allegedly needed, they would both be very expensive. To become economically feasible, a large proportion of their costs would have to be written off to flood control, which is nonreimbursable. The Corps of Engineers has always been vested with the authority to compute the flood-control benefits of federal dams. It was, as expected, quite generous in computing the flood-control benefits of Dos Rios; it was inexplicably niggardly in the case of English Ridge. David Shuster, who would become the operations manager for the Central Valley Project—and who would later be hounded out of the Bureau for being too fair-minded about Jimmy Carter's water projects "hit list"—would later insist that English Ridge was a perfectly feasible project. "There was nothing wrong with English Ridge," Shuster remembered. "A lot of the water could have gone to municipalities in the North Bay and to grape growers. Repayment-wise, it was in sound shape. The thing that killed us was the Corps wouldn't give us the flood-control benefits they gave to Dos Rios. It was like they were using two separate formulae."

Flood-control benefits for both dams, it would turn out, were largely a fraud. Even huge Dos Rios Dam would have reduced the thirty-five-foot crest of the monstrous 1964 flood by less than a foot—a fact which the Corps took pains to camouflage, but which its enemies, especially a local Dartmouth-educated rancher named Richard Wilson, who led the opposition, managed to bring out and make stick. It was Wilson, and Ronald Reagan—who, as governor, refused to approve the project—who ultimately killed Dos Rios Dam, but it was the infighting between the agencies that set the stage for its defeat—and for the ultimate collapse of the whole carefully orchestrated development push. By 1981, not a single one of the thirteen North Coast dams on the Corps' and the Bureau's priority lists had been built. In that year, moreover, all the major North Coast rivers were added to the federal Wild and Scenic Rivers system by the Carter administration—which, in theory, puts them forever off-limits to any dams.

The single cooperative achievement since the Marysville summit meeting a decade and a half earlier was the erection of New Melones Dam on the Stanislaus River. There, the Marysville formula was finally tried: the Corps built the dam, the Bureau gets to market the water—if it can. By 1985, seven years after the dam was completed, not a teacup of New Melones water had been sold, which was especially infuriating to those who despaired at watching the last wild stretch of the lower Stanislaus River drowned. One reason no water had been sold was lack of demand, no matter what the Bureau said about demand "being there." The other reason was that no canals had been built to carry water from the reservoir to the farmers' fields. Why had they not been built? The conspiracy theorists—who, by now, include a lot of people who have watched the progress of California water politics from the losers' side—thought they had an answer. As long as New Melones water remains unsold, it simply runs out, in carefully regulated and fully usable flows, to the Delta, where it can be sucked up by the State Water Project's battery of ten-thousand-horsepower pumps and conveyed either to the big San Joaquin growers or to Los Angeles. It is hard to argue with people who insist that this was the intention all along.

The relatively small yield from behind New Melones Dam, however, is scarcely enough to supply the Chandler family's Tejon Ranch or to satisfy two years' worth of subdivision growth in southern California. As a result, the State Water Project seems destined to remain

chronically undersupplied, unless Californians do a remarkable about-
face and approve $10 billion or $20 billion worth of new water development. The state, the Bureau, and the Corps have all heaped blame
on environmentalism and "selfish" northern Californians for the fact
that so little has been built, but if anyone was selfish it was the Bureau
and the Corps, who coveted too much, and cooperated too little, for
their own good. As for Pat Brown and Bill Warne and the California
water lobby, they appear to have schemed themselves into a dry hole.

No one will ever know how many ill-conceived water projects *were*
built by the Bureau and the Corps simply because the one agency
thought the other would build it first. What is clear, thanks to long-
hidden files from the Bureau that have come to light, is that the Corps
of Engineers has kept a full-court press on the Bureau since it moved
on the Kings and Kern rivers forty years ago. And it was during this
period that by far the most objectionable projects were built.

A May 19, 1962, memorandum from Bruce Johnson, the Bureau's
regional director in Billings, Montana, to Floyd Dominy offers a vivid
illustration of how far things could go. Johnson's memo discussed a
series of potential conflicts between the Bureau and the Corps in the
upper Missouri Basin. One project which the Corps was talking about
building at the time was Bowman-Haley, a dam on the Grand River
in North Dakota—which is not much of a river, despite its name. "They
will build [Bowman-Haley] if they get the money," Johnson warned
Dominy. "I predict the state will see to it that they get the money
unless steps are taken to have the Secretary of the Interior [that is,
the Bureau] authorized to build it."

Would it make sense for the Bureau to build it, in that case? "We
have reported on Bowman-Haley, always unfavorably, at various
times for some thirty years," Johnson wrote. ". . . If we take this on
we will be building another tributary dam with little to show in the
way of repayment contracts. Benefits and repayment are based on a
delivery of 3,000 acre-feet per year. However, some years delivery of
this amount of water will not be possible." It was, in short, a perfectly
miserable Reclamation project, a project whose yield was not only
pitiable, but impossible to guarantee. But it made no better sense,
Johnson quickly added, for the Corps to build the dam instead. Flood
control was a poor justification; the damsite was so near the river's
headwaters that most of the floods were raised downstream. That left
municipal water supply as the sole conceivable justification. But "most
of the municipal and industrial [water-supply] benefits," he continued,
"are anticipatory of urban and industrial growth." And not only was

North Dakota the one state in the union that was *losing* population, but the little town of Haley—the town that would presumably get the water—is so small it wasn't even listed in the American Automobile Association's road atlas for 1976. The water might be piped to Bowman, a considerable distance away, but even Bowman, a relative metropolois, had only thirteen hundred people. If the dam were justified on the basis of local water supply, then, it would give Bowman and Haley about two and a half acre-feet per person—twelve times as much water as average per-capita use.

It was difficult to conceive of a more worthless project, but in the 1950s and 1960s projects as dubious as Bowman-Haley had a way of getting built. The agency that ended up building it was, indeed, the Corps of Engineers; authorized by Congress in 1962, Bowman-Haley was finally completed eight years later. Seeing it there, on a piddling river snaking through the drought-bleached rises and swales of western North Dakota, one needn't be a hydrologist or an engineer to fathom why Bruce Johnson was right. The dam itself is huge: more than a mile across and seventy-nine feet high from the base, it has nearly half the bulk of the smaller of the Corps's main-stem Missouri dams. The reservoir, by contrast, is tiny and shallow, a puddle as reservoirs go; it holds only 19,780 acre-feet, while the smallest of the main-stem Missouri reservoirs holds ninety times more. A lot of tax money had gone for a thimbleful of water.

The Bureau's main problem throughout the Missouri Basin, Johnson added in a footnote to his secret memorandum to Dominy, was the indefatigable opportunism of the Corps of Engineers. "They [the Corps] will build projects that we may find unacceptable from a financial standpoint. The states are aware of this. . . . The Corps will gladly give us their 'bad' project proposals on the tributaries but do not intend to refuse Congress if money is appropriated to build either 'good' ones or 'bad' ones. I do not think they believe that the Memorandum [an informal division of responsibility Johnson and his counterpart in the Corps had recently signed] ends the historic game of the states playing the Army against the Bureau to get what is locally desired."

It was an incredible admission—although it was obviously not intended as such—since neither the Corps nor the Bureau would assert publicly that *any* federal water project, anywhere, had ever been a waste of tax dollars. As Johnson intimated, however, the ultimate blame for the bad projects had to be laid at the feet of the "local interests," the contractors and irrigation farmers and patriotic Chamber of Commerce types who haven't the slightest compunction about wasting the taxpayers' money on pointless dams. A perfect example was offered by the Bureau's area engineer in Salem, Oregon, John H. Mangan, who wrote a confidential letter—what the Bureau calls a blue-envelope letter—dated January 22, 1965, to Harold Nelson, the regional director in Boise, recalling a conversation he recently had with a member of the Oregon State Water Resources Board. "He expressed his feeling," Mangan wrote, "that the Corps of Engineers working through the Public Works Committee did not have the difficulties Reclamation has. . . . He did not feel that the Public Works Committee was concerned with legislation such as Public Law 9032 of the last Congress relative to reimbursement of recreation and fish and wildlife functions." Mangan said he told the man that any such environmental protection provisions would likely apply to the Corps as much as to the Bureau. But his derisive response, according to Mangan, was that he should watch what happened during the upcoming Congressional session. "If the Corps is able to secure rapid authorization of a number of projects and the Bureau is having trouble getting their projects authorized by the Interior and Insular Affairs Committee, [the man said], 'perhaps we should have the Corps building all our projects.' "

Harold Nelson forwarded Mangan's letter to Floyd Dominy, adding a postscript of his own. He had just spoken confidentially with the head of a local pressure group organized to support a new Bureau project in eastern Oregon. Nelson's confidant, a Mr. Courtright, said he was finding considerable sentiment that the group should switch its allegiances and push for rapid authorization by the Corps instead. "Courtright . . . stated quite frankly that the argument which they are having the greatest difficulty to counter is the one that authorization through channels available to the Bureau will be much more difficult and time-consuming than through Public Works Committee channels." Actually, Courtright told Nelson, he knew the real cause of the Bureau's difficulties. "He attributes [them] to field representatives of the Oregon Water Resources Board and to the Corps of Engineers" itself.

As Harold Nelson intimated, an unholy alliance of local economic interests and a powerful member of Congress was something the Bureau was at pains to resist. In 1967, the Johnson administration, preoccupied with the war in Vietnam and the chronic inflation Johnson's policy was creating, requested only a minuscule appropriation for Auburn Dam in California. Robert Pafford, the regional director, wrote a memo to Dominy discussing the options the Bureau had. The obvious one was to slow down the construction schedule on the dam itself, but this was "quite inconsistent with the urgent needs for flood control and power." Another was temporarily to stop work on the irrigation and conveyance facilities—the Forest Hills development and the Auburn-Folsom South Canal. "However," Pafford wrote, "[Congressman] Bizz Johnson has made it quite clear that he wants Forest Hills moved rapidly, and I am sure you know how unhappy the East Side Association [the main local pressure group] is that we are not moving the Folsom South Canal even faster—they and [neighboring Congressman] Bernie Sisk would react violently if we cut the canal out in fiscal 1967." Pafford proposed a more palatable alternative: "[O]ur soundest course will be to reprogram Auburn funds internally to handle the urgently needed preconstruction program, and reduce our right-of-way program accordingly. . . . By reducing land acquisition from $900,000 to $135,000 we will be able to carry out a preconstruction program suitable to Denver's needs for design data. This will provide for some additional land acquisition, although not nearly as much as would be desirable."

The remarkable thing about this suggestion was that, first of all, it scorned the will of Congress, which had specifically allocated money for land acquisition and expected it to be used that way. Secondly, its effect could only be to put the squeeze on landowners who sat in the path of the reservoir. It was critical to keep the land-acquisition program moving because of the rapid inflation in California land values, but now Pafford was proposing to do that with one-seventh of the money the Bureau had deemed necessary. This could only mean that people would be offered less money to sell out, and might well accede, since the Bureau could always hold the threat of condemnation over their heads. But it was typical of the way the Bureau operated. If it had a cash-flow problem, the losers would be the people who had had the bad judgment to own property in the valleys it wanted to flood with its reservoirs.

One might be tempted to feel a little sorry for the Bureau of Reclamation. It was, after all, operating at a great disadvantage compared to the Corps, which was unencumbered by social legislation and ostensibly built its reservoirs with the holiest of motives in mind, controlling floods. The available evidence also suggests that the Bureau was not quite as committed to self-perpetuation and self-promotion nor as inclined to trample its opposition. Under several Interior Secretaries—Ickes, Udall, Andrus, even Nixon's Walter Hickel—it had environmental constraints imposed on it that the Corps needn't have bothered with. But one's sympathies might be tempered if one were told that the Bureau, over the intense opposition of a local town, and on a pristine stretch of river up for inclusion in the Wild and Scenic Rivers system, was perfectly capable of proposing a dam which, by its own admission, was completely useless.

The fact that the Yellowstone River was one of the four or five remaining rivers of any size in the American West without a single major dam on it had made it attractive to the Bureau since the 1920s. At one point, according to a former director of the National Park Service, Horace Albright, it had even toyed with the idea of damming the river's outflow from Yellowstone Lake and turning the jewel of Yellowstone Park into a regulated reservoir, and Albright had ordered his rangers to take the drastic step of hiding the Park Service boats so the Bureau couldn't come in and survey. The original Pick-Sloan Plan included a dam lower down on the Yellowstone, which is a major tributary of the Missouri, but in twenty years of trying the Bureau hadn't been able to justify it. The farmers along the river had already built a number of small-scale diversion projects without the Bureau's help; there was plenty of irrigation going on. Flood control wasn't a good enough reason, either, since the damaging floods were all on the lower Missouri, and by the 1960s the Corps had that river completely under control. Power potential didn't amount to much, weighed against the cost of a dam. By 1965, the river had survived six and a half decades of the Bureau and nearly two centuries of the Corps without being dammed—a noteworthy feat of sorts. At about the same time, the conservation movement awoke to the fact that a river of great beauty and substantial size still flowed wild through the northern Rockies and plains, and began to push for official protection in the Wild and Scenic Rivers system. There seemed to be no earthly reason for the Bureau to resist such status—but it did.

The person assigned to take a last, long look at the Yellowstone River, in the light of the conservationists' effort, was dour, impassive Gil Stamm, a future commissioner, who had just been promoted to assistant commissioner by the man he admired and had served so well. In a long blue-envelope letter to Dominy, dated February 3, 1965, Stamm delivered his report. In general, Stamm wrote, "No storage regulation in the Yellowstone River is required . . . as Yellowtail Dam, now under construction, will provide regulation of the Bighorn River and this will insure dependable supplies [of water] below the mouth of the Bighorn," where most of the irrigation was. The only residual interest the Bureau could rightfully claim was "to provide electric power and flood control to the city of Livingston."

The problem was compounded by the fact that the Mission site, where the dam was originally planned, was now occupied by several miles of Interstate 90, which went right along the river below Livingston. Relocating the highway would cost more than it was worth. That left three other sites to select from. The best of them, in beautiful Yankee Jim Canyon above Gardiner, Montana, would back water into Yellowstone Park; Stamm decided to rule it out. The Wanigan site was more expensive to develop and, therefore, "can barely show a [benefit-cost] ratio of one to one." That left the Allenspur site, which was practically in the town of Livingston.

"There is intense local opposition to storage on the upper Yellowstone and particularly the Allenspur site," Stamm cautioned. "The dam would be very close to Livingston, in effect inundating valuable farm and ranch properties and a reach of outstanding stream fishing with national reputation. . . . [Both] ranchers and conservationists have expressed strong opposition to any storage development above the town of Big Timber, which is about 35 miles downstream from Livingston. . . . Findings of the Bureau of Sport Fisheries and Wildlife and the National Park Service show that a dam and reservoir in this area would be detrimental to both fishery and outdoor recreation." The reservoir, Stamm said, would inundate thirty miles of Class I trout fishery—8 percent of the outstanding trout habitat left in Montana. On top of that, it would create an ideal habitat for goldeneye, a rough fish highly competitive with trout; there was "a definite threat of eventual invasion of the streams of Yellowstone National Park by this generally unwanted fish."

As if that were not enough, Stamm said that "a single-purpose flood-control reservoir at Allenspur"—which is essentially what the Bureau was left with—"would cost more than presently estimated benefits." Designing it as a power project wouldn't help; "if the power were to be evaluated realistically in the light of present-day power values . . . Allenspur power would not be very attractive." But adding a hydroelectric plant might be necessary to win authorization, because "the only support for the potential project is from a few public power supporters."

In short, a miserable project: without irrigation benefits, without worthwhile power benefits, without demonstrable flood-control benefits, certain to ruin a long reach of the most productive trout river in the West (if not the entire country), and opposed by virtually everyone who stood to benefit from it—for once, by ranchers and conservationists alike. On top of this, an expensive project, projected to cost at least $128 million—say half a billion dollars today. Stamm's letter reads like an argument *for* giving the upper Yellowstone Wild and

Scenic status—a conservationist couldn't have said it much better himself. But would the Bureau make such a recommendaton? Would it at least not *oppose* such a recommendation?

Only if it was allowed to build the Allenspur Dam. "[F]uture events such as a disastrous local flood possibly could change local attitudes," Stamm concluded. Therefore, his recommendation to Dominy was that the Bureau try "to get the wild river determination altered . . . to accommodate the potential future construction of the Allenspur Unit." By doing so, it would ensure that "all foreseeable desireable future water resource developments would be protected." The Bureau was prepared to accept Wild and Scenic River status for the Yellowstone, in other words, as long as it could someday build the dam that would largely destroy it as a wild and scenic river.

Behind such nearly pathological unwillingness to let go of even *one* river stood, of course, the lurking shadow of the U.S. Army Corps of Engineers. The only conceivable justification for a dam on the Yellowstone was flood control. For now, the Bureau held the authorization to build the project. If it demurred, the Corps might waste no time in trying to build it instead.

If, by the late 1960s, the rivalry between the Bureau and the Corps of Engineers had degenerated into an ongoing squabble over needless projects instead of necessary ones; if each agency was reaching farther afield from its original mandate—the Bureau now talking about building a single-purpose flood-control dam, the Corps incessantly trying to steal the loyalty of the Bureau's irrigation constituency; if they were trying to move into geographic territory where they had no business being—the Bureau into the swamps of Louisiana (there are internal memos suggesting that even this wet state should perhaps be brought into the Bureau's orbit, per request of Senator Russell Long), the Corps into the middle of the Central Valley Project's sevice area—if all of this was true, then it was entirely fitting that the climactic battle between the Bureau and the corps should be fought in, of all places, Alaska.

On April 7, 1961, Daryl Roberts, the head of the Bureau's Alaska District office, wrote a blue-envelope letter to Commissioner Dominy reporting on a luncheon conversation he had just had with C. W. Snedden, the publisher of the Fairbanks *Daily News-Miner*. Snedden, Roberts wrote, had told him that "the Corps of Engineers was cutting my throat and brainwashing the local people in favor of Rampart Dam." Snedden reported that the Corps had "held two meetings with the City Council, had met with the Chamber of Commerce, the National

Resource Committee and others to sell them on holding off on the Devil's Canyon Project until the Corps completes their Rampart study." This news had so upset Roberts that he made a proposal to Dominy that, in all likelihood, no one had ever made before: the Bureau should enlist the same conservationists who had just defeated one of its most beloved dams, Echo Park, in a joint effort to make war on the Corps of Engineers.

What was ironic about the Bureau and the Corps staging their climactic battle in Alaska was that, strictly speaking, neither of them had any business being there. Alaska has very little agriculture—about the only place one can grow anything is in the Matanuska Valley north of Anchorage—and its few farmers employ little irrigation, if any. Besides, the state has more groundwater than one can dream of, most of it a few feet beneath the surface of the earth. The only navigable inland waterway is the Yukon River, and what the Corps was proposing to build would have put an end to that. Anchorage, Fairbanks, and the tiny towns along the Yukon sit on bluffs; only once did Fairbanks have a serious flood, and the city was expanding up the hill, away from the Tanana River. Irrigation, flood control, navigation—none of those applied; yet those were the principal assignments of the Bureau and the Corps. Everything else—recreation, power, fish and wildlife "enhancement"—was supposed to be incidental to those activities. In Alaska, however, such "incidental" benefits were the only rationale they could come up with to build dams. And the dams they wanted to build were too monumental to pass up.

The Corps' dream project, Rampart Dam on the Yukon River, was, at last, an opportunity to show the world what it could really do. It wasn't its size that was so breathtaking—although, with a speculative height of 530 feet and a length of 4,700 feet, it had the dimensions of Grand Coulee—as the size of the reservoir that would form behind it. Lake Rampart would become the largest reservoir in the world. It would cover 10,800 square miles, making it almost exactly the size of Lake Erie. And it was the *power*—five million kilowatts of it, two and a half times more than the initial output of Grand Coulee. Rampart was, by far, the grandest virgin hydroelectric damsite under the American flag; there were only a dozen like it in the entire world.

The Bureau's project, Devil's Canyon Dam on the Susitna River, was, by contrast, almost invisible. But it was still huge: a high plug in a great canyon on the river which ranked sixteenth in the United States in terms of annual flow, Devil's Canyon would produce hundreds and hundreds of megawatts of power, depending on how high it was built. In Alaska, it was second only to Rampart as a hydro-power site.

The Bureau's dam would drown Devil's Canyon, a remote stretch of almost unbelievable wildwater rapids about a hundred miles north of Anchorage. Even fish couldn't navigate those rapids, and no sane person would try—although in the mid 1970s, a group of kayakers led by Dr. Walt Blackadar, a fifty-three-year-old surgeon from Salmon, Idaho, did, and succeeded, at least in the sense that none of them died. Devil's Canyon's value was mere spectacle, even if it was the greatest spectacle of whitewater on the North American continent.

Rampart Dam, however, was an ecological disaster probably without precedent in the world. It would drown the entire Yukon Flats, a sightless plain of marshes, bogs, and small shallow lakes that nurtures more ducks than *all of the United States* below the Canadian border. In its report on the project, the U.S. Fish and Wildlife Service stated, "Nowhere in the history of water development in North America have the fish and wildlife losses anticipated to result from a single project been so overwhelming." At least a million and a half ducks were contributed to the North American flyways by the Yukon Flats, besides 12,500 geese, thousands of swans, an estimated ten thousand little brown cranes, eagles, sandhill cranes, osprey, and moose—thousands and thousands of moose, to which such boggy habitat was pure paradise. The ducks, the moose, the geese, and the swans all required drowned lands, shallow wet habitat, and the Yukon Flats were the greatest continuous expanse of it in North America.

There were salmon. More than a quarter of a million salmon passed through Rampart Canyon every year, some of them destined to go through two time zones to spawning tributaries all the way across Alaska and into Canada. A high dam would end their migration, irrevocably. The Corps' plan to lift them out and carry them across the 250-mile reservoir in barges wouldn't help, because the tiny fry couldn't possibly navigate such a vast body of slack water on their way back to the sea.

There were also furbearing animals—wolverines, lynx, weasels, martins, muskrat, otter, mink, beaver—animals which were the livelihood, to greater or lesser degrees, of most of the Yukon people. Some forty thousand pelts, accoridng to the Fish and Wildlife Service, could be taken from the area to be covered by the reservoir on a sustained-yield basis every year.

And there were people—twelve hundred of them in the taking area, another eight or nine thousand whose livelihoods would be drastically affected, by either the drowning of animal habitat or the end of the salmon runs. Many of those people were Canadian citizens, many others were American Indians and Eskimos who had been promised, by treaty, a land that could sustain them forever. The Corps was promising jobs building the dam, jobs in the tourist industry, jobs in the lake trout fishery that was supposed to replace the salmon. Those were promises; what was already there had sustained their ancestors for five hundred generations.

The whole idea behind Rampart Dam was to turn Alaska, overnight, into an industrial subcontinent. Five million kilowatts were enough to heat and light Anchorage and ten other cities its size, with power left over for a large aluminum smelter, a large munitions plant, a couple of pulp and paper mills, a refinery, perhaps even a uranium-enrichment facility tucked safely away in the wilderness—and even then, about half of the power would be left over for export. But that was the problem. Export where? The dam made sense only if all of the power could be immediately sold.

Realistically speaking, the dam made no sense at all. Neither did Devil's Canyon Dam. The last thing Alaska had to worry about was an energy crisis. It had 300,000 inhabitants; its population could fit inside a few square blocks of Manhattan. Even then, before the gigantic North Slope oil field was discovered, it had proven oil reserves estimated at 170 million barrels (the North Slope was to increase the figure by some ten billion more). It had 360 million board-feet of timber; the driftwood floating down the Susitna River seemed enough fuel to fulfill Anchorage's needs. It had, right around Anchorage, some of the most dramatic tidal variations in the world; the difference between high and low tide approached twenty feet, and a single tidal project taking advantage of similar conditions at DeRance, France, was producing hundreds of megawatts, more than Anchorage (which held half of Alaska's population) would need for decades. Mainly, though, it had plenty of smaller hydroelectric sites scattered about, some of them practically at Anchorage's doorstep. They should be developed first—that was the "orderly" water development the water planners were always talking about.

The problem was simply that Alaska might have to build those itself.

Behind their fiercely independent stance, Alaskans, in the 1960s, were a people completely dependent on Washington, D.C. Their major industry, after fishing, was the U.S. military; their third major industry was the rest of the U.S. government. Alaskans spoke of their state as a "colony," but as colonies go they had themselves a pretty good thing, and they exhibited all the character traits of colonial people—which is to say that they wanted to exploit "their" resources for themselves, but expected the federal government to pay the cost.

Senator Ernest Gruening, formerly a governor of the state, was the main booster of Rampart; he lobbied for it with a zeal that bordered on the fanatic. Behind him were pressure groups like Yukon Power for America, or the more picturesque North of the Range association, which said in its brochures that Alaska's future depended on "coming forward with both guns blazing." What mattered most to the boosters was that Rampart was an opportunity—the first real opportunity— to leave mankind's mark on a place that held it in magnificent contempt. George Sundborg, Gruening's administrative assistant, dismissed the area to be drowned by the dam as practically worthless; there were "not more than ten flush toilets in it." Gruening went further: it was totally worthless. "[T]he Yukon Flats," he wrote, "—a mammoth swamp—from the standpoint of human habitability is about as worthless and useless an area as can be found in the path of any hydroelectric development. Scenically it is zero. In fact, it is one of the few really ugly areas in a land prodigal with sensational beauty."

And, since these were the 1960s, and since this was the army that wanted to build the project, there may have been a further consideration working behind Rampart Dam. Ernest Gruening had, he said, recently returned from Russia, where he had seen "hydroelectric power dams larger than the largest in America." The dam, then, was to be a monument against Communism; and if it made it any easier to build it, one might as well note that the ducks whose habitat would be drowned were Communist ducks—many of them migrated to Siberia. Did it make sense, a director of Yukon Power of America asked, "to mollify these feathered defectors"? It is hard to judge whether or not he was serious.

It is also hard to say, in retrospect, how close Rampart Dam ever came to being built. The odds are, moderately close. But Floyd Dominy killed it.

If the dam was built, the Bureau would have no future in the last place where there were still plenty of big damsites left. Congress wouldn't authorize another dam there for decades; the power probably wouldn't be needed for two hundred years. That was the argument Dominy used, and used brilliantly. With Stewart Udall's enthusiastic blessing, Dominy had the Bureau turn all of its guns on Rampart Dam—the planning division, the hydroelectric division, the demographics branch: everyone who had some expertise that could cripple Rampart's chances was enlisted in the cause. In 1967, the Interior

Department produced its Rampart report, a document nearly a hundred pages long, complete with appendices and reams of supporting documentaton in the files. The report demonstrated that Rampart power had to be sold immediately or the project would be a financial fiasco of the first order. But it also showed that the power market projected by the Corps and the local boosters couldn't possibly develop within the state—not in twenty years, not in fifty, perhaps never. Shipped to the nearest market—the Pacific Northwest—the power couldn't possibly sell at competitive prices; the cost of *transmission alone* would be more than people were paying, more even than nuclear electricity.

In the end, Dominy was asked to testify on Rampart Dam, and it was one of the most brilliant performances of his career. Without anger, without malice, he tore the Corps' justification to shreds. Even the pedestrian rhetoric of his successors—of a Gil Stamm or a Keith Higginson—might have demolished Rampart's prospects, but Dominy spared nothing in his presentation. When he was finished, Rampart Dam lay pretty much in ruins. The project surfaced a few more times during the 1970s, then floated under and hasn't been seen since.

Devil's Canyon Dam, however, was seemingly dead, too—at least as far as the Bureau was concerned. (Late in the 1970s the state of Alaska announced plans to build the dam itself; it will be interesting to see whether it can without falling into the kind of financial hole that the $18 billion Itaipu Dam dug for Brazil.) And there the irony of the whole long fight between the Bureau and the Corps of Engineers came full circle. Had they really cooperated—as General William Cassidy had stated they would, and must—there is no telling what they might have built. Their rivalry prevailed, and grew more intense, during one of history's truly unique periods—a time when we had the confidence, and the money, and, one might say, the compulsion to build on a fantastically grand scale. The money invested in the dozens of relatively small projects each agency built—in many cases because the other threatened to build first—would have sufficed to build the great works they insisted were necessary, but which required extraordinary determination, cooperation, and raw political clout in order to be authorized. Fifty million here, eighty million there, a hundred million here, and soon one was talking about real money. In the 1960s, Dos Rios Dam could have been built for $400 million; today it might cost $3 billion or more. A diversion from the Columbia River to the Southwest could have been built for $6 billion or so in the sixties, and there was so much surplus energy in the Northwest that a few million acre-feet of water removed from a river that dumps 140 million acre-feet into the sea might not have been missed. Today the cost seems utterly prohibitive, and Washington and Oregon would probably resist the engineers with tanks. The opportunity was there. But the Corps of Engineers and the Bureau squandered their political capital and billions in taxpayers' money on vainglorious rivalry, with the result that much of what they *really* wanted to build does not now exist, and probably never will.

CHAPTER SEVEN

Dominy

When Emma Dominy, writhing and shrieking, finally evicted her son Floyd, the doctors dumped him on a scale and whistled. Floyd Elgin Dominy, ten pounds, four ounces, at birth. Floyd Elgin Dominy, larger than life. All of Floyd's siblings were born huge. His brother Ralph weighed twelve pounds. Emma's six giant babies were a cross she was to bear through the rest of her life. Her uterus became distended, causing her horrid pain. She developed a nervous condition. Her temper became explosive, her outbursts hysterical. Strong-willed, French-Irish, and beautiful, Emma May Dominy was a handful anyway. Charles Dominy and his wife fought day and night. They had what is referred to as a "difficult" marriage, cemented precariously by children, religion, and a pious wheatbelt condemnation of divorce. Life, remembers Floyd, was like living on an earthquake fault. There was never any peace. "They fussed and fumed from morning to night. We'd lie awake at night and listen to them tearing into each other." He is seventy when he says this, but his childhood is still a bad memory; you can read it in the turned-down corners of his mouth. "I remember what a relief it was to get away from home. It bugged me right through college. When everyone else was having nightmares about missing exams, I was having nightmares that my parents were murdering each other."

Hastings, Nebraska, is a long way from paradise: Libya in the summer, Siberia in the winter; too wet for the Bureau of Reclamation, too arid for trees. Hard up against the hundredth meridian, Hastings occupies America's agricultural DMZ. Neither God nor government has taken it under its wing. Disaster is Hastings's stock-in-trade—that and dullness. "The capriciousness of nature is the one thing that livens that place up," says Dominy. "When they aren't talking crop prices or tattling on their neighbors, all anyone talks about is the weather." Hastings is tornado country (one of the few double-funneled tornadoes ever seen was photographed near there), baseball-size-hail country, banshee blizzard country, drought-without-end country. The region's whole economy can be drained by a summer's drought, dashed by an afternoon's hailstorm. The anarchy of nature may be one reason why most of Hastings's residents—Republican or Democrat, dry farmer or irrigation farmer, city dweller or country dweller—devoutly believe that man should exercise as much dominion over the earth as he can. Hastings, Nebraska: birthplace of Floyd Dominy, future Commissioner of Reclamation.

Floyd was headstrong and impulsive—"an independent cuss from the beginning." He was an above-average but somewhat uninterested student, and his intelligence was more obvious than evident in his grades. His distinguishing characteristic was self-reliance. Floyd had great confidence in himself. At the age of eleven, he could manhandle a neighbor's two-thousand-pound Belgian draft horses as if they were a pair of pygmy ponies. He fixed things, ran things, organized things. Other children respected and feared him. To most children, the home is a refuge from a dangerous world; in Floyd's case, it was the other way around. Compared to home, shadowed by gloom and rumbling with thunder, the world was a sunlit place.

"I always felt there was a contradiction between my parents' fussing and fuming and their Christian piety," he says. "It seemed inconsistent to me. As a boy, I was very moral. I was president of my Sunday school class. I thought money was the root of evil. If someone had offered me a job paying $300 a year for life, I would have taken it. When I first married Alice, I made her take off her lipstick if we went out for the evening.

"I'm an enigma, even to myself."

At seventeen, Floyd fell in love. Her name was Alice Criswell. She was sweet, demure, and very pretty, a little heroine out of Willa Cather. They met at a state convention; he was Master Counsellor for the Order of DeMolay, and she was the Queen of Job's Daughters. Alice's family lived in western Nebraska, near Chappell, a good two hundred miles away. Floyd was mad for her, but his father refused to let him borrow the car. Floyd had $30 to his name. He spent $25 of it on a beat-up one-cylinder motorcycle that, with luck, would take him to Alice. "It was a helluva trip out there. The roads were all dirt in those days. I wore out a pair of boots balancing that one-lunger, but I made it. When I got ready to go back home, the damn thing wouldn't fire up. Alice's father looked at it and said, 'Your magneto's shot.' I said, 'Can we fix it?' He spent two hours trying, but the sonofabitch was beyond repair. I had to sell it for what I could get, which was five bucks, and start hitchhiking home. Hitchhike, hell. You hardly saw a car in western Nebraska in those days. I'd walked about three miles when I came upon an old guy with his head stuck under the hood of his truck. I said, 'What's the matter?' and I looked in and saw that *his* magneto was shot. Well, in the last two hours I'd learned about magnetos. I took his apart, saw right away what was wrong with it, and fixed it then and there. That old geezer was so impressed that he offered me a job on the spot. I never went home again."

Floyd and Alice married secretly in Georgia, where Floyd had gone after two years at Hastings College to work on a gas pipeline being built across the South. They spent their three-day honeymoon in Florida. Floyd signed in ahead for three days of work and they took off. A supervisor, his heart warmed by a young couple in love, covered for him. "I was nineteen," Floyd says. "I think that was the first lie I ever told in my life."

When his stint in Atlanta was up, Floyd and Alice went back to Hastings. For $15 a week, he drove a truck between Hastings and Lincoln. Driving anything—a team of horses—was a dream job to many a farm boy, but Floyd found it excruciatingly dull. "I finally said to myself, 'Hell, $15 a week is nothing. I'll go out to western Nebraska with Alice.' I got myself a job on Fred Smith's place. Man, that was a badly run operation. They had new weeding tractors and their wheat fields were still being run over by weeds. They only ran the tractors during the daytime—they were too lazy to run them at night. This land was dry-farmed, and those weeds were using precious rainfall that was needed by the wheat. There were lights on the tractors. They should have been running the goddamned machines twenty-four hours a day. So I finally said, 'This is a helluva way to run a farm!' Fred Smith thought I was quite an upstart. He said to me, 'How would *you* run it?' and I said, 'I'll show you.' I climbed on one of those tractors and I ran it till ten o'clock at night. Then I went to bed, got up at three in the morning, and finished the job by four the next afternoon. Cleared out every weed on that farm. I was hell-for-leather. I didn't stop to take a leak. Old Fred Smith came up to me later as I was changing clothes and said, 'With that kind of drive, you're wasting yourself. You ought to go back to college.' "

The sensible thing for a mechanically gifted farm boy who didn't particularly like farming to major in was engineering. At Hastings College, Dominy had given it a brief go and quit. "I didn't like the preciseness," he says. In 1930, he entered the University of Wyoming at Laramie, choosing economics as a major. He was captain of the hockey team. He stayed on and won a master's degree in 1933. By then the country's economy was in a screaming nosedive and the West was five years into the Great Drought. The ranchers around Laramie couldn't sell their cattle—first because no one had money to buy them, second because the cattle weren't worth buying anyway. They were thirsty and starving, vacant-eyed beasts with bellies bloated from hunger and protruding ribs. Stupefied by the intensity of the disaster, Wyoming's people were in the same condition, mentally if not physically. Campbell County, two hundred miles north of Laramie, was typical of the places that had plummeted through FDR's safety net of relief. Roosevelt couldn't launch a federal dam project there because Campbell County had no river worth a dam. It had no highway project because no one went there and it hardly had cars. It had no writers' projects, no hospital projects, no dog census. All it had was the cattle liquidation program. The Agriculture Department's county agent paid the ranchers $8 a head for their scrawny cattle, then shot them. The farmers took the $8 and spent it on horse feed and rifle shells, then headed into the uplands in search of deer and rabbits. During the Depression, Campbell County reverted substantially to the hunter-gatherer existence of the Crow and northern Cheyenne who had forfeited the territory. The two things it had going for it were reasonably abundant herds of game and the county agent, Floyd Dominy.

At eleven o'clock one morning in the spring of 1980, Dominy, floating on three gin and juices and powered by two cigars, was in a mood to talk about his Campbell County days. "We had a drought, grasshoppers, crickets. I tell you it was something else. It looked as if nothing could live. Under the federal regulations, five thousand cattle were to be bought in the whole state of Wyoming. Fifty thousand were dying in Campbell County alone. I called up Washington and said, 'This is worse than you can believe. Send me another vet, dammit.' They sent me three vets. That got me some attention. The range improvement program, though, really put me on the map. That took creativity and force. The government was paying farmers fifteen cents a cubic yard to move dirt. Hell, I wasn't going to pay fifteen cents if it cost ten. I said to those ranchers, 'I'm gonna pay you cost—nothing more.' Naturally, they bellyached. But with my relief allotment stretched further I could build a lot more dams."

Campbell County is drier than crisp toast, but it does get some rain. There are mountains around that produce orographic clouds, and some of them produce rain—not much, but enough to make it worth trying to store the runoff that occasionally pours down the creeks. "I said to myself, 'It's stupid to let a drop of that stuff escape. We've got to capture that water.' I'd take these ranchers out to where I wanted them to build a dam, some godawful-looking dry creek somewhere, and they'd say, 'A dam's no good. There's no water to take.' And I'd say, 'Goddamn it, a ten-minute downpour in this devegetated moonscape and you'll see a nice little surge come through here.' The one good thing about Wyoming is there's not enough groundcover to soak up the rain where it falls. I said to the farmers, 'You capture that water and at least your cows won't die of thirst. You get a little extra for irrigation and you can grow some grass on it. What do you want to do—just sit here and starve?'

"So I got them building dams. I practiced myself with a little four-horsepower Fresno scraper. The county surveyor and I developed our own set of regulations. We said it's got to have ten-foot width and five feet of freeboard. The federal regulations said the Soil Conservation man had to approve the damsite. The Forest Service guy was supposed to have his say-so, too. I said to hell with it. I cut all that red tape. The extension director and the Wyoming dean of agriculture finally got wind of what I was up to. They said to me, 'Floyd, you can't *do* that. You've got to play by the *rules*.' I said, 'The Democrats would have a really black eye if they announce a program that doesn't work.' "

Dominy took a swig of gin and juice, leaned back in his black easy chair, and chuckled. "That was the end of 'prior approval.' Henry Wallace took the phrase right out of the law.

"We built three hundred dams in my county. That was more than in the whole rest of the West. I was a one-man Bureau of Reclamation. We were moving! I was twenty-four years old, and I was king. Campbell County was my demesne. They still talk about me out there. I saved a lot of cattle from dying and a lot of farmers from going on relief. After that, I started getting job offers from Washington. But I had already psychoanalyzed myself as a strong starter who got bored easily. I figured I'd have to watch that if I wanted to succeed in life. So I had made up my mind to stay in Campbell County five years."

For Floyd and Alice, the first two and a half years in Campbell

County meant a life-style a cut above that of his ancestors when they arrived in Nebraska in 1873. They lived in a stone dugout built into a hillside; they had a gasoline lantern and a coal-burning stove, but no windows. "The place had been abandoned for thirty years. It was vandalized. The house had a leaning chimney and big holes in the floor. I was being paid $130 a month, plus five cents a mile for the car. The guy who owned the hovel was named Mr. Bartles. He was as bald as a billiard ball. I said, 'What's the rent?' He said, 'You're crazy wanting to live there in the first place. I'm not going to let you live there *and* charge you rent.' "

Dominy didn't quite achieve his goal of staying five years in Campbell County; he finally succumbed to an offer from the Agricultural Adjustment Administration to help administer the nation's increasingly complex farm program, working as a field agent for the western states. In 1942, he transferred to the Inter-American Affairs Bureau, working under Nelson Rockefeller. The war effort demanded immense quantities of bauxite, rubber, and cinchona, most of it coming out of the Caribbean and South America. Tens of thousands of miners and loggers were dumped in the middle of the jungle without enough to eat. Instant farms became Dominy's specialty. He set them up in nine Central and South American countries, and, later, on the islands of Saipan, Tinian, Iwo Jima, and Peleliu as they were recaptured from the Japanese.

In March of 1946, Dominy was back from the Pacific. Reviewing his career on a homebound ship, he decided that nothing had been as satisfying as building all those dams in Campbell County. It was one thing to hack a farm out of a jungle clearing—that was brutal and monotonous work, requiring neither brains nor talent. It was quite another thing to build a dam, store the water, and make the desert bloom. That, in a small way, was changing the order of the universe. On the same day he returned to Washington, Dominy went to a phone booth and put in a call to the Bureau of Reclamation. He had a job in three hours.

As a land-development specialist for the Bureau, Dominy proved his mettle quickly. His experience helped, as did his prodigious energy, but Dominy also had something a great many of the Bureau's engineers lacked—a knack with people. "It was two things," he says. "First, I cared about making these projects work. The engineers would build the dam and the irrigation features and walk away from it. They felt the projects were supposed to work out by themselves. When I got there, we had projects failing all over the place. The Bureau would send a threat out to the farmers to shape up, then forget about them for five years. No one took us seriously. Well, by God, they took *me* seriously. I was tough, but they saw I cared about their problems. That was number two. I proved myself right away. One of our early projects in big trouble was Milk River in Montana. The regional director, Ken Vernon, had revised the repayment contract under political pressure and it was a complete giveaway. I had moved up to Allocation and Repayment then, and I sent him a blistering letter about it. Vernon was several ranks above me and he couldn't believe it. He called up Goodrich Lineweaver, my superior, and made himself hoarse chewing him out. 'Who is this goddamned upstart?' Lineweaver thought he could put me in my place by sending me to negotiate a better deal. He was sure I'd fail. So I went out to Montana. I saw these old farmers lined up in a room like a country church. They were hostile as hell. I demanded that tables and chairs be brought in. I gave them all pencils and a scratch pad and something to drink. Now they could put their feet under something, light up a smoke, and we could have a serious goddamn discussion. We got a whole new package out of this."

Floyd Dominy's rise to power in the Bureau of Reclamation was astonishingly fast. From dirt sampler to waterlord of the American West took just thirteen years, and he might as well have been commissioner during the last three. Like a chess master, Dominy leaped and checked his way to the top, going from Land Development to an entirely different branch, Allocation and Repayment, then sidelong to Operation and Maintenance, then to the Irrigation Division, and finally to assistant, associate, and full commissioner. His strategy was simple. He would settle in a branch with a weak man as chief and learn as fast as he could. Then he would flap up to the ledge occupied by the chief and knock him off. The first to go was Bill Palmer, who headed Allocation and Repayment and was there largely because he was a Mormon and had an influential constituency. "Mike Straus was totally unsatisfied with Palmer," says Dominy, "so I told Lineweaver that they ought to replace him with me. He said, 'I can't do that.' I said, 'Well, what *can* you do?' Lineweaver said, 'We can make you acting director and not tell Palmer about it.' I said, 'How long acting?' He said, 'Well, I don't know, until we can work something out.' I said, 'Let's make it sixty days.' Lineweaver mumbled and grumbled, 'I don't know, Floyd, that's awfully short.' I said, 'It's long for me.' Well, I got him to agree. There I was, 'acting director,' and Palmer doesn't even know it. The first thing he does is start making a fuss about having to train me, because he'd just trained some other guy. So I walked into his office late one day and said, 'Bill, I think you've got a bad attitude. I hear you've been complaining about having to train me. Well, you don't have to. Dominy can train Dominy.' He looked up at me and said, 'What do you mean by that, Floyd?' I looked him *cold* in the eye and said, 'I mean I'm about to run this division, Bill. It's you or me, and I can guarantee you it's going to be me. So maybe what you ought to do is request a transfer. Maybe you should go out West.' " Mimicking his tone of voice then, Dominy sounds like a Mafia shakedown artist running a recalcitrant store owner out of the neighborhood. "Well, he took my cue. Next thing I know Bill Palmer is requesting a move to Sacramento and I'm chief of Allocation and Repayment. It took exactly sixty days, just like I said. I brought him back, though. Ultimately, I made him an assistant commissioner. Bill was a good man."

In his new position, Dominy had an opportunity to learn anything he wanted about the three-hundred-odd Reclamation projects in existence. He read every project history, reserving for special attention the "bad elements"—the projects that were failing. "Half of our projects were insolvent. I was fascinated: why some and not others? I said to myself, 'Whoever figures this out and starts to haul Reclamation out of this financial ooze is going to be the next commissioner.' The reasons were complicated. In the early days, Reclamation made some bad mistakes—we miscalculated water availability, we laid out canals that didn't work right, we had drainage problems that we should have anticipated. Soil, altitude, crop prices, markets—they all made a difference. On top of that, there were practically no requirements. Straus and Warne let any idiot get into a Reclamation project. You didn't have to demonstrate that you had capital, farming skills, anything. Any fool could sign up and get on a Reclamation farm and use whatever intelligence he had cheating the government. When the projects began to go bankrupt, Straus and Warne were afraid to expose them. They covered the goddamn things up and that got us in a hell of a lot of trouble with Congress. We were illegally delivering water all over the place. Payments were way in arrears and no one was doing a damn thing about it. I think we were violating the law at least as often as we were not violating it."

Dominy approached the problem in a somewhat schizophrenic way. Privately, he was appalled by the lassitude of the Reclamation program, by the indifference of the engineers to its problems, and by the hypocrisy of members of Congress who voted for bad projects as special favors to colleagues and then griped about the money they were losing. At the same time, he was, in public, the program's most belligerent defender after Mike Straus. His defenses were so eloquent he even came to believe them himself.

Once a prominent Senator from South Dakota, Chan Gurney, sent Straus a copy of an article that was witheringly critical of the Belle Fourche Project in his own state, implying that he agreed with it. For years, Belle Fourche had been perhaps the Bureau's preeminent fiasco. Streamflow calculations and reservoir carryover capacity were based on nine months of gauging during a wet year; when the drought of the 1930s came, the reservoir was dry within months. No investigation had been made of the need for drainage, which was turning out to be a terrific problem the farmers could not begin to pay to solve. Farmers settling the project were not selected on the basis of character, aptitude, or available capital, and the vast majority of them were bankrupt within a few years. Even with the Bureau forgiving almost all their obligations, many of the farmers were going broke. They were still receiving water, however, so the project was technically in violation of the law. Congressmen hostile to the Reclamation program loved to crucify Belle Fourche at appropriations time; it was like stoning a flightless auk. Even blustery Mike Straus was going to send Gurney a milquetoast letter in response. When he reread the draft that had been prepared by an aide, however, he couldn't bear to do it. So Dominy volunteered.

Of course the project was in deep trouble, Dominy wrote. It was planned at the turn of the century, one of the first large-scale irrigation ventures since the Fertile Crescent. There was hardly any experience to go on. Records of North America's climate scarcely existed. But it was *Congress*, not the Bureau, that had been especially anxious to get the Reclamation program underway—that was the main reason Belle Fourche was undertaken on such a paucity of data. It was *Congress*, not the Bureau, that had established impossibly short repayment periods, that had failed to appropriate funds for demonstration projects. It was *Congress* that demanded projects in areas where the value of agriculture wasn't worth the cost of irrigation, making subsidies inevitable. The point was the project was there. Thousands of South Dakotans depended on it; they had helped feed the country when the state's dryland farmers were utterly ruined. What would the Senator do? Shut it down? Tear down the dam? Kick defaulting farmers off their lands and onto the relief rolls? Or would he help the Bureau come up with solutions to put the Reclamation program on a sound foundation? After all, if anyone was embarrassed by the Belle Fourche Project, it was the Bureau. Did the Senator believe that the greatest amalgamation of professional talent in the government was *glad* when its projects became financial disasters? "Straus read that letter and loved it so much he read it twice again," Dominy chuckled. "He didn't change one word. I was in thick with him from that point on. We really blew smoke up that Senator's ass."

Dominy had the instincts of a first-rate miler. He could pace himself beautifully, moving on the margin of recklessness but always with power in reserve. He knew when to cut off a runner, when to throw an elbow, when to sprint. He also knew that there was nothing like a grudge to make him run harder.

If Dominy harbored a lifelong grudge, it was against engineers. Away from their drafting tables, he thought, engineers could be inexcusably stupid. On the other hand, they had a mystical ability to erect huge structures along exact lines, using bizarre formulas he could not even read. They could map a river basin, analyze some abutment rocks, measure the streamflow, and build a dam of precisely the shape, size, and structure to suit it. They had labored through the trigonometry, the calculus, the chemistry, the topology, and the geology that he had backed away from—the one time in his life he had given up on anything. The problem was, they couldn't explain their own work or its importance, couldn't understand human relations, couldn't see a political problem about to smack them in the face. He could do all of that—brilliantly. Dominy needed them, and he knew it, and they needed him—and didn't know it. It made him furious. In the mid-1950s, after mastering Operation and Maintenance and Repayment and Irrigation, Dominy felt he should move on to the second most important job in the Bureau—the assistant commissioner for legislative liaison. He should be the one working Congress—explaining new projects, justifying the problem ones, tantalizing members with grandiose plans, horse trading, cajoling, threatening. After all, if the Republicans held to their "no new starts" policy, the Bureau would soon have nothing to do.

The position, however, had never gone to a non-engineer, and the person Commissioner Wilbur Dexheimer wanted to appoint was Ed Neilson. Dominy had warned Dexheimer about Neilson. He was, he told Dex, just like him: good-natured, somewhat bumbling, uninterested in politics, and therefore inept. Neilson was the last person who should be sent up to explain the Bureau's work to Congress. "He had already admitted that he didn't even know the names of most of the projects, and if someone mentioned one to him he wouldn't be able to say what state it was in. For Christ's sake!"

The Public Works Subcommittee of the House Appropriations

Committee, which authorized every penny the Bureau spent, had been reorganized after the 1954 election in a way that was profoundly inauspicious for the Bureau. Only two Congressmen sympathetic to Reclamation still sat on it, and one of them, Mike Kirwan, was from Ohio, whose farmers were beginning to raise hell about subsidized competition from Reclamation lands. Everyone else on the subcommittee was hostile or indifferent to the Bureau.

The Appropriations Committee hearings began in April of 1955, and, as Dominy had predicted, the roof caved in. "Dexheimer had gone off for two weeks to watch an atomic bomb test in Nevada. It was utterly inexcusable. The assistant commissioners, Neilson and Crosthwait, and the regional directors were all there, but they were the most tongue-tied bunch of engineers you ever saw. They muffed answers to the simplest questions. It was the biggest fiasco. But Neilson and Crosthwait kept telling me my presence 'wasn't required,' because the subcommittee was only allowing five witnesses to be present at one time. Actually, they were scared I would upstage them. On the tenth day, I was invited to lunch by Senator Gale McGee of Wyoming. Word was getting around about how unbelievably inept Reclamation's witnesses were, and like every other member from the West, he was concerned. He said, 'Floyd, can you do something?' See, I already had a reputation as the most knowledgeable person in the Bureau. After lunch, I called in for my messages.

"My secretary told me I'd gotten a telephone call from Neilson up on the Hill. 'He needs you desperately,' she said. I was madder than hell. I stalked into that hearing room and went up to Neilson and said, 'You got your chestnuts burned pretty good and now you want me to pull them out of the fire.' You should have seen the look on his face. He said, 'Are you being insubordinate?' I said, 'Hell, no, I'm being loyal. I'm here to save your can. But you introduce me first.'

"Rudy Walters, the regional director from Denver, was up there at that moment testifying about the Kendrick Project. I knew all about the Kendrick Project—it was in Wyoming. Rudy was totally tongue-tied. You could read the exasperation on those committee members' faces. Neilson ran up to the front of the room and said, 'Mr. Chairman, Mr. Chairman, Floyd's here.' 'Floyd's here.' No introduction, no last name, nothing. I was mad as a bull with a spear in his back, but I know how to channel anger. I walked to that witness dock and said, 'Mr. Chairman, my name is Floyd Elgin Dominy. I am not an engineer. I'd be happy to tell you about the Kendrick Project. In the first place, the Kendrick Project would never have been built if it hadn't been for Senator Kendrick. If our engineers had been left solely with the decision, they probably wouldn't have built it.' That kept them from dozing off. Then I told them everything they wanted to know.

"For the first hour I was standing up, resting my hands on the chair of the official reporter. Neilson didn't even give me a goddamned seat in his pew. Then the committee wanted me to testify about some other projects, and the chairman directed Neilson to make room for me. I went on all afternoon, and they invited me right back the next day. I ended up testifying for a week. The committee publicly reprimanded the Bureau for inexcusable lack of preparedness and unwillingness to provide facts, but they specifically mentioned *Dominy* as the one exception. From then on, if a Congressman wanted to know anything about Reclamation, he came to *me*. Before long, they were asking me about the Corps of Engineers projects, too. I became the person they trusted. I wasn't afraid of any of them, either. I chased one out of my office once.

"What I did on Fred Smith's farm got me my start in life. What I did in Campbell County got me to Washington. Those hearings made me commissioner."

"I liked Floyd. I trusted him. I thought he would be loyal to me as secretary."

"I liked Stewart. He was a bad administrator, but he had marvelous instincts. He also had guts. He wouldn't bite a chainsaw, but he had guts."

"Dominy despised Stewart Udall, and Udall regarded him like a rogue elephant. Dominy used to come storming out of Udall's office and say, 'Who does he think he is?! The Commissioner of Reclamation?' "

"Dominy was the most able bureaucrat I've ever known."

"I was amazed by him. He had the constitution of a double ox. He'd be dead drunk at a party at three A.M. and he'd be testifying at eight-thirty the next morning and you couldn't tell."

"He was merciless to the people around him. He could be hell on his assistant commissioners. He was horrible to some of the regional directors. If you made a stupid mistake he was all over you and he wouldn't quit."

"When we went on tours abroad, Dominy was treated like the President of the United States."

"He was a magician with Congress. His friends there would do anything for him. They believed every word he said."

"When he testified he spouted numbers like a computer. He spoke with absolute self-assurance. It was all hogwash. If he didn't know a number, he made one up."

"When Dominy was ousted the Bureau of Reclamation fell apart. It will never recover. The disarray over there now is ridiculous."

"When you worked for Dominy you were always terrified of the page-eight syndrome. If you handed him a memorandum and page eight was missing, he'd call your supervisor and say, 'Get that asshole off the job. Put him in a hole someplace.' Guys ruined their careers because they stumbled on the rug when they entered his office."

"Basically, he was a terrorist."

"All the wives were disgusted with him. Some of them refused to come to parties when he was going to be there, because he'd start propositioning them all."

"We played a game of golf once. Floyd was a below-average golfer and I'm an above-average golfer, but he beat me with psych. On the second or third hole, I sliced a ball. He spent the rest of the game ridiculing my slice. I didn't know whether I was madder at him or at myself. He got me all worked up and nervous. Ordinarily, when one grows up and becomes successful, one learns not to let silly mistakes or ridicule become bothersome. But I was so bothered I felt like a little kid on the verge of tears. He psyched me out. He won the game."

"He was one of the best gamblers I ever saw. I was on an airplane with him once and watched him play a game of high-stakes bridge. He won $1,200 in a couple of hours. He took the money and bought himself a tractor."

"If Dominy were commissioner today, he'd be killed."

Nominally, the Bureau of Reclamation is a part of the Interior Department. The commissioner is, in theory, directly responsible to the
Interior Secretary and the President, and carries out the wishes of
whatever administration occupies the White House—whether that
administration appointed him or not. Actually, everyone who has
watched the Bureau in action over the years knows it doesn't work
that way. The Bureau is a creature of Congress, and most Presidents
have not been able to control it any better than they could control the
weather or the press. The role of the Bureau vis-à-vis the White House
and Congress might be likened to that of a child placed in a foster
home by a doting pair of unstable parents. The child may tell lies,
throw tantrums, wreck the house, and eat everything in the icebox,
but if his foster parents finally decide to give him a thrashing, his real
parents materialize out of nowhere and wrest the paddle from their
hands. Jimmy Carter lost the momentum of his presidency, and a
chance at a second term, through a hapless effort to bring the Bureau
and the Corps of Engineers under control. Eisenhower, Johnson,
Nixon, and Ford all tried to dump or delay a number of projects the
Bureau and Corps wanted to build, and failed in almost every case.
Congress simply tossed the projects into omnibus public-works bills,
which would have required that the President veto anything from
important flood-control projects to fish hatcheries to job programs in
order to get rid of some misbegotten dams.

The peculiar relationship between the Bureau and the two leading
branches of government—in which it can defy the wishes of the branch
that supposedly runs it and is largely subservient to the wishes of the
other—is something relatively new. Mostly it is a development of
the postwar era. In the past, the President often had to champion the
Reclamation program *against* the objections of an eastern-dominated
Congress, which found the whole idea a waste of money. Teddy Roosevelt, Franklin Roosevelt, and even Herbert Hoover all fought with
Congress over Reclamation dams they wanted built. As the dams oc-
tupled the population of the West, however, and as long-lived members
of Congress from the South and West rose into important committee
chairmanships, the character of Congressional leadership changed,
and its attitudes followed. With Wayne Aspinall and Carl Hayden
running the Interior and Appropriations committees, Ike could no
more enforce a "no new starts" policy than Jimmy Carter could bounce
a $40 million Corps of Engineers dam whose sole beneficiary was to
be a private catfish farm in the district of an influential Congressman
from Oklahoma. As far as public works were concerned, by the 1950s

it was Congress, not the White House, that ran the government. We had become a plutocracy of the powerful and entrenched.

No one in government recognized this earlier, or exploited it more brilliantly, than Floyd Dominy. Dominy cultivated Congress as if he were tending prize-winning orchids. Long before he became commissioner, on almost any day you might find him eating lunch with some powerful or promising Congressman or Senator who needn't necessarily represent a western state. Not only would Dominy have lunch with him, but often Dominy would pick up the tab. If a Congressman broke his toe, he might receive a nice letter of condolence. Dominy sent out reams of condolence letters, often to acquaintances who could only be described as casual, though he didn't write too many himself; much of his underlings' work had nothing to do with dams. Favored Congressmen like Mike Kirwan (an easterner) might receive an expensive, custom-crafted set of bookends in the shape of Flaming Gorge and Hoover dams, which they could use to contain the public works bills that were flooding the country in a tide of red ink.

Dominy was a meticulous list-keeper. In his files he kept lists of the Bureau's friends on Capitol Hill, arranged in categories: close friends, reliable supporters, occasionally wayward supporters. Those on the "A" list were handsomely rewarded. "Dominy yanked money in and out of those Congressmen's districts like a yo-yo," says a former associate assistant Interior Secretary who admired Dominy so much he was assigned to tell him he was fired, and whose name was James Gaius Watt. "If some Senator was causing him trouble, money for his project could disappear mighty fast. It went right into projects for the politicians who were Dominy's friends." All Dominy had to do was order his engineering department to say that it simply couldn't spend the money any faster. A memorandum dated April 10, 1967, from Dominy's chief of public affairs, Ottis Peterson, put together, at Dominy's request, a list of Senators whose terms were about to expire and whom, in Peterson's words, "we should make a particular effort to protect and give as many news breaks as possible." The list of thirteen names—among which were McGovern of South Dakota, Morse of Oregon, Church of Idaho, and Magnuson of Washington—was for "very special attention and protection," although "we can fatten our batting average by taking care of everyone to the best of our ability." Small wonder that George McGovern became so blindly wedded to the Bureau's Oahe Diversion Project that his constituents voted him out of office thirteen years later when they turned against it.

Dominy's power and influence with Congress were so extraordinary that all he usually had to do to change his superiors' minds—

whether they were contemplating his dismissal or merely a stretch of Wild and Scenic River where he wanted to put a dam—was make a few phone calls to Congress. At worst, he simply had to threaten to resign.

Talk of resignation was Dominy's ace in the hole. "Dominy threatened to resign so many times I lost count," says his onetime regional director in Sacramento, Pat Dugan. Early in the 1960s, Stewart Udall's Under Secretary, Jim Carr, a voluble pro-Californian who loathed Dominy at least half as much as Dominy loathed him, ordered Dugan to fire his chief of planning, Pat Head, for allegedly causing delays in the preconstruction work for Auburn Dam—delays that Dominy may very well have instigated himself. Dugan was in Washington at the time, and he and Dominy went out to lunch. After they had consumed two big steaks and several belts of whiskey, Dugan told Dominy about Carr's order, and suggested self-effacingly that maybe *he* had better resign, since he was Pat Head's superior. Dominy was enraged. "Hell, let's *both* resign!" he boomed in a voice that stopped conversation cold. And, in fact, he made his customary threat, which wouldn't have worked so well if Udall hadn't suspected that he was mercurial enough to carry it out. But it did work, and neither Dominy, nor Dugan, nor even Pat Head left his job, and Jim Carr died without watching a bucket of concrete poured for his favorite dam. Small wonder Dominy used the threat of resignation so much—after all, it had made him commissioner.

Floyd Dominy was furious when Dexheimer failed to appoint him assistant commissioner, and he believed in carrying a grudge. After Dexheimer's designee, Ed Neilson, failed so miserably before the Appropriations Committee in 1955, only to be rescued by Dominy, the chief of the Irrigation Division went to see the commissioner after he returned from watching his atomic bomb blast. "Today I told the Commissioner that in eighteen years on government payroll . . . I had never seen an agency perform so ineptly," Dominy confided in his diary on June 7, 1955. "I went on to tell him that I thought it was a crime to personally absent himself from the City through practically all of the hearings. . . . I concluded that I was prepared to move up to strengthen the front office . . . I had made my speech and if he wished to think it over I would be available. With this I terminated the discussion." Contempt dripped from every word. Obviously, Dominy no longer thought he should be assistant commissioner; he thought he should be commissioner. Over the next several months, he lobbied assiduously on his own behalf with Congress. He was only forty-five, and he had been in the Bureau less than half as long as others who were eminently qualified to replace Dexheimer. Not that this was about to deter Dominy—after all, they were merely engineers.

The campaign worked. Dominy fastidiously made a notation in his diary every time he won a Congressman over. Once, after going to see Congressman Keith Thomson of Wyoming only to find him preparing to pay a visit to Interior Secretary Douglas McKay, Dominy wrote approvingly, "His purpose in seeing McKay was to urge the appointment of Floyd E. Dominy as Commissioner." By 1957, Fred Seaton, who had replaced McKay as Secretary, was so besieged with requests to make Dominy commissioner that he had to do something. Seaton's solution was to appoint Dominy "associate commissioner"—a position that, as Seaton conceived it, would be about as meaningful as Vice-President. It had never before existed in the Bureau, and it has never existed since. Seaton, however, thought Dominy would be satisfied with a fancy title, and there he badly misjudged the man. Dominy wanted power. When, after several months, he still didn't have enough of it to suit him, he began making his wish plain to his friends in Congress—and threatening to quit. Fortunately, his wish was their wish, too. One day, Seaton called Dominy into his office for a chat. "The Secretary . . . advised me that he had been getting almost unanimous demands from Senators and Congressmen that I be put in charge of the Bureau's budget presentation and other works with the Congress," Dominy typed in his diary. "He went on to make it plain that he desired to carry out these changes in the Associate Commissioner role with as little discomfort to Commissioner Dexheimer as possible. He asked me to guard against any reaction that would tend to belittle the Commissioner. . . . I assured him that I would be as careful as possible in that connection."

That was hardly the way it was to be. "The whole thing was pathetic to watch," says an old Interior hand who was there. "Dexheimer was like an old bull who's been gored by a young contender and has lost his harem and is off panting under a tree, licking his wounds." The associate commissioner was now in substantial charge of the Reclamation Bureau—Dominy knew it, Dexheimer knew it, nearly everyone in the Bureau could see it. But Dexheimer had nowhere else to go. His whole life had been dams, and now he had reached the pinnacle of the dam-building profession. Any move would have been a step down, a terrible loss of face. One could hardly blame the commissioner for absenting himself as much as possible to deal with "important business" abroad. It was during one such trip—a month in Egypt— that Dominy decided to make his move. The day Dexheimer returned,

Dominy walked into his office and demanded all the authority he had been asking for. If he didn't get it, he would resign. Dexheimer said he would "think about it," and in the weeks that followed he continued to hedge and waffle, relinquishing as little power as he could but relinquishing it anyway, afraid that his popular associate commissioner would really deliver on his threat. Dominy was effectively in command when Congress put poor Dex out of his misery. A number of higher Reclamation officials—Dexheimer included—had been moonlighting at consulting jobs, and when the news reached Congress some members were furious about it. (These were the days when Cabinet members still resigned over ethical transgressions which, today, would be considered almost innocent.) When the commissioner refused to produce a list of offenders, Congress demanded that Eisenhower force him out. On May 1, 1959, Dexheimer, "for personal reasons," announced his resignation as Commissioner of Reclamation. "My decision was not arrived at easily," he said. Floyd Dominy landed in his seat a few days later with a terrific thump.

Most Commissioners of Reclamation were dull, pious Mormons—or, if not Mormon, and pious, then at least dull. Floyd Dominy was a two-fisted drinker, a gambler; he had a scabrous vocabulary and a prodigious sex drive. In interviews, Bureau men tend to be careful, guarded, and obviously suspicious of reporters. Dominy was candid and amazingly open. Most commissioners like to operate within carefully defined parameters, always going by the book. Dominy was freewheeling and reckless, racing yellow lights and burning rubber in three gears. He could be methodical, he worked incredibly hard, he always did his homework—those were the qualities that sustained him through four successive administrations. But he had a self-destructive impulse, a violent temper, and a compulsion to tempt fate. He could, for example, make a lifelong enemy of a very powerful politician over lunch.

The governor of Utah during the early 1960s, George Dewey Clyde, personified, as far as Dominy was concerned, the hypocrisy of conservative Mormons—a faith he privately detested—where the Reclamation program was concerned. Clyde wanted the government to build as many dams as there were sites in his state, but he wanted private utilities to be able to sell the power. Dominy knew the Bureau needed the power to make the projects appear feasible, and besides, he was a Harry Truman Democrat—a warm, if not quite passionate, public-power man. At the National Reclamation Association's annual convention in Portland in 1962, Clyde gave a ringing speech calling for unity among the western states in support of the Reclamation program. He deplored the fact that 40 percent of the members of Congress from the seventeen western states had failed to vote for two big projects the Bureau wanted built. However, Clyde said, the West had a duty to veto "counterfeit" reclamation projects—dams whose purpose was not irrigation but public power. He then went on to single out "a current example in a state neighboring Utah, where a project continues to be pushed by public-power interests which has no reclamation values, whatever." The project which he alluded to, but did not name, was the Bureau's Burns Creek Project in Idaho, which would occupy a hydroelectric site that the company of which Clyde was a puppet, Utah Power and Light, wanted to own itself.

Clyde might as well have impugned the morals of Dominy's daughter. Edward Weinberg, the Interior Department's solicitor, was sitting with Dominy as Clyde spoke. "Dominy just turned maroon," Weinberg recalls. "He said, 'Eddie, you keep me out of jail, but I gotta attack this guy.' Over lunch, he hunkered in a back room redrafting his prepared speech. He showed it to me after lunch, and I said, 'Jesus Christ, you can't say that! They'll crucify you!' 'Let them try' was all he said."

By the time Dominy was scheduled to give his speech, the three thousand conventioners already had an inkling that something portentous was likely to occur. "The title of my speech is 'Crosses Reclamation Has to Bear,' " Dominy began in a sarcastic voice. After making some desultory remarks about the Bureau's routine difficulties, he turned with relish to the subject at hand. "Only yesterday, my good friend, Governor Clyde of Utah, preached the gospel of unity to this association. He warned the West that if it did not unite, the cause of reclamation was in danger. I want to underscore the governor's warning. It is timely and it is true, but apparently the governor's warning fell on some deaf ears. Among those deaf ears, I regret to say, were those of Governor George D. Clyde of Utah." Dominy then tore into Clyde for attacking the Burns Creek Project—"a counterfeit reclamation project," he said acidly, "that was first proposed by those well-known foes of private power, Dwight Eisenhower and his Secretary of the Interior, Fred Seaton." As Clyde sat in the audience redfaced, Dominy's attack became more and more bitter. The delegates were absolutely stunned. "This is the Burns Creek Project which Governor Clyde considers false and a masquerade," Dominy was now shouting. "Is it any wonder that Reclamation's position in the Congress is threatened when the Governor of one of our own Western States attacks a project not even located in his state?"

Nineteen years later, Weinberg was still shaking his head. "No one could believe it," he said. "George Dewey Clyde sat there like he'd been hit by a Buck Rogers ray. Dominy just stood up there smiling serenely. I've never known such nerve. It took the audience thirty seconds to decide whether it dared applaud him at the end of his speech.

"You'd probably have to go back to Andrew Jackson's administration," said Weinberg, his tone full of wonder, "to find another instance where a bureaucrat attacked a sitting governor like that."

Going after a sitting governor was one thing. Going after an entire profession was another, especially if it was a fraternity to which 95 percent of your immediate colleagues belonged. But Dominy was quite capable of that, too.

When the American Society of Civil Engineers held its annual meeting in 1961, they asked Stewart Udall to be the keynote speaker. Udall had a prior engagement and had to decline, and the natural person to speak in his stead was Floyd Dominy. This was the same society, however, whose president had twice written a letter to the President asking that Dominy not be appointed Commissioner of Reclamation— first when Eisenhower appointed him, then when Kennedy reconfirmed his appointment. The reason was both simple and gratuitous: Dominy was no engineer. "When Udall said I should speak in his place," Dominy remembers, "I told him, 'The hell I will!' I wasn't going to speak to a bunch of people who didn't think I deserved my job. I told Stewart, 'You make them send me a personal invitation to give the address. Then I will *consider* whether my schedule permits me to appear.' I didn't think they'd invite me, but damned if they didn't."

When he was introduced and took the lectern, the assembled engineers should have known what was coming. "I'm never fully at ease before so large a group," Dominy began, "but in this one instance I am at ease. I'm at ease because *I* know that *you* know that *I* know that I would never have been appointed commissioner if two Presidents had listened to your organization's advice. Be that as it may," Dominy went on, "I'm here to offer you gentlemen a little edification. I think that both you and your honorable president should go back and read the Reclamation Act, the document that has provided so many of you with jobs. I've read the act many times, and nowhere do I see evidence that it was set up as a job security program for engineers. The act is a land settlement program, and if land settlement were left solely to engineers I think we would still be hunters and gatherers, because it's a lot sexier to design a better mace than it is to plant a garden.

"I'll make you a solemn vow here tonight," Dominy concluded after another few minutes of this. "I promise never to refuse to promote anyone in the Bureau of Reclamation just because he happens to be an engineer."

A few weeks after his speech, Floyd Elgin Dominy was inducted as an honorary member into the American Society of Civil Engineers.

If attacking the governor of Utah took nerve, if taking on the entire engineering profession took gall, then waging ceaseless war against one's superiors would have to be regarded as slightly nuts. But Dominy continually attacked and defied all three of his immediate superiors in the Interior bureaucracy—the Secretary, Under Secretary, and Assistant Secretary—and won nearly every time.

Stewart Udall, who served as Interior Secretary during the Kennedy-Johnson reign, was an enigmatic man. A jack Mormon—a lapsed member of the faith—who hailed from a desert state but assumed office on the threshold of the conservation era, he spent his entire term trying to reconcile his conflicting views on preservation and development, especially when it came to water projects. A smooth politician, handsome, vigorous, and diffident, he was a favorite of Jack Kennedy and a darling of the press; he was continually getting his picture in the papers. There was Stew Udall rafting rivers, Stew Udall climbing Alaskan peaks, Stew Udall and his sometime friend Dave Brower trekking through one of the National Parks. This was the same Stew Udall who wanted to build a nuclear-powered desalination plant off Long Beach to slake Los Angeles' giant thirst; the same Udall who secretly plotted aqueducts carrying water from the Columbia River to the Southwest; the Udall who gave his official, if not private blessing to plans to dam the Grand Canyon. However, what was to Udall a delicate reconciliation of divergent instincts was to Dominy—who held the conservation movement in contempt—a Hamlet-like ambivalence or, even worse, outright capitulation to "posy-sniffers."

To make a strained relationship worse, Udall appointed as his Under Secretary James Carr, a brash, opinionated young Irish Catholic from California who could not help inflaming the ire of a brash, opinionated, and older Floyd Dominy, who happened to be a Celtic-Irish Protestant. To make matters still worse, Udall appointed as his Assistant Secretary for Water and Power a big, dour South Dakota Norwegian named Kenneth Holum, a man whose very essence and style found their exact opposites in Floyd Dominy.

Dominy's battles with Udall were, for the most part, due to dis-agreements on issues; personally, when neither had the other's goat, they liked each other tolerably well. On the other hand his battles with Holum and Carr had more to do with the fact that Dominy de-spised them both as much as they loathed him. Carr had been the legislative assistant of someone else Dominy hated: Congressman Clair Engle of California, who tried repeatedly to get him removed from his job for not favoring California enough. (When Engle died of brain cancer, Dominy told his inner circle, half seriously, that he was re-sponsible. "That cancer in his head was something I put there. He got it arguing with me all the time." Twenty years later, the commissioner still loved to tell about the time he booted the Congressman out of his office.) Personal dislike soon escalated into all-out war: Holum was trying to prevent Dominy from giving a speech; Carr was ordering him not to make a trip; Carr and Holum were trying to give the com-missioner a new secretary who Dominy suspected was their personal spy. By late 1962 or 1963, the feud had grown so intense that it kept the denizens of the Interior building coming to work just to see what would happen next. Before long, Dominy, to the amazement and ex-asperation of Udall, had established a firm policy on dealing with Holum: the commissioner would no longer walk downstairs to speak with the assistant secretary. If the big dumb sonofabitch wished to speak with the commissioner, he could walk upstairs to see *him*. "As his superior I simply had to rein him in from time to time," muttered Holum during a telephone interview, and declined to discuss the sub-ject further. The truth was, however, that Dominy made a fool of Holum much more frequently than Holum made a fool of him. The one time he did—when he and Carr managed to freeze the commis-sioner off the Presidential airplane during one of Kennedy's western tours—Udall returned to his office only to find powerful Congressman Wayne Aspinall on the other end of the telephone, waiting to chew off his ear. After that, Dominy not only got to ride on Air Force One, but he had his *own* fancy aircraft—and his own building.

For years, the world's great amalgamation of engineering talent had been housed in a complex of warehouses, military depots, and glorified barracks outside Denver known today as Federal Center. Then, it was simply known as the Ammo Depot. Thrown up hastily during the war, the Bureau's headquarters, a two-block-long hangar called Building Fifty-six, had neither air conditioning nor many windows. The only source of heat was some undersized radiators spaced many yards apart. Chunks of ceiling calved like icebergs; water dribbled from a hundred leaks. The plumbing sounded as if a team of Russian weight-lifters were banging wrenches against the pipes.

Mike Staus and Dexheimer had tolerated this travesty of a head-quarters, but Dominy would not. He wouldn't keep his cows in there. He was going to get Congress to appropriate money for a new build-ing—a new building that would, in time, become known as the Floyd E. Dominy Building. Under his tutelage, the Bureau's public relations department produced a picture book called *Inside Building Fifty-six*. In it were photographs of rusting pipes, of rotting ceilings suspended over bowed heads, of huddled secretaries typing in overcoats. Accom-panying the pictures was a text that might have described the Sheraton Maui. It was, especially from engineers, a high-class piece of wit. The results, however, were negligible. Udall was frightened of a new build-ing's cost; a few Congressmen even wondered out loud why such a brochure should be produced at public expense. That was enough to make Dominy mad, but not half as mad as he was when he learned that the General Services Administration, run by a close friend of James Carr—the same Jim Carr who had told Dominy that the Bu-reau's headquarters were adequate—erected a new building next door to house the complex's garbage cans.

The federal code stated things plainly enough: the construction of new federal edifices, unless Congress voted otherwise, was left to the discretion of the GSA. Dominy asked his lawyer, Eddie Weinberg, to give him the exceptions to the rule. There were none, Weinberg said—except that, obviously, the GSA had no say-so over the Bureau's dams. "Well, then, it's simple," he told Weinberg, "we'll get the goddamned thing authorized as a dam."

It was a quintessential Dominy solution, brilliant in its simplicity, splendid in its insolence. The building would be authorized as a dam. The Senate Appropriations Committee—Carl Hayden, chairman—would approve money for Dominy Dam, and the dam would meta-morphose into a building. Then it was only a matter of getting the House to agree.

Fascinated by the outcome of this thing, Weinberg was finally per-suaded to go along. Later that year, there was Dominy, with Hayden's blessing already in hand, testifying before his counterpart on the House Appropriations Committee, chairman Clarence Cannon of Missouri. Dominy was eloquent in his blunt Harry Truman style. "I've got a building where icicles practically form in winter," he complained, "and a plane where ice *does* form, right in the carburetor. My people need a decent place to work, and I need a plane that isn't going to fall out of the sky so I can live to see them enjoy it."

Cannon asked, "Do you have any idea when your plane might fall out of the sky?"

"Probably on the very next flight," said Dominy.

"Well, you let me know, then, when you plan to arrange it," said Cannon. "I've got a list of passengers for you."

Then, without further questioning, Cannon approved both of Dominy's requests.

When Carr's friend, the GSA administrator, found out that Dominy had sneaked a new building into a bill that nominally authorized only dams, he was apoplectic. When Carr found out soon thereafter that Dominy had immediately signed a $250,000 design agreement without his approval, *he* was beside himself. Carr forgot, however, that Dominy had been clever enough to make a friend in every strategic place; and there was no more strategic place in the Interior Building than the mailroom.

Stewart Udall was out of town, making a speech, but he was indignant when he learned from Carr how Dominy had operated behind his back. With the Secretary's approval, Carr wrote and signed a letter agreeing to hand the $250,000 back to the Treasury. "When I found out about that," says Dominy, "I called my man in the mailroom. I said, 'I'll take the rap and you'll keep your job—don't you let that letter out of the building.' He promised me he wouldn't. Then I called up Udall that night in his hotel room. I dialed him every fifteen minutes so he wouldn't get away from me. When I got through to him, I said, 'Stew, dammit, you can't do that. It's not $250,000 cash. It's $250,000 credit with the Appropriations Committee. I promised them I'd save that amount of money in the rest of the program. It's their money, not yours. You do this and you're going to run smack into Senator Carl Hayden and Congressman Clarence Cannon.'

"That did it," Dominy chortled. "There was nothing he could do. I got my building. I got my airplane, too. When the GSA chief found out the building was going to be a high-rise, he really squeaked. He sent me three letters of complaint. I didn't bother to answer one."

For years, the Dominy Building—a name it has not yet officially received—was the only high-rise anywhere around Denver. You could see it from far across the Platte River, rising significantly behind the thrusting skyline of downtown. Without knowing what it was, you knew it was a monument to something or someone powerful. "I want it functional, dammit!" Dominy barked at his architects. "I want a building like a dam." What he got is a lot worse. Square as a cinder block, thuddingly banal, it is done in the Megaconglomerate style of the 1960s and 1970s—a J. Edgar Hoover Building without the grotesque semicantilevered overhangs. Despite the cold, the heat, and the feeling of marcescence, Building Fifty-six had a refreshing air of purposefulness, a MASH-like crisis atmosphere. The Dominy Building, by contrast, is fixed, solid, and sealed, as impervious to a rose's scent as to a typhoon—rather like a dam. When it was finished, thousands of Bureau engineers could leave their climate-controlled suburban homes, climb into their climate-controlled cars, and drive to their climate-controlled, windowless new offices, never once encountering the real world.

It is probably pure coincidence that, at about the same time, the mid-1960s, the Bureau—especially its chief—began losing touch with other types of reality.

In the early days, Floyd Dominy had been something of a crusader, if only because he hated being pushed around by politicians and big farmers. Bureau water was by far the cheapest in the West, sold at a fraction of its free-market worth, and if you could manage to irrigate enough land with it you could not only prosper, you could grow rich. Legally, under the Reclamation Act, you could irrigate 160 acres and no more. "We didn't even want them to irrigate that much land," says Dominy. "The law was created to pack as many farmers as possible in a region with limited water. If they could make a living on forty acres, we gave them water for forty. We were talking about subsistence." However, many farmers in Bureau projects were irrigating 320 acres, the result of a liberal interpretation of the act that permitted joint ownership and irrigation of 320 acres by a man and wife. (Married men, it was discovered, made more reliable farmers than bachelors.) In all but the highest and coldest regions of the West, you could make a good living on 320 acres irrigated by subsidized water. If you were in California and raised two cash crops a year with water that cost a quarter of a cent per ton, you could make more money than a lawyer. In 1958, the Fresno Chamber of Commerce published a brochure whose purpose was to lure more farmers to the Central Valley, and which estimated the number of irrigated acres one had to plant in various crops to support a family. The figure for oranges was twenty to thirty acres; for peaches, thirty to forty acres; for grapes and raisins, forty to fifty acres; for figs, sixty to eighty acres. Even a hundred and twenty acres of cotton and alfalfa, comparatively low-value crops, could support a family if you had Reclamation water.

Rumors abounded, however, of corporate farmers illegally irrigating thousands of acres with the super-subsidized water—by inventing complicated lease-out lease-back arrangements, by controlling excess land through dummy corporations, by leasing from relatives, and so on. It is unclear how much the Bureau knew about this and how exact its knowledge was; what *is* clear is that it did little or nothing to end it. Even a self-proclaimed populist like Mike Straus was afraid to tangle with the giant California farming corporations and the politicians they helped elect. "Straus huffed and puffed about the acreage limit," Dominy said later, "but he didn't do a damn thing to enforce it." (This is largely but not completely true. One of Straus's worries, which turned out to be well founded, was that the Corps of Engineers, unencumbered by social legislation or much of a social conscience, would gladly step in and replace the Bureau as the major water developer of the West if the Bureau began cracking down too hard on violators.)

At first, Dominy was self-righteous about enforcing the Reclamation Act. In 1954, when the Corps of Engineers, with the acquiescence of Interior Under Secretary Clarence Davis, tried to do exactly what Mike Straus feared—let water from its two biggest California reservoirs run free of charge onto the lands of two gigantic farming corporations, the J. G. Boswell Company and the Salyer Land Company— he was apoplectic. "Special Assistant Frye showed the [Under Secretary's] letter [of acquiescence] to me confidentially," Dominy wrote in his professional diary on February 4, 1955. "I blew my top and stated emphatically the detrimental effect that would have on Reclamation's ability to conclude repayment contract negotiations . . . with other groups of water users. [A] very plausible legal basis can be made that Congress has directed that irrigation water available as a result of Army construction should be sold pursuant to Reclamation law."

Later, Dominy, now chief of the Irrigation Division, paid a visit to the Boise regional office and learned, he wrote in his diary, "that there is apparently a rather widespread evasion of the incremental land provisions of the Columbia Basin Project Act." (According to the incremental-land-value provisions of the act, beneficiaries newly supplied with Bureau water are supposed to sell their excess lands at a price reflecting their worth *before* the Bureau water arrived. Otherwise, speculation would be as rampant as in the old days of the Homestead Acts; people with an insider's knowledge of future projects could buy land in the project area for $10 or $20 an acre and sell it later for fifty times as much.) "I made it plain," Dominy wrote, "that it was the Bureau of Reclamation's responsibility to either (a) energetically enforce the law or (b) ask Congress to repeal it." When Assistant Interior Secretary Aandahl privately expressed extreme reluctance to prosecute the violators, Dominy wrote, "I am happy to report that this is the first time in my 24 years of Government work that I have heard a top administrator say that he was unwilling to take action to enforce a law which he was sworn to uphold and which comes under his jurisdiction." Ultimately, there was an FBI investigation, a prosecution, and a conviction in the Columbia Basin case. The sentence was a fine of $850. "The sentence made you feel like a fool," says Gil Stamm, who worked on the case with Dominy and was ultimately to succeed him as commissioner.

It did gross injury to Floyd Dominy's image to be made to look like a fool. That may be the main reason why, as commissioner, his indignation over violations of the Reclamation Act appeared to evaporate like a summer cloud. Under Dominy's tenure, the one serious example of enforcement in the Bureau's career did take place: the breakup of the huge DiGiorgio Company holdings in California after it was proved that the lands were illegally receiving subsidized water. But the main instigator in that action was not Dominy but Frank Barry, the first Interior solicitor under John Kennedy. And though it is true, as Dominy insists, that the record of enforcement during his reign was at least as good as any other commissioner's, that isn't saying much, because the record of enforcement over eighty years has been almost nil. Not only that, but the violations had become more frequent and worse by the time Dominy was appointed. It wasn't until the administration of Jimmy Carter that a serious attempt was even made to find out how bad the violations were. The conclusion was that they had multiplied considerably after the Second World War and reached their apogee about the time Dominy became commissioner. As it happened, the Carter investigation found that the vast majority of illegalities were occurring in California and Arizona. But the senior Senator from Arizona was Dominy's best friend in Congress, Carl Hayden. In California, the Congressmen who represented the region where most of the acreage violations were taking place were three Dominy stalwarts: Bizz Johnson, John McFall, and Bernie Sisk. None of those gentlemen ever showed much interest in enforcing the acreage limitations of the Reclamation Act. They did, however, display a passionate interest in new dams, and their attitudes became Floyd Dominy's attitudes the longer he remained in office. He had begun as a crusader, a person who at least appeared to possess a sense of fairness and justice, a non-engineer whose outlook was basically agrarian. He ended his term as a zealot, blind to injustice, locked into a mad-dog campaign against the environmental movement and the whole country over a pair of Grand Canyon dams.

The fact is that Dominy *knew* that scandalous violations of the acreage limit were occurring right around Los Angeles—for example, that the Irvine Ranch, one of the largest private landholdings in the entire world, was illegally receiving immense amounts of taxpayer-subsidized Reclamation water—and did absolutely nothing to stop it. When he was shown the list of violators, compiled during a months-long secret investigation, he put it in his desk drawer and never looked at it again. Though he went to great lengths to try to disprove it, Dominy knew that the Bureau was opening new lands for crops which farmers were paid not to grow back east—cotton being the prime offender. The Bureau could easily have refused to supply new water to a region until it could demonstrate that its crop patterns would not make the nation's agricultural surpluses worse, but its response, under Dominy, was to launch a belligerent campaign to deny that the problem existed.

When Dominy appeared *not* to realize was that these three syndromes, often occurring at once—farmers illegally irrigating excess acreage with dirt-cheap water in order to grow price-supported crops—were badly tarnishing the Bureau's reputation. By the 1960s, the Reclamation program was under attack not only from conservationists but from church groups (who objected to its tacit and illegal encouragement of big corporate farms), from conservatives, from economists, from eastern and midwestern farmers, and from a substantial number of newspapers and magazines that had usually supported it in the past—even from the Hearst papers in California. Dominy was not so blind that he didn't see this; his fatal mistake was in believing that the protest and indignation amounted to sound and fury signifying nothing: Dominy had a peculiar adeptness at denying reality. And the conservation movement was the reality he liked least of all.

Throughout its history, the conservation movement had been little more than a minor nuisance to the water-development interests in the American West. They had, after all, twice managed to invade National Parks with dams; they had decimated the greatest salmon fishery in the world, in the Columbia River; they had taken the Serengeti of North America—the virgin Central Valley of California, with its thousands of grizzly bears and immense clouds of migratory waterfowl and its million and a half antelope and tule elk—and transformed it into a banal palatinate of industrial agriculture. The Bureau got away with its role in this partly because its spiritual fathers, John Wesley Powell and Theodore Roosevelt, happened to be two of the foremost conservationists of their day—a heritage which, in the right hands, might have all but immunized it against more modern conservationists' attacks.

The Bureau's response to the rising tide of conservation, however, was to let them eat cake. It might have learned some valuable lessons from the Corps of Engineers, which at least knew how to build a Trojan horse. While the Corps was preoccupied with such mightily intrusive wonders as the Tennessee-Tombigbee Waterway and its county-size reservoirs in the South, it was proclaiming the 1970s the "Decade of the Environment," publishing a four-color magazine devoted to wild rivers and fish and swamps, and holding regular palavers with its environmental adversaries to throw them off guard. General John Woodland Morris, who became chief of the Corps in 1970, is regarded by many conservationists as the most brilliant and effective adversary they ever met. Some of the same adjectives are used to describe Dominy—tough, brilliant, formidable—but it is odd how seldom anyone refers to him as "effective."

Dominy's problem stemmed from a fatal sin—pride—and a fatal misjudgment: that his despised adversary, David Brower, was the corporeity of the conservation movement—its unanimous voice, its unified soul. To Dominy, anyone who objected to any single thing the Bureau wanted to do was "a Dave Brower type." He failed utterly to understand that Brower had always been a fringe figure in the conservation movement—respected, admired, but not necessarily followed or trusted or believed. Jack Morris of the Corps understood that, as a rule, conservationists enjoyed widespread public respect—that an endorsement from one conservation organization was worth the endorsements of a hundred Chambers of Commerce. He knew that when it came to a conflict between nature and civilization, millions of Americans automatically turned to the conservation groups for guidance. If such an organization endorsed a compromise proposal, general opposition could die like a puff of wind.

But the last thing Floyd Dominy was going to do was seek a compromise with conservation groups. If he went out of his way at all, it was to antagonize them. On February 13, 1966, he gave a speech in North Dakota lambasting the principle that certain rivers, or portions of rivers, ought to be set aside as "wild and scenic." Calling the undammed Colorado River "useless to anyone," Dominy harrumphed, "I've seen all the wild rivers I ever want to see." The speech elicited a testy letter from the state's fish and game commissioner (who was hardly a Dave Brower type) to Stewart Udall, suggesting that Dominy badly needed some edification about changing American values—not to mention the importance of rivers and wetlands to waterfowl. "Floyd, it seems to me that Commissioner Stuart has a point," Udall wrote in a short memo with a copy of the letter attached. "My Secretary's becoming a Dave Brower type," Dominy sneered to his comrades in arms. A few months later, ignoring his advice, Congress passed the Wild and Scenic Rivers Act.

Under Dominy, the Bureau lost touch with reality so completely that it developed an uncanny knack for snatching defeat from the jaws of victory. At the northern end of Lake Havasu, a few miles south of Needles, California, it had inadvertently created a large freshwater wetland known as Topock Marsh. Migrating ducks and geese that were evicted from the Central Valley soon discovered the marsh and descended on it by the tens of thousands during their winter sojourns. By the late 1940s, Topock Marsh had become one of the most important man-made attractions on the Pacific Flyway, and the Bureau, had it had any sense, would have graciously accepted its share of credit and basked in it. The grasses and duckweed, however, were phreatophytes, and consumed valuable water that could have been sold to Imperial Valley farmers for $3.50 an acre-foot. As a result, the Bureau began trying to dredge the marsh in 1948; when at first the dredging didn't work, it spent millions of dollars and stepped up its efforts and pursued them so relentlessly that by the 1960s about 90 percent of the food grasses were gone. The marsh's visiting waterfowl soon diminished from forty or fifty thousand a year to a few hundred or thousand at most.

Dominy's Bureau regarded the operation as a "success," failing utterly to recognize the public relations catastrophe into which it had happily stepped. Even Imperial Valley farmers, who had so much water to waste that some of them applied ten or twelve feet per year to their crops, were opposed to the dredging because they liked to shoot ducks. Ben Avery, a widely read outdoor columnist for the *Arizona Republic*—a newspaper never known to oppose water development unless it was California's—adopted Topock Marsh as his personal crusade and made a point of savaging the Bureau several times a year. In June of 1966, one of his columns finally caught Dominy's attention. "I believe we will have to take Avery on," he wrote to his regional director, Arleigh West, "or face up to the realities [sic] that there is a great deal of truth in what he is saying." In other words, Dominy *knew* Avery was right. He *knew* that Topock Marsh was pitiful compensation for all the habitat the Bureau and Corps had ruined. He *knew* that the marsh would reappear unless the Bureau continued to spend millions of dollars trying to annihilate it. But which course of action did he choose? The Bureau, he decided, was going to deny everything Ben Avery said and continue demolishing the marsh.

Stewart Udall was upset over the Topock Marsh situation, and since the marsh was being eradicated for the sake of California—not Arizona—he ordered Dominy to do something about it. In typical fashion, Dominy's response was to try to make an end run around Udall, through the Congress. Though he was nominally Dominy's boss, Udall didn't like tangling with his two-fisted commissioner; that was the reason he had John Carver on his staff. Small, tough, and profane, built like a bantamweight prize-fighter, Carver had been hired to be Udall's all-purpose troubleshooter. Manhandling Dominy, however, was turning into his full-time job.

"The summit meeting was to take place in Udall's office," remembers John Gottschalk, who was then the director of the Fish and Wildlife Service. "It was a good choice—the Secretary was absent, but the trappings of authority would impose themselves. I was a little late in arriving, and as I was walking down the hall I could already hear Carver and Dominy at each others' throats. God Almighty! Were they screaming at each other! When I walked in they were standing at opposite sides of Udall's desk just like a couple of football players facing off. They were pounding the table with their fists. Dominy's face was beet-red. I remember him yelling, 'What do you want me to do? Resign my fucking job?' And Carver was shouting back, 'We want you to get on the *team*, Floyd! We don't want you to resign. We want you to stop throwing tantrums and get on the goddamned *team*!'

"I just stood there transfixed," says Gottschalk. "I didn't know whether to try to break it up or slink out the door. It went on like that for another fifteen minutes until Dominy gave up. I remember exactly what he said. He yelled, 'You realize you're asking me to go against every sound precept of water management for a bunch of goddamned birds and fish!' And then he barged out the door like a Sherman tank."

By the mid-sixties, Dominy finally had realized that the conservation movement was a serious enough threat to the Reclamation program that he would have to acknowledge not only its existence, but its political power. At first he had paid it as much attention as he would a flea, but now he began to go after the flea with a hydrogen bomb. In one issue of *Audubon* magazine—which had a circulation far smaller than it does today—the magazine's bird-watching columnist, Olin Pettingill, made a derogatory reference to the Bureau in an article which, for the most part, was about curlews and gallinules. Pettingill remarked that the Bureau's Nimbus Dam, on the American

River east of Sacramento, "has ruined what once were spawning grounds for salmon and steelhead rainbow trout"—an observation that happens to be entirely true. That was the sum total of Pettingill's criticism: one sentence in a two-thousand-word article about birds. However, as far as Dominy was concerned, the magazine was guilty of delivering "a gratuitous slap in the face." He wrote to his regional director, "We think it would be opportune and worthwhile to work with the Sacramento newspaper in the development of a feature story on the lengths to which Reclamation has gone . . . to enhance the fishery and wildlife resources of the Central Valley. An ideal situation would be for such a story to be used in the *Bee* on the opening day of the Audubon Society convention in Sacramento, to be followed up by an editorial."

Two interesting questions are raised by Dominy's response. One is whether he really had enough influence with the Sacramento *Bee* to enlist it in an orchestrated campaign to perfume the Bureau's reputation. One also wonders what he had in mind when he spoke of Reclamation projects "enhancing" fish and wildlife habitat in the Central Valley. By the mid-1960s, nearly 90 percent of the valley's wetlands habitat was gone, almost entirely because of irrigation farming, and wetlands were by far the most important natural feature in all its five-hundred-mile length; the valley was once the winter destination of a hundred million waterfowl cruising the Pacific Flyway, and now their numbers were reduced to five or six million, jammed onto refuges or forced to scrounge a meal in unwelcoming farmers' fields. The Sacramento–San Joaquin river system once had six thousand miles of salmon-spawning streams, but by the mid-1960s there were perhaps six hundred miles left, and it was the Bureau's dams, cemented across rivers low down in the foothills, that blocked the salmon most effectively. So what had the Bureau done to "enhance" fish and wildlife resources? At best, it had created a series of slack-water reservoirs that were host to such rough fish as catfish, crappie, and bass, plus some trout and an occasional landlocked salmon. The reservoirs were useless to ducks and geese, which couldn't feed in their deep waters and would be driven mad by the powerboats anyway.

Those reservoirs, however, were the only thing Dominy could have had in mind, unless he had completely lost touch. To him, it seemed, nothing in nature was worthwhile unless it was visited by a lot of people. If it was a pristine river, accessible only by floatplane or jeep or on foot, navigable only by whitewater raft or kayak or canoe, populated by wily fish such as steelhead that were difficult to catch, then it was no good. But if the river was transformed into a big flatwater reservoir off an interstate highway, with marinas and houseboats for rent—then it was worth something after all.

There was, for example, Lake Powell. Before Glen Canyon Dam had been built, that stretch of the Colorado River was one of the remotest, most inaccessible places in the United States. Only a few thousand people had seen it. Utterly unlike the turbulent reaches of the Grand Canyon, Glen Canyon was a stretch of quiet water drifting sinuously between smooth, rainbow-colored cliffs. Labyrinthine and cool, some of the canyons were as lush as a tropical forest, utterly incongruous in the desert. All of this was drowned by Lake Powell, but to demonstrate how nature had actually been improved, Dominy decided to publish a book called *Lake Powell: Jewel of the Colorado*. He even decided to take the photographs and write the text himself. "Dear God," he wrote on the inside cover, "did you cast down two hundred miles of canyon and mark: 'For poets only'? Multitudes hunger for a lake in the sun." He went on:

> How can I describe the sculpture and colors along Lake Powell's shores? Over eons of time, wind and rain have carved the sandstone into shapes to please 10,000 eyes. The graceful, the dramatic, the grand, the fantastic. Evolution into convolution and involution. Sharp edges, blunt edges, soaring edges, spires, cliffs, and castles in the sky. . . . Like a string of pearls ten modern recreation areas will line Lake Powell's shores, with names that have the tang of the Old West. . . . Feel like exploring? Hundreds of side canyons—where few ever trod before the lake formed—are yours. . . . You have a front-row seat in an amphitheater of infinity. . . . Orange sandstone fades to dusky red—then to blackest black. . . . There is peace. And a oneness with the world and God. I know. I was there.

Dominy's war against the conservationists may have given him some satisfaction, but, from his point of view, it was hardly time well spent. No public figure would be as hated by the environmental movement until James Watt came along a decade later. His blind insistence on building dams in the Grand Canyon—not just dams, but cash register dams whose purpose was to generate income to build *more* dams—won him the wrath of *Reader's Digest* and *My Weekly Reader*; his habit of making end runs around federal laws and regulations by begging special relief from Congress did not endear him to those whose laws he was circumventing; and hundreds of well-placed officials in Washington, many within his own building, were laying for him.

Despite all this, in the late 1960s Dominy was as entrenched as any bureaucrat in Washington. The main reason was his relationship with Senator Carl Hayden of Arizona, the chairman of the Appropriations Committee, the most powerful man in legislative government. It was the relationship of a fawning nephew and a favorite uncle—the kind of relationship young Lyndon Johnson enjoyed with Sam Rayburn—and it gave Dominy an authority, an insolence, an invulnerability scarcely anyone else enjoyed.

When Carl Hayden was in his late eighties, senile, half blind, half deaf, confined to a hospital bed half the time, Floyd Dominy all but served as chairman of the Appropriations Committee when dam authorizations came around. He managed this by telling Hayden exactly what he wanted him to say—by actually writing dialogue for the two of them to recite. He would go to Hayden's office, sit down with his legislative aide, Roy Elson, and write the questions he wanted Hayden to ask him; then he would go back to his own office and write the answers. It is unclear whether he did the same for other witnesses. The Hayden-Dominy scripts were of dubious enough ethical propriety for Dominy to keep them locked in the Bureau's sensitive files, their existence known to only a handful of aides. Old, frail, and sick as he was, Hayden was still a man no one wanted to cross, and Dominy, knowing this, basked as long as he could in his failing light. "When you walked into Dominy's office," says John Gottschalk, "the first thing you saw was a huge framed picture of Hayden and Dominy getting off a plane in Hawaii all decked out in leis. Hayden's inscription went something like this: 'As this photograph was being taken I was thinking to myself that Floyd Dominy is the greatest Reclamation Commissioner who ever lived.'

"It was powerful medicine," says Gottschalk. "There's no member of Congress today who's nearly as powerful as Hayden was then. You'd walk in there to complain about something the Bureau did and see that picture and say to yourself, 'How the hell am I going to go up against this man and win?' "

Dominy was, of course, much too canny to put all of his eggs in Carl Hayden's basket. In the House, he maintained the most cordial of relations with Wayne Aspinall, the chairman of the House Interior Committee Aspinall, a former schoolteacher from Palisade, Colorado, with a nasty disposition and a religious conviction that only the Bureau of Reclamation stood between the West and Armageddon, would say that Floyd Dominy was "not only the best Reclamation Commissioner

I have ever known, but the *only* good Reclamation Commissioner I have known." Besides cultivating the powerful, Dominy, for the most part, did a marvelous job of concealing his political prejudices from the world. He could get on famously with Frank Church, the liberal Senator from Idaho, and get on just as famously with William Egan, the right-wing governor of Alaska. If a Congressman didn't get on famously or even politely with him, Dominy had little compunction about taking revenge: a dam project in his district might suddenly become unfeasible, a weather modification program might move somewhere else. "He pulled money in and out of those Congressmen's projects like a yo-yo." Loved by some, feared by many, respected by all, Dominy seems to have had only one enemy of consequence in the whole Congress—Senator Henry Jackson of Washington. But Jackson knew better than to take his enmity too far.

And Dominy could be jovial, amusing, a lot of fun. Reclamation parties were legendary in Washington—hardly what one would expect in a hotbed of Mormon engineers. He could beat the conservationists at their own game. When the Sierra Club and the Wilderness Society and others complained bitterly that a finger of Lake Powell would extend to Rainbow Bridge, a spectacular natural arch in Utah, leaving a stagnant, fluctuating, man-made pool of water under one of the nation's scenic wonders, Dominy went to see the place himself—on foot, with a mule. It was a grueling twenty-mile hike in desert heat to the arch, a trek so tough the mule almost didn't make it. Later, he flew a bunch of conservationists in by helicopter so they could see it themselves, taking care to ask each one whether he had been there before. Almost none had. Dominy used that fact to great advantage in testimony before Congress. Not only had they never seen what they so passionately wanted to protect, he said acidly, but they wanted him to erect a *dam* to keep the waters out. A dam! After regaling the committee with his story, Dominy got a special exemption from the federal law prohibiting significant man-made intrusions in national monuments. Today Rainbow Bridge is visited mainly by overweight vacationers clambering out of houseboats and trudging up to stare briefly at the arch.

He had a politician's way with names. On visits to the Bureau's dams, he greeted maintenance people whom he had met briefly years before, and he even knew the names of people he had never met. When the University of Wyoming awarded him an honorary degree, he was invited to dinner at the home of Gene Gressley, the director of its American Heritage Center. He had never met Gressley, nor his family, but when he walked in the door he knew all of Gressley's children by name. When, during an interview, I reminded Dominy of the incident and told him how impressed Gressley said he had been, his response seemed somehow predictable: "Who's he?"

One of his former aides said Dominy liked people the way we like animals—we like them, but we eat them. His employees laughed at his antics, admired his guts, profoundly respected his abilities, and were scared half to death. He could be sadistic, and he would carry a grudge to his grave. As soon as he became commissioner, he tried to fire all of his regional directors—not on the basis of incompetence, necessarily, but because they had been appointed by Dexheimer. But he couldn't dislodge the one whose head he wanted most, Bruce Johnson in the Billings office, because Johnson had strong political support. The reason he wanted to fire Johnson so badly is that he had refused to arrange a "date" for Dominy with his secretary, whom Johnson was courting himself. Unable to depose him, Dominy tried to hound Johnson out—ridiculing him mercilessly, intimidating him, humiliating him. Johnson took it for several years and finally quit.

He hated weakness, but he needed a weak person to serve as his whipping boy, and he had one in Arleigh West, his regional director in Boulder City. "Arleigh was his Sancho Panza," says Pat Dugan, one of the few whom Dominy didn't cow. "He had a rough life. He brought out everything that was sadistic in Floyd." When West was in Washington, Dominy commandeered his hotel room as his trysting spot, and there were evenings when poor Arleigh found himself out window-shopping, waiting for Dominy to finish. He had someone in Denver—another weak man, a top-level aide—whom everyone referred to as the "Official Pimp." His responsibilities went beyond procurement. When a public relations flack leaked the story of how Dominy had gotten Congress to give him a new airplane, thinking he was doing Dominy a favor—after all, he was always telling those kinds of stories on himself—the Commissioner was beside himself. He was in the middle of a meeting with some Colorado bureaucrats at the time. "You fire that son of a bitch," he yelled to the Official Pimp in the presence of the astonished bureaucrats. "We can't fire him," said the Pimp, "he's civil service." "You fire him," roared Dominy, "or I'll can your goddamned ass, too!"

It wasn't his blindness, his stubbornness, his manipulation of Congress, his talent for insubordination, his contempt for wild nature, his tolerance of big growers muscling into the Reclamation program—in the end, it wasn't any of this that did Dominy in. It was his innate self-destructiveness, which manifested itself most blatantly in an undisguised preoccupation with lust. His sexual exploits were legendary.

They were also true. Whenever and wherever he traveled, he wanted a woman for the night. He had no shame about propositioning anyone. He would tell a Bureau employee with a bad marriage that his wife was a hell of a good lay, and the employee wouldn't know whether he was joking or not. He preferred someone available, but his associates say he wasn't above paying cash. "The regional directors were expected to find women for him," says one former regional director. "It always amazed me how he carried on in the light of day. He was opening himself up to blackmail, but somehow he always seemed immune." The Bureau airplane was known, by some, as the "Winged Boudoir in the Sky."

As he bullied weak men, Dominy preyed on women whom he considered easy marks. According to one regional director, Felix Sparks, the head of Colorado's Water Conservation Board, was married to a woman who occasionally overindulged, so Dominy went right after her. In time, an indignant Mary Sparks refused to attend any party where Dominy threatened to show up. Sparks, one of the most decorated veterans of World War II, might have been expected to punch Dominy in the jaw. Everyone, however, seemed to humor him. "He's just being *Floyd*," they would say. "You know how *Floyd* is." "He's just a little drunk. Ignore him."

Alice Dominy must have known. Her life was insulated, she rarely went with him on trips, but for years everyone suspected that she knew. And there came a day when she had to find out for sure. She drove into town to the hotel where according to the rumors, he liked to conduct his trysts. She took the elevator upstairs, mustered her courage, and knocked on the door. A woman opened up. Floyd Dominy, her husband, was in the back of the room. "He just told her to go home and mind her own business," says one of Dominy's confidants. "And she was of that era where that's what women did. I don't know how he rationalized it. He probably said, 'Well, lots of people commit adultery.' He had a talent for rationalizing anything.

"Alice was sweet. She was a dear lady. It broke your heart to see her treated that way."

Dominy did not even aspire to discretion. He bragged about his exploits. He taunted his assistants with remarks about their wives. He ordered them to find him women. It seemed as if he simply couldn't help himself. He could testify before Congress on a half bottle of bourbon and two hours of sleep, he could throw Representative Clair Engle out of his office, he could learn more about the Reclamation program than any person alive—he was tough, ferociously disciplined, indomitable. But he was also compulsive, addicted, a fool for lust—and exposed himself quite recklessly to full view. "I'm not sure what Dominy is better remembered for," says one Washington lawyer who knew him well, "having been Commissioner of Reclamation or having been the greatest cocksman in town."

"I've tried to psychoanalyze him," says Pat Dugan, "and I don't think he ever believed that his playing around would get him in real trouble. He got away with so much that after a while he must have decided he was immune."

But he wasn't.

The man assigned to tell Floyd Dominy that he was fired was a young, intense, middle-level Interior bureaucrat barely thirty years old, a fire-breathing evangelical Christian from Wyoming named James G. Watt. The order came directly from the newly inaugurated President of the United States, Richard M. Nixon. At Nixon's behest, the FBI had run its customary investigation of top federal officials to look for improprieties and had come back with a file on Dominy that was inches thick. ("The FBI knows every woman I've ever fucked," Dominy once confessed to me.) "He didn't act surprised when I told him," Watt remembered. "I think he knew it was coming. We had decided to let him stay on a while longer so his pension could vest, and he acted grateful about that. I was in awe of this man. Everyone was. I was half his age. But he took the news very mildly. I can remember feeling very, very relieved."

When Dominy was himself relieved, he retired to his cattle ranch in the Shenandoah Valley, leaving his twenty-five-year Reclamation career behind him as if it had never occurred. "When I quit something," he said, "I really quit it." Once in a while he could be enticed into a lucrative consultancy—in 1981, he was hired by Egypt to help draft a solution to the grotesque drainage problems created by the Russian-built Aswan High Dam—and he drove to Capitol Hill now and then to testify against the likes of a Hells Canyon National Monument (which would preclude more dams on the lower Snake River); mostly, though, he preoccupied himself with enshrining his reputation and with his cows. In 1979, he was named Virginia Seed Stockman of the Year, a fitting title: he had been proclaimed the state's preeminent stud expert.

Dominy's reputation and legacy are more problematical—at least as complex as the man himself. In *Encounters with the Archdruid*, John McPhee portrays him as a commissioner who led Reclamation on a terrific binge, plugging western canyons as if they were so many basement leaks. His reputation, even today, is outsize; he is often talked

about in Washington, and in the conservationists' annals of villainy
he remains a figure as large as, if not larger than, Ronald Reagan's
Interior Secretary, the same James Watt. Watt, however, hopped
around so much with his foot in his mouth that he didn't really have
a chance to do much that the environmental movement regarded as
awful. But Dominy presided over Glen Canyon Dam, over Trinity Dam,
over a dozen other big dams, over the federal partnership with California in that state's own water project, which dammed the Feather
River and allowed Los Angeles' explosive growth to continue, and with
it its appetite for even more water. Those enamored of such giant
engineering works were at least as sorry to see Dominy go as the
conservationists were thrilled; no successor, they believed, could ever
hope to equal him as a master tactician in Congress, as a fiercely
committed believer in the cause of reclaiming the arid West.

On balance however, Floyd Dominy probably did the Bureau of
Reclamation and the cause of water development a lot more harm
than good. That, at least, is Daniel Dreyfus's assessment. Brilliant and
hardheaded, the Bureau's house intellectual—and a native New
Yorker—Dreyfus was the only person it had who could sit down with
an influential Jewish Congressman from New York City, trade some
urban banter and rabbi jokes, and convince him that he ought to vote
for the Central Arizona Project. He left, in part, because of Floyd Dominy. "You could take so much of him," Dreyfus remembered one day
in 1981, sitting in his office at the Senate Energy Committee, where
he had gone to become staff director. "He got to be like a stuck record.
The same damn stories about himself, the same fights with the same
people over and over again. The mood of the country was changing,
but Dominy refused to let the Bureau change. You got the feeling that
you belonged to the Light Brigade." The loss of Dreyfus was especially
ironic, because the chairman of the Senate Energy Committee was
Henry Jackson, Dominy's one powerful enemy from a western state.
In Dreyfus, Jackson had acquired the one person on earth who knew
as much about the Bureau and its work as its commissioner.

Jim Casey, the Bureau's deputy chief of planning, worked under
Dreyfus and also left in disgust. Like Dreyfus, Casey had become cynical about the whole Reclamation program, but he couldn't help retaining his loyalty to the Bureau. Once, in the early 1970s, when a
friend sent a young engineering graduate over for job advice, Casey
suggested that he apply at the Bureau, and the young man made a
sour face. "He told me that the Bureau of Reclamation was a disgrace,"
Casey remembered. "And I got mad at him for saying that, but here
was a guy fresh out of one of the top engineering schools—the kind of

guy who once would have loved to work for the Bureau—and he said it was nothing but a bunch of nature-wreckers out to waste the taxpayers' money. It was Floyd Dominy who gave it that reputation. You couldn't convince him that the Bureau's pigheadedness about things like Marble and Bridge Canyon dams was turning the whole country off. After he'd told me his Rainbow Bridge story for the seventh time and how he'd licked the conservationists, I said, 'Well, you won that one, but you haven't won too many others lately.' He said. 'What haven't I won?' And I said, 'Well, they licked you pretty good on Marble and Bridge Canyon.'

"You know what his answer was? 'My Secretary turned chickenshit on me.' The man was blind. He went completely blind."

These are mere opinions, but the record speaks for itself. The Central Arizona Project which Dominy finally managed to build is a medium-sized dwarf compared with the Pacific Southwest Water Plan he had planned, and he had to sacrifice the last years of his career to the effort to get it authorized. Today, few of the other grand projects conceived under him exist. There is no Devil's Canyon Dam on the Susitna River, no Texas Water Plan, no Auburn Dam, no Kellogg Reservoir, no English Ridge Dam, no Peripheral Canal, no additional dams in Hells Canyon on the Snake River, no Oahe and Garrison diversion projects. Dominy wanted to move the Bureau's activities into the eastern United States, because he came to believe that irrigation often makes better sense in wetter regions than in emphatically dry ones, and also because he wanted to invade the Corps of Engineers' domain in order to retaliate for the Corps having encroached on the Bureau in the West. But all of those plans—for irrigation projects in Louisiana, for a series of reservoirs in Appalachia set around new industrial towns—came to naught. The legacy of Floyd Dominy is not so much bricks and mortar as a reputation—a reputation and an attitude. The attitude is his—one of arrogant indifference to sweeping changes in the public mood—and it is probably the foremost obstacle in the Bureau of Reclamation's way as it tries to play a meaningful role in the future of the American West.

Actually, there is one more legacy, one of flesh and blood. In Dominy's office at his Shenandoah farm, next to his huge commissioner's desk, is a photograph of him with his son on a boat speeding across Lake Powell, arms around each other. Remove the film of thirty years and Floyd could be Charles Dominy's twin—they look that much alike. In the 1980s, Charles was the chief of the southeastern district of the Army Corps of Engineers. He was turning the Savannah River into a continuous reservoir, channelizing countless miles of meandering

streams and creeks, draining the last wild swamp and forest lands of
the wet Southeast for soybean farms. He was also plotting to revive
the cross-Florida barge canal—a casualty of the same administration
that deposed his father.

A couple of hours earlier Dominy had been lambasting the Corps,
saying it "has no conscience." As he saw his guest look at the photograph, however, he broke into a proud grin. He said, "That boy is going
to be Chief of Engineers someday."

CHAPTER EIGHT

An American Nile (II)

Nineteen twenty-eight, the year the Hoover Dam legislation was passed, was a milestone year in Arizona in another sense. The population went past 400,000—the largest number of people who had lived there in approximately seven hundred years.

The original 400,000 Arizonans (that is an outside estimate; the number may have been somewhat smaller) were, for the most part, members of the Hohokam culture, a civilization that thrived uninterrupted near the confluence of the Gila, Salt, and Verde rivers for at least a thousand years, until about 1400, when it disappeared. The confluence of Arizona's only three rivers occurs in the hottest desert in North America, a huge bowl of sun now occupied by modern Phoenix and environs. Average summer temperature is 94 degrees; average annual rainfall is just over seven inches. There are far more hospitable places in the state, such as the cool Ponderosa-clad Mogollon Rim, but archaeologists surmise that the inhabitants of Arizona's higher and wetter regions drifted down to join the Hohokam in the latter days of their realm; something about the desert proved irresistibly attractive. The lure was probably food, which the Hohokam rarely lacked. They were the first purely agricultural culture in the Southwest, if not all of North America. Midden remains, well preserved by the desert's dryness and heat, suggest that the Hohokam rarely hunted, or even ate meat; their copious starch and vegetable diet was supplemented only occasionally by a bighorn sheep, antelope, raven, or kangaroo rat. Sometimes they ate sturgeon. That sturgeon bones have been found amid the Hohokam ruins suggests a Gila River considerably fuller and more constant than the ghost river whites have known—a river that, even before its headwaters were dammed, usually ran underground. And this, in turn, suggests a possible reason for the Hohokam's demise: that the climate was considerably wetter during the centuries their civilization flourished, then turned suddenly dry.

Whatever happened, the Hohokam, by A.D. 800, had already established a civilization that rivaled the Aztec, Inca, and Maya farther south. They were good builders, using rafters for houses and I-beams to create ancestral skyscrapers four stories high. They lived in small cities; the ruins of one of them, Pueblo Grande, occupied a large piece of land just about where downtown Phoenix is today. Superb flint and stone masons and excellent potters, they also worked beautifully with shells; they may have traded with people living on the Mexican coasts. For sport, they built enclosed ball courts very much like those of the Maya, who probably gave them the idea.

When it came to irrigation, however, the Hohokam were in a league by themselves. The largest of the canals they dug was fifteen miles long and eleven yards wide from bank to bank; like the other main canals, it had a perfectly calibrated drop of 2.5 meters per mile, enough to sustain a flow rate that would flush out most of the unwanted silt. There were dozens of miles of laterals and ditches, implying irrigation of many thousands of acres of land. Because of the dry climate and the provenance of the irrigated land, the Hohokam should have enjoyed good health; they made superior weapons; they were more populous than any culture around. Why then should they disappear? It is hard to imagine a civilization covering thousands of square miles and comprising hundreds of thousands of people just vanishing, but according to Emil Haury, an archaeologist who became fascinated by their demise, they apparently did. "We are almost totally ignorant of Hohokam archaeology . . . after 1400," writes Haury in *Snaketown*, an archaeological record of the impressive Hohokam artifacts he and his colleagues unearthed. The relatively few Pima Indians whom whites found living in central Arizona in the 1800s were presumably descended from the Hohokam—which in Pima language, means "those who have gone"—but they offered no explanation as to what happened to them. Drought remains a possibility—perhaps a twenty-year drought the likes of which they had never seen—but an equally plausible explanation is that they irrigated too much and waterlogged the land, leading to intractable problems with salt buildup in the soil, which would have poisoned the crops. In either case, the mysterious disappearance of Hohokam civilization seems linked to water: they either had too little or used too much.

And that is exactly the problem that Arizona faces today.

When Franklin Roosevelt came out to dedicate Hoover Dam on September 30, 1935, the one important dignitary who refused to attend the ceremony, which drew some ten thousand people, was the governor of Arizona, B. B. Moeur.

Though the dam had been built to safeguard the future of the entire Southwest—that was what FDR said in his speech—Moeur, like many Arizonans, looked on it more with trepidation than with satisfaction and awe. The Colorado River Compact hadn't really given Arizona anything; it had just promised the lower basin 7.5 million acre-feet. In passing the Boulder Canyon Project Act, Congress had implied that Arizona's share was at least 2.8 million acre-feet, but this, Moeur felt, was only a paper guarantee. For one thing, the guarantee had probably been jeopardized, in a legal sense, by Arizona's refusal to sign the compact. Even if it wasn't, Arizona's water rights would become exceptionally vulnerable the moment the Bureau of Reclamation completed its giant canal to the Imperial Valley and California built the mammoth aqueducts headed for the Coachella Valley, San Diego, and Los Angeles. Southern California was growing much too fast to be satisfied with 4.4 million acre-feet of the river's flow—its compact entitlement. In all likelihood, its demand for water would overtake its allotment in another twenty years. Suppose, then, that California began "borrowing" some of Arizona's unused entitlement, which it could probably do. Would Arizona ever get it back, if millions of people depended on it? For the foreseeable future, Arizona was in no position to use its share of the river, because most of the people and most of the irrigated lands were in the central part of the state, nearly two hundred miles away. Rich, urban Los Angeles had the money to build an aqueduct that long, but Arizona, still mostly agricultural, did not. And yet California had vowed to blockade any effort by Arizona to have a federal aqueduct authorized unless the major issue that still divided the two states—the Gila River—was resolved in California's favor.

The Gila, with its tributaries, the Salt and the Verde, was Arizona's only indigenous river of consequence. In the historic past, it evaporated so quickly as it meandered through the scorching Sonoran desert th all that reached the Colorado River at Yuma was an average flow of 1.1 million acre-feet. However, the Salt River Project, by erecting damns in the mountain canyons east of Phoenix, had increased storage and reduced evaporation enough to give the state 2.3 million acre-feet to use. Which of those figures ought to be deducted from Arizona's 2.8-million-acre-foot share of the Colorado watershed? Arizona said neither, or, at most, 1.1 million acre-feet, which was the historical flow. California said 2.3 million acre-feet—the amount which the dams effectively conserved for Arizona's use. If California's reasoning prevailed, Arizona would be left with a paltry 500,000 acre-feet of compact entitlement, which was hardly enough to sustain growth. But if Arizona's reasoning prevailed, California had vowed that a Central Arizona Project would never be built.

To Moeur, a showman politician in the grand carnival style, California's threats were worse than an outrage. In the arid West, denying one's neighbor water was a virtual declaration of war. But Moeur had his own response to such a challenge. He would begin waging a *real* war.

The advance expeditionary force consisted of Major F. I. Pomeroy, 158th Infantry Regiment, Arizona National Guard, plus a sergeant, three privates, and a cook. Their instructions, issued personally by the governor, were to report "on any attempt on the part of any person to place any structure on Arizona soil either within the bed of said river [the Colorado] or on the shore." Moeur knew full well that such an attempt had already been made, for the Bureau was doing some test drilling at the site of Parker Dam—a smaller regulation dam downriver from Hoover—from a barge, and the barge was secured against the current by a cable whose eastern end was anchored in Arizona soil.

When the newspapers caught wind that an army had actually been dispatched, they were ecstatic. The Los Angeles *Times* promptly inducted its military correspondent to cover the hostilities. He made it to the Parker Dam site on his state's fast macadam roads before the expeditionary force even arrived. When it did, exhausted from the heat, dust, and twelve fords across the ooze of the Bill Williams River, Major Pomeroy requisitioned a ferryboat from the town of Parker, and the force was instantly renamed the Arizona Navy. After a full inspection ng cable, Pomeroy tried to steam up the Colorado to the Bill Williams to reconnoiter, but the ferry was too high r the cable, and it got hung up. It was a harbinger of re to turn out that the occupants were finally delivered te by the Los Angeles Department of Water and Power's ch.

Pomeroy stayed at the site for seven months, sending daily dispatches to the governor by radio. When the Bureau finally began to lay a trestle bridge to the Arizona shore, Moeur decided to demonstrate that he meant business. He declared the whole Arizona side of the river under martial law and sent out a hundred-man militia unit in eighteen trucks, some with mounted machine guns. According to residents of the town of Parker, who were watching a good joke turn sour in a hurry, the guardsmen seemed eager for a fight. By now, however, the imbroglio had became national news and a source of embarrassment to Interior Secretary Harold Ickes. Well aware that Arizona had at least a moral case to make, Ickes ordered construction halted on the dam while the dispute was settled in the courts. To its own surprise as much as everyone else's, Arizona, which had already lost twice in the Supreme Court in its efforts to block Boulder Dam, was upheld. Parker Dam, ruled the Court, was technically illegal because it had not been specifically authorized by Congress. (That the Bureau could begin to put up a big dam without even asking Congress for formal approval says a lot about how far it had come in the intervening years.) Four months later, however, California's Congressional delegation pushed a bill through Congress that specifically authorized the dam, and Arizona was left without recourse, unless it wanted to declare war on the United States.

A few years later, in 1944, Secretary of State Cordell Hull formally promised Mexico the 1.5 million acre-feet that had been set aside for it by the Colorado River Compact. Feeling itself the odd man out, Arizona finally gave in and signed the compact in disgust; it also signed a contract with the Interior Secretary to purchase 2.8 million acre-feet of water. A few years after that, in 1948, the upper-basin states apportioned their 7.5 million acre-feet among themselves, and only two major issues involving the river remained unresolved: how much of her allocated 2.8 million acre-feet Arizona could take out of the main-stem Colorado, and whether California could invoke prior appropriation and deny Arizona most of that. These, at any rate, were the last *legal* issues left to be resolved. The *real* issues had much more to do with nature and economics than with law, and they were just beginning to make themselves felt.

The 1940s and 1950s were boom years in Arizona. Phoenix—population in 1940, 65,000; population in 1960, 439,000—grew overnight from outsize village to big city. Between 1920 and 1960, the state's population doubled twice, and millions of irrigated acres came into production. One of the revelations of the postwar period was that, given the opportunity, people were happy to leave temperate climates with cold winters for desert climates with fierce summers, provided there was water to sustain them and air conditioning to keep them from perishing (Phoenix, in the summer, is virtually intolerable without air conditioning). Not that the migrants had bothered to ask whether there was enough water before they loaded their belongings and drove west. They simply came; no one could stop them. How they were to fill their pools and water their lawns was Arizona's problem.

Arizona's solution was the same most other western states relied on; it began sucking up its groundwater, the legacy of many millennia, as if tomorrow would never come. By the 1960's, despite the Bureau's big Salt River Project—which captured virtually the entire flow of the Gila drainage—four out of every five acre-feet of water used in the state came out of the ground. The annual overdraft—the difference between pumping and replenishment by nature—went past 2.2 million acre-feet a year, which was more than the historic yearly runoff of all the rivers in the state. In dry years, it approached four million acre-feet. In the early days, artesian wells flowed around Phoenix. By the 1960s, some farmers could drill to two thousand feet and bring up nothing but hot brine. Parts of Maricopa County, which included metropolitan Phoenix, literally began to subside as the water below was pumped out and the aquifers collapsed. Drivers heading toward Tucson on Interstate 10 learned to watch for fissures opening in the highway as a vast block of land sank several inches and the one next to it stayed put. Arizona had reversed the pattern of some western states— it had fully developed its surface water first, and *then* began to overdraft its groundwater. Except for its Colorado River entitlement— whatever it was—it literally had nothing left.

Arizona's politicians reserved their most grandiloquent and apocalyptic imagery for speeches about the state's water dilemma. "Without more water," Congressman John Rhodes liked to say, "we are all going to perish." Morris Udall, Rhodes's ideological opposite on most matters, sounded no less like John the Baptist. Arizona, he said, was "returning to desert, to dust." As far as Senator Carl Hayden was concerned, "the survival of our dear state is at stake."

In 1952, when Los Angeles built a second battery of pumps at the head of its aqueduct and California's diversion climbed toward 5.3 million acre-feet—900,000 more than its entitlement—Arizona, in desperation, went to the Supreme Court a third time to try to get the issues resolved. The case, *Arizona v. California*, was to become one of the longest-running lawsuits in the annals of the Court, and the Justices appointed a "Special Master," the New York lawyer Simon Rifkind, to review the case. (Rifkind found the lawsuit and the constant commuting to San Francisco, where the trial was held, so taxing that he suffered a heart attack halfway through; he was still in his early fifties.) The performance of California's chief attorney, Northcutt Ely, is still studied by lawyers interested in the high art of dilatory obfuscation; one expert witness complained that Ely spent three days cross-examining him about a matter that could have been settled in a minute and a half. California, of course, had a vested interest in delay, since each year of irresolution meant 300 billion more gallons of water for the state.

In 1963, the Supreme Court finally ruled. To California's astonishment, it upheld Arizona on virtually all counts. The Salt-Verde-Gila watershed was exclusively Arizona's except for a small portion that belonged to New Mexico. Its use of that water would not be counted against its 2.8-million-acre-foot main-stem Colorado entitlement, which remained intact. That Los Angeles counted on hundreds of thousands of annual acre-feet it might never see—that a California-born Interior Secretary, Ray Lyman Wilbur, had contracted to sell it 5,362,000 acre-feet of water—mattered not in the least. The real zinger, though, came at the end of the decision, and had nothing to do with the immediate issues at hand. If, during a natural calamity or a drought, the river could not begin to satisfy all the claims on it, then, according to the Court, it was up to the Interior Secretary to decide who got how much. From that moment on, the genealogy of each Secretary became a matter of high importance to all the basin states. But there was another matter of even greater importance. The one exception to the rule it had just established, said the Court, was when someone had water rights that predated the Colorado River Compact. Those rights had to be satisfied first, no matter what.

There was an exquisite irony in this. Most of the Indians of the Southwest were hunter-gathers when whites arrived; a purely agricultural culture such as the Hohokam no longer existed. When the whites came and killed off the buffalo and antelope and ran the Indians onto reservations, their old way of life perished, and they had no choice but to become farmers or wards of the state. The reservation land they got, however, was, for the most part, land no one else had wanted. Much of it was terrible farmland, too sandy or infertile or high in elevation to grow anything well. Because it was such poor land, it required a lot of irrigation water, and the government had implicitly attached large water rights to it—rights that were confirmed in 1908 by the Supreme Court under the *Winters* doctrine. The Navajo Reservation in Arizona carried implicit rights to nearly 600,000 acre-feet, about one-fifth of the natural runoff of the state. Now, according to the Supreme Court, the Navajo could use every drop of that water during an extended drought even as people in Phoenix and Tucson were being allocated five gallons per day, even as millions were fleeing Los Angeles and leaving it the largest ghost town in the world. It probably wouldn't come to that, but the Indians, where water was concerned, clearly had the upper hand. The white man's cavalry had made beggars of them; now his courts had made them kings.

Things looked pretty bleak for southern California after the Supreme Court decision. At some point it would presumably have to give up the 600,000 acre-feet of Arizona's entitlement it was diverting, enough water for the city of Chicago. But things looked bleak for Arizona, too, because the Central Arizona Project, which was supposed to deliver the water to Phoenix and Tucson and the dying farmland in between, was neither built nor even authorized, and California could be counted on to try to achieve politically what it hadn't been able to achieve in court. Only the Indians were satisfied with what they had won. As it would turn out, however, things were even worse for California and Arizona—white man's Arizona—than they looked.

The 17.5-million-acre-foot yield that the Compact negotiators had ascribed to the Colorado River was based on about eighteen years of streamflow measurement with instruments that, by today's standards, were rather imprecise. During all of that period, the river had gone on a binge, sending down average or above-average flows three out of every four years. Not once had the flow dropped below ten million acre-feet, as it had repeatedly during the Great Drought of the 1930s. But all it takes to make statisticians look foolish is a few very wet or very dry years. In San Francisco, precipitation records have been kept for more than a hundred years—a log which, one might think, is good enough for a highly accurate guess. But 1976 and 1977, two unprecedented drought years, lowered the average rainfall figure from 20.66 to 19.33 inches. In a marginal farming region such as the Great Plains, an inch less precipitation can mean all kinds of trouble. In a desert region such as the Southwest, utterly dependent on one river, a difference of a couple of million acre-feet can spell disaster.

The first serious doubts about the 17.5-million-acre-foot figure were raised by Raymond Hill, a distinguished hydrologic engineer, at a conference in Washington, D.C., in 1953. "The discharge of the Colorado River at Lee Ferry [near the Arizona-Utah border]," Hill told his disbelieving audience, "has averaged only 11.7 million acre-feet since 1930." As Hill pointedly noted, the Colorado Basin states had not only been counting on 17.5 million acre-feet per year; they had been building and planning as if they thought that figure was *conservative*. But during the period from 1930 to 1952, the river's annual average had fallen nearly six million acre-feet shy of the accepted safe yield. He didn't need to tell his audience that this was enough water for thirty million people or a couple of million acres of irrigated farmland, maybe more.

As it would do on innumerable occasions, the Bureau refused to believe any expert who told it what it didn't want to hear. Three years later, it was frantically lobbying the Colorado River Storage Project through Congress, as if it considered Hill's figures bunk (if he was right, some of the upper-basin reservoirs it wanted to build might never fill). Then, despite mounting evidence that Hill was more right than wrong, it began planning the Central Arizona Project, which would divert another two million acre-feet from the lower basin to Phoenix and Tucson and the sinking farmland in between. Even as it continued to hold forth for 17.5 million acre-feet, however, the Bureau was beginning to develop some serious internal doubts—doubts which it would attempt to conceal for several more years, but which, in the meantime, would lead it on the most ambitious quest for water in U.S. history.

On August 18, 1965, the Bureau's resident expert on the Colorado River, Randy Riter, forwarded a long letter to Commissioner Floyd Dominy by blue envelope. Blue-envelope mail was meant to be seen by only the commissioner, the regional directors, and a small handful of top assistants. It was the Bureau's version of a diplomatic pouch, and the contents usually meant trouble.

Riter, a hydrologist and a bishop in the Mormon church, had just attended a meeting of the Colorado Water Conservation Board, a group whose purpose is to prevent a single drop of water from leaving that state's borders without first having been put to beneficial use. The featured speaker at the closed-door meeting was Royce Tipton, a consulting hydrologist in whom the Bureau placed considerable stock. Tipton's reluctant conclusion, Riter told Dominy, was that "there is not enough water in the Colorado River to permit the Upper Basin to fully use its apportionment of 7.5 million acre-feet and still meet its compact obligations to deliver water at Lee Ferry." Tipton's estimate of the river's flow was a lot more optimistic than Hill's had been, but even he felt that it should be set no higher than fifteen million acre-feet. In that case, if one divided the shortage equally between the two basins, each would be left with 6.3 million acre-feet. After you deducted another 1.5 million acre-feet or so for evaporation, and another 1.5 million acre-feet for Mexico, you had a figure low enough to throw seven states into panic.

The implications were enough to make the Bureau panic, too. The Colorado River Storage Project, which it had begun to build—Glen Canyon Dam was already completed—and the Central Arizona Project, which it dearly wanted, were both predicated on the availability of 7.5 million acre-feet to each basin. What if it invested billions in both projects only to find that there wasn't nearly enough water in the river to operate them? The upper-basin projects, in particular, were critically dependent on the full volume of water flowing through the dams; that was the only way the Bureau could generate enough hydroelectric income to give them the illusion of being economically "viable." A shortfall of nearly two million acre-feet could initiate a chain of bankruptcies among thousands of farmers or else force the Bureau to appeal to Congress for rescue. It would also open up a ghastly can of worms involving water rights. Would the shortages come equally out of each basin's hide, or would the earlier projects invoke seniority and try to keep their water under the doctrine of appropriate rights? Obviously, the new figures could knock the whole painstakingly constructed edifice of the Colorado River Compact into rubble. And what would happen when someone discovered that the Bureau had been ignoring warnings such as Hill's and Tipton's for years?

As far as Riter was concerned, there was only one way to face it. "It is futile to argue about an inadequate water supply," he wrote to Dominy. "[F]uture development in the Colorado River Basin is dependent upon the future importation of water to augment the dependable supply in the basin." He suggested that, "as a minimum," the Central Arizona Project legislation pending before Congress be rewritten to contain "a conditional authorization of an import plan of at least 2.5 million acre-feet." Riter didn't say where 2.5 million acre-feet of water from outside the basin should come from. But he knew, and Dominy knew, that there were only a few places where it *could* come from. That much unappropriated water couldn't be found within eight hundred miles. It could come from the rivers of far-northern California. It could come from the Pacific Northwest. Or it could come from Canada.

The idea of relocating distant rivers into the depletion-haunted Colorado Basin—"augmentation" was the euphemism of choice—was really nothing new. One of its earliest and most relentless proponents was William E. Warne, who was brought into the Bureau by Mike Straus and later built the California Water Project under Governor Pat Brown. As a young boy, Warne had moved from Indiana—average precipitation, thirty-six inches per year—to the Imperial Valley of California—average rainfall 2.4 inches per year—and Warne seems never to have gotten over the shock. A smooth, handsome, genial sort (though even some of his fellow Bureau men considered him water-mad), Warne in his later years would raise his voice to shouting pitch over just one issue: the "ridiculous waste" that was condoned by continuing to allow the rivers of northern California to spill practically unused out to sea. It was unconscionable, Warne would say, that those rivers were so near—"within striking distance," as he put it—and still undammed. Warne was haunted not only by the desert, but by the desert's growth. As a boy in the Imperial Valley, he heard stories about how it had been when not a soul lived there, ten years before. By the time his family arrived, forty thousand people had already moved in. Five years later, another forty thousand had come, and the valley was appropriating about 20 percent of what was then considered to be the Colorado's flow. In the same period, the population of Los Angeles had gone from 100,000 to 500,000 people. "It was the wonder of the world," Warne mused, "how that city grew." By the time he became Assistant Secretary of the Interior for Water and Power in 1949, Bill Warne had developed an obsession: rerouting the fabulous amount of water that spilled into the Pacific from Eureka on north.

The engineering study that would determine how best to do it was called the United Western Investigation. It is, to this day, the best-kept secret in the history of water development in the West; people who have been in the business all their lives have never heard of it. Since it would involve the movement of unprecedented flows of water over unprecedented distances at unprecedented expense, the investigation would need someone of unusual vision and character to lead it, and Mike Straus and Bill Warne would have to go outside the Bureau to locate him. The found him in Bogotá, Colombia, building dams for the descendants of conquistadors. His name was Stanford P. McCasland.

Stan McCasland had worked in the planning division of the Bureau for some years. He quit, evidently, because the predictable tedium of designing small projects on small rivers was something which he considered beneath him. In South America, where you could find unnamed tributaries of the Amazon bigger than the Colorado, he at last found the landscape of his dreams. Like a lot of Bureau engineers, McCasland had only a faint interest in irrigation; it was damming rivers that got his juices flowing. An irascible Scot, he viewed rivers not so much as challenges or opportunities but as willful monsters to be beaten into submission. In the likes of him, the rivers had an unlikely foe. "He was as skinny as a rail," says Pat Dugan, a longtime Bureau engineer who worked with him, "and he had a shock of flaming red hair that made him look as if the top of his head had ignited. He always wore a tweed suit that was about three sizes too big. He looked like he could turn around 180 degrees in that suit while the suit stood still."

The investigation took two years to complete. Its conclusions filled several volumes with descriptions, economic analyses, appendices, and maps. To Clarence Kuiper, a young engineer recruited from the Corps of Engineers, "it was the closest I ever came to feeling omnipotent. We were looking at ideas even Mike Straus hadn't thought of yet." The UWI team raced around the Pacific rim like Rommel's army, concocting schemes to put deserts to flight. They dinged rock samples out of canyon walls. They traced future reservoir basins by air. They floated rivers and explored by jeep. They spread contour maps across the floors of rooms and built tunnels and aqueducts with pencils. They spread oceans of theoretical water over horizons of potential farmland; on paper, they turned half the Southwest green. "Straus told us to look at every possibility," Kuiper would recall years later. "He said, 'Don't you laugh at a goddamm thing.' Well, we didn't. We looked over every harebrained idea that ever came up. We looked at an undersea pipeline from the mouth of the Columbia to Los Angeles. Lord, we found every conceivable place where you could divert the Columbia. We looked into jumping the Willamette out of its bed at Oregon City and turning it right around in an aqueduct to California. Southern Oregon is a big mess of mountains, so we plotted a tunnel that would have been 135 miles long. We had one guy in the Bureau who thought you could keep wall-to-wall tankers moving between the mouth of the Columbia and L.A., so we looked at that, too."

If anything, the United Western Investigation suffered from a surfeit of choices. "Numerous possibilities exist for the interbasin transfer of supplies into water-deficient regions," wrote McCasland in the cover document, which bore the splendidly militaristic title *United Western Investigation, Interim Report on Reconnaissance, Report of the Chief.* You could, for example, take a few million acre-feet out of the Snake River at Twin Falls, Idaho, pump it up the south side of the Snake River plain in fifteen-foot siphons, and drop it into the Humboldt River, the only constant river in the state of Nevada, meandering small and forlorn beside Interstate 80 for three hundred miles until it disappears in shallow, salty Humboldt Lake. Then you could move the suddenly prodigious Humboldt straight across seamless desert to the Owens Valley, two hundred miles away. You could tunnel thirty miles under the White Mountains and just dump it in. Then you could quadruple the size of everything Los Angeles had built to divert the Owens River, and move the mingled waters of the Snake and the Humboldt and the Owens to L.A., to San Diego, to the Mojave Desert, and dump the surplus in the Colorado to satisfy our treaty with Mexico, leaving the other basin states with the whole Colorado to hoard for themselves.

Alternatively, you could build a whole series of dams at a more or less equal elevation on the bigger rivers of coastal Oregon and, at a level approximate to the elevation of the upper Sacramento Valley, run a gravity-diversion aqueduct from reservoir to reservoir, picking up half-million-acre-foot increments as a bus picks up passengers, then run the aqueduct beneath the Siskiyou Mountains and plop the water into Shasta Lake, then lead it south from there. You could take millions and millions of acre-feet out of the Pend Oreille in Washington, an obscure river bigger than the Wabash or the Hudson or the Sacramento, and move it by gravity—aqueduct-tunnel-aqueduct-tunnel-aqueduct-tunnel-from Albeni Falls, near the Canadian border, across the deserts of eastern Washington and Central Oregon all the way to California, passing by the Rogue and the Illinois and picking up some surplus flows, with the end result that California's developed water yield would be increased by nearly one-half. "The total length of the aqueduct . . . would be about 1,020 miles, of which about 290 miles would be tunnel and 40 miles in siphon. No estimates of cost were made for this plan because the necessary length of aqueduct causes it to appear unattractive." Most or all of these diversions, the *Report of the Chief* implied, would *have* to be built, sooner or later. "Regardless of magnitude, scope, and timing of the undertaking, if it can be shown that moving surplus waters of the Northwest to water-deficient areas elsewhere is in the realm of sound public interest, it is, in Reclamation's opinion and half century of experience, only a matter of time before exhaustion of nearer water supplies forces the undertaking of a suitable project for that purpose." All of this, however, might still be fifty years in the future. For now, the immediate need was in the Colorado Basin and its parasitic appendage, southern California, and the obvious river of rescue was the Klamath.

Remote, wild, half-forgotten, the Klamath was a perfect example of how God had left the perfection and completion of California to the Bureau of Reclamation. The second-largest river in the state—three times the size of the third-largest river—it was imprisoned by moun-

tains and hopelessly remote from Los Angeles. Spilling out of Klamath
Lake in southern Oregon, a huge shallow apparition cradled between
mountains and desert, the river drops across the California border and
bends its way westward toward the coast. Then it dips suddenly southward toward populated California, and, as if recognizing covetous
intent, immediately doubles back on itself and flees to Oregon through
the plunging topography of the Siskiyou Range. Diverting the Klamath
would be easy along the first half of its course, but it doesn't contain
much water yet. A hundred miles from the Pacific, however, rainfall
shoots up to a hundred inches, the Trinity and Scott and Salmon rivers
pour in, and the Klamath is suddenly huge. On a random day late in
February of 1983, after a week of rain, the Klamath was flowing at
four thousand cubic feet per second below Klamath Lake and at
148,000 cfs near its mouth, a Niagara-size flow in a canyon you can
bat a ball across. Small tributaries were tumbling oven-size rocks like
ice cubes. To the Bureau, the Klamath's huge and reliable winter
surges were only its second greatest attraction—the first was its availability. The Klamath was wasting twelve million acre-feet to the sea
with hardly a claim on it. Its principal appropriators were salmon,
steelhead, and bears.

To capture the Klamath, you had to dam it twelve miles from the
Pacific, then move the water in reverse across, or under, a hundred
miles of the most rugged topography in the United States. The dam,
which would be called Ah Pah, would occupy the river's last gorge. It
would stand 813 feet high. The Pan Am Building in New York City
stands 805 feet high. A man-made El Capitán, it would pool water
seventy miles up the Klamath and forty miles up the Trinity to form
a reservoir with 15,050,000 acre-feet of gross storage. (The reservoir
that obliterated Johnstown, Pennsylvania, held fifty thousand acre-
feet.) The Klamath, both forks of the Trinity, and the Salmon River
would, for all practical purposes, disappear; 98 percent of the salmon
and steelhead spawning grounds would be lost; at least seven towns
would vanish, including the main settlements of the Hoopa tribe, from
whose language the dam's name was borrowed, and whose reservation
it would drown. "Only minor improvements [i.e., towns] exist in this
[the reservoir] area," said the United Western report. The site, in a
dense metasandstone formation, was presumed to be safe, although it
"probably contains minor faults."

Trinity Tunnel, which would spin water out of the bottom back
side of the reservoir and carry it to the Central Valley, would be sixty
miles long, Its shape would resemble a horseshoe, and its diameter
would be thirty-seven feet. There would be no tunnel remotely like it

anywhere in the world. The Delaware Aqueduct, stretching from the Catskill Mountains to Westchester County, is eighty-five miles long, but its diameter is only fourteen feet. Trinity Tunnel could hold four passenger trains operating on two levels. It alone would cost nearly half a billion dollars, in 1951, and it was merely the longest and biggest of numerous tunnels. The Tehachapi Tunnel, forty miles long, would move the water through the Transverse Ranges, which cordon off Los Angeles from the rest of the state. Seven known fault zones would be crossed along the aqueduct route, and one of them, the San Andreas Fault, would present a fracture zone at least two miles wide in the middle of the Tehachapi Tunnel which could pose "unusual construction problems." An ordinary tunnel would shear if the fault slipped—it slipped nearly twenty feet in the 1906 quake—leaving Los Angeles unwashed and unquenched. "Extra-heavy supports would be required throughout this zone."

The cost of everything—Ah Pah Dam, the other dams, the tunnels, the aqueducts, the pumps, the canals, the receiving reservoirs, and an item called a Peripheral Canal, which would be built to carry the Sacramento's greatly swollen flow around California's Delta—would be $3,293,050,000. It was an incredible bargain; today, a couple of nuclear power plants cost much more than that. Had the Bureau reckoned how expensive life was going to become, the Klamath Diversion might well have been built. "In those days, almost everything the Bureau proposed was being built," Kuiper says. "But Straus decided to move cautiously on this one. If you read the report, you'll see that we were always talking about 'orderly development.' That's code talk for building at a deliberate pace, taking care to butter everyone's bread, instead of going gung-ho, which is what they did on the Missouri. In California, you had two choices: you could build a lot of little projects on tributaries of the Sacramento and the San Joaquin, or you could built one huge project on the Klamath. I don't know whether it made better sense to do one or the other. What I do know is that Mike Straus was constitutionally incapable of seeing that the clock was already running out on these tremendous projects. He thought it was the other way around—that the Bureau would keep building bigger and bigger things. The only way Mike knew how to think was bigger and bigger. He and Bill Warne were sure the Klamath Diversion was going to be built someday, so they didn't try to railroad it through."

McCasland didn't help. Contentious and prickly, he may have been a fine engineer, but he was the public relations equivalent of Sherman's march to the sea. Without asking clearance from the commissioner's office, he wrote an article describing the Klamath Diversion for *Civil*

Engineering in 1952. Northern California's thirty-five years of passionate opposition to southern California's diversion plans can be traced
directly to that article. McCasland would not even say that these Pan-
tagruelian waterworks would take care of the southland's need for all
time: "The plan described as the Northern California Diversion," he
wrote, "would not by any means constitute a complete water supply
for the Southwest. It would meet the most imminent demands . . . but
it would more probably constitute the initial stage of a large plan to
serve much wider markets a future economy might dictate." Clarence
Kuiper says, "What I remember about that time is the phone ringing
off the hook with reporters from Oregon and Washington asking me
if it was true that the Bureau was planning to divert the Columbia
River." But the outrage cascading down from the Pacific Northwest
was the least of the Bureau's problems. Its main problem was that
Los Angeles, for whose benefit the Klamath Diversion had mainly been
conceived, was unalterably opposed to it, too.

The idea of the city it was trying to save vilifying the project it had
planned in order to save it left the Bureau of Reclamation speechless.
Had its engineers accepted a thing or two about law and psychology,
however, they wouldn't have been the least surprised. Los Angeles, in
the middle of an epic feud with Arizona over Colorado River water
rights, saw the Klamath Diversion as a ploy to encourage it to relin-
quish its claim on the share of the river that it wanted to consider its
own. In fact, if any Californian even *mentioned* the idea of going north
for water, Los Angeles came down on him like Thor. When Republican
Congressman Richard J. Welch of San Francisco did just that, the Los
Angeles *Daily News* denounced his idea as "the kiss of Judas." "This
San Franciscan," it fumed, "is trying not to succor but to sucker us."
As Carey McWilliams wrote in *California: The Great Exception*, "To
suggest that Colorado River water was not the only water which might
be made available in southern California was, of course, an act of
treason, a betrayal." The Republican Party of the state, with its center
of power in Los Angeles and Orange County, went so far as to mount
an effort to excommunicate poor Welch, who, as bewildered as the
Bureau by then, said he was only trying to help.
    The Bureau was flummoxed. Copies of the UWI report were buried
in the archives in the regional office in Salt Lake City, where they sat
under lock and key. Before he could do more damage, McCasland was
transferred to a desk job in Washington, and young Kuiper was left
with the job of repairing the wreckage his boss had created. The Klamath Diversion, potentially the greatest engineering scheme of all

time, was dumped on the scrapheap of human dreams. "The whole thing kind of backfired on them," said Kuiper in 1981, still wryly amused after all those years.

The Eisenhower era put transbasin diversions into the Colorado on hold for at least another eight years. Ike's Interior Secretary, Douglas McKay, a Chevrolet dealer from Portland, Oregon, followed the honorable Republican tradition of using the office as a vending machine for timber and minerals, but recoiled at the idea of an activist government marketing water and power. Ike's water-development policy, announced shortly after his inauguration, was that there would be "no new starts" during his administration, especially if the production of power was involved. His own Republican allies from the western states would soon make him eat his words, but his immediate problem, after his inauguration, was finding a Reclamation Commissioner who would do his bidding—an exact opposite of Mike Straus. Since the Bureau had been stuffed with liberals, public-power advocates, and super-engineers of the McCasland ilk during the previous twenty years of Democratic reign, he wouldn't be easy to find within the Bureau—whence commissioners traditionally came. After leaving the post vacant for several months, the Republicans finally came up with Wilbur Dexheimer, the Bureau's assistant chief construction engineer. Dexheimer was handsome, amiable, and a competent engineer, but he was, as Winston Churchill said of Clement Attlee, a modest man with much to be modest about. He had spent his entire career in the Denver engineering headquarters, and he was an ingenue at politics, which was the breath of life to Mike Straus. Dexheimer was the first to admit that he was the consummately wrong choice for the job. As soon as he was appointed, he called his regional commissioners to Washington, gathered them in his office, and blurted out, "I don't have to tell you guys that I'm the least likely person in Creation to be sitting at this desk. I'm ignorant as hell about what goes on in this town, but by God they made me commissioner and here I am and now you've got to follow my orders even if you and I think they stink."

To the routed myrmidons of the New Deal, the golden age of water development seemed truly over. But Republican principles would prove to be no match for the stark imperative of the American desert. In 1956, Ike would end up signing the Colorado River Storage Project Act against his better judgment, and the budgets of both the Bureau and the Corps of Engineers would increase dramatically during his administration. In the lower Colorado Basin, however, Eisenhower had an excuse to do nothing. Until the Central Arizona Project was given final shape—and that couldn't happen until the legal battle had ended and it was determined who had rights to what water—the river's looming deficit would remain an inconsequential fact. Once the lawsuit was settled, however, the Bureau would face two seemingly insuperable problems at once: how to build the most expensive water project of all time; and, even worse, how to authorize an even *more* expensive augmentation scheme that would give the Colorado Basin enough water for everything it was planning to build.

On a map of Arizona, the Colorado River can be seen making a wide circle around the northern and eastern half of the state. At every point along that six-hundred-mile sojourn, the populated center of the state is walled off from the river by mountains. In the north, the river flows in a bottomless canyon, a mile below its southern rim; to lift it out of there and lead it to Phoenix would be out of the question—even though the water, once out of the Grand Canyon, could flow downhill all the way. Closer to its mouth, the river escapes its canyon confines and flows across broad sandy wastes, but numerous ranges stand between the river and Tucson and Phoenix—the Aquarius Cliffs, the Black Mountains, the Maricopa Mountains, the Saucera Range. Regardless of where one located the point of diversion, to move a portion of the Colorado River to Tucson and Phoenix would involve a pump lift of at least twelve hundred feet. Pumping irrigation water there would be like taking it out of the Hudson River and lifting it over the World Trade Center in order to water lawns on Long Island. The CAP was to be, first and foremost, an *irrigation* project, a rescue project to save the dying farmlands between Phoenix and Tucson; the cities would also get some water, but the farmers would receive the overwhelming share. Hardly anywhere on earth, however, is water lifted that high in order to irrigate crops, unless the water flows nearly as far downhill somewhere along its route as it was lifted uphill, so that much of the energy required to lift it can be recovered. Even then, the Second Law of Thermodynamics exacts a heavy toll: for every hundred units of energy expended to lift the water, only seventy or so can be recovered on the way back down. Using the most optimistic predictions—high-value crops, high crop prices, dirt-cheap power from preexisting dams—the Central Arizona Project was still likely to need more public welfare than anything the Bureau had built.

A simple matter of physics, then, made the Central Arizona Project even worse, in an economic sense, than the Colorado River Storage Project. But politics demanded that it be built, and in the 1960s, Arizona had power. Barry Goldwater was the presidential candidate of the Republican Party; Carl Hayden was the chairman of the Senate

Appropriations Committee. He could, if he wanted, hold up every other
water project in the country until his state was satisfied. And there
was the issue of equity. California had its water, Nevada had its water,
the upper basin was developing its water, and Arizona still had nothing. What were a couple of billion dollars in the face of these other,
more important concerns?

Still, something would have to be done about the project's horrif-
ically poor economic rationale. And something would ultimately have
to be done about the fact that the river now seemed certain to dry up
if the CAP was built. Something—but what? The obvious answer was
a couple of big cash register dams that could generate enough power,
and enough money, to give Arizona's irrigation farmers the 90-percent
subsidy they would probably need. If the dams were big enough, there
might be enough revenue left over to begin a fund that, in the future,
could help build the gigantic augmentation project that the basin
would require.

But where could one locate the dams? There were no sites for big
dams left in Arizona, and besides, the Gila River system didn't have
nearly enough water to develop the kind of power the Bureau had in
mind. California still had a lot of undeveloped hydroelectric potential,
but it wouldn't think of allowing dams to be built within its borders
whose revenues would allow Arizona to divert water it was then using.
The Colorado River Storage Project was cementing dams in all the
best hydroelectric canyons in the upper basin. New Mexico's rivers
had neither the sites nor the water flows. There was only one place in
the entire Southwest where reliable water flowed through a section of
river with a thousand-foot drop—the Grand Canyon.

The proposal for Grand Canyon dams was officially revealed on January 21, 1964, with the release of something called the Pacific Southwest Water Plan. One had only to read the title to see that, now that
another New Deal Democrat was enfranchised in the White House
(Lyndon Johnson was about to beat Barry Goldwater with 60 percent
of the popular vote), the Bureau had happily returned to the mode of
thinking prevalent during the FDR and Truman years. The plan was
majestic. It contemplated two huge new dams on the Colorado River
in Marble Gorge and Bridge Canyon, at opposite ends of Grand Canyon
National Park. Both had been carefully situated so as not to flood the
park itself—except for what the Bureau called "minor" flooding that
would drown lower Havasu Creek, the canyon's most beautiful side
stream, and submerge Lava Falls, the river's most thunderous rapid.
But the park would sit inside a dam sandwich: Bridge Canyon Dam

would back up water for ninety-three miles below it, entirely flooding the bottom of Grand Canyon National Monument, and Marble Gorge Dam would create a reservoir more than forty miles long right above it. The dams had one purpose—hydroelectric power—and a single objective: lots and lots of cash. They would not conserve any water, because there was none left to conserve; in some years, they would cause a net *loss* to the river through evaporation. They were there only to take advantage of the thousand feet of elevation loss between Glen Canyon and Hoover dams. Together, they would generate 2.1 million kilowatts of peaking power, marketable at premium rates. Later, the power revenues would finance an artificial river of rescue; for now it would pay for the other features of the plan.

One of those features—actually, it was the centerpiece of the plan— was a pair of big dams on the Trinity River, in far-northern California, and a long hard-rock tunnel that would turn their water into the Sacramento River, where it would begin its journey to Los Angeles. That city and its burgeoning suburbs would thus receive a huge surge of high-quality water from northern California to replace the salty Colorado. The San Joaquin Valley would siphon off a considerable portion along the way; it was going to be rescued, for the *third* time, from its suicidal habit of mining groundwater. New Mexico would get Hooker Dam, which would inundate yet another scenic monument— the Gila Wilderness—and Utah would get two more projects. In the middle of the list, camouflaged under "water salvage and recovery programs," was the most expensive item of all: the Central Arizona Project. It was the same multibillion-dollar shell game that the United Western Investigation had proposed: new water from northern California would take care of southern California's needs so that the Colorado could be conserved for the upper basin and Arizona.

Curiously, the United Western Investigation did not even rate a passing mention in the report, though it dwelled at some length on earlier plans to solve the Southwest's water dilemma. Evidently, the UWI was still so closely associated with a raid on the Columbia River that the framers of the Pacific Southwest Water Plan would rather have pretended that it never existed. Another name hard to find in the report was that of the Interior Secretary, Stewart Udall. There were three possible explanations for this. One—the one conservationists wanted to believe—was that the plan did not really have Udall's support. He was, after all, being described by them as the best Interior Secretary since Harold Ickes. How could the best Interior Secretary since Harold Ickes wish to inundate the most stunning feature of the American landscape? How could he talk about the "minor" intrusion of a reservoir into a national park? Another explanation was that Udall, as a native of Arizona, felt that he had to distance himself from a plan whose ultimate purpose was to deliver a couple of million acre-feet to his home state. A third explanation—the one conservationists least wanted to believe—was that Udall supported the plan but didn't want to admit it.

The most interesting curiosity about the plan, however, was the obvious discrepancy between the amount of new water the Trinity River could deliver and the looming shortfall in the Colorado River. At the moment the plan was released, the second-largest reservoir in California, Clair Engle Lake, was beginning to fill on the upper reaches of the Trinity. Its capacity of 2,448,000 acre-feet was not much less than the river's annual flow of 3,958,000 acre-feet. Clair Engle Lake was a main feature of the Central Valley Project; its water, therefore, was exclusively for California's use. According to the Pacific Southwest Water Plan, only 1.2 million acre-feet would be left in the Trinity to augment the Colorado River—and that was assuming the Trinity, one of the world's great salmon and steelhead rivers, would be bled virtually dry before reaching the sea. But the shortfall which the Bureau was projecting in the Colorado Basin, privately if not publicly, was at least 2.5 million acre-feet. Where would the other 1.3 million acre-feet come from? The Pacific Northwest Water Plan said nothing about it. It only hinted that it "does not provide an overall solution for the region's water needs," then failed to mention what such an ultimate solution would be. The Bureau's maps had other reservoirs all over the place, drawn in gray—several on the Eel River, one on Cache Creek, the huge Ah Pah reservoir on the Klamath—but referred to these as "alternative storage possibilities," as if they might substitute for, but not augment, the Trinity dams. Where, then, was water for six million people to come from?

In the Pacific Northwest, there was a lot of suspicion that the Pacific Southwest Water Plan was merely a smokescreen for a much larger plan, long a gleam in the Colorado Basin's eye, to tap the Columbia River. Such paranoia was inflamed by occasional speeches delivered to sympathetic ears by some of the Bureau's engineers, insisting that this was the final solution that would someday have to be built to allow continued growth in the parching Southwest. Officially, however, the Interior Department went to great lengths to reassure the Northwest that it had no such designs. Udall publicly scoffed at the notion of diverting the Columbia, and Floyd Dominy, the Bureau commissioner, sharply reprimanded his underlings if they even mentioned the idea. But the truth of the matter was that the Pacific Southwest

Water Plan *was* a smokescreen. The Columbia was on Udall's and Dominy's minds the entire time.

On December 15, 1964, less than a year after the Pacific Southwest plan was revealed, a four-hour-long meeting quietly took place at the regal new offices of the Metropolitan Water District of Southern California. (Built on a hill at one end of Sunset Boulevard, the MWD headquarters had a splendid view of the immense sprawl and traffic congestion it had helped create—four freeways converged right below its windows—but it was walled off from same by a forest of fountains and, fittingly, a moat.) The participants in the meeting were Udall, Dominy, Interior solicitor Edward Weinberg, Los Angeles Congressman Chet Holifeld (whose twin passions were water diversion and nuclear power), and seven carefully selected members of the MWD. Officially, this was a meeting that never took place, but as the chairman of the MWD, Joe Jensen, enthused in a "Confidential Report to MWD Directors," it was "one of the most constructive conferences we have attended." Udall, he reported, "expects to discuss with Senator Jackson of Washington a feasibility study and the eventual taking of ten to fifteen million acre-feet of water from the Columbia River. He would hope to have Senator Jackson lead off with the statement that the export program would be possible according to such guidelines as Jackson felt necessary for such an export program."

"Mr. Dominy," Jensen continued, "explained that a group in Denver had been working for thirty days on a preliminary study to bring water from the Columbia River, and that by March he should be able to give a definite answer as to the route and the general features of the project; as well as a comparison of cost of this project and the cost of delivering water from California and desalting. . . . Washington may need a stepped-up reclamation program," Jensen quoted Dominy as saying, "in order to offset the adverse effects of closing down several federal installations in that state."

"The Secretary stated two courses appeared to be possible at the present time," Jensen wrote his fellow board members. "(1) Have a study made and defer action on authorization while the study is done right; (2) Introduce a bill which would authorize the import program, Bridge and Marble Canyon Dams, Central Arizona Project, and a few of the other projects. By March more definite information should be available and it should be possible to have the Committee report a bill to authorize the study and authorize the construction of the import program. . . . Dominy indicated the first six months of any presidential term was the best time to hit Congress. *He stated that the Bureau had never had any trouble getting funds once a project had been authorized,* but it frequently had trouble getting projects authorized" (emphasis added).

Udall and Dominy, in other words, wanted to study the feasibility of the Columbia diversion *after* it was already authorized, on the assumption that even if it wasn't economically sound, it would be too late to stop it. Their real concern seemed to be lining up the political firepower that would let them succeed. And the plan, as Jensen described it, included so many gifts to so many states that it certainly *ought* to succeed. It contemplated numerous new irrigation projects in both Oregon and Nevada, some more projects in the upper Colorado Basin, and the stepped-up reclamation program in Washington that would make up for the mysterious "facilities" that might have to be shut down. Even then, Jensen said, "up to 7.5 million acre-feet of [Columbia River] water" would still reach Lake Mead every year. The plan, then, had to be far more expensive and ambitious than anything ever contemplated—more so, by far, than the Pacific Southwest Water Plan, more so even than the Klamath Diversion studied by the Bureau twelve years earlier. That was remarkable enough. What was *really* remarkable, however, was that the water would be available *"at the present price of Colorado River water."*

To charge no more for Colorado River water delivered to Los Angeles or Arizona than was being charged for water from nearby Hoover Dam would be a feat as astonishing as Moses' bifurcation of the Red Sea. The water would have to come a thousand miles by aqueduct; Hoover water came only a couple of hundred miles, and the immense power output of the dam subsidized the big pump lift to L.A. Hoover Dam was financed with Depression-era interest rates and built by workers earning $4 a day; this project would be financed by Vietnam-era interest rates and built by unionized labor earning at least $6 an hour. There would be little, if any, hydroelectric power produced, but a lot of power might be required for pumping; the water had to go over or through two major mountain ranges! The difference in cost, per acre-foot, ought to be at least 800 percent, probably much more. But, according to Joe Jensen, Stewart Udall was offering it at the same price the MWD paid for water from Lake Mead. Somewhere, there was an immense subsidizing engine, but where?

In his memorandum, Jensen merely hinted at an answer; it may have sounded so good he didn't believe it himself. "The cost of such Colorado River supply," he wrote, "is to be paid out of the first power revenues. The remaining power revenues would be available for assisting in the payment of the main program or project, and water revenues would pay part of the cost." Apparently, Jensen was promised by Udall and Dominy that the power revenues generated by the dams that would be part of the import scheme were going to subsidize the price of water before they even began to pay back the cost of the facilities! Before the dams were paid for; before the aqueducts were paid for; before the tunnels were paid for; before the siphons and canals were paid for—before a penny went to all of that, the power revenues were going to go directly into the pockets of water consumers in southern and central California and Arizona, subsidizing the price of their water. How else *could* Udall be promising the Southwest water—water that probably wouldn't be available until the 1980s—at 1935 prices? If this was what it took to get the Central Arizona Project built—and Jensen, a leading foe of that project, did not say an unkind word about it in his memorandum—Udall and Dominy were just about prepared to give the water away.

In an economic sense, what the backers of the Pacific Southwest Water Plan were proposing was unprecedented. It violated every principle of economics, even the fast and loose principles of Reclamation economics. If the lion's share of the power revenues were going to subsidize not only irrigation but *municipal* water costs—municipal water whose revenues had usually subsidized *irrigation* in the past— the project could not possibly be paid back for hundreds of years, if ever. The cost, which had to be in the many billions, would simply be borne on the backs of the taxpayers. From a national perspective, it was a stunningly ill-conceived idea; but from a regional perspective, it was a wonderful idea—an offer none of the basin states could refuse. At a price guaranteed to be affordable—not only affordable but dirt-cheap—the yield of the Colorado River would be increased by one-half. Oregon would get a slew of new irrigation projects, as would Nevada. California's irrigators would be relieved of their most desperate worry, the self-inflicted groundwater overdraft. And it could all be accomplished by taking a mere 10 percent of the flow of the Columbia River and turning it southward.

Behind the proposal was a dramatic gamble—that Congress and the public would go along with the idea; or, even if they didn't, that the Southwest had the political power to persuade them to. But how could one sell the public on a program that was supposed to remain a tightly guarded secret? On December 31, 1964, two weeks after learning of the Columbia plan, Joe Jensen sent his New Year's greeting to Stewart Udall, expressing "very great appreciation" for Udall's decision to support the project that had always been Jensen's dream. Then he added, "Since your program is to be kept confidential there is little that we can do except give you assurance of our support and

our desire to assist in every way." It must have been frustrating for
Jensen. The Metropolitan Water District had the mightiest propaganda
apparatus in the entire West, and he didn't dare push the button to
fire it up.

Maintaining a self-enforced silence about the proposal was actually
the least of the proponents' problems. By 1965, the war in Vietnam
was consuming an ever-larger bite of the federal budget, and LBJ's
antipoverty programs also promised to cost a tremendous amount. No
price had been put on the Columbia diversion, but the Trinity River
version of the Pacific Southwest Water Plan was expected to cost
$3,126,000,000; going as far as the Columbia for much more water
could easily cost three times that much. The federal budget in 1965
was only $118.4 billion; to persuade the Congress to authorize perhaps
$10 billion for a single water project would take some doing. But the
biggest and most unyielding obstacle would not even be the enormous
cost. It would be the man who, Udall foolishly felt, he could persuade
to lead the bill through Congress—a pugnacious, five-foot-ten-inch,
third-term Senator and fellow Democrat from Washington state
named Henry Jackson.

In June of 1965, with no discernible opposition, Senator Henry
Jackson tacked an innocent-looking rider onto an innocuous-seeming
bill that established standardized guidelines for the allocation of costs
to fish and wildlife enhancement. What the rider did, in a couple of
brief sentences, was prohibit the Bureau of Reclamation from undertaking feasibility studies that Congress did not approve in advance.
The effect of the maneuver, which few recognized at first, was the same
as if Jackson had strung a six-hundred-volt electrified fence along the
entire south bank of the Columbia River. Without a feasibility study,
the Bureau couldn't approach Congress for authorization. Without a
Congressional authorization, it couldn't build. Explaining his amendment to a couple of reporters who were smart enough to see what it
meant, Jackson made no mention of the Columbia River. He was an-
noyed, he said, by the Bureau's habit of "working up local interest
and enthusiasm for projects in the field before presenting its case to
Congress." Such tactics, he said, put Congress in a "take it or leave it
position" when the Bureau came to authorization hearings with a
gaggle of local politicians and noisy project boosters in tow. His
amendment was nothing for the Bureau to get upset about; "the Corps
of Engineers has operated under similar provisions for many years."

It was true that the Corps operated under a similar restriction; it
was also true that it rarely paid much attention to it. But Jackson's
rider had made illegal the feasibility study that Dominy had quietly

ordered on the Columbia diversion. Jackson, who obviously had heard rumors of the secret plan, was out to kill it in its embryonic state. The Northwest had water to spare, but it no longer had power to spare, and nearly all of its electricity came from dams. To remove ten million acre-feet from the Columbia River meant a reduction of several billion kilowatt-hours in power output, unless one diverted the water below the dams. The Bureau would undoubtedly want to do that; but suppose the pumping cost of a diversion from low elevation would add tremendously to the project's cost, and it made much more sense to divert above the dams? If the enormous momentum that could develop behind the diversion scheme really got rolling, the Northwest would look awfully selfish refusing to part with some of its superabundant water just because it insisted on paying one-fourth the average national price for electricity.

"I told Jackson that we ought to let them study the idea," recalled Daniel Dreyfus, who was then the Senator's closest aide. "There was no way it was going to be economically feasible. Twenty years earlier, maybe. In the sixties, absolutely not. 'Let's let 'em study it,' I told Jackson. 'Study the damned thing and it will slay itself. It's a crazy idea.' But his reasoning was that there'd been other crazy projects that got built just because they were studied. I still never thought it could get built, but he was right on that point."

Without a feasibility study—which Jackson, as chairman of the Senate Interior Committee, would never allow—the Columbia diversion was stillborn. What is more interesting is how quickly the Trinity Diversion died with it, even though Jackson had not publicly opposed it. One reason may have been that Los Angeles viewed it, as it had viewed the United Western Investigation, as a threat—an implied source of water that wasn't the Colorado River (it didn't mind the Columbia because that source was *really big*). But another and better reason was that it didn't make any economic sense. The Trinity River offered too little water at too great an expense. No matter what the cost or opposition, the Colorado Basin had to get its hands on the Klamath, the Snake, or the Columbia; those were the only rivers left in the American West that were worth thinking about.

It was actually the upper Colorado Basin states, not the lower, that were pursuing the water importation idea with particularly feverish interest. California wasn't worried. The Imperial Valley had so much water it was almost drowning in it, and Los Angeles had more on the way from the State Water Project, then just being built. Through the CAP, Arizona might soon receive most of its entitlement to the river through a single diversion. The upper-basin projects, however, were small and spread all over the map, and few of those authorized by the Colorado River Storage Project had yet been built. Several, in fact, had been denied startup funds in Congress—partly because their backers lacked the awesome Appropriations Committee clout of the California delegation or a Carl Hayden, partly because they were beginning to be regarded by some members of Congress as a scandalous waste of taxpayers' money, especially with a war going on. Floyd Dominy had told Joe Jensen that he always got funds for projects that had been authorized, but the upper basin was learning that, indeed, this was *not* always the case. At the languid rate its projects were being built, the upper basin would be the last to develop its full entitlement to the river. And when the overappropriated river was played out, the compact might not mean a thing. Whoever was using the most water would end up keeping the most water; the various Congressional delegations—especially the powerful one from California—would see to that. No one was going to turn off the spigot to Los Angeles, Arizona, or the Imperial Valley for the sake of a few marginal irrigation projects in the upper basin—especially if they hadn't even been built.

Exactly how adamant the upper basin was on this issue became apparent, for the first time, at a secret summit meeting attended by representatives of the four states at Denver, Colorado, on January 18 and 19, 1966. The subject of the meeting was the CAP legislation that Dominy, Udall, and the Arizona and California delegates had coalesced behind, HR 4671. HR 4671 was a drastically trimmed-down version of the Pacific Southwest Water Plan. It authorized only the Central Arizona Project, Bridge Canyon Dam (Marble Gorge, the other Grand Canyon dam, had been dropped because its more meager power output didn't seem worth the inevitable fight), and a new aqueduct to Las Vegas. The bill also authorized something called a "development fund"—a receptacle for revenues from the power dams that, in the future, would help finance the augmentation scheme everyone knew would be needed. The legislation, in other words, authorized the projects that would ensure the Colorado River's early exhaustion; it also authorized the means of financing the basin's rescue. What it did not authorize—what it didn't even *mention*, let alone describe—was the importation plan itself. Udall and Dominy had evidently concluded that the development fund would be enough to mollify the upper basin. Only after Dominy's regional director in Salt Lake City, Dave Crandall, sent him his report on the Denver meeting did they see how utterly wrong they were.

Of the four upper-basin states, the one that seemed most intent on a specific authorization for the rescue project was Colorado, within whose borders half the river's flow originates. This, from Udall and Dominy's point of view, was most unfortunate. Colorado's delegation was headed by Felix Sparks, the head of its Water Conservation Board. Sparks had won the Medal of Honor in World War II, among many other medals, for single-handedly storming a machine-gun nest with a sidearm and a jacketful of grenades and killing half a platoon of Germans. According to those who knew him, he was not afraid of God, man, or the devil. He was also stubborn, vindictive, and a bully, but in Colorado, where water was concerned, he was king.

According to Dave Crandall, Sparks had terrorized the Colorado delegation into asserting that "a feasibility study of import must be a part of the [CAP] bill, otherwise they would not support it. They would prefer an authorization of import but recognize the impracticality of seeking such authorization at this time." Wyoming, he said, took a similar view: "It feels that import is an absolute necessity for their future development and protection and they desire conditional authorization. . . . Studies of importation are an absolute minimum and anything else would result in opposition to the bill." Utah was slightly less adamant than Colorado and Wyoming, but not much. New Mexico would accept a bill that only authorized a feasibility study, but nothing less than that. Between them, as Dominy well knew, the four states had the power in Congress to kill any bill they didn't like. Wayne Aspinall of Colorado was the autocratic chairman of the House Interior Committee, which would have to report out the bill in order for it to reach the floor of Congress; he could bottle it up forever if he desired. Clinton Anderson of New Mexico had similar power in the Senate, Carl Hayden notwithstanding. And there were plenty of others in both houses to be reckoned with.

However, for all their insistence on an augmentation project that might be viewed as something akin to a military invasion by northern Columbia or the Pacific Northwest, the upper-basin representatives were curiously ambivalent about the one item *already authorized* in HR 4671 that would generate the billions that would allow such a rescue project to be built—Bridge Canyon Dam. "New Mexico observes that its inclusion could be untimely and unwise," reported Crandall to Dominy. Even the choleric Felix Sparks, he wrote, was inclined "to defer to the lower-basin states on this question." Wyoming's and Utah's positions were "not materially different than the position of Colorado and New Mexico." And yet if Bridge Canyon Dam were not built, with its promise of huge amounts of high-priced peaking power, how could the rescue project they insisted on be self-financing? It couldn't. But no Reclamation project had ever been built that didn't at least create the *illusion* that it was self-financing.

For the moment, however, the upper-basin states were not worried about that. They were much more worried about a former magazine editor and amateur lepidopterist from Berkeley, California, named David Brower.

David Brower's passionate opposition to dams has its origins in his teeth. Brower's childhood, spent in that most tolerant of American cities, had not been happy. He had an awkward case of shyness and, to boot, a row of missing teeth, and his schoolmates taunted him mercilessly about both. In his midteens he departed for the only place in California where he felt he would be left alone or at least find better company: the Sierra Nevada. In those days—the late 1920s—backpacking and mountaineering were considered the oddest of preoccupations, the province of slightly deranged British peers. The Sierra Nevada, which is invaded by so many hikers today that it feels like a zoo, was virtually devoid of humanity. The rapture Brower experienced there transported him to a mystic state; it became a dependency, a drug. He had food and supplies cached all over the place; he could return to one weeks after laying it in and it would still be there. Like his hero John Muir, Brower grew intimate with vast portions of that range. He would return to Berkeley, work at odd jobs for a while, make enough money to quit, and leave for the mountains again. By his late twenties, Brower had become the sort of person the water-development lobby cannot fathom: someone who puts unspoiled nature above the material aspirations of mankind. For his part, by the time he became the first paid executive director of the Sierra Club, in 1952, Brower had decided that no work of man violated nature as completely, as irrevocably, as a dam.

Relatively late in life, Brower had discovered the sublime emptiness of the plateau and red canyon country of the Colorado River Basin. It was the same terrain that had enchanted John Wesley Powell eighty years before, and it was almost as unpeopled and unspoiled as it had been then. Brower loved everything about it: the bottomless dry wind-sculpted canyons, beginning suddenly and leading nowhere; the rainbow arches, overhangs, and huge stately monoliths (an expert rock climber, Brower had pioneered the route up the most impressive of them, Shiprock in New Mexico); the amphitheater basins ringed by great hanging rock walls; the chiaroscuro desert sky, with its promise of rain that rarely came. Above all, he loved the desert rivers. Brower's favorite place in the Colorado Basin was Echo Park. Near the confluence of the Green and the Yampa rivers, Echo Park was a pure indulgence in the most austere of deserts. In autumn, its groves of cottonwood and yellowing willow gave it a New England air. In the spring, the swollen Green would flood the canyon bottom and leave lush meadows as it went. Echo Park was probably the most beautiful canyon flat in all of Utah, part of Dinosaur National Monument. It was also an ideal site for a dam.

Echo Park Dam was to have been a part of the Colorado River Storage Project—one of the first of the giant cash register dams. David Brower loathed it as he had never loathed something before. Brower had no training as an engineer, but he was the son of an engineer, and he led the fight against Echo Park Dam in the late 1950s, going after the Bureau with its own favorite weapon—statistics. Brower liked to quote Disraeli about the three kinds of lies: lies, damned lies, and statistics. The Bureau had confidently proclaimed that Echo Park would conserve 165,000 acre-feet of water over any alternative site; Brower demonstrated convincingly that it would conserve nineteen thousand acre-feet at most. The Bureau said it would add to the basin's water supply; Brower argued, with evaporation figures, that the basin might well lose water if Glen Canyon, the other big cash register dam, was also built. He demonstrated that a coal-fired powerplant would produce power for less money. It would be a great mistake, he told an incredulous Congressional subcommittee composed mainly of westerners, to rely on the Bureau's figures "when they cannot add, subtract, multiply, and divide." The Bureau reacted to such challenges with a mixture of bafflement and contempt, especially after Brower admitted that he had only made it through the ninth grade. But he had been secretly coached by Walter Huber, then the president of the American Society of Civil Engineers—for someone who had limited skill with people, Brower had an amazing ability to marshal expertise—and his calculations were largely supported by General Ulysses S. Grant of the Corps of Engineers. (Sometime later, the Bureau's regional director in Salt Lake City, Olie Larson, was presented with a rubber slide rule by a group of fellow engineers; it was his award for stretching the truth at Echo Park.)

In the end, Brower and a handful of conservationists managed to bring about the biggest defeat the western water lobby had suffered until then: a denial of funds to build Echo Park Dam. To pull it off, though, they had had to compromise; for the sake of victory at Echo Park, they had agreed to leave Glen Canyon Dam alone. Later, when

the dam was already under construction, Brower floated this then
almost inaccessible reach of the Colorado River in a dory much like
Major Powell's. He was astonished by the beauty of the place, as were
most of the handful of people (a few thousand perhaps) who managed
to see Glen Canyon before it was drowned. When the reservoir filled,
Brower's friends actually wondered whether he might shoot himself.
In the forward to a Sierra Club book called *The Place No One Knew*,
he flagellated himself over the loss. "Glen Canyon died in 1963," he
wrote, "and I was partly responsible for its needless death. So were
you. Neither you nor I, nor anyone else, knew it well enough to insist
that at all costs it should endure." Never again, Brower vowed, was
he going to compromise over such a dam.

The battle over the Grand Canyon dams was the conservation
movement's coming of age. Only the upper basin had wanted Echo
Park built; the lower-basin states had either remained neutral or opposed it. But now everyone knew the river was overallocated, and
everyone wanted to see it replenished by water from somewhere else,
so all the basin states were in favor of the Grand Canyon dams. Never
before had conservationists challenged the collective will of seven
states. Brower and the Sierra Club led the fight. As in the Echo Park
battle, he managed to recruit heavyweight expertise. Luna Leopold,
one of the country's leading hydrologists and the son of Aldo Leopold,
the famous ecologist, was willing to take a swipe at the Bureau's flow
calculations. Brower found some nuclear engineers from M.I.T. and
Bechtel who were eager to demonstrate why nuclear reactors were a
cheaper alternative. (Brower would later become one of the leading
opponents of nuclear fission.) His most valuable discovery, however,
was an utterly unknown thirty-year-old mathematician from New
Mexico named Jeffrey Ingram. Ingram was a self-described fanatic
about two things: the Grand Canyon, and numbers. He loved playing
with figures, and above all he loved exposing figures as frauds. In
particular, pyramid schemes fascinated him, and in the Bureau's pay-
back scheme for the Pacific Southwest Water Plan, he thought he had
discovered the greatest pyramid scheme anyone ever saw.

In order to finance the CAP and Bridge and Marble Gorge dams,
Ingram discovered, the Bureau planned to capture the revenues from
Hoover, Parker, and Davis dams, after their power sales had paid them
off in the late 1980s, and reroute them into the new projects. For one
thing, the Bureau, under Reclamation law, had no business doing this.
All surplus power revenues were supposed to revert to the Treasury,
in order to compensate the taxpayers for having forgiven interest obligations on the irrigation features of the projects. But that was not

the half of it. The whole rationale for the Grand Canyon dams was that the river would have to be augmented someday, and power dams were the only means of raising the money for an importation project. The new dams, however, would be terrifically expensive compared to their predecessors. Hoover had been built for the incredibly low sum of $50 million; Bridge Canyon would likely cost close to $1 billion. Because of their enormous cost, the new dams would see their revenues tied up for years, for decades, repaying their own costs and subsidizing the CAP—a subsidy that was crucial if the Bureau was to find anyone to buy its water. Even if revenues from Hoover, Parker, and Davis dams were added, all of that money would be consumed for a seemingly endless period paying for the new works and the CAP subsidy. It would be financing the *depletion* of the river; there would be no money for *augmentation* until long after the basin was predicted to run out of water. In fact, according to Ingram's calculations, if you *didn't* build the Grand Canyon dams, money would start flowing into the development fund sooner than if you *did*. Ultimately, the dams would generate a lot of money—perhaps enough to finance most of the cost of diverting a distant river, if one could be found. But by then it would be too late. The Colorado River would have long since run dry.

It was a formidable argument, and it forced the Bureau of Reclamation to redirect its creative energies toward convincing the Bureau of the Budget that it wasn't really so. In the meantime, David Brower was free to do what he did best: publicity. With the help of two San Francisco advertising men, Jerry Mander and Howard Gossage, the Sierra Club took out full-page advertisements attacking the dams in the Washington *Post*, the New York *Times*, the San Francisco *Chronicle*, and the Los Angeles *Times*. One of the Bureau's arguments for building the dams, an argument which it would later regret, was that tourists would better appreciate the beauties of the Grand Canyon from motorboats. "Should we also flood the Sistine Chapel," asked one advertisement, "so tourists can get nearer the ceiling?" The response was thunderous. Dan Dreyfus was still with the Bureau of Reclamation then, in charge of planning projects like the CAP. "I never saw anything like it," remembers Dreyfus. "Letters were arriving in dump trucks. Ninety-five percent of them said we'd better keep our mitts off the Grand Canyon and a lot of them quoted the Sierra Club ads."

"Jerry Mander and Howard Gossage were both geniuses," Brower would later reminisce. "We did a split run of one ad. I wrote one, which went, 'Who Can Save Grand Canyon—An Open Letter to Stewart Udall.' Jerry Mander's said, 'Now Only You Can Save Grand Canyon from Being Flooded for Profit.' We arranged to have a split run because I thought my ad was saying the right things and he thought his ad was. The upshot of it all was that Jerry Mander's ad outpulled mine two to one. The Sistine Chapel line was suggested by a Sierra Club member from Princeton. I wasn't sure about it. Jerry Mander jumped at it. He was right. That ad was dynamite. It was the ad the Internal Revenue Service cited when they revoked our tax-deductible status."

Who persuaded the IRS to revoke the Sierra Club's tax-deductible status is a question still debated today. Brower is convinced that Congressman Morris Udall, Stewart's brother, was behind it. He insists Udall even confessed to him once in an unguarded moment. Others suspect Stewart. Everyone wanted to lay the blame with Dominy, but private memoranda from Dominy's files suggest that he was as perplexed as everyone else; he wanted to locate the culprit so he could congratulate him. It was, obviously, a purely political strike. Other tax-deductible groups were at least as active in trying to influence legislation as the Sierra Club, and nothing happened to them. Whoever was responsible, the Sierra Club suddenly found itself tilling fund-raising soil as dry as the West's, and had a close brush with bankruptcy. Brower, for his part, would soon find himself without a job, fired by the club's board of directors for fiduciary irresponsibility. But, in the end, none of it was to make much difference. The ads had been published; the public was outraged; the Grand Canyon dams were doomed to defeat. Everyone knew it except Floyd Dominy, the Bureau of Reclamation, and the Colorado Basin states.

At the same time that Lyndon Johnson was telling himself and anyone who would listen that the opponents of his war in Vietnam were a handful of draft-dodgers, the proponents of the Grand Canyon dams were telling themselves that their opposition was limited to the Sierra Club. The real problem, Wayne Aspinall, Carl Hayden, and Floyd Dominy would fume, was Dave Brower's "lies." Once people understood that Bridge Canyon Dam would only flood Grand Canyon National Monument, and not the park itself, they would come around and support the dams. They believed, in other words, that the fate of the dams hinged on a technicality. They couldn't fathom that a sea change in public feeling toward the natural world was taking place, one of those epochal shifts that guarantee that things will never be the same. But it was, and people didn't care whether the dams flooded the monument or the park, or whether they drowned a mile or a hundred miles of the canyon, or whether they submerged the bottom fifty feet or the entire chasm. They wanted no dams—period.

In 1966, the National Reclamation Association held its annual meeting in Albuquerque, and Brower, to his considerable surprise, was invited to speak. To the NRA's surprise, he showed up. So did Wayne Aspinall, the chairman of the House Interior Committee, and when a photographer spied them twenty feet apart he tried to arrange a picture. Aspinall glared at Brower and shouted, "No picture of mine is being taken with that liar!" When a reporter asked the Congressman how Brower had lied, he responded that the dams would in no way flood the national park. They would merely flood 120 miles of the Grand Canyon. As far as Aspinall was concerned, that was a distinction of the utmost significance.

As for Dominy, facing the prospect of a major defeat for the first time in his life, he not only believed that Brower was a liar—he was convinced he wasn't worth worrying about. "If you even suggested to Dominy that Brower was winning," says a former Bureau man, "he would have fired you on the spot." Finally, when even his allies in the Southwest began to have misgivings about Bridge Canyon Dam, Dominy began to take his nemesis a little more seriously. He ordered employees to stalk Brower, showing up at his speaking engagements to report on what he said and get in a little heckling on the side. But Dominy's men either were poor judges of audience response or were so afraid of their chief that they told him exactly what he wanted to hear. "Mr. Brower's talk . . . was highly emotional," wrote a Bureau man in a blue-envelope report on a Brower address. "It was completely lacking in any kind of substantiating data, and he appeared a far less formidable opponent than anticipated. It is my opinion from this encounter that the Bureau should encourage face-to-face discussions with Mr. Brower before unbiased audiences because any reclamationist, armed with basic facts, could adequately defend the Bureau's position against his pure emotionalism."

In that particular speech, Brower had said that he wouldn't mind dams in the Grand Canyon as long as the Bureau built a comparable canyon somewhere else. He received a standing ovation—in Denver.

The handwriting was on the wall by March of 1966, when the *Reader's Digest* ran an article attacking Marble Gorge and Bridge Canyon dams in a tone that could almost be described as enraged. "Right after the *Reader's Digest* article, *Life* ran a big goddamned diatribe," remembered Dan Dreyfus. "Then we got plastered by *My Weekly Reader*. You're in deep shit when you catch it from them. Mailbags were coming in by the hundreds stuffed with letters from schoolkids. I kept trying to tell Dominy we were in trouble, but he didn't seem to give a damn. It was kind of surprising, because Dominy could be very

flexible when it came to the smaller projects. He made some big conces-
sions here and there and wasn't bothered by it. On this one he was an
utter maniac. In a way you can't fault the man, though, because even
though Dominy was a good liar when he had to be, here he was a
prisoner of his own intellectual honesty. A lot of people figured that
no one was going to let the Southwest run out of water, and if the
time came when it wanted more the country would just pay for it,
whatever the cost. I mean, New York City was full of immigrants,
criminals, minorities, so who gave a damn if it went bust? But Phoenix
and irrigated farmland—that was America! So it may have been a
correct assumption. But Dominy said, 'No way—this project is going
to include those dams.' "

By 1967, it had become obvious to everyone but Dominy and Carl
Hayden that the Grand Canyon dams would have to go. Rescue for
the Colorado Basin might never come without them, but the Central
Arizona Project would never be built with them. The problem, for
Stewart Udall, was how to sneak the amended legislation past Hayden
and Dominy. Hayden might not be too much of a problem; he was old
and senile and in the hospital half the time, and he was desperate to
see the CAP authorized before his death, which might come at any
time. It was Dominy—bullheaded, willful, obsessed with defeating
Brower—who somehow had to be handled. The opportunity came
fairly soon. With the Bureau now helping to build dams all over the
world, the commissioner had to make an annual global inspection of
projects-in-progress; it was a condition imposed by the Agency for
International Development, which was pumping billions of dollars
into dam construction, and even as the Colorado River battle raged
away Dominy had to absent himself for a few weeks. In early 1967,
the commissioner grabbed his hat and was gone. Almost as soon as
his plane left the runway at Dulles Airport, Udall was telling his Assistant Secretary, Ken Holum, to take Bridge Canyon Dam out of the
CAP legislation and come up with an alternative before Dominy returned. The main objective was to find enough power to pump the
water to central Arizona. The means of financing a rescue project
would simply have to be put off. A Dominy representative would, of
course, have to sit on the task force, and Udall had just the person in
mind—Daniel Dreyfus. Publicly, Dreyfus could write a good rah-rah
speech for Dominy about Marble Gorge and Bridge Canyon dams.
Privately, he believed neither in them nor, for that matter, in the CAP.
He wasn't even sure he believed in the Bureau of Reclamation
anymore.

"The hardest part for me was getting the regional commissioners to go along," Dreyfus would recall in his Senate office in 1981. "Dominy had them all so scared that when I told them what we were up to, they wanted to crawl in a hole. 'Oh, no, *Floyd's* got to be here!' 'You know what *Floyd* would think of this.' '*Floyd* will shit a brick.' One regional director was so terrified I had to fly out to Phoenix to put some fiber in his backbone. The solution itself was kind of clumsy, but it was simple. We decided to buy a share of the Navajo Powerplant in northern Arizona. For the first time, the Bureau was going to own something it always hated—a piece of a great big smoke-belching coal-fired powerplant. It didn't solve a damn thing except that it gave us the power to pump water to central Arizona. The fact is we were licked. The conservationists and the press and ultimately the public licked the Bureau of Reclamation, and the last person in the world to admit it was Dominy. He wouldn't admit it, but I can't believe he didn't know what was coming. By the time he took off to go overseas he was fighting a rearguard action, and he knew it. Maybe being out of the country was a way for him to save his honor. When he returned, I was the one who had to go see him with a copy of the agreement we'd worked out. I thought he was going to go through the roof, but Dominy always had a way of catching you off guard. His reaction was complete and total lack of interest. He already knew all about it. He just said, 'I don't even want to hear about it,' and told me to get the hell out of his office. He didn't even look up from what he was reading on his desk."

Like the westbound wagons that had to jettison furniture, food, even water in order to plow through the desert sands, the Central Arizona Project was finally light enough to move. The Colorado River Basin Project Act was signed into law by Lyndon Johnson on September 30, 1968—the most expensive single authorization in history. Besides the CAP, it authorized Hooker Dam in the Gila Wilderness of New Mexico, the aqueduct from Lake Mead to Las Vegas, the Dixie Project in Utah, and the Uintah Unit of the Central Utah Project—the first piece of a water-diversion scheme that promised to be nearly as grandiose as the CAP. It also authorized the San Miguel, Dallas Creek, West Divide, Dolores, and Animas La Plata projects in Colorado, and it authorized a Lower Colorado Development Fund, still penniless, to build an augmentation project that hadn't yet been defined, let alone approved. Almost unnoticed alongside everything else, the bill made deliverance of Mexico's 1.5 million acre-feet of water—of tolerably sweet water—a *national* responsibility, whatever that meant. Loosely interpreted, it might mean a pipeline from Lake Superior to Mexicali.

The five Colorado projects—which could easily add a cool $1 billion to the cost of everything else—were an object lesson in the workings of the Congressional pork barrel. They were put into the bill at the insistence of Wayne Aspinall, the black-eyed former schoolteacher with a testy principal's disposition who had climbed from a little western Colorado town to become the chairman of the House Interior and Insular Affairs Committee. Aspinall distrusted urban, expansionist California with all the recondite loathing of a small-town mind, and he didn't trust Arizona much more. The overallocated river ran right under the window of his expensive home on Aspinall Drive in Palisade, Colorado, and he figured that Colorado had better extract every drop of its rightful share or California and Arizona would take it and never give it back. If the CAP was to get past the chairman of the House Interior Committee, Colorado was going to be satisfied first.

The problem was that by 1968, there wasn't a single irrigation project left on the West Slope of the Rockies that was economically feasible. The best ones—or, to put it more accurately, the least senseless ones—had already been authorized by the Colorado River Storage Project Act in 1956. If Colorado had a need for more water, and a place where a new project might actually make sense, it was on the eastern plains, where both the growing cities of the Front Range and the farms atop the Ogallala aquifer were facing water famine thirty or forty years down the road. One of the Bureau's most successful projects, Colorado–Big Thompson, was already delivering Colorado River water across the Continental Divide through a tunnel to the East Slope; the power produced by the steep drop down the Front Range was enough to justify the expense of the tunnel, and the additional water diverted from the upper Colorado to tributaries of the Platte River was welcomed by everyone from canoeists to whooping cranes to irrigators in Colorado and Nebraska. There was no reason why another such trans-basin diversion project couldn't be built. No reason, that is, except Wayne Aspinall. The eastern plains were in someone else's district.

During an interview in 1979, Felix Sparks, who selected the five projects at Aspinall's behest, conceded as much. "Twenty years ago, we already saw urbanization as inevitable," Sparks said. "So I looked around for a place where we could keep a viable agricultural industry going. *We didn't want* to let cities and industry have the water. We picked those projects on the basis that it would be impossible, physically impossible, for Denver to get its hands on that water." It was an extraordinary admission. All that Sparks failed to mention was the fact that he was likely to benefit personally from new projects on the West Slope. Though a modestly paid public servant, Sparks was a fairly wealthy man, the result of some shrewd and highly secretive business investments across the Front Range. He was widely rumored to own a large interest in a food-processing plant on the West Slope— a plant that could use a fresh supply of locally grown fruit nurtured on taxpayer-subsidized water. Of course, Felix Sparks, like a lot of western farmers, didn't believe in such a thing as federally subsidized water. "This business of federal Reclamation subsidizing irrigation water," he snorted, "is absolute, utter, unmitigated crap."

Subsidy, however, was exactly what Aspinall and Sparks's five projects would require, subsidy on a scale that made even the Bureau cringe. It fell to Dan Dreyfus, the Bureau's house magician, to invent enough benefits to make them pass muster. "Those projects were pure trash," said Dreyfus in an unusually candid interview in 1981, as he prepared to retire from public service. "I knew they were trash, and Dominy knew they were trash. The way they got into the bill was, Aspinall called up Udall one day and said, 'No Central Arizona Project will ever get by me unless my five projects get authorized, too.' When Udall passed the word on to us, we were appalled. The Office of Management and Budget had just bounced Animas–La Plata. Now we had to give it back to them and make them reverse themselves. I had to fly all the way out to Denver and jerk around the benefit-cost numbers to make the thing look sound.

"As a last resort," Dreyfus continued with a grim smile, "Dominy and I went to see Aspinall and tried to talk him out of it. Dominy said, 'Look, Congressman, these projects won't work as irrigation projects. We can't afford to pump water from the reservoirs to the irrigable lands because we haven't got any surplus power in the river, and the alternative is to follow the land contours with canals that are going to be ungodly long and expensive. They'll cost so much you might run into some real problems getting appropriations for these things.' What Dominy suggested was to build the dams and forget the rest. He said, 'What you really want is to capture your entitlement. The dams alone will do that. California will never see that water, and you'll cut the cost in half.'

"Dominy could be the most persuasive man I ever met," Dreyfus said, "but Aspinall wouldn't budge. He liked to think of himself as almighty principled, so he got huffy and said, 'The Reclamation program knows no such thing as a project without beneficiaries. The answer is no.' "

Those kinds of principles usually end up costing the taxpayers a lot of money, but in this case they may have cost Aspinall his projects. Why would California's and Arizona's Congressional delegations, which outnumbered Colorado's ten to one, vote for appropriations for five projects which would mean surrendering water their own constituents were using? Since the projects made so little sense, and were so expensive, the rest of Congress might follow their lead. Aspinall, however, had already succumbed to the twin delusions that affect so many committee chairmen—that he would be reelected forever, and that he would live nearly that long. As long as he sat in his committee chair, he could deny California and Arizona whatever he pleased unless they voted in favor of his projects. It was a reasonable argument, until he was bumped out of office four years later by a virtually unknown law professor named Alan Merson—a candidate who had campaigned heavily on the environmental principles that Aspinall often scorned. By 1987, only the Dolores Project was close to being finished—it alone would end up costing $450 million, and the water promised to be so expensive that the farmers were anxiously trying to back out of their contracts. "We were dumb and greedy," said one Junior Hollen. "If they force us to buy the water now, it will bankrupt us."

Meanwhile, twelve years and more than $2 billion after the passage of the Colorado River Basin Project Act, the Central Arizona Project was nearly built too. It would be a Dolores Project on a far, far grander scale.

A political mirage for three generations of Arizonans, the Central Arizona Project is now a palpable mirage, as incongruous a spectacle as any on earth: a man-made river flowing uphill in a place of almost no rain. To see it there in late 1985, just being filled, induces a kind of shock, like one's first sight of Mount McKinley or the Great Wall. But it is an illusion that works both ways. Up close, the Granite Reef Aqueduct seems almost too huge to be real. Where will all the water come from? From the air, however, the aqueduct and the river it diverts are reduced to insignificance by the landscape through which they flow—a desert that seems too vast for the most heroic pretensions of mankind. The water the aqueduct is capable of delivering is more than Cleveland, Detroit, and Chicago consume together. Pour it on Arizona, however, and it would cover each acre with two hundredths of an inch. In the summer, when the temperature reaches 135 degrees at ground level, that much water would evaporate before you had a chance to blink.

To build something so vast—an aqueduct that may stretch eventually to 333 miles, pumps that will lift the water 1,249 feet, four or five receiving reservoirs to hold the water when it arrives—at a cost that may ultimately reach $3 billion, perhaps even more, would seem to demand two prerequisites: that there be a demand for all the water, and that it be available in the first place. In Arizona, all of this has been an article of blind faith for more than half a century. Build the CAP, and the aqueduct will be forever filled because of Arizona's Compact entitlement; fill the aqueduct, and the water will be put to immediate use—that is what every politician who ever aspired to sainthood in Arizona has said. But there are a number of reasons why this will not be the case—perhaps not remotely the case. If anything, the Central Arizona Project may make the state's water crisis *worse* than ever before.

When the Colorado River Basin Storage Act was bottled up in the House Interior Committee in the mid-1960s, it wasn't just the Sierra Club and the Grand Canyon dams that were responsible. The dams, it was feared, might drag the bill down to defeat on the floor of Congress, but it had to get out of committee first, and the bill's major hurdle there—a hurdle that seemed about fifty feet high—was California. California had five members on the committee and a powerful ally in John Saylor, the senior Republican committee member, who was from Pennsylvania. Saylor was as antagonistic toward the Bureau of Reclamation as anyone in Congress; he especially loved to pick a fight with Floyd Dominy; and he was unalterably opposed to the CAP. He was so valued in office by California that tens of thousands of dollars poured into his campaign coffers from that state to keep him there.

What California demanded as the price for acquiescence was simple—devastatingly simple. Before Arizona received a drop of its entitlement, it wanted its full 4.4-million-acre-foot entitlement guaranteed. As far as California was concerned, there would be no equitable sharing of shortages, no across-the-board cuts in times of drought; it wanted satisfaction no matter what. In fact, what it was really asking was a legislative reversal of the lawsuit it had lost in the Supreme Court. It was an outrageous demand from Arizona's point of view, and few believed that its Congressional delegation would swallow it. But, in the end, they did.

"How do I explain it?" asked Sam Steiger, then a junior committee member from Arizona, repeating the question just asked of him. "I can't. Obviously a deal was struck. I was too junior to be in on it. Mo Udall, Stewart's brother, and John Rhodes were the ones in a position to do it. *Why* did they do it? The only answer I can think of is that they didn't really believe the river was overallocated—that, or else they really believed we were going to get an augmentation project, even without Bridge Canyon Dam. The Bureau of Reclamation wasn't running around Capital Hill crying, 'The river's overallocated! The river's overallocated!' I don't know what figures they were using, but we sure as hell weren't hearing the ones that came out a few years later. They made like there was plenty of water for everyone."

And so, before a real fight even developed over California's imperious demand, the CAP legislation became saddled with what is known as the California Guarantee: 4.4 million acre-feet or bust. Come drought, come calamity, California must be satisfied first.

A few years later, the Bureau was finally forced to admit that its estimate of 17.5 million acre-feet a year was a convenient fiction, and amended it to around fifteen million acre-feet. A few years after that, even the latter figure looked optimistic; independent hydrologists were putting the Colorado's average flow at somewhere around thirteen million acre-feet, perhaps a little more. Southern California was diverting its full 4.4 million acre-feet as it had for years. The upper basin had a diversion capability that had moved past 3.6 million acre-feet and was still building moderately. Evaporation varies from year to year, but averages close to two million acre-feet from all the reservoirs on the main stem and tributaries; and Mexico must get its 1.5 million acre-feet.

Work these figures out and the Colorado River is almost used up if its flow is as low as some say. If the higher estimates are used, there are two to two and a half million acre-feet left. Now consider the projects that are authorized and, in some cases, nearly built or being built. The Central Utah Project. The Animas–La Plata Project. The Dolores Project. The Fruitland Mesa Project. The West Divide Project. The Dallas Creek Project. San Miguel, Savery Pot Hook, Paonia, Florida, and the largest of them all, the CAP. Three or four of these could send the Colorado River into "deficit"; the rest will merely make the deficit hopeless. Everything has turned out exactly as could have been predicted twenty years ago—everything, that is, except the rescue project that was supposed to save the basin states from a Sumerian fate.

The prospects that an augmentation project would be built were already dim in the mid-1960s, before double-digit inflation, before double-digit interest rates, before environmentalism, before federal deficits four times larger than the federal budget was then. Meanwhile, northern Californians have grown so jealous of their "underused" rivers that in a 1982 referendum they emphatically refused to release more water even to the desperate supplicants in the southern half of their own state. The Klamath River alone has nearly as much water in it as the Colorado, and flows to the ocean almost entirely unused, and one could build a reservoir on it two-thirds the size of Lake Mead, but the odds of the Klamath River being rerouted to southern California so the Colorado Basin states can have more water are about the same as the odds of being bitten by a rattlesnake while crossing the street in Washington, D.C. If that is unthinkable, then the odds that Oregon's rivers will be turned southward are even less so. As for the Columbia River diversion, it still has at least one champion, a Los Angeles supervisor named Kenneth Hahn who introduces a resolution calling for it every year, but his resolution cannot even make it past the board of supervisors of one of the most water-starved cities in the world, and that, with luck, is about as far as it will get.

The Colorado Basin, then, is a few years away from permanent drought, and it will have to make do with whatever nature decrees the flow shall be. If the shortages were to be shared equally among the basin states, then things might not be so bad for Arizona. But this will obviously not be the case; there is that fateful clause stipulating that California shall always receive its full 4.4-million-acre-foot entitlement before Phoenix and Tucson receive a single drop. What began as an Olympian division of one river's waters emerged, after fifty years of brokering, tinkering, and fine-tuning according to the dictates of political reality, as an ultimate testament to the West's cardinal law: that water flows toward power and money.

Despite one of the most spellbinding and expensive waterworks of all time, Arizonans from now until eternity will be forced to do what their Hohokam ancestors did: pray for rain. During wet cycles, when Lake Mead and Lake Powell are sending water down the spillways as they were in 1983, the Granite Reef Aqueduct may be delivering something close to peak yield. During drought cycles, the aqueduct may run half empty, if that, and the odds are extremely high that it will run progressively more empty as the years go by. It would be foolish, at this stage, to surmise that all or even most of the upper-basin projects are going to be built, but a few of them are likely to be, and each one will cut into the CAP's supply. The Colorado River, to which Arizona decided to marry its future hopes, will prove no more trustworthy than a capricious mistress, delivering a million acre-feet one year, 400,000 the next.

And this, in turn, raises a bizarre possibility, as unthinkable to modern Arizona as it was to the planners of the CAP: the people of Arizona may not even *want* the modest amount of precious water this $3 billion project is able to deliver.

In early 1980, Phoenix experienced a series of damaging winter floods. The Salt River goes through the center of town and is usually an utterly dry bed of pebbles and rocks; therefore, city streets are laid right across the river, as if it had long since gone extinct. In 1980, however, it rolled cars like boulders—cars whose owners were so used to driving through the riverbed that, despite repeated warnings on the radio, they didn't bother to detour and cross on a bridge as the waters began to rise. Even if they had, it wouldn't have done them much good. Only two of Phoenix's bridges were designed to withstand a flood flow greater than twenty-five thousand cubic feet per second. In February of 1980, the Salt River peaked at 180,000 cfs.

Phoenix owes its existence to this ephemeral desert river, but even so it doesn't seem to hold the Salt in high esteem. On both banks, the floodplain is encroached on by industrial parks, trailer parks, RV parks, but no real parks. The flood channel itself has been developed to a degree, playing host to establishments which are, by nature, transient: topless-bottomless joints, chop shops, cock-fighting emporia. Paris built its great cathedral by its river, Florence its palaces of art; Phoenix seems to have decided that its river is the proper place to relegate its sin. When the Bureau of Reclamation performed a cost-benefit analysis for Orme Dam, once a central feature of the Central Arizona Project, it included as a benefit the flood protection the dam would offer to the cock-fighting and striptease establishments downstream.

That particular dam—a $400 million structure intended to store Colorado River water shipped over in the Granite Reef Aqueduct, and to hold back the occasional flood surge—was one of the main topics of conversation in 1980. On February 27, just after the biggest flood hit, the *Arizona Republic* ran a huge editorial that read, "Are you fed up sitting in traffic, creeping to work, because floods have taken out all but two of the major bridges crossing the Salt River? Are you fed up with reading stories about a new study and more hearings into whether construction of Orme Dam would interrupt the nesting habits of bald eagles . . . of this community playing second fiddle to high-and-dry special pleaders who shed tears over nesting eagles, but can't find compassion for the thousands of families who endure hardship, fear, and ruin as flood waters rampage through the valley?

"I'm mad!" continued the editorial, which was signed by the *Republic*'s editor-in-chief, Patrick Murphy. "I'm mad as hell that high-and-dry Washington bureaucrats have been dilly-dallying for at least ten years over approval of Orme Dam. . . . Now, dammit, give us our dam!"

The "special pleaders" Murphy referred to numbered among them the Yavapai Indians, whose remnant population of three hundred or so lives on a reservation near the confluence of the Salt and Verde rivers, and who would lose their homes to the reservoir. The Yavapai, who appear to be some of the most peaceful, sweet-natured souls on earth—many of them are old and still weave baskets for a living—had won a lot of well-placed sympathy, which was apparently what Murphy was complaining about. Cecil Andrus, then the outgoing Secretary of the Interior—and someone who spent a good part of his term trying to stop the Bureau from carrying out its plans—vowed that the tribe would be relocated over his dead body, and one local attorney who was preparing to fight Orme Dam on their behalf was Stewart Udall—the man who, as much as anyone, had made the CAP and Orme Dam possible. (In later years, Udall, unlike the Bureau, was to rue much of what he said and did in the 1960s; he even spoke at a testimonial dinner, in 1982, celebrating the seventieth birthday of his old nemesis, David Brower.) The dam was in the news so often that one could almost imagine the dancers in the bars debating the pros and cons between acts. What was most striking about the debate, however, was that practically no one seemed to be asking the more fundamental question about Orme Dam. As a $400 million flood-control structure, it made little economic sense; it would be much cheaper to move the relatively few threatened structures and reinforce the bridges. Only if it received and stored a substantial amount of Colorado River water—which implied not only a decent flow in the river but a demand for the water, and an ability to pay for it—did Orme Dam make any sense. Would the water arrive, and arrive predictably and often enough, and be economical enough, so that anyone would want to buy it?

In 1980, one of the few people in the state who seemed to be asking this question was William Martin, an economist at the University of Arizona at Tucson. For having done so, and answering negatively, Martin had been accused in local newspapers of being a paid agent of California, where he was born. The dean of his department denied him merit raises for eight years, and even led a campaign to discredit his academic qualifications, though he wouldn't go quite so far as to try to have him fired outright.

Large and bearded, inclined toward jeans, cowboy boots, and western shirts, Martin looks as if he would feel more at home in the cockpit of a Peterbilt than at a professor's desk, even if his writings are nationally known. His first notoriety came in 1973, when he and a colleague, Robert Young—who was so wounded by the hounding he got that he opted to leave the state—published a book called *Water Supplies and Economic Growth in an Arid Environment*, an innocuous-sounding little tract which, in Arizona, was almost as revolutionary as *Das*

Kapital. They first asked, as a matter of speculation, what might happen if the Central Arizona Project was *not* built. The underground aquifers, Young and Martin reckoned, would undoubtedly be depleted as the farmers continued to pump them out (in the 1960s, the rate of overdraft—use over replenishment—climbed as high as four million acre-feet per year). As pumping costs rose due to the dropping water table, some farmers would begin to go out of business. But there was still enough water so that the decline would be very slow. Arizona's farm income, by Young and Martin's calculations, would be reduced by about one-fifth of 1 percent per year. The reason the decline in income would be so modest was self-evident: as pumped water got more expensive, the farmers would conserve it better and switch to higher-value crops, and they would do more with less. The way to see if the Central Arizona Project Was worth building, then, was to see if each acre-foot of water it brought in would be cheaper than the value (in lost farm income) of each disappearing acre-foot from the aquifers. Martin and Young figured that every acre-foot that was being mined was causing a loss of $5.35 in farm income—a conservative estimate, as far as they were concerned. Could the CAP deliver water cheaper than that? By the Bureau's own calculations, CAP water would cost at least $10 per acre-foot—without even figuring the cost of distributing it. As a result, the farmers would make *more* money if they continued pumping groundwater than if they bought water from the CAP. In fact, if the price of distribution systems—which the farmers would presumably have to build themselves—was as high as it promised to be, buying CAP water might be a ticket into bankruptcy.

Twice since then, Martin has repeated the analysis, and his results confirm his earlier conclusions—only far more emphatically. By 1977, the projected canalside price of CAP water had reached $16.67 per acre-foot. Add the cost of a distribution network, and farmers growing any kind of low-value crops—alfalfa, small grains, perhaps even the state's main crop, cotton—could not afford it. In 1980, he and another colleague from the University of Arizona, Helen Ingram, did a detailed study, region by region, of the likely cost of distribution systems, and were amazed by what they found out. In one irrigation district, Maricopa-Stansfield, the price of the distribution system—hundreds of miles of canals and laterals, headgates, and people to operate them— would likely come to $160 million, leaving each farmer a bill of $100 per acre-foot of water *per year* just for distribution. The Bureau's canalside estimate for CAP water had, by then, risen to around $30 per acre-foot, per year. The price of pumped groundwater *was nearly $100 less* per acre-foot at Maricopa-Stansfield—around $39. It was an extreme case, but Ingram and Martin couldn't find a single irrigation district where CAP water promised to be cheaper than groundwater. In most of them, it would cost half again or twice as much, sometimes more. One of the main arguments the farmers had always made for the CAP was that they couldn't all switch to high-value crops as the groundwater table went lower and pumping costs became intolerable. The American consumer, they said, could only eat so many lemons and oranges. But if Martin's figures were right, farmers who signed contracts to buy CAP water might not even be able to raise oranges on it. In 1980, about the only crop you could raise with water that cost $130 per acre-foot was marijuana.

But that was the *good* news. The bad news was that during periods of drought, with California guaranteed its full entitlement before Arizona received a drop, this incredibly expensive water might often not arrive. The Bureau's own projections showed "firm" CAP water dwindling from 1.6 million acre-feet at the beginning to 300,000 acre-feet or less in fifty years; only during wet years, or if the upper-basin projects are never built, will there be more. To think of the Central Arizona Project as salvation, then, is not just to stretch things a bit. For those groundwater-dependent farmers who will have to build distribution systems, at least—and there are a lot of them—the Central Arizona Project could spell economic ruin.

Did Arizona's farmers realize any of this? One of William Martin and Helen Ingram's graduate students, Nancy Laney, traveled around the state to find out. To her astonishment, most of the farmers didn't. One of the farms Laney visited was the Farmers' Investment Corporation, a huge pecan-growing operation south of Tucson that is about as far from the diversion point on the Colorado River as one can be. (Why pecans, which are native to the Mississippi Delta, should be grown on subsidized water in a desert state is another matter entirely.) If it arrives, CAP water will have surmounted a lift of well over a thousand feet and traveled more than three hundred miles to get there. Meanwhile, there is still plenty of water immediately under the farm, less than two hundred feet down. Despite the huge subsidies written into the CAP—as with any Reclamation project, the farmers are excused from paying interest costs—the groundwater is certain to be much cheaper, at least until the aquifer drops several hundred more feet. (The worst areawide decline in Arizona's water table has been around two hundred feet, and that took decades to happen.) But the farm manager at Farmers' Investment expressed to Laney his unalterable belief that "CAP water will be cheaper than pumping." "Water is essential," he said with religious conviction, adding that he "would back any plan where more water would be available." He had no idea what CAP water would cost him, but planned to sign contracts to buy it anyway. His state of knowledge and level of blind faith were not unusual. One farmer thought that the water was going to arrive by gravity instead of being pumped many hundred of feet uphill. One believed that there was still enough surplus water in the Colorado River to turn the entire Grand Canyon into a reservoir—something he devoutly wished. Only two of the farmers Laney interviewed seemed to have a sense of things as they really were. One realized that Arizona's Colorado River water was jeopardized and thought it was high time we "took" Canada's surplus water to replenish it. The other said that even if it turned out he couldn't afford CAP water, he was going to sign a contract to buy it anyway, because "contracts are made to be broken."

"Contracts are made to be broken." There, in a simple phrase, was perhaps the worst legacy of the Bureau of Reclamation's eighty years as the indulgent godfather of the arid West. The irrigation farmers not only had come to expect heavily subsidized water as a kind of right, allowing them to pretend that the region's preeminent natural fact— a drastic scarcity of that substance—was an illusion. They now believed that if it turned out they couldn't afford the water, the Bureau (which is to say, the nation's taxpayers) would practically *give* it away. These farmers were about the most conservative faction in what may be the most politically conservative of all the fifty states. They regularly sent to Congress politicians eager to demolish the social edifice built by the New Deal—to abolish welfare, school lunch programs, aid to the handicapped, funding for the arts, even to sell off some of the national parks and public lands. But their constituents had become the ultimate example of what they decried, so coddled by the government that they lived in the cocoonlike world of a child. They remained oblivious to what their CAP water would cost them but were certain it would be offered to them at a price they could afford. The farmers had become the very embodiment of the costly, irrational welfare state they loathed—and they had absolutely no idea.

In 1984, Congress began to demonstrate why the farmers might not be so foolish after all. Early that year, it voted to lend them $200 million to help build distribution systems—an interest-free loan, as one might have expected, but the sum was only about half of what they would need, and there was a lot of resistance to lending them the rest. But they still weren't out of the woods. For one thing, the Indian water-rights issue was still substantially unresolved. There was a good chance that the white farmers would have to lease water from the

Indians, who could well end up with most of the water in the CAP. The Ak Chin and the Papago tribes had recently settled with the Interior Department for 300,000 acre-feet, about the consumption of Phoenix. The Papago tribe's water will come directly out of the Tucson Aqueduct—water which the farmers, most of whom had conveniently ignored the Indian water-rights question, had always expected to get. More and more, the CAP was metamorphosing from an agricultural rescue project into an expensive atonement for travesties visited on the Indians, and, perhaps, into a municipal water supply project for Phoenix and Tucson—if *they* feel they can afford it.

"The cities in Arizona are going to get hit even worse than the farmers," Bill Martin told an interviewer in 1984. "The farmers at least get the interest-free subsidy, which is worth a fortune to them. They also get interest-free loans on things like the distribution systems. The cities get none of that. They pay full fare.

"Here in Tucson, we're already drawing groundwater out of neighboring basins because we've depleted ours, and we pay around $430 per household, which seems like a lot. But most of that, I'd say around $400, is to pay off the water mains, the infrastructure, the bureaucracy. It's a distribution cost. It only costs us $30 or a little more to pump the water. But to pump CAP water all the way from the Colorado River to Tucson is going to cost at least $250 per acre-foot; that's what the water is worth when you get rid of all the interest subsidies and so forth. Add $250 to $400, the distribution cost, and people are going to be paying $650 for water. There are families around here who only earn ten or fifteen times that much in a year. So what's obviously going to happen is people are going to conserve, and use a lot less water, and there will be less and less of a need for the CAP.

"It's already happening," Martin continued. "We've all gotten water-conscious, even if we weren't before. Tucson uses a third less water than Phoenix, because up there they still get cheap water from the Salt River Project. Once Phoenix starts paying $600 a year, though, they're going to conserve just like we are."

But if the farmers can't afford the water, and the cities can't afford the water, then who is going to buy it and justify the whole expense?

"Damned if I know" was Martin's response.

I f it seems implausible that Arizona's farmers will buy here-today, gone-tomorrow water that costs three times what farm economics suggest they can pay; if it seems implausible that cities will want to waste millions of dollars a year buying turbid, alkaline water from the Colorado River when they can pump cheaper, fresher groundwater instead—if all of this seems unlikely, what is even *less* likely is that Arizona and the Bureau of Reclamation will permit a giant concrete aqueduct to sit empty in the desert, a ruin before its time. For the aqueduct to remain full as long as it can, however, the farmers must receive most of the water; their collective thirst is much greater than that of the cities. (And in 1985, work on the extension aqueduct to Tucson, the only big potential urban buyer besides Phoenix, had not even begun.) What this in turn implies is subsidy, more than is already there in the form of cheap electricity and interest exclusion—subsidy on a rather heroic scale. The question is, how will it be done?

One person who thought he had it figured out was Sam Steiger, a former Congressman from Prescott, a small city up in north-central Arizona. In the 1960s, Steiger was a prototypical Arizona Republican—crew-cut, jut-jawed, archconservative—who nonetheless had little trouble voting for the CAP. "Of course I was for it. Any Arizona politician who wanted any kind of political future had to be for it. Besides, I was on the Interior Committee, which authorized the thing—one of two Arizonans versus five Californians on the committee. If I had voted against it, I would probably have been shot." In the 1980's, however, Steiger, no longer in office, had gotten into the water-brokering business, which was becoming a cottage industry in western states whose laws permitted some degree of free market in water rights. Suddenly Steiger had an economic interest in the very condition the CAP would pretend to relieve—scarcity—because he was earning a living helping people with good water rights—mainly farmers—sell those rights to people who could pay top dollar for them—usually subdevelopers and cities. If the CAP suddenly brought in a big volume of water to be sold at vastly subsidized rates—or if CAP water was somehow forced on cities that didn't really want to buy it—it would create an artificial glut and hurt his business. But that was exactly what Steiger thought would happen: subsidy and political coercion were going to create a "demand" for CAP water which, even in this third-driest state in the country, would otherwise not exist.

"In the first place," Steiger said during an interview in 1985, "we passed a strict groundwater law here in 1980, one that was supposed to have been passed ten years earlier. The CAP legislation we passed in 1968 demanded it—what was the point of approving the project if the farmers kept sending the aquifer down to hell anyway? When the Carter people threatened to withhold funds for the CAP until the law was passed, it finally went through the state legislature. What that law does, besides restrict pumping, is demand that any developer who sells a new home guarantee the buyer a hundred-year supply of water. Otherwise, he can't sell. Hell, I can sell you a home and guarantee you that in a hundred years I'll give you desalted water from the ocean. I'll be dead then anyway—that's how ridiculous the provision is. But the way it's being interpreted by the Department of Water Resources is this: no developer gets his certificate unless he's signed up for CAP water, and without that certificate he can't sell his house. The odds that there'll be water in the Granite Reef Aqueduct in a hundred years are probably lower than the odds we'll be getting water from the ocean, but the developers are stuck. So are the cities. If a city wants to grow, it has to buy water from the Central Arizona Project."

That, by Steiger's reasoning, was how the cities would be forced into the hand. The farmers, he felt, would be corralled by the new law's restrictions on groundwater pumping; at some point, they would have to rely more on surface water, and the only available surface water would be the CAP. The problem with the farmers, though, is that their demand is, to use that economists' word, inelastic: charge them too much and they'll go belly up. So the farmers, according to Steiger, will be brought in with the carrot rather than the stick. In 1984, the first fifty-year contracts for cheap Hoover Dam power expired—the dam was finally paid off. The new contracts negotiated by the Interior Department didn't raise the rates much, but they did tack on a surcharge of four mills per kilowatt-hour which is to go as a direct subsidy to the CAP. Four mills per kilowatt-hour—a few cents per day—may not sound like much, but multiply it by a couple of million users and it is a fair piece of change: millions of dollars per year. It is an almost poetic irony that most Hoover power is sold in southern California; at last, Arizona was going to get its pound of flesh from California, after involuntarily "loaning" that state water for so many years. A similar, smaller subsidy applies to power sold by the Navajo Power Plant. On top of *that*, the Central Arizona Conservancy District—the imperium created to receive and distribute CAP water—is permitted, by law, to buy cheap Hoover power and resell it at market rates, funneling the profits directly into the project to subsidize the water.

"Add all this nonsense to Congress's interest-free loan for distribution systems and some other things they're bound to cook up, and it's all of a piece," Steiger said with palpable disgust. "They'll skin the cat twenty ways if they have to, but they're going to make the water affordable. Congress will go along, because it will be goddamned embarrassing for Congress to have authorized a multibillion-dollar water project when there's no demand for the water because no one can afford it. The CAP belongs to a holy order of inevitability. Will Congress bail out the big banks that pushed all those loans on Latin America, when the countries finally default? Of course. Will it make water affordable for Arizona's farmers? Of course.

"The sensible thing would have been for the farmers to move," Steiger said. "There are hundreds of thousands of acres of good farm- land right along the Colorado River where you'd only have to build short diversion canals and maybe pump the water uphill a few hundred feet. But the farmers got established in the central part of the state because of the Salt River Project. The cities grew up in the middle of the farmland. The real estate interests, the money people—they're all in Phoenix and Scottsdale and Tucson. They didn't want to move. So we're going to move the river to them. At any cost. We think."

CHAPTER NINE

The Peanut Farmer and the Pork Barrel

At the restaurant in the Dillard Motor Hotel in Clayton, Georgia, a little town in a mountainous northern corner of that state, a yellowed old newspaper clipping has been posted by the telephone for years. The story includes a photo showing two men in an open canoe going through Bull Sluice, a Class V rapids on the Chattooga, one of the South's preeminent whitewater streams. According to the official classification system of the American Whitewater Affiliation, a Class V rapids consists of "extremely difficult, long, and very violent rapids with highly congested routes which nearly always must be scouted from shore. Rescue conditions are difficult and there is significant hazard to life in event of a mishap." In the photo, the man in the stern of the canoe looks scared to death, but the man in the bow has a look of grim, Annapolis determination on his face—as if he were smoking out a nest of wasps. According to the story, which is dated sometime in 1972, this was the first run of Bull Sluice in an open canoe, ever. Others have their doubts about that—which is, of course, to be expected on a river with this sort of reputation—but most everyone acknowledges that even if they were not the very first, they were among the first.

The man in the stern is Claude Terry, an expert local river runner. The man in the bow is the governor of Georgia, Jimmy Carter.

The lore of the South could not survive without rivers any better than the human body could survive without blood. Rivers wind through Twain's and Faulkner's and James Dickey's prose; they flow out of Stephen Foster's lyrics. Yet it is the South, more than any region except California, that has become a landscape of reservoirs, and southerners, more than anyone else, are still at the grand old work of destroying their rivers. With one hand they dam them; with the other they channelize them; the two actions cancel each other out—the channelized streams promote the floods the dams were built to prevent— and the whole spectacle is viewed by some as a perpetual employment machine invented by engineers.

The reasons behind the South's infatuation with dams was somewhat elusive. Precipitation in the South is uniformly ample, the rivers run well and often flood, and good damsites are, or were, quite common. But the same applies to New England, and there the landscape contains relatively few dams. There are water-supply reservoirs and small power dams, but only a handful of mammoth structures backing up twenty-mile artificial lakes, which are encountered everywhere in the South. Whatever the reasons, it is an article of faith in the South that you send a politician to Washington to bring home a dam. The first southern politician of national stature who went on record opposing one may have been Jimmy Carter.

Carter's misgivings about dams appear to have been rooted in metaphysics, flintiness, and a sense of military honor. As a businessman, a state legislator, and the chairman of the Middle Flint River Planning and Development Council, he was at first enthusiastic when the Army Corps of Engineers announced plans to erect Spewrell Bluffs Dam, a $133 million structure on the Flint, which is one of Georgia's larger rivers. However, some of Carter's personal friends belonged to the state's environmental community, and at about the same time he was running for governor, they introduced him to canoeing and river rafting, a sport with which he immediately fell in love. Caught between political expediency—many of the state's business and labor interests were equally in love with Spewrell Bluffs Dam—and the appeals of close friends and his own changing values, Carter decided to make up his mind purely on the facts. He got a copy of the Corps' general plan and environmental statement, closeted himself in a room, and, displaying that passion for detail that was to contribute to his political undoing, read it from cover to cover. He cross-checked its assertions

CADILLAC DESERT 308

with a number of experts; he did his own math; he graded the Corps' hydrology (Carter had graduated from Annapolis as an engineer). In the end, he wrote a blistering eighteen-page letter to the Corps accusing it of "computational manipulation" and of ignoring the environment; then, exercising his gubernatorial discretion, he vetoed the dam. According to friends, Carter was deeply incensed by the Corps' reliance on deception to justify the dam; as an Annapolis graduate, he didn't believe a military unit would do such a thing. And, perhaps because he did go to Annapolis instead of West Point, he took it personally. "The Corps of Engineers lied to me," he told his friends. He said it as if a stranger had wandered into his house, eaten everything in the icebox, and then, on leaving, chopped down his favorite tree.

Carter also possessed something rare among American politicians—a sense of history—and, according to those close to him, he began to wonder what future generations would think of all the dams we had built. What right did we have, in the span of his lifetime, to dam nearly all the world's rivers? What would happen when the dams silted up? Fixed, huge, and permanent, dams were also oddly vulnerable. What if the climate changed? What if there were floods which the dams, their capacity drastically reduced by silt, couldn't hold? What if there were terrible droughts, and farms and desert cities that owed their existence to dams faced economic ruin? Besides, having already built fifty thousand of them, what were we getting for our investment now? By the time Carter became President, the cumulative federal debt was approaching a trillion dollars and inflation had already visited the double digits, but the federal water bureaucracies were still going through $5 billion every year. One of the first things he was going to chop out of the federal budget was dams.

To a degree that is impossible for most people to fathom, water projects are the grease gun that lubricates the nation's legislative machinery. Congress without water projects would be like an engine without oil; it would simply seize up. If an influential southern member of Congress didn't much like a program designed to aid a certain part of the Northeast, then it would not be unheard-of for the Congressional delegation from that region to help him get a dam built in his state. If a Senator threatened to launch a filibuster against a particular program, perhaps the program's advocates could muster support for the Senator's favorite water project.

In the Congress, water projects are a kind of currency, like wampum, and water development itself is a kind of religion. Senators who voted for drastic cuts in the school lunch program in 1981 had no compunction about voting for $20 billion worth of new Corps of Engineers projects in 1984, the largest such authorization ever. A jobs program in a grimly depressed city in the Middle West, where unemployment among minority youth is more than 50 percent, is an example of the discredited old welfare mentality; a $300 million irrigation project in Nebraska giving supplemental water to a few hundred farmers is an intelligent, farsighted investment in the nation's future.

Among members of Congress, the intricate business of trading favors is commonly referred to as the "courtesy" system, or, more quaintly, the "buddy" system. Among its critics—a category that extends to include anyone who has not yet benefited from it—it is called log-rolling, back-scratching, or, most often, the pork barrel. (The phrase "pork barrel" derives from a fondness on the part of some southern plantation owners for rolling out a big barrel of salted pork for their half-starved slaves on special occasions. The near riots that ensued as the slaves tried to make off with the choicest morsels of pork were, apparently, a source of substantial amusement in the genteel old South. Sometime in the 1870s or 1880s, a wag decided that the habitual efforts by members of Congress to carry large loads from the federal treasury back to their home districts resembled the feeding frenzies of the slaves. The usage was quite common by the late 1880s; and in 1890 it showed up in a headline in the New York *Times*, assuring its immortality.) Members of Congress who believe in the system—there are many who fervently do, and probably an equal number who dislike it but go along—argue that it benefits the nation as a whole by distributing public-works money to all the fifty states in more or less equal proportion. It doesn't. Anyway, to say the Congress cannot function without the "courtesy" system is to say that it cannot conduct its business without indulging in bribery, extortion, and procuring.

Ideology is the first casualty of water development. Senator Alan Cranston of California, who is well out on the left of the Democratic Party, spearheaded the successful effort to sextuple the maximum acreage one could legally own in order to receive subsidized Reclamation water. Having accomplished that, Cranston, heavily financed by big California water users, launched his presidential campaign, railing against "special interests." Senator Ernest Gruening of Alaska, who built a reputation as one of the most ardent conservationists in Congress, also campaigned mightily for Rampart Dam, which, if built, would have destroyed more wildlife habitat than any single project ever built in North America. In 1980, Steve Symms of Idaho, a right-wing small businessman, ran against and defeated Senator Frank Church, one of the Senate's most respected liberals; the one thing they ever agreed on was that the Bureau of Reclamation ought to build Teton Dam.

"New Age" politicians who strive to disassociate themselves from the old Left or the old Right seem to fall into the same old habits where the pork barrel is concerned. In 1984, Senator Gary Hart of Colorado ran for president as a neoliberal and a self-proclaimed expert on how to trim the federal budget; he also supported, consistently, a couple of billion dollars' worth of unbuilt Colorado reclamation and salinity-control projects, most of them sporting costs far greater than benefits. Former Governor Edmund G. Brown, Jr., of California flew to London at his own expense to attend the funeral of his hero E. F. Schumacher, who wrote *Small Is Beautiful*, then returned to promote what could turn out to be the most expensive single public-works project ever built, the expansion of the California Water Project.

Politicians beach themselves in such ideological shallows for various reasons: the power of money, the selfishness of their constituents, or their own venality. The system thrives as it does, however, largely because of the power and nature of the committee system in Congress. The leadership of the appropriations and public-works committees that approve and fund water projects traditionally comes from the South and West, where water projects are sacrosanct. In 1980, for example, Congressman Jamie Whitten of Mississippi was chairman of the House Appropriations Committee; Congressman Tom Bevill of Alabama was chairman of its Subcommittee on Public Works; Congressman Ray Roberts of Texas was chairman of the House Public Works Committee; Jennings Randolph of West Virginia was chairman of the Senate Environmental and Public Works Committee; Mike Gravel of Alaska was chairman of its Subcommittee on Water Resources; Mark Hatfield of Oregon was chairman of the Senate Appropriations Committee. In that same year, 1980, 288 individual projects were included for funding in the omnibus Public Works Appropriations bill. Only eight got more than $25 million. All but one of the eight were located in the South or West. The most expensive item on the menu was the $3 billion Tennessee-Tombigbee Waterway, which was to receive $243 million—in a single year. The waterway is in the districts of Bevill, Whitten, and the immortal John Stennis, who was second in seniority on the Senate Appropriations Committee that year.

Together, the House and Senate committees and the water-development agencies run a remarkably efficient operation. They work in concert, rewarding those who vote for water projects and punishing those who do not, sometimes to the point of stopping virtually any federal money from going into their districts. They would, of course, much rather use the carrot than the stick. In 1978, before he had even set foot in Washington, Senator-elect Alan Simpson of Wyoming was paid a special visit by three high-ranking officers in the Corps of Engineers asking if there was anything they could "do" for him. Once in Washington, Simpson was approached again, this time by the leaders of the appropriate committees, who made him the same offer. Every freshman Senator and Congressman got the same treatment, even Bob Edgar. "The old-boy network comes to you," says Edgar, who was elected to the House of Representatives in 1974, at the age of thirty-one. "They say, 'You've got a water project in your district? You want one? Let us take care of it for you.' Then they come around a few months later and get their pound of flesh. You actually risk very little by going along. You get a lot of money thrown into your district for a project that few of your constituents oppose. In return, you vote for a lot of projects your constituents don't know about or care about. Not many of my constituents are going to base their vote for or against me on whether or not I supported Stonewall Jackson Dam in West Virginia. Then everyone wonders why we're running such big federal deficits, and they cut the social programs, which must be the culprit."

As it turned out, Edgar did not support Stonewall Jackson Dam in West Virginia, nor did he support dozens of other projects earmarked for funding in the Appropriations Committee that year. He has even made a concerted effort to have them taken out, year after year. For this, Edgar has become a virtual pariah among his colleagues and a hero among conservation groups. By general consensus, no one among the 535 members of Congress has been quite as willing to risk his political career attacking the pork-barrel system. The reason may have something to do with the fact that Edgar is a former Methodist minister who became a Congressman almost by accident. Well-built, handsome, a picture of rectitude in repose, he was, in the early 1980s, perhaps the most stubbornly principled person in that legislative body, a distinction that has worked against him at every turn. "Some of my colleagues come up to me and say, 'Bob, I wish I had your guts,' " says Edgar. "Then they attack me on the floor." Actually, Edgar has a built-in advantage in his district. He represents suburban Philadelphia, and it would be difficult for the Corps of Engineers to tantalize his constituents with a water project—where would one build one in the suburbs?—and then see to it that the appropriations committees deny him funds (a strategy which, according to a number of Congressional staff aides, has been used on numerous occasions, with good results). Still, federal public-works money has, in recent years, tended to detour around Edgar's district. His colleagues have also subjected him to threats. "Tim Lee Carter of Kentucky came up to me once after I fought to remove Paintsville Lake from the appropriations bill," says Edgar. "He was blazing mad. He punched a finger in my chest and said, "I know nothing about the Philadelphia shipyard, *but I will.*' Another Congressman told me he hopes I *am* successful in knocking off his project, because then hundreds of his constituents will walk into my district and work for my defeat."

After a while, it is difficult to remain principled in such an atmosphere, let alone be effective. "Congress as an institution is pretty sick," says Bob Eckhardt, who was a liberal Congressman from Houston until his defeat in 1980. "It has two diseases: special interestitis and parochialism. My opponent made a big issue out of the fact that I was too generous to the Northeast. He said I voted to guarantee New York City's loan when the money could have been spent in Texas. He boasted about *not* being a candidate with a national perspective. New Yorkers are just as parochial in their own way. Liz Holtzman of New York feels the question of the Concorde landing at Kennedy Airport is as important as the Equal Rights Amendment. People like Pat Moynihan [the Democratic Senator from New York] oppose western dams but want to waste even more money on a crazily expensive project like Westway. If New York City *had* gone bankrupt in 1975 it would have been a terribly serious blow to the bond markets of many other cities, including places like Boise, Idaho, and Jackson, Mississippi. I didn't detect that many members recognized that fact, or cared about it if they did. They mainly didn't want to be accused of spending their constituents' money on a lousy place like New York."

"We are a tyranny presiding over a democracy," says Edgar. "Congressman Floyd Fithian of Indiana has a water project planned for his district which he doesn't want. He wants it out of the bill, deauthorized. I don't know whether a majority of his constituents support him or not, but that should be his problem and their problem. He should be able to take a project out of his own district and if his constituents don't like it they can vote him out of office. But he hasn't been able to remove the project from the appropriations bill. Congressman John Myers sits on the Appropriations Committee and its Energy and Water Development Subcommittee. He has some big construction people in his district, which is next door to Floyd's, who would get some big contracts if the project is built. So every time Fithian tries to remove the project, Myers puts it back in.

"It's pathetic to watch what can happen to grown men here. One guy had a good project—I thought it was good—in the 1978 appropriations bill, but Ray Roberts yanked it out because he was upset over a couple of votes the guy had cast. He had the poor Congressman crawling up to him on his hands and knees for a year. He finally got his project back. Ray jerked him around like a beaten dog."

It was against this system that Jimmy Carter, a rube from Georgia who had never been elected to public office outside the state, decided to declare war.

Carter's appointments alone probably got him off on the wrong foot; in their own way, they were like Ronald Reagan's chemical-industry people taking over the EPA. His Interior Secretary, Cecil Andrus, had been governor of Idaho and, before that, a sawmill owner; but Andrus was a stranger to Washington, and he had made a reputation in Idaho as an unusually conservation-minded governor from a state full of millionaire sheep ranchers and irrigation farmers. Andrus's Assistant Secretary, Guy Martin, looked like a bearded logger, but he was a lawyer and made a reputation as a politically canny resources director under another conservationist governor, Jay Hammond of Alaska. The first head of Carter's Council on Environmental Quality was Charles Warren, probably the most active conservationist in the California legislature. One of the other members, later chairman, was Gus Speth, a lawyer from the archconservationist Natural Resources Defense Council. Speth was a Yale-educated Rhodes Scholar from Orangeburg, South Carolina, who had a dense drawl, resplendent southern charm, and Carter's ear on water projects and nuclear energy, which he had fought relentlessly at NRDC. Katherine Fletcher, a scientist with the equally archconservationist Environmental Defense Fund, became a natural resource specialist under Stuart Eizenstat, the head of Carter's domestic policy staff. In the Environmental Protection Agency and the Interior Department were a dozen more high- and middle-level appointees pulled off the environmental organizations' staffs. All of the conservation groups were, of course, beside themselves with glee to lose so many people to Carter. In view of the astonished anger that greeted the appointments among the entrenched committees in Congress, however, they may have been one of the worst mistakes Carter made.

Long before the inauguration, Carter's domestic policy staff, under Eizenstat, was working up alternatives to the Ford budget it had inherited for fiscal year 1978. Since Carter's most dramatic campaign promise had been to balance the federal budget by the end of his first term, he needed to make substantial cuts right away; besides that, like many new Presidents, he wanted to inaugurate his term by doing something bold. In a series of memoranda, Eizenstat gave him his options. There weren't many. Most of the budget was soaked up by defense and the entitlement programs, and it seemed impossible to touch the discretionary part of the budget without ruffling the feathers of some large interest group. In February of 1977, on a working weekend, Carter flew to Georgia for the first time aboard the "Doomsday plane"—the jet from which the President is supposed to run the country, or what is left of it, in the event of a nuclear war. His reading material was the Eizenstat issue paper on water projects. Sitting there, imagining himself running an incinerated nation from an airplane, Carter worked himself into a negative mood. As he flipped through Eizenstat's memo, which was written largely by Kathy Fletcher, Carter began to smolder. "There is no coherent federal water resources management policy," he read. ". . . extensive overlap of agency activities . . . several million acres of productive agricultural and forest land and commercial and sport fisheries [have been ruined] while [other] large expenditures have been made to protect these resources . . . overlapping and conflicting missions . . . large-scale destruction of natural ecosystems . . . 'the pork barrel' . . . obsolete standards . . . self-serving . . . pressure from special interests." By the time he returned from Georgia, according to one of his aides, he knew how he was going to make his big splash. He called up his chief lobbyist, Frank Moore, and told him to put Congress on notice that he wanted to cut all funding for nineteen water projects. That same day, Cecil Andrus, who knew nothing of this, stepped on a plane and flew off to Denver for a western governors' conference on that year's severe drought.

The incident demonstrated a characteristic that was to plague the Carter administration for the rest of its term—a capacity for mind-boggling political naiveté. That the news of the hit list got out before Andrus was even notified was soon attributed to a "leak" within the White House, and the culprit was identified, by sly innuendo, as Kathy Fletcher. According to one of Carter's own legislative aides, however, the source of the news was none other than Carter himself. "He told Frank Moore to put the Hill on notice that he wanted those projects cut," says the aide. "The projects had been selected at a meeting attended by Andrus, but he didn't know they were actually going to go ahead with the idea. He was opposed to it from the start."

Whatever the case, the timing was miserable. It was 1977 and California was in the midst of its driest year on record—the year before had been the third-driest—and Auburn Dam was on the hit list.

Though Auburn's existence would hardly have helped the state a bit, no one was about to notice that during a drought. Colorado, whose mountains were so bereft of snow that many of the ski slopes were closed in February, had three projects on the list, the most of any state. None of them would have helped much, either, but reason is the first casualty in a drought. The Central Arizona Project was already half-built, but it, too, was on the hit list. The western governors, who saw, by Andrus's own embarrassed and baffled reaction, the hopeless disarray of the Carter administration, milked the incident for all it was worth. Governor Richard Lamm of Colorado had to plumb the depths of his emotions to convey properly his deep and profound sense of outrage and shock. "We're not going to be satisfied," Lamm shouted at a huge crowd of scribbling reporters, "until we get our projects back." Governor Raul Castro of Arizona was "stunned and angry." The ever-opportunistic Jerry Brown of California, who had won over the state's powerful environmental community by publicly opposing the only two federal dams then being built in California—Auburn and New Melones—made one of his deft about-faces and said, "We want to build more dams."

The reaction from Congress was even stronger. Congressman Morris Udall of Arizona immediately dubbed the incident the "George Washington's Birthday Massacre," a term that stuck. Interestingly, Udall was one of several dozen Congressmen who had written a much publicized letter to Carter only five days earlier, saying, "During your campaign you stated many times that as President you would halt the construction of unnecessary and environmentally destructive dams . . . We support . . . your efforts to reform the water-resources programs of the Army Corps of Engineers and the Bureau of Reclamation." Reminded of this, Udall was gracious enough to admit that "one man's vital water-resources project is another man's boondoggle." His colleagues were not so gracious. Words like "infamous," "dastardly," "incredible," "incomprehensible," and "mind-boggling" peppered the pages of the *Congressional Record*.

If Carter was counting on help from anyone, it was the press. After all, newspapers had been criticizing other regions' public works projects since the nation's founding, and the national press was nothing if not cynical about Congress. The press, however, found Carter a better target than the projects themselves. Even principled David S. Broder wrote in the Washington *Post*, "That Carter would let something like the Red River Project put him at odds with the man [Senator Russell Long] whose cooperation is essential for passage of all the vital economic, energy, health, and welfare legislation on the administration's

CADILLAC DESERT   316

agenda is so unlikely that some observers conjured up a theory that
made the President seem much shrewder." Evidently Broder couldn't
fathom a stand of principle on something as inconsequential as a new
$900 million artificial waterway a few jumps away from the Missis-
sippi River. *Newsweek* and *Time* made a desultory effort to explain the
projects to their readers, then implied that people, not surplus crops
(as was the case), were using most of Arizona's water. *Time* unques-
tionably accepted Morris Udall's prediction that without the CAP,
"Tucson and Phoenix are going to dry up and blow away." There was
good coverage in *Science, National Journal*, and *Congressional Quarterly*, but those were publications few read.

   The intensity of the reaction from Congress and the affected regions
was so white-hot that Carter had to move much more quickly than he
had reckoned toward conciliation. In a letter to Congress, he chastised
its members for authorizing projects that made so little sense, but
promised regional hearings on every project in question and invited
the leadership to the White House for a talk. It was hardly the kind
of talk he had in mind. "All they did was tell him what an idiot he
was for doing this," said Carter's House lobbyist, Jim Free. "It was
like a lynch mob. He was the sheriff throwing calm facts back at them,
but they kept yelling at him to release the projects. One Congressman
kept banging his fist on the table. They compared him to Nixon—the
Imperial Presidency line. They were rude. They interrupted him. And
most of them belonged to his own party."

   Despite its best efforts, Congress couldn't budge Carter. He may
have been naive, but he was adamant. Seeing this, Congress, as the
New Republic remarked, began "breaking out the high-minded rhetoric
that Congressmen reserve for their grubbiest and most cynical undertakings." Majority leader Jim Wright of Texas, for example, wrote
a letter to his colleagues urging them "to help defend the Constitutional prerogative of Congress. The White House," Wright said, "in
trying to dictate [budgetary] line items, is reaching for powers never
granted any Administration by Congress." (This was the same Jim
Wright who was one of the key backers of the constitutionally dubious
Gulf of Tonkin resolution; it was the same Jim Wright who, in defiance
of his own constituents—who had decisively rejected a bond issue to
help finance the proposed Trinity River Project—kept sticking money
for it back into the public-works appropriations bills.) Senator Edmund Muskie of Maine picked up Wright's Imperial Presidency line
in the Senate—the same Edmund Muskie who was pushing the Corps'
$800 million Dickey-Lincoln Dam on the St. John River even as it was
opposed by both the president and minority leader of the Maine senate,

by Maine's two U.S. Representatives, by most of the local newspapers, and, according to several opinion polls, by a majority of the people in the state. Senator Robert Byrd of West Virginia said, "A project is not 'pork barrel' to someone who has to shovel black mud . . . or see his home swept away." The most recent flood disaster in Byrd's state, which killed more than sixty people, was caused by the collapse of a dam, and the West Virginians most immediately threatened by flooding were the homeowners who lived in the valley behind Stonewall Jackson Dam.

Notwithstanding Congress's threats, Carter continued to move his water reforms along. Simply applying a reasonable discount, or interest, rate of 6¾ percent—still too low, but reasonable—the hit list easily swelled to eighty projects. Vice-President Walter Mondale, who regarded the hit list as a terrible idea from the start, told Carter that a stand against eighty projects would be his last. With reluctance, he and his water-policy staff began a deliberate effort to winnow it down. The Tennessee-Tombigbee Waterway would devour more money, for a more illusory purpose, than anything on the list, but it had to be left alone; even the NAACP was for it. The Red River Project was also to survive; Carter had evidently read David Broder's column. Animas–La Plata in New Mexico and Colorado offered something to the local Indian population; it would survive. In most cases, Carter was going against his own deeper instincts when he let a project slip by. Once, in the midst of a string of rank political judgments, he called Charlie Warren of the Council on Environmental Quality over to his office. "He spent the first half hour telling Charlie about how outrageously wasteful and harmful some of the projects were," says one of Warren's aides. Then, together, he and Warren reduced the final hit list to eighteen projects.

On April 18, Carter announced his final, unalterable decision on the projects. It was obvious to anyone that the administration had tried to steer around states from where powerful committee chairmen came; nonetheless, it couldn't help crashing into some formidable egos and interest groups along the way. There were three projects in Colorado—Dolores, Fruitland Mesa, and Savery–Pot Hook—which was home to the second-largest Congressional delegation in the West and a Democratic governor, Dick Lamm, who hadn't hesitated to attack Carter before. The Dayton, Plainsville, and Yatesville projects were all in Kentucky, a swing state in an election year. There were Cache Basin in Arkansas, Grove Lake in Kansas, The Harbor Project and the Bayou Boeuf, Chene, and Black Channel were both in Louisiana, Russell Long notwithstanding. There was Dickey-Lincoln in Maine; Merremac Park in Missouri; Lukfata Lake in Oklahoma—peanuts as such projects go, but irresistible because the only real beneficiary of a $39 million investment would be a private catfish farm. And then, to make the whole effort financially worthwhile, there were five immense projects, none of them worth less than $500 million, two of them likely to end up costing six or seven times that much, all conceived by the Bureau of Reclamation: Garrison in North Dakota; Oahe in South Dakota; Auburn Dam in California; the Central Utah Project; and then—one could almost sense the administration crunching the bullet between its teeth—the most expensive project the West had ever seen, the rival of Tennessee-Tombigbee itself, the Central Arizona Project. Carter said he wanted all of the projects terminated. Not just unfunded—terminated.

As Carter had by then come to expect, the decibel level was highest from within his own party. Republicans, of course, stood up for their own threatened projects, but the Minority Leader in the House, Congressman Robert Michel of Illinois, said privately—and sometimes not so privately—that he thought the hit list was a pretty good idea. It was the Democratic leadership, their values and spending habits unchanged since the New Deal, that gave Carter fits. In a lectern-thumping floor speech, Jim Wright said that Carter was carrying his environmental ideas so far he threatened to become "a laughingstock." Then, to show that he, too, was an environmentalist, Wright help up a glass of water to extol its goodness. Public Works Committee chairman Ray Roberts said Carter was a captive of "environmental extremists and budget hackers." House Speaker Tip O'Neill took the highly unusual (and, for Carter, embarrassing) step of arranging a meeting with the New York *Times* to complain that Carter was "not listening" to Congress. Senators Gary Hart and Floyd Haskell of Colorado began to pepper the administration with Freedom of Information Act requests, ostensibly to learn how their projects were selected. ("They implied that we were practicing some kind of secret skulduggery," a Carter staff member complained bitterly later on. "The skulduggery was when the Bureau justified those dams, not when we reevaluated them.") Even Mondale began undermining Carter's effort—whether he knew it or not—by going around the country privately assuring Democrats that it was all a phase, that Carter meant well, of course, but that he was certainly subject to reason.

On June 13, the House Appropriations Committee, studded with Democrats, reported out its own version of the 1978 Public Works Appropriations bill. If Carter had hoped it would heed his request and delete the eighteen projects, he was mistaken. The committee bill represented not only outright but vindictive defiance of his wishes. Only one of the projects he wanted to abandon—Grove Lake in Kansas, which lacked firm support even in the district where it was to be built—was omitted. Everything else was generously funded, some with minor conditions attached (Auburn Dam wouldn't receive more money until there was a better idea whether or not an earthquake would destroy it). On top of that, money was included for a dozen new projects nowhere to be found in the administration's budget. And on top of *that*, there was a section of the bill that rejuvenated the Cross-Florida Barge Canal, which was anathema to environmentalists, and which Richard Nixon himself decided to halt in 1971.

Publicly, Carter said nothing. Privately, he was seething. "The only way now is a veto," one of his aides was quoted as saying. "We're in a game of chicken." A quick head count, however, showed that the Senate could muster the two-thirds majority required to override a veto. If he was serious about vetoing the bill, Carter would have to shore up his support in the House. With moral support from the administration, and perhaps some rewards—to his chagrin, Carter was learning that he might have to resort to the pork barrel to win his fight against the pork barrel—the House was a distinct possibility. It would only take one branch of Congress to win.

Carter's lobbyists, Frank Moore and Jim Free, worked the House furiously, joined by the railroads (which were being undercut by competition from federally subsidized barge traffic), lobbyists from the conservation groups, and every dissident farmer, businessman, rancher, and mayor from a project region whom they could get to come to Washington to help them. Vote by vote, the frailest of margins was stitched together. On the straight head count, Carter would surely lose; the problem was holding Congress's margin below the two-thirds necessary for an override. Many Congressmen, especially those whose support would take great political courage—South Carolina's Butler Derrick, for example, who had opposed Richard Russel Dam in his own district, or Philip Burton of California, who leaned heavily on labor support—demanded absolute assurances that Carter would veto the bill. If they voted not to override and he signed it anyway, their embarrassment would be acute. Meanwhile, the administration was fighting insubordination within its own ranks. The Bureau of Reclamation was widely suspected of feeding numbers to Capitol Hill that made the administration's figures appear suspect. The Corps, which had more than once disregarded the wishes of its commander in chief, was suspected by Carter's people of doing the same thing. Once, as Jim Free was passing by the Public Works committee room, he noticed several high-ranking officers of the Corps talking with Ray Roberts. Free stopped and eavesdropped long enough to capture the gist of the conservation. "They were laughing about how they were going to beat us at our own game," he says.

By fall, as the showdown approached (the Senate had already passed a close equivalent of the House bill), Moore and Free were finally convinced they had the votes to stop an override in the House. Tip O'Neill, the House Speaker, who wanted to avoid such an outcome at all costs, was apparently sure of it, too. At the last minute, he decided to play his trump card."Tip called Ham Jordan," the President's top aide, remembers Free, "and made him a bargain. Something would be worked out on Clinch River [the demonstration breeder reactor which Carter wanted to stop even more than the water projects]. A few projects would be deleted, and Tip would help the President get a reform process going.

"It was a nice piece of work," Free grudgingly admits. "They went right to Hamilton because he was the closest thing we had to a good ol' boy. He was also in a little trouble for not returning people's phone calls and things like that. If he worked out a compromise, it would make him look good, and they knew it."

O'Neill's offer was actually far less than it seemed. Although he had gotten Tom Bevill to agree to take nine projects out of the 1978 bill, he had not secured a firm promise that he would not put them back in next year. The same applied to Clinch River: the compromise might slow it down, but there was no commitment to stop it, even for a couple of years. Bevill had also agreed to a 3-percent across-the-board cut in funding, but that did not affect the ultimate cost of the projects; if anything, it made them more expensive in the long run.

No one knew exactly what had been discussed except O'Neill and Jordan and Carter themselves. Had O'Neill promised that the projects were out for good, or had Carter simply accepted that on faith? Did he really believe he had stopped the Clinch River reactor? No one who was intimately familiar with Bevill, or with Congress, believed they were in a mood to make such an offer. Andrus and Guy Martin were still urging Carter to veto the bill; now that he had gone this far, they argued, he couldn't abandon the fight unless he got nearly everything he had asked for. There was no indication from the White House that Carter felt otherwise. "Up until the last moment," says Free, "I was being told, and was telling everyone, that he was going to veto." Then, with no advance word to anyone, Carter signed the bill.

Carter's allies in Congress were thunderstruck. No one had been forewarned. Butler Derrick, according to his staff, was white with anger. Silvio Conte, the one senior Republican member on the House side who vociferously supported the administration, said that he would never trust Carter again on anything. His own lobbyists were furious. Even Andrus, who had opposed the hit lists from the beginning, was mad. Free, a young Tennessean, had had a local-boy-makes-good profile published abut him in his hometown newspaper, which happened to be in the district where Columbia Dam was to be built, and his parents had received so much verbal abuse because he was lobbying against the dam that they unlisted their phone number and took their name off their mailbox. "It hardly seemed worthwhile after that," Free said dejectedly.

Even though Carter protested that the compromise was a good one—it was still unclear exactly what it meant, and would remain so for over a year—one thing was becoming abundantly clear: Carter was already in a mood to retreat. He had underestimated Congress's passion for dams and overestimated his ability to move the rest of his legislative program forward. In January of 1977, Cecil Andrus told the New York *Times*, "Thank God, there'll be no more hit lists." A lot of fence-mending was obviously being done. Later that month, Lou Cannon, the Washington *Post*'s correspondent in San Francisco, could write that "the West's Democratic governors have been offered unconditional surrender by the Carter administration, [which] has backed away from nearly every position" on water projects. An "options paper" drafted shortly thereafter and leaked, to Carter's chagrin, to the environmental groups made no mention of several of the main water-policy reforms Carter had spoken of earlier.

Having reversed himself once, however, Carter was perfectly capable of reversing himself again. In October of 1978, his second big challenge on water projects came around. The fiscal year 1979 public-works appropriations bill that emerged from the House and Senate conference committee did exactly what most of Carter's advisers said it would. To begin with, it restored money for every one of the nine projects deleted the previous year. Carter, in his innocence, evidently believed that the projects had been killed for good, and he was livid. On top of that, the bill contained money for a number of new starts, despite the fact that inflation was well into double digits, interest rates were topping 15 percent, and a balanced budget was slipping out of Carter's grasp.

Once again, Jim Free began making his rounds on Capitol Hill, urging support of a presidential veto when the vote came—even though the administration's allies were still seething over Carter's performance with the previous year's bill. Whatever doubts they had about Carter's courage, however, were soon stilled. A few days later, after making a terse, angry statement denouncing it, Carter vetoed the entire appropriations bill.

The timing of the veto, as it happened, coincided neatly with the passage of Proposition 13 in California, a draconian measure which effectively held the annual increase in property taxes to about 1 percent. Everyone knew the public was fed up with government spending; this was the first sign that it was *really* fed up. The main sponsor of the measure, a real estate lobbyist named Howard Jarvis, instantly became something of a celebrity. And though the rest of the country felt that California was more than slightly daft, everything that happened there had an odd way of spreading eastward.

One of the people who realized this right away was Larry Rockefeller, the nephew of Nelson and son of his elder brother Laurance. Rockefeller, who was then a staff attorney at the Natural Resources Defense Council, was thirty-six, almost neurotically shy, and a strikingly gifted propagandist and politician. Almost single-handedly, he pieced together the Alaska Coalition, the vast umbrella organization that was responsible, two years later, for passage of the Alaska Lands Act—which created, in an instant, as much federal parkland as the country had set aside in more than a hundred years. The full-page advertisements run by the Alaska Coalition were written and often paid for by Rockefeller, and they were astute; mostly, they talked about how much resource development and fabulous economic growth the Alaska Lands Act would still allow.

The Alaska campaign was based on persuasion. To make Congress sustain Carter's veto of the appropriations bill, however, a campaign would have to be based on fear. There was too little time to try any other tack, and fear seemed to be the one universal motivator on Capitol Hill. At that particular moment, Rockefeller reasoned, Congressmen feared no one more than Howard Jarvis.

Getting Jarvis's cooperation was surprisingly easy. Although the value of real estate in his hometown, Los Angeles, depended entirely on aqueducts bringing water from three directions, they were already built. Besides, an opportunity take on Congress was more than the feisty old man could resist. Rockefeller recalled some of Jarvis's speeches, shut himself in his office, and imagined what sort of advertisement Jarvis might write. When he finished a draft, he read it to him over the phone. Jarvis was stunned. "That's just what I would have said," he answered.

On the morning of October 5, with a vote to override Carter's veto just hours away, four hundred-odd members of the House opened their

copies of the Washington *Post* and the New York *Times* and saw the
scowling visage of Howard Jarvis staring back at them. "IT'S AN OUT-
RAGE," he croaked. "THE PUBLIC WORKS APPROPRIATIONS BILL IS THE BIG
TAX, BIG GOVERNMENT, BIG SPENDING, BIG WASTE BILL OF THE YEAR." During the debate that day, the "spirit of Howard Jarvis" was invoked
several times. When the vote was taken, the attempt to override Carter's veto had barely failed.

As the dam saboteurs in Carter's Administration were to discover,
however, victories over the Congressional pork-barrel system tend to
be short-lived. They are especially short-lived if they come thirteen
months before an election year.

In July of 1979, a group of California's wealthiest irrigation farmers, many of them from the Westlands Water District, played host to
Rosalyn Carter at a big Democratic fund-raiser in Fresno. Soon thereafter, a number of big growers from the nominally conservative San
Joaquin Valley were making hefty campaign contributions to the
Carter-Mondale reelection campaign. Their reward was a new water
contract obligating them to pay only $9.10 an acre-foot—well below
cost, and a subsidy worth $60 million over the term of the contract.

Westlands, which the Bureau had illegally expanded back in the
1960s at the behest of the farmers, was the one place where Carter
could put one of his most ballyhooed reforms, realistic water pricing,
to work, because the illegal expansion had technically voided the original contract. He not only failed to do that, but, by caving in on an
issue he could easily have won—Westlands had no other source of
water except groundwater, which was running out, and therefore had
little choice but to accept the administration's terms—he sent a signal
to Congress that he was prepared to do business with them.

It was just the beginning. Carter had entered office convinced that
the 160-acre land limitation in the Reclamation Act was a sound principle. But in Congress there was talk of removing the acreage limitation
altogether, and of allowing unlimited leasing (which was, in effect, the
same thing as removing the limitation). The more "moderate" proposals called for a limit of 1,260 acres, an eightfold expansion. Most
of Carter's advisers were telling him that he had to hang tough on the
acreage issue: if subsidized water suddenly became available to the
biggest growers in the West, it would not only be an outrageous subsidy
of the wealthy, but it would intensify pressure for even more projects.
Assistant Interior Secretary Guy Martin, the administration's canniest
strategist on western water policy, says he recommended a revised
acreage limit of perhaps six hundred acres—a compromise which, he

felt, the administration could sell. By late 1979, however, Martin's boss, Cecil Andrus, was suddenly agreeing with Jerry Brown, another lapsed champion of the 160-acre limit, on a new limit of 1,260 acres. (It wasn't clear whether that meant 2,520 acres for a man and wife.) In California, with 1,260 acres and subsidized water costing between $3.50 and $9 per acre-foot, a halfway ambitious farmer could become a millionaire—which was not exactly the intention of the Reclamation Act.

And then, on top of everything else, there was Tellico Dam.

Tellico was a dam the Tennessee Valley Authority had conceived as early as the 1930s and hadn't gotten around to proposing seriously until the 1960s—which was mute testimony to the kind of project it was. The dam itself would produce no power—it would merely raise and divert the Little Tennessee River about a mile from its confluence with the main Tennessee so some extra water could be run through the turbines of nearby Fort Loudon Dam. The result would be twenty-three megawatts of new power, about 2 percent of the capacity of one of the nuclear and coal plants the TVA was simultaneously building. There were no flood-control benefits; there were hardly recreational benefits (the region had more reservoirs than it knew how to fill with boats); there were no fish and wildlife benefits. On the other hand, the Little Tennessee was about the last fast-flowing coldwater stream in the state. It was dammed only once upriver, while most tributaries of the Tennessee were dammed several times. It had a large and healthy population of trout. It was a splendid canoeing stream. It flowed through a beautiful valley, one of those happy places that contain both farms and bears. The Cherokee Indian Nation had had its pick of all the rivers of central Appalachia, and it chose the valley of the Little Tennessee as its home. There were hundreds of archaeological sites, some probably yet to be discovered. With its pretty white clapboard houses and its well-tended little farms, the valley was a beautiful anomaly, a place more at home in the nineteenth century. Tellico Dam would put all of this under eighty feet of water.

After wrestling with its lack of a *raison d'être* for a while, the TVA decided that the only way it could justify the new dam was to change the whole character of the region in which it would be. The solution, it finally decided, was to create an entirely new town around the reservoir, a chrome-and-steel headquarters for a major branch of the Boeing Corporation which would go by the somewhat ironic name Timberlake. (Actually the TVA may have come up with the idea because the Bureau of Reclamation had thought of it first. In the 1960s, it was no secret that the Bureau, boxed out of much of its historical domain by the Corps of Engineers, was looking to expand its activities eastward, and Appalachia was the first place it planned to give things a try, building exactly the kind of sterile, reservoir-centered new towns of which Timberlake would be a first example.) It was like deciding to put a fifty-thousand-seat Superdome in the middle of Wyoming and then building a city of 150,000 people around it to justify its existence. And there was no real guarantee by Boeing that it would establish itself there; it had merely expressed interest in the idea. But that was enough to get the project moving. By 1969, Tellico Dam was well on the way to being built.

As it was going up, however, two entirely new hurdles were thrown in Tellico's path. One was the National Environmental Policy Act of 1969, which requires an environmental-impact statement and a discussion of alternatives before any major federal project can proceed. (The TVA claimed it was exempt from NEPA and had to be taken to court before it complied.) The other hurdle, which no one paid much attention to at first, was the Endangered Species Act of 1973.

In that same year, 1973, a professor of zoology from the University of Tennessee was snorkeling around in the Little Tennessee when a small fish, resembling a dace, darted out from under a rock in front of his face and gulped a snail. The zoologist, whose name was David Etnier, followed the fish until he could get a good look at it. He had never seen one like it before. After some taxonomic investigation, the fish was identified as a snail darter—a species that appeared to inhabit only a portion of the Little Tennessee, mainly the taking area of Tellico Dam. Its numbers estimated to be in the low thousands, its habitat apparently confined to one place, the darter seemed eligible for classification as an endangered species. Before the act, that would have meant merely that the fish was probably doomed. Now it meant, by law, that "protection of habitat . . . critical to [its] continued existence" was federal government's number-one priority.

The TVA tried to get around the act by attempting, without much success, to transplant the darter to other nearby streams. Meanwhile, instead of suspending construction, it redoubled its efforts to complete the dam in a hurry, a time-honored strategy employed by the public-works bureaucracies—but one which, this time, resulted in its being hauled into court by the Environmental Defense Fund. The federal district court essentially found for the EDF, but ruled that the Endangered Species Act was never intended by its framers to stop a project which was already 80 percent built. On appeal, however, the district court's decision was overturned, and completion of Tellico was stopped cold.

The national media, which had covered the story with yawning lack of interest up to then, were suddenly tearing each other's clothes trying to get onto the Tellico site. Half the newspapers in the country seemed to run the story on page one, under some variation of the same headline: "Hundred-Million-Dollar Dam Stopped by Three-Inch Fish." In most cases, the coverage went little deeper than that. Some editorial writers couldn't even see humor in the impasse; the Washington *Star* harrumphed that it was "the sort of thing that could give environmentalists a bad name."

Had the editorialists and reporters taken a longer look, they might have seen that the big story was not the dam at all but the TVA itself, an agency that had evolved from a benevolent paternalism into the biggest power producer, biggest strip miner, and single biggest polluter in the United States. Unaccountable to the public, largely unaccountable to Congress, the TVA was an elephantine relic of the age of public works; it had undoubtedly done its region some good, but by the 1970s it had passed the uncharted point in an agency's career—twenty years, thirty years, sometimes much less—when it confronts new challenges with barnacled precepts and, in a sense, turns on the constituency it was created to help. Had they looked around them, the reporters might have seen that Appalachia, the godchild of this benevolent agency for four decades, still looked socially depressed; physically, it looked horrifying. The single most important reason was the TVA's purchase of immense quantities of strip-mined coal. It still clung to the discredited notion that the salvation of Appalachia lay in cheap power, and strip-mined coal was the cheapest fuel. But the strip-mining, besides eliminating thousands of jobs in deep-mined coal, was creating a scene of gruesome devastation. The denuded mountains seemed covered with a reddish-brown rash, and rivers that were once pristine were running with what looked like old blood. Meanwhile, the TVA's older coal-fired power plants were creating pollution traps in the valleys where they were situated, and its newer ones, with smokestacks a thousand feet high, were wafting sulfur and nitrogen oxides up to New York State and Canada, where they fell as acid rain.

This same obsession with cheap electricity had, of course, resulted in the TVA's having built thirty-odd major dams in the Tennessee Basin over the course of thirty-odd years. The dams, mostly built during the Depression and the war era with low-interest money and by workers earning a few dollars a day, were the cheapest source of power around, and TVA's rates were as low as those in the Northwest. As in the Northwest, a complement of energy-intensive industries had moved in—aluminum, uranium enrichment, steel—and now the TVA was afraid they would move right back out if it raised its rates. It was a fear whose end result, rational or not, was the Tellico Dam.

In June of 1978, the Supreme Court upheld the injunction against the dam on the basis of the Endangered Species Act, as written. Legally, the Court had little choice, even though, by then, the dam was more than 90 percent built. Chief Justice Warren Burger, who wrote the decision, was clearly offended by the whole situation, and all but invited Congress to amend the act. Congress required no such prompting. The legislative hopper began to spin with amendments to weaken or gut the act. Through the leadership of Senator John Culver of Iowa, however—and of Senator Howard Baker of Tennessee, whose only real interest was completing the dam—a less drastic amendment was passed, by which an endangered species review committee would be created to resolve any case where a major project such as Tellico ran up against the act. It was to be a Cabinet-level committee, composed of the Secretaries of Interior, Agriculture, and the Army, in addition to the administrators of the Environmental Protection Agency and the national Oceanic and Atmospheric Administration, the chairman of the Council of Economic Advisers, and a representative from the affected state. According to the language of the amendment, the committee, which some began to call the God Squad, could grant exemptions to the act where no "reasonable and prudent" alternative exists, where the project is of national significance, or where the benefits of building it "clearly outweigh" any other course of action.

The makeup of the interagency committee suggested a predisposition toward completing stalled projects, especially in the case of a dam. At best, Tellico's opponents were hoping for a four-to-three split in favor of construction, which might seem like enough of a hung jury to let them try another tack. They were wondering what such a tack might be when the committee's decision was announced. No one was prepared for the outcome: a unanimous decision that held for the snail darter and against the dam. In so doing, the committee skipped over metaphysics, transcendentalism, and evolutionary philosophy and ruled solely on the basis of economics. Tellico was a terrible investment—even worse, if the committee was to be believed, than the environmentalists had said. "Here is a project that is 95 percent complete," said Charles Schultz, the chairman of the Council of Economic Advisers, "and if one takes just the cost of finishing it against the benefits . . . it doesn't pay." Cecil Andrus added, "Frankly, I hate to see the snail darter get the credit for delaying a project that was so ill-conceived and uneconomic in the first place." God, in his new bureaucratic incarnation, had spoken. Tellico was a loser—it didn't deserve to be finished.

The dam's two main Congressional defenders, Senator Howard Baker and James Duncan, a Republican Congressman whose district encompassed both the dam and the TVA's headquarters, still tried to blame everything on the snail darter. "Should a worthless, unsightly, minute, unedible minnow outweigh a possible injustice to human beings?" groused Duncan on the floor of the House—ignoring the injustice to the thousand-odd people who would be evicted from their homes. Nonetheless, the finale had been written. Baker and Duncan had been beaten, fair and square—beaten, through some oddly poetic reprise, by Howard Baker's own amendment. The only thing for them to do was to accept defeat gracefully.

June 18, 1979, was a dull day on the floor of the House, even duller than most. Little was going on, so hardly anyone was there. Bob Edgar was one of the many who were absent, and he still hates himself for it. He was one of the few Congressmen who might have been suspicious enough to stop what was about to take place. "Duncan walked in waving a piece of paper," Edgar recalls. "He said, 'Mr. Speaker! Mr. Speaker! I have an amendment to offer to the public-works appropriations bill.' Tom Bevill and John Myers of the Appropriations Committee both happened to be there. I wonder why. Bevill says, 'I've seen the amendment. It's good.' Myers says, 'I've seen the amendment. It's a good one.' And that was that. It was approved by voice vote! No one even knew what they were voting for! *They were voting to exempt Tellico Dam from all laws.* All laws! They punched a loophole big enough to shove a $100 million dam through it, and then they scattered threats all through Congress so we couldn't muster the votes to shove it back out. I tried—lots of people tried—but we couldn't get that rider out of the bill. The speeches I heard on the floor were the angriest I've heard in elective office. For once, a lot of my colleagues were properly outraged. Senator Baker and Representative Duncan couldn't have cared less. They got their dam."

A few days later, the House passed the appropriations bill with the Tellico rider still in it. The Senate followed suit, 48-44, despite two earlier votes against the dam. "That," said Edgar with sardonic disgust, "is the democratic process at work."

There was, of course, still the possibility of a presidential veto. If anything, it seemed inevitable. Here in the case of one dam, was everything that was rotten in Denmark: a bad project proposed by a dinosaurian bureaucracy; needless destruction of one of the last wild rivers in the East; usurpation of a quiet valley; and a cynical Congress sneaking around one of its own laws. Guy Martin and Cecil Andrus were both urging a veto in the strongest possible terms. Gus Speth, by then chairman of the Council on Environmental Quality, was privately talking of resignation if Carter backed down. Few in Carter's conservationist constituency even entertained the possibility that he wouldn't veto the bill. Congress, however, had taken care of everything. Carter was in the midst of negotiating a treaty that would give the Panama Canal back to Panama, and he was meeting stubborn resistance in Congress. The votes were lined up closely enough to put the President in a position of wretched vulnerability. The threats were quite naked. If Carter vetoed the bill, there would be no treaty; his education bill might suffer the same fate. In both cases, his embarrassment would be extreme—worse, perhaps, than if he swallowed the Tellico exemption. The gulp was almost audible. On the night he signed the bill, the President telephoned Zygmunt Plater, the young law professor from the University of Tennessee who handled the case before the Supreme Court, and performed a *mea culpa*. Plater was taken aback. He was, in fact, speechless, and he wasn't even sure why. Was it having the President on the other end of the phone, or was it the fact that a dam was not dictating foreign policy?

When the gates closed on Tellico Dam a year or so later, Carter's humiliation was just about complete. Not a vestige of his water-policy reforms remained. Everything he had asked for was out; everything he wanted out was in. Congress had made a mockery of one of its own laws, and even of an amendment weakening that law, for the sake of a water project so bad it made better sense to abandon than to finish it. The Tellico vote was one of the things that prompted the normally restrained Elizabeth Drew, the *New Yorker*'s Washington correspondent, to write a devastating series on Congress's capitulation to money and power. To those familiar with water projects, though, it was nothing new.

With the benefit of hindsight, some of Carter's own people are scathingly critical of how the administration handled the water-projects issue. Guy Martin, his Assistant Interior Secretary, is one. "He blundered from the word go," says Martin. "He might as well have gone up to the Hill with a six-pound codfish and slapped it across their faces. Andrus begged him not to come out with a big long hit list, but he did it anyway, and from that point on the merits of the whole issue got lost. It became 'Congressional prerogative,' the 'Imperial Presidency.' He was his own worst enemy. He had a great big chip on his shoulder about water projects, that was his problem. It made him focus way too much on the environmental issue, when the only way he could win was with the economic one. Most Congressmen don't really care about wild rivers. The New Deal mentality is entrenched up there—even the right-wingers believe in it. Carter loved wild rivers, and in the end they thought he was just plain kooky.

"What Carter *could* have done," Martin continued, "is pick the three or four worst projects instead of nineteen, or thirty-two—that was another problem, he kept changing the numbers on them—and get rid of them. He could have done it. In war, you don't take two dozen beachheads on the same day. You can't, for God's sake. But he could have won some big ones. Auburn Dam, for instance. If that dam failed, it would be the worst peacetime disaster in American history. He had them there. Garrison and Oahe were awful. The farmers didn't even *want* Oahe. The Tug Fork Project is so ridiculous it strains belief. I can't help believing that if Carter had focused on a few he could have eliminated them. Then he would have had a small victory, but a real one. Then there's next year."

Having said all this, Martin added, almost apologetically, "Carter was right, though. The projects are as bad as he said, most of them. The environmental damage is bad. The economics are bad. The politics are bad. The agencies are out of control. If the Corps and the Bureau built everything they wanted to, we'd hardly have any flowing water left. His instincts were good."

Many western members of Congress, not to mention the water lobby and the bureaucracies, were overjoyed when Ronald Reagan was elected President after Jimmy Carter. Reagan might talk like a fiscal conservative, but surely he wouldn't be against water development. After all, he was a westerner. He owned a ranch in dry country. His Interior Secretary, James Watt of Colorado, was the environmentalists' anti-Christ. Most of his other key domestic advisers were westerners, too—James Baker, Ed Meese, William Clark, Paul Laxalt. All of them, and Reagan, too, certainly *talked* as if they believed in water development.

No sooner had Reagan taken office, however, than his budget director, David Stockman, was talking about recovering 100 percent of the costs of new navigation projects from the beneficiaries—not just the capital costs, but the operating costs, too. (In 1985, the Corps of Engineers spent around $1 billion on project operation and maintenance alone.) There was also talk of forcing states to pay a large share of the costs of flood-control dams—something Carter had never seriously proposed. Even Watt was suggesting that the states should contribute to Reclamation projects—up front. It wasn't exactly clear how large a share the administration had in mind, but privately Watt was suggesting that 33 percent might be a reasonable amount. Since that would preclude practically all new water development, the water lobby didn't know quite how to react. Jan van Schilfgaarde, the director of Agriculture Department's Salinity Control Laboratory, was speaking one day with William Johnston, the assistant manager of the Westlands Water District in California, and he asked, "Why do you think Reagan is your friend if he wants you to pay a third of the fare? Carter only wanted 10 percent." As van Schilfgaarde recalls it, Johnston was silent for a moment, then said, "Well, Reagan understands us." "You can get cheaper understanding from a psychiatrist" was van Schilfgaarde's response.

As expected, Reagan's original proposals were slowly nibbled away by Congress, but meanwhile, year after year, no new authorization bills managed to clear the floor—partly because the federal government was suffocating under its own mass, but also because Reagan, like Carter, was threatening to veto. In 1984, the entire $20 billion water-projects authorization in the public-works bill—three hundred projects' worth—was taken out due to such a threat. A year later, when an almost identical bill reached the floor of the House, environmentalists, who had formed a discreet alliance with Stockman and other fiscal conservatives in the administration, had managed to sneak in amendments and conditions requiring local cost-sharing on the order of 10 to 30 percent—even for flood control. If the amendments and conditions stayed in the bill, only a handful might get built; when a state sees that it has to put up $50 million toward construction of a dam, its enthusiasm is apt to wilt like a plucked blossom. As for the Bureau, one of its largest projects, Central Utah, had been burdened with a supplemental repayment contract that absolutely guarantees recovery of all costs before the CUP receives any further funding. That provision, which could stop the project dead in its tracks, also had Reagan's private blessing. No one could predict how much of this would remain in this or successive bills when they cleared Congress and reached the President's desk—and the Tellico experience led some to think, not much—but the pork barrel seemed finally to have lost its anchorings, and to be adrift on the very thing it helped produce: an uncontrollable tide of national debt.

CHAPTER TEN

Chinatown

Everyone knows there is a desert somewhere in California, but many people believe it is off in some remote corner of the state—the Mojave Desert, Palm Springs, the eastern side of the Sierra Nevada. But inhabited California, most of it, is, by strict definition, a semidesert. Los Angeles is drier than Beirut; Sacramento is as dry as the Sahel; San Francisco is just slightly rainier than Chihuahua. About 65 percent of the state receives under twenty inches of precipitation a year. California, which fools visitors into believing it is "lush," is a beautiful fraud.

California is the only state in America with a truly seasonal rainfall pattern—stone-dry for a good part of the year, wet during the rest. Arizona is much drier overall, but has two distinct rainy seasons. Nevada is the driest state, but rain may come at any time of year. If you had to choose among three places to try to grow a tomato relying on rainfall alone, South Dakota, West Texas, or California, you would be wise to choose South Dakota or West Texas, because it rains in the summer there. California summers are mercilessly dry. In San Francisco, average rainfall in May is four-tenths of an inch. In June, a tenth of an inch. In July, none. In August, none. In September, a fifth of an inch. In October, an inch. Then it receives eighteen inches between November and March, and for half the year looks splendidly green. The reason for all this is the Pacific high, one of the most bewildering and yet persistent meteorological phenomena on earth—a huge immobile zone of high pressure that shoves virtually all precipitation toward the north, until it begins slipping southward to Mexico in

October, only to move back up the coast in late March. More than any other thing, the Pacific high has written the social and economic history of California.

Actually, San Francisco looks green all year long, if one ignores the rain-starved hills that lie disturbingly behind its emerald-and-white summer splendor, but this is the second part of the fraud, the part perpetrated by man. There was not a single tree growing in San Francisco when the first Spanish arrived; it was too dry and wind-blown for trees to take hold. Today, Golden Gate Park looks as if Virginia had mated with Borneo, thanks to water brought nearly two hundred miles by tunnel. The same applies to Bel Air, to Pacific Palisades, to the manicured lawns of La Jolla, where the water comes from three directions and from a quarter of a continent away.

The whole state thrives, even survives, by moving water from where it is, and presumably isn't needed, to where it isn't, and presumably is needed. No other state has done as much to fructify its deserts, make over its flora and fauna, and rearrange the hydrology God gave it. No other place has put as many people where they probably have no business being. There is no place like it anywhere on earth. Thirty-one million people (more than the population of Canada), an economy richer than all but seven nations' in the world, one third of the table food grown in the United States—and none of it remotely conceivable within the preexisting natural order.

For all its seasonal drought, its huge southern deserts, and its climatic extremes, there is plenty of water in California for all the people who live there today. If, God forbid, another twenty-five million arrive, there will still be plenty for them. The only limiting factor will be energy: to get to where the people are likely to settle, a lot of the water has to be lifted over mountains. Take any ten of the largest reservoirs— Shasta, Bullard's Bar, Pine Flat, Don Pedro, New Melones, Trinity, a few others—and you have enough water for the reasonable needs of twenty-five million people; enough for their homes, their schools, their offices, their industries, even (in all but the driest times) their swimming pools and lawns. As for the other 1,190 California dams and reservoirs, their purposes are threefold: power, flood control, and, above all, water for irrigation. What few people, including some Californians, know is that agriculture uses 81 percent of all the water in this most populous and industrialized of states.

California's $18 billion agricultural industry—and it is a gigantic, complex, integrated industry—is the largest and still the most important in the state, Silicon Valley notwithstanding. That figure, $15 billion, only begins to convey what agriculture really means to California. A great proportion of its freight traffic is agricultural produce. A disproportionate amount of the oil and gas mined in the state is used by agriculture. California agriculture supports a giant chemicals industry (it uses about 30 percent of all the pesticides produced in the United States), a giant agricultural-implements industry, an unrivaled amount of export trade. Because it relies on irrigation—and therefore on dams, aqueducts, and canals—there is a close symbiotic relationship with the construction industry, which is why politicians who lobby hard on behalf of new dams can count on great infusions of campaign cash from the likes of the Operating Engineers Local No. 3 and the AFL-CIO. And, more than any other state, California has been a source of opportunities for the Bureau of Reclamation and the Corps of Engineers.

All of this production, all of these jobs, all of these concentric rings of income-earning activity nourish California's awesome $485 billion GNP. It is a gross *state* product, obviously, but everyone seems to refer to it as the "California GNP," as if the state were a nation unto itself—which it really is, and nowhere more so than in the example of water. California has preached and practiced water imperialism against its neighbor states in a manner that would have done Napoleon proud, and, in the 1960s, it undertook, by itself, what was then the most expensive public-works project in history. That project, the State Water Project, more than anything else, is *the* symbol of California's immense wealth, determination, and grandiose vision—a demonstration that it can take its rightful place in the company of nations rather than mere states. It has also offered one of the country's foremost examples of socialism for the rich.

In the 1850s, when the California gold rush was at full flood, the Great Central Valley traversed by the miners on the way to the mother lode was an American Serengeti—a blond grassland in the summertime, a vast flourishing marsh during the winter and spring. The wildlife, even after a century and a half of Spanish settlement, was unbelievable: millions of wintering ducks, geese and cranes, at least a million antelope and tule elk, thousands of grizzly bears.

The winter of 1861 and 1862 was the beginning of the end for this scene of wild splendor. Relatively few of the Forty-niners found enough gold to pay their fare back home, let alone retire in the style of which they dreamed. California in the 1850s was full of broken men, searching for whatever day labor they could find. Many of them, having given up on returning home, decided to make a try at farming or ranching in the Central Valley. Most of the pioneers who followed the miners in wagon trains had farming on their minds, too, and by the 1860s the Central Valley was already a vista of cows. Because of the rainless summers, no important crop except wheat could be raised without irrigation, which was an alien form of agriculture to Americans. But the valley's chronic scarcity of moisture was suddenly reversed in 1861 and 1862. A series of vast drenching Pacific storms bashed the state for weeks on end; in January of 1862, San Francisco recorded twenty-seven inches of rain, half again what it usually gets in a year. The floods would have been bad anyway, but their destructiveness was greatly intensified by the incredible amount of spoil—whole sides of mountains—which hydraulic mining had sent down the rivers to the lowlands. The beds of the valley rivers were raised several feet, and could not begin to contain the torrential runoff; downtown Sacramento was under seven feet of water. The 1862 flood marks the beginning of the valley's obsession with bringing the rivers under control. Meanwhile, farther south, in the San Joaquin Valley, Henry Miller was using the same flood to acquire hundreds of thousands of acres of ephemerally drowned lands under the Swamp and Overflow Act. Miller's acquisitive nature, combined with the serendipity (in his case) of the flood, made him enough money to construct a large dam, and he soon had his hundreds of thousands of acres under irrigation. Before he died, he was likely the richest farmer in the United States.

When the valley ranchers saw how rich one could become through irrigation farming, they began to switch from cows and dryland wheat to crops. Few had Miller's ambition or wealth, however, and even when organized into irrigation districts they couldn't duplicate his dam, so they irrigated with primitive sluiceways cut from the rivers, much as did the farmers along the Nile. As for the state and federal governments, they wanted nothing to do with publicly financed irrigation projects, which were widely regarded as socialistic.

Everything changed with the invention, shortly after World War I, of the centrifugal pump. Suddenly able to draw hundreds of gallons per minute out of the valley's shallow aquifer, the irrigation farmers no longer had to worry about building expensive canals, about cleaning them of silt; they no longer had to dream of regulating the rivers with dams to ensure summer flows. By the mid-1920s, thanks to irrigation pumping, California had surpassed Iowa as the richest agricultural state in the country; the Central Valley was the largest semicontinuous expanse of irrigated farmland in the world. The aquifer, which had collected over many thousands of years, was prodigious; before pumping began, it may have held three-quarters of a billion acre-feet. With the expansion of irrigation farming from a few thousand to millions of acres, however, the water table began to drop sharply. By the end of the Great Drought of the 1930s, the farmers had so badly depauperated the groundwater that the depletion curves were precipitous. Twenty thousand acres had already lost their groundwater and gone out of production; hundreds of thousands more overlay a groundwater table that was becoming dangerously low. Suddenly, the valley's reserve of groundwater, which had so recently seemed limitless, had only a few more decades of economic life.

The farmers could look in two directions for help: Sacramento and Washington, D.C. The Bureau of Reclamation, which was just completing Hoover Dam, had such a hold on the public imagination and the Roosevelt administration that it could build almost anything it pleased. On the other hand, it was supposed to create new subsistence farms in the West, not rescue the farmers who were already there from the consequences of their short-sighted avarice. Besides, Hoover Dam had been a great gift to California, and the other western states were waiting in line.

Sacramento, then, was the better bet, even if it couldn't dip into the federal treasury to finance the farmers' rescue. In 1933, the state legislature succumbed to heavy lobbying from the growers—who had become its biggest source of campaign contributions—and passed the Central Valley Project Act. The legislation was a striking display of ambition for a single state, proposing as it did the control, through dams and reservoirs, of the largest and third-largest rivers in California. The project bonds, however, could not be sold in the middle of the Depression, so the state was forced to let the Bureau of Reclamation take over the Central Valley Project; it was such a gargantuan scheme that the completion of its main features, including four big dams, required eighteen years.

The Central Valley Project was without question the most magnificent gift any group of American farmers had ever received; they couldn't have dreamed of building it themselves, and the cheap power and interest exemption constituted a subsidy that would be worth billions over the years. It is interesting, therefore, that originally many of the farmers *hadn't wanted* the Bureau to build the CVP.

The wedding between the Bureau of Reclamation and the Central Valley farmers was never more than a marriage of convenience, and, like many such marriages, it was soon on the rocks. As a starlet trades a virile but impecunious husband for a wealthy old tycoon, the farmers had, in effect, traded whatever hope they had of becoming agricultural grandees like Henry Miller for a secure supply of water. The Reclamation Act, which would apply, in theory, to the CVP even though it only delivered supplemental water to most, required a farmer owning more than 160 acres of land (320 for a man and wife) to sign recordable contracts to dispose of the excess holdings in order to continue receiving subsidized water. Since a great many farmers owned far more than that, the CVP looked as if it might become the first real land-redistribution device in U.S. history. Leasing acreage above and beyond the 320-acre limit was also prohibited under the act, and all excess holdings were supposed to be sold at their pre-project worth—which, in a valley where no crop could be raised without irrigation, was very little. On top of all this, farmers receiving Reclamation water were required to live on their land, not farm from Fresno or San Francisco, as many of them did. (The Bureau stopped enforcing the residency provision in 1916, but a federal court later determined that it was still valid.) The whole idea was to keep speculators away, and to open up arid land to as many new farmers as possible. "We weren't even supposed to give them 160 acres if they could make a living on less," says former commissioner Floyd Dominy. "And in warm states like California you could make a living on a lot less. We were talking about subsistence—nothing more."

The CVP, in short, was fundamentally different from every earlier Reclamation project. It did not create many new irrigated farms. It rescued thousands of farms that were already there, including a good many that were far larger than the law allowed. One of the "farmers" whose land lay within the service area of the Central Valley Project, and who was scheduled ultimately to receive its water, was the DiGiorgio Corporation, whose lands grew more commercial tomatoes than any state except Florida. Another was the Southern Pacific—not a mere railroad, but the largest private landowner in California, and the eventual owner of 109,000 acres in the Westlands Water District, which was scheduled to become the largest single recipient of CVP water. The roster of landlords within the San Joaquin Valley was a Who's Who in corporate agriculture. Figures for 1946, published in a Senate report on the acreage limitation, reveal that Standard Oil owned 79,844 acres in the probable CVP service area; Will Gill and Sons owned 29,926 acres; the Bellridge Oil Company owned 30,120 acres; the Tidewater Associated Oil Company owned 25,554 acres; the Richfield Oil Company owned 10,718 acres; the Anderson and Clayton Company owned 19,144 acres; and the J. G. Boswell Ranch Company, which, among others listed, was already receiving Kings River water virtually free courtesy of the Army Corps of Engineers, owned 16,760 acres—part of a worldwide land empire later estimated at some 860,000 acres minimum. If such growers availed themselves of the Bureau's water, which they would doubtless want to do, the law was quite clear about the disposition of their cases: they would have to sell all lands in excess of 160 (or, more likely, 320) acres that received the subsidized water. The Reclamation Act's chief sponsor in the House, Frank Mondell, had said on introducing the bill that this divestiture provision "was drawn with a view to breaking up any large landholding which might exist in the vicinity of Government works." It was hard to imagine it stated more emphatically than that.

The threat of divestiture gave the big growers in the CVP service area fits, even if the Bureau was far more interested in building more dams than in trying to enforce such an unpopular law—especially when the Interior Department's lawyers, few of whom were legal stars, had to go up against some of the craftiest legal talent in the state. One modest example of how the farmers managed to deceive the Bureau was provided by the case of Russell Giffen, one of the big landowners in the Westlands district. A Fresno rancher who stitched together seventy-seven thousand acres of valley property—about seven times the acreage of Manhattan Island—Giffen was the largest cotton grower in the world: nationally, he also ranked just behind Boswell and one other farming company in the combined federal farm subsidies he received. In the 1970s, Giffen decided to clean up his estate for probate, and sold most of the land for $32.5 million. One of the buyers was a New York–based company called Jubil Farms, in which a Bakersfield couple, William and Judith Rogers, owned an 80 percent interest. The Rogerses, five other couples (most of them Rogers employees), the trusts of four Rogers children, and a mail-order denture company took title to 1,812 acres, all of it in parcels of 160 acres or less. All the new landowners then leased their property back to Jubil Farms. Financing for the whole deal, in the amount of $3.5 million, was provided by the Nissho Iwai American Corporation, the subsidiary of a Japanese conglomerate, which happened to own the other 20 percent of Jubil Farms.

On paper, and in the Bureau's recordable contracts file in Sacramento, the requirements of the Reclamation Act were satisfied. In reality, the whole business was a translucent sham. One company, Jubil Farms, with its headquarters in New York City, was farming eleven times as much California land as the law allowed, with water it bought from the government for a few dollars per acre-foot—probably one-tenth of its worth on the free market, had there been such a thing. But this phony transaction, cynical as it was, was at least a

gesture of compliance with the Reclamation Act. Other farmers chose
to stonewall the Bureau in court, moving into compliance a centimeter
at a time. Any self-respecting lawyer could drag such a case out for
years, while his client continued to receive subsidized water the whole
time. Others were being granted special exemptions by the Interior
solicitor's office. No one has ever produced hard evidence, but there
has been speculation that such exemptions bore a more than casual
relationship to the size of a campaign contribution—and these were
growers who could easily contribute $50,000 to a candidate's coffer.
Rita Singer was a lawyer in the Interior solicitor's office through the
1960s and early 1970s, until she resigned and joined the legal staff of
California's Department of Water Resources. "We'd be working on a
case for months," Singer recalled during an interview in 1984, "and
then my supervisors would send down an interpretation of the law
that nullified our cause of action. Some of the subterfuges would be
allowed. Others would be disallowed. There wasn't any rhyme or reason. In most cases we never got an explanation. It was legal hairsplit-
ting. The solicitor's office would recognize 'distinctions' in cases that
were identical.

"In effect, we were telling the growers, 'Go ahead. Do whatever
you want.' When we moved for enforcement, it was always inconsistent. We never gave them a serious message that we meant business."
In public, Singer says, the growers cursed the Bureau, calling it "dic-
tatorial" and using epithets far stronger than that. "In private, they
regarded the federal government as a laughingstock."

The Bureau knew full well that numerous violations were taking
place in California. In 1964, Interior Secretary Stewart Udall ordered
Commissioner Floyd Dominy to investigate the number of violations
occurring within the service area of the Metropolitan Water District
of Southern California—presumably to use the information as a
weapon to force the Met to drop its campaign of divide and delay
against the Central Arizona Project. Dominy's regional director in
Boulder City, Arleigh West, hired an investigator from Phoenix named
Ralph O. Baird to conduct a surreptitious hunt for violators, learning
what he could through deed records and word of mouth. According to
a December 30, 1964, blue-envelope memo sent to Dominy by West,
"extreme caution was required"—apparently Baird thought he had
some reason to fear for his safety. In three months, he managed to
document ninety-nine violations of the excess-lands provision, totaling
105,229 acres. Several growers were irrigating thousands of acres with
federally subsidized water wholesaled to them by the Met; the largest
of them was the Irvine Ranch, which, in 1980, was the eleventh-largest

landowner in California, with 28,257 acres of cropland, 82,344 acres all told, and $140 million in annual sales, according to the California Department of Corporations and Dun and Bradstreet. The list of violators has apparently been destroyed; not a trace of it could be found in Dominy's files or Bureau records in Boulder City. But West would admit in retirement that the violations had indeed occurred, that they might still be occurring, that in his estimation it was a clear-cut illegality under Reclamation law, and that—for reasons he "wasn't privy to"—nothing was done. "I didn't even dare mail the list to Dominy," he said. "I hand-carried it to him on a plane. He looked it over and put it away. He told me never to talk about it—and he said it in that tone of voice of his that meant you'd better obey—and I never saw it again. There were some pretty powerful people on that list."

In eighty-two years, the Bureau would see the breakup of only one major illegal landholding through to the end. That was the DiGiorgio Company, and it was stripped of its excess lands only because John Kennedy's Interior Department solicitor, Frank Barry, was relatively serious about enforcing the act. Later, when Jimmy Carter began making noises about enforcing the letter of the law again, the growers managed to lobby through (in 1982) the most extensive and, in the view of those who had watched in frustration as large growers evaded the Reclamation Act not just for years but for decades, the least justifiable revision of the law in its eighty-year history. The 160-acre limit was raised to 960 acres, and the leasing and residency restrictions were eliminated. In return, the growers are supposed to pay "full cost" for water delivered to all lands beyond the 960-acre limit. (In 1987, however, the Reagan administration delivered an Interior Solicitors opinion allowing subsidized water to be sold to unlimited 960-acre "paper farms" owned by relatives and trusts in the same family—the same fraud, on a much larger scale, that had gone on before the Reclamation Act was "reformed.")

A Carter administration investigation conducted a couple of years earlier before the reform—the first serious effort to gauge the degree of compliance with the law—had established that more than 90 percent of the acreage violations were occurring in California and Arizona, whose hot climate permits high-value crops and two-crop seasons— exactly the kind of climate where the original 160-acre limit is eminently fair. In Colorado, or Montana, or Wyoming, where most farms are at altitudes of at least 4,000 feet, where the freeze-danger period runs to eight months, and where farmers are lucky to raise one good crop of low-value corn or wheat, a revision of the acreage limit was probably in order. But California's farmers, having received the gift of subsidized water not long before, were now awarded with a so-called reform whose chief result was to legalize wholesale noncompliance with federal law.

Even if the farmers sensed that, ultimately, the government would cave in, the Reclamation law was at the very least a nuisance to the big growers in the CVP service area. And to many of the *really* big growers who owned huge acreages toward the southern end of the valley, between Fresno and Bakersfield, the Central Valley Project meant nothing at all. Nearly all of its water deliveries stopped at Fresno, and most of it went to the valley's east side. The biggest owners were south of Fresno, and a number were on the west side, where they had amassed fiefdoms of dirt-cheap scrubland, which they were either irrigating or hoping to irrigate someday. Not a single substantial stream drains the lee side of the Coast Range south of San Francisco—precipitation is barely six inches a year—so most of the west-side growers were utterly dependent on groundwater. It was fossil water, water that had accumulated over hundreds of thousands of years but which, at the rate it was being pumped, evaporated, and transpired by plants, would barely last another fifty, if that.

The Central Valley Project was, in fact, to have an interesting—a startling—effect on the groundwater table of the San Joaquin Valley. In Tulare County, at one test well, the aquifer dropped sixty feet between 1920 and 1960, the year the first CVP water arrived. Thanks to the flood of new surface water, the water table then rose twenty feet in nine years. Just three years later, however, it had dropped another thirty-three feet. In Kern County, where the depth to groundwater is much greater, farmers who had pumped from 275 feet during World War II were pumping from 460 feet by 1965. The reason was obvious: the CVP and the Corps of Engineers projects on the Kings, the Kaweah, the Tule, and the Kern had delivered a lot of surface water throughout the valley, but they had encouraged so much agricultural expansion that they hadn't really relieved the pressure on the aquifer at all. For a while things were better; *then the projects actually made things worse.* Half the agricultural water used in the state was still coming out of the ground—even farmers who got cheap federal water continued to pump from their own wells in order to irrigate as much land as possible—and with three times as much irrigated land in production as there had been thirty years before, the big projects, besides depriving San Francisco Bay of half of its historical outflow, were just encouraging more pumping.

If there were no controls over groundwater pumping, a lot of farming in the southern half of the San Joaquin Valley faced extinction.

By the late 1950s, the land was producing the greatest agricultural bounty in the world. Four counties—Fresno County, Kings County, Kern County, Madera County—that were consistently among the six wealthiest agricultural counties in the nation now looked as if they might topple like a row of dominoes. The farmers were like addicts, oblivious to their self-destructive ways; they were making so much money they wouldn't think of groundwater regulation, and any politician who so much as uttered the phrase was instantly marked as a threat. (A hand-picked Fresno legislator named Ken Maddy once referred to groundwater regulation as "World War III.") The only answer, then, was to try once more to have the citizens of the nation's richest state build them a huge project to bring in more water from somewhere else.

A Himalaya of obstacles, a series of seemingly insurmountable crests, stood between the San Joaquin Valley and its goal. Cities could afford to build dams and aqueducts, because urban water was at least ten times more valuable than irrigation water. And urban property was worth much more than agricultural land—the richest acre of valley land couldn't be traded for a ten-by-ten-hundred plot in Beverly Hills—so a big urban aqueduct would have billions in assessed valuation standing behind its bonds. But without the fabulous subsidies written into the Reclamation Act—the "ability to pay" clause, the exemption from interest, the hydropower profits shoveled right back to the farmers—few irrigation projects could be built anywhere. The only feasible ones were at perfect sites—where a first-class river with a first-class gunsight canyon lay right above some first-class irrigable land. If one had to build a huge dam on a middling river or an aqueduct hundreds of miles long, or if the water had to be pumped uphill, any nonfederal project was out of the question.

Unfortunately, the San Joaquin Valley had every one of those problems. Much of the land in need of rescue was second- or third-class, even fifth-class, with vast depths to groundwater or drainage problems or alkaline deposits in the soil. Some of the barren acreage held for speculative purposes by oil companies at the southern extremity of the valley had no usable groundwater at all. The big rivers were all in the north, so an aqueduct hundreds of miles long would have to be built. And since the San Joaquin Valley slopes imperceptibly upward as one travels south, most of the land lay several hundred feet above sea level. The water would have to go to sea level in order to cross the Delta, in the middle of the state; then it would have to be pumped three to five hundred feet uphill.

One thing was clear: the growers, rich as they were, could never finance such a project themselves, as cooperative irrigation districts had financed a few smaller projects on the east side. The state would have to build it. But California had become highly urbanized since World War II; the votes had shifted toward the cities on the coast. Those urban voters would be crucial in getting the project through the legislature. In fact, they would probably demand a public referendum, and a referendum cannot be bought as easily as an act of legislation. The urban voters would obviously have to subsidize the growers, too. Between the astronomical cost of building such a project *and* the cost of pumping the water uphill, the farmers could never afford it—not as long as CVP water was being sold to farmers next door for $3.50 an acre-foot. Not as long as their cotton-farming competitors in Georgia and Texas and Louisiana (cotton was the main crop in the southern San Joaquin) got their water free from the sky. And that meant only one thing: urban Californians would have to get some of the water. If they didn't, they wouldn't vote for the project.

Only one major city could logically be tied into the project, and that was Los Angeles. Water on its way from northern California to Los Angeles would, of course, pass right through the San Joaquin Valley. With its meager and erratic rainfall, Los Angeles had always been haunted by drought; the thought of more water always set off a Pavlovian response. On the other hand, the metropolitan region didn't really *need* the water. The city of Los Angeles proper was getting virtually all its needs fulfilled by its Owens River Aqueduct, and its countless suburbs, together with San Diego, had recently gotten the first of their 550,000-acre-foot entitlement from the Colorado River. By the early 1950s, Los Angeles was extending its aqueduct into Mono Basin, where it planned to divert the streams tumbling out of Yosemite that feed Mono Lake. Meanwhile, the Metropolitan Water District, the area-wide water imperium serving most of southern California, was already planning a *second* aqueduct to the Colorado River, which would double that supply. (This was water that southern California planned to "borrow" from Arizona's entitlement for as long as Arizona—stymied by southern California's Congressional delegation—was unable to build the Central Arizona Project.) Six million new people could settle in southern California before a water famine developed.

What made matters worse was that in order to deliver northern California water to Los Angeles, you would have to contend with the Tehachapi Mountains, which separate southern California from the San Joaquin Valley. Either you had to tunnel through that brutish, barren summit, or you had to pump the water up and over, two-thirds of a vertical mile. Since the Tehachapis sit on two major active earth-quake faults, the Garlock and the San Andreas, tunneling would be risky. An earthquake could crush the aqueduct inside the mountains and shut off the water for months or years. That meant you would have to pump the water uphill, and the energy requirements would simply be awesome.

Why, then, would Los Angeles, which had most of its water arriving entirely by gravity from the Owens River, and the rest of the South Coast region, which got its water pumped by subsidized electricity from Hoover Dam, vote for a project that would·sell them expensive water they wouldn't need for decades?

There were two possible reasons. One was Arizona's lawsuit against California over its Colorado River entitlement. If California lost, and the Central Arizona Project was built, southern California would have to forfeit a vast quantity of water, on whose promise much of its expected growth was based—water enough for three million people. With such stakes, its smug confidence that it would win the lawsuit had to be at least *somewhat* shaky.

The other reason southern California might go along was simply that opportunities to find water did not arise every year. Ten or twenty years would be required to complete the project; by the time it was finished, if the region continued its spellbinding growth, there would be millions of new people there. Los Angeles was growing so fast that it might not want to pass up *any* opportunity to find more water—whether it made good sense or not.

If one thought about it this way, and thought about it long enough, it all began to seem inevitable. Los Angeles would resist, it would drag its feet and fret, but once the project began to roll through the legis-lature it would climb aboard. Since southern California was, finan-cially speaking, the key to the whole plan, it simply *had* to be dragged along. Southern California would sign on—out of fear, out of simple ignorance if nothing else. And southern Californians would get some of the water. But not too much.

During the winter of 1955, California was hit by the biggest floods since the monumental deluges of 1861 and 1862. After weeks of almost continuous rain, the rivers of the Sierra Nevada and the North Coast were tumultuous. The Eel River in the coastal mountains, which nearly dries up during the late summer and fall, was carrying the flow of the Yukon, the St. Lawrence, and the Missouri combined. The flood that spilled out of the mouth of the Eel—550,000 cubic feet per second—could have driven a fleet of battleships to Japan. The Sacramento

River, despite the enormous bulk of Shasta Dam in its path, also rose to monstrous heights. But it was the Feather River, the Sacramento's main tributary in the northern Sierra, that was the killer.

At the end of December, as a series of huge, slow-moving cloud-masses wrung themselves out against the western wall of the range, the Feather River rose with hurricane suddenness. Swelling toward a crest of 250,000 cubic feet per second, it burst out of its canyon and flooded over Yuba City and Marysville, two small cities on the flood-plain below, near the confluence with the Sacramento. Within hours, a parade of houses, some wrecked and some nearly intact, was floating toward San Francisco. Yuba City was substantially destroyed, first by water and then by mud. More than twenty people died.

The San Joaquin growers would never have admitted to feeling relief, but the Marysville and Yuba City disaster was the best news they had heard in years. If there had to be floods, the Feather River's wrath was a serendipitous one, for it had already been chosen as their river of rescue.

The origins of the rescue project went back to the Bureau of Rec-lamation's United Western Investigation, the two-year study of trans-continental water-diversion schemes, completed in 1951, that had been the swan song of Commissioner Mike Straus. Having looked at the possibility of diverting the Columbia, the Snake, and all the larger rivers of the Northwest to the desert Southwest, the Bureau had settled on the Klamath, which it wanted to run in reverse, through a sixty-mile tunnel, back into the Sacramento River and then south. The plan had collapsed under the weight of its own ambition, and the Eisen-hower administration had administered the *coup de grâce* by firing Straus. But the idea of a transbasin water diversion had quickened the pulse of California's state engineer, A. D. Edmonston, an unrecon-structed, gung-ho, New Deal water-development type. In 1951, Ed-monston, backed by the agricultural lobby, persuaded the legislature to give him enough money to undertake an "inventory" of the state's water resources—where water was in surplus, where it might be needed. What emerged three years later was something else entirely. The "inventory" had metamorphosed into something called the Cal-ifornia Water Plan—a scheme for moving water southward that vir-tually duplicated the Bureau's plan. Only two things were different. There would be no Martian aqueduct leading from the Klamath River to Lake Mead; the remote Klamath was, in fact, out of the picture, replaced by the smaller but much more accessible Feather River. The other distinction was that the plan, as envisioned by Edmonston, would not be a federal project in any sense. It did not come right out and say so (perhaps because it hoped to get some federal help), but if one read between the lines, the state, or at least Edmonston, was now contemplating something this monumental on its own, just as it had originally planned the CVP. In fact, no sooner was the California Water Plan released than a new agency, the Department of Water Resources, was created out of a jumble of fifty-two agencies that had previously dealt with water, and given administrative powers to match.

Edmonston's scheme was mesmerizing. The largest water project ever built by a state or local government was New York City's Delaware water system, completed during World War II. The Delaware Aqueduct was eighty-five miles long and entirely underground—by far the longest hard-rock tunnel in the world. But the California Water Plan, in its first phase alone, contemplated the movement of four times more water over a distance six times as long. What was even more startling was that most of the water would go to irrigation. The Delaware Aqueduct had left New York, a Babylon of wealth, up to its ears in debt. But since each average household paid around a hundred dollars an acre-foot for water, and because the city had a huge commercial and industrial sector sharing the cost, the bonds would be paid off unless the city, for some reason, saw its growth curve go drastically into reverse. In the mid-1950s, the most that irrigation farmers could pay for water was, by a generous average, about $15 an acre-foot—less than a fifth as much as New Yorkers paid. And New York City's water arrived by gravity; California's farmers would have to pay for several hundred vertical feet of pumping, and Los Angeles would have to buy water pumped more than three thousand feet if the aqueduct went over instead of through the Tehachapi Mountains. (The United Western Investigation had already concluded that tunneling was too risky because of earthquake hazards.) How could anyone afford it?

What passed for an answer provides an insight into the thinking of Edmonston and the water lobby and a good many politicians at the time. It was also as remarkable a statement as any certifiably sane person ever made. "It is believed that the cost of water *will not be a limiting factor* in ultimate development of the water resources of California," Edmonston's report read. "It is indicated that urban communities will always be able and willing to pay the cost of water to meet their municipal needs. Furthermore, it is considered probable that under pressure of future demands for agricultural produce, *the water necessary for greatly expanded irrigation development will be provided, at whatever cost may be required.* . . . Many works financially infeasible today will undoubtedly be financed and constructed in the future" (emphasis added).

If anyone found such a statement preposterous—it was really like saying that, because of population pressure, we were bound to settle Mars—he kept his opinions to himself. The nearest thing to a publicly expressed doubt was the somewhat timorous suggestion of the Stanford Research Institute, which was asked to comment on the report, that a "definite price policy" would be required for "more realistic estimates of probable water sales," and that these, in turn, might well decide "the financial outcome of the project"—that is, whether or not it would end in the greatest bankruptcy of all time. The prevalent mood was more accurately reflected in a remark by the director of California's new Department of Water Resources, Harvey Banks—a remark he used in a great many of the speeches he gave to drum up support for the plan. "We must build now," Banks would say, "and ask questions later."

Meanwhile, the financial foundation of this most recklessly ambitious of plans was quietly being laid.

In the 1940s, some petroleum deposits were discovered off the southern Californian coast, near Long Beach. A few years later, when several major oil companies announced that they planned to begin exploiting the reserve, California decided to impose a severance, or extraction, tax, and agreed to give the revenues to Long Beach. After all, the money wouldn't amount to all that much, and Long Beach would need it to enlarge its harbor and cope with the mini-boom that would inevitably result. But after the tidelands oil revenues had been promised to Long Beach, in a contract duly signed by the city and state, the amount of oil offshore was discovered to be far greater than the initial estimates had indicated. The severance tax, if these estimates were correct, would amount to hundreds of millions of dollars over the years. As a result, the attorney general of California decided that there was only one sensible course of action: he nullified the contract.

The attorney general, whose name was Edmund G. Brown, was at the time a politician of less than starlit promise. Of middle height, a little squat, Pat Brown was a cheery Irish ward-heeler kind of politician—hale, earthy, utterly lacking in the complexity and awkwardness of his future rival, Richard Nixon. At about the same time he voided Long Beach's tidelands oil contract, Pat Brown developed an obsession, one that would remain with him for the rest of his life: water. As his water czar, Bill Warne, was to describe it later on, Bob Edmonston, the state engineer, had corralled him one day in the capitol and implored him to do something about "the water crisis." Brown, who grew up in San Francisco, said he wasn't aware there was any. Hadn't Los Angeles built its Colorado River aqueduct? Hadn't the Bureau just built the Central Valley Project? Yes, answered Edmonston, and that was precisely the problem. When you added a couple of lanes to a freeway or built a new bridge, cars came out of nowhere to fill them. It was the same with water: the more you developed, the more growth occurred, and the faster demand grew. California was now hitched to a runaway locomotive. At the rate the state was growing in both population and irrigated agriculture, it ought to be developing 750,000 new acre-feet each year. It was developing nothing. It had no major plans. Even if it started today, it would take twenty years to get a big project authorized, financed, and built. By then, California could have another seven or eight million people. "When we finally come to our senses," Edmonston told Brown, "the biggest bandwagon in history is going to come rolling through with water written all over it. If you want to be elected governor, you jump on it early—now."

It was a moment of epiphany, Brown told his friends. The thought of all those people arriving to no water, perhaps even to a Biblical drought, suddenly left him staggered. He would never be the same. Edmonston was right—water was worth developing at whatever cost. Nearly twenty-five years later, in 1979, he still believed it. In an interview he granted to the University of California's Oral History Program, Brown said, "No, I don't think it [cost] has any validity because you need water. Whatever it costs you have to pay it. It's like oil today. If you have to have oil, you've got to pay for it. What's the value of oil? What's the value of water? If you're crossing the desert and you haven't got a bottle of water, and there's no water anyplace in sight and someone comes along and says, 'I'll sell you two spoonfuls of water for ten dollars,' you'll pay for it. The same is true in California."

In 1958, after campaigning for and winning the governorship of the state, Pat Brown turned to the task of building his new dream, Edmonston's water plan, with an energy few of his friends had ever seen. He wheedled, cajoled, and mule-traded like a home-grown Lyndon Johnson, trying to accomplish something which, in its own way, was as daunting as Johnson's Great Society agenda: uniting a state divided into wet and dry parts, into sophisticated cities and hundreds of mean little farm towns, on a breathtaking agenda of water development. An Irish Catholic, Brown came across like a missionary preaching to the damned when he spoke to Californians of their water crisis. But he was also ruled, at times, by a Catholic's impulse to confess, and later he would tell an interviewer about his other, more prosaic motivation. "I loved building things," he blurted in an unguarded moment of candor. "I wanted to build that goddamned water project. I was *absolutely determined* I was going to pass this California Water Project. I wanted this to be a monument to *me*."

It must have been frustrating for Brown that the most implacable opposition did not at first come from northern California, as expected. It came from the corner of the state whose cooperation was essential if the project was ever to be built: metropolitan Los Angeles.

The stubborn resistance of the Metropolitan Water District of Southern California to a plan that would give it more water, at one stroke, than it had ever received was perfectly understandable from its point of view, even if it was baffling on its face. The water it had been counting on to meet its future growth was water that Arizona felt it rightfully owned, and was at issue in a seemingly endless lawsuit then before the Supreme Court. The Met's case, which was based largely on Arizona's initial refusal to sign the Colorado River Compact, was somewhat flimsy; it wasn't so much a legal argument as a game of chicken with the Supreme Court. In effect, the Met was daring the Court to take away water for three million people just as they were coming to depend on it. Because of the weakness of its legal position, southern California had at least as great a stake in thwarting bills that would have authorized the Central Arizona Project—something which its Congressional delegation had accomplished for twenty years. But the key to victory, in Congress if not in the Supreme Court, too, was demonstrating that the contested water was crucial to its growth, if not its very survival.

From the Met's point of view, then, the Feather River Project, which it ought to have viewed as salvation, was in a more immediate sense a threat. If it was built, it could wash away the strategic foundation of its legal and moral argument. It was an absurd position to be in, but the Met was committed—it had to pretend that no water was available from anywhere else.

As a result, the chairman of the Met's board of directors, Joe Jensen, decided to oppose the Feather River Project at all costs. The Met also disliked the idea of subsidizing the growers in the southern San Joaquin, who would receive half of the water but pay less than a third of the cost, and that was the argument it trundled out for public consumption. "If an urban area is to help carry this agricultural load," Warren Butler, the Met's vice-chairman, told the Los Angeles *Times* on August 10, 1960, "the urban area of Kern County should." (As Butler

well knew, that urban area—Bakersfield—couldn't possibly afford to.)
If any project bringing water to the South Coast was going to be built,
the Metropolitan Water District was going to build it on its own. While
Pat Brown thumped his Feather River Project up and down the state,
Joe Jensen was talking about water from the Eel River, from the Trinity, from the Columbia—in due time (which was to say, *after* it had
won its lawsuit with Arizona). While Brown talked of water famine in
apocalyptic tones, the Met board issued a statement that "these fore-
casts of disaster are without foundation in fact." To the utter con-
sternation of the growers, who were frantically lobbying for the project
under the auspices of the Feather River Project Association, the Met
went after the idea hammer and tongs, arguing against it on every
conceivable ground: cost, need, feasibility, practicality, even morality.
In 1957, the board of directors staged an opulent victory dinner in
honor of several legislators who had successfully crushed the project's
hopes in the last legislative session. "They refused to listen to reason,"
Bill Warne, Brown's water chief, would recall. "I must have gone down
to talk to them a dozen times, but all they could think about was that
they might weaken their case before the Supreme Court. I didn't think
they would. As a matter of fact, I didn't think they had much of a case
to begin with. But *they* thought they did."

Pat Brown was wise enough to see that eventually the Met would
be brought into the fold. "I remember Norman Chandler saying he
was going to oppose the project in the Los Angeles *Times* unless we
went along with the Metropolitan's viewpoint," Brown recalled later
in an interview. "I told Norman, 'Then you just oppose the project,
Mr. Chandler. The people will look at you with scorn as the years go
on.' So he walked out and I didn't know whether he was going to
support it or not. . . . But they *had* to do it. I knew we had them. I
knew that if they didn't get this bond issue over, they'd never get water
in southern California."

Actually, though, the Met's opposition wasn't Pat Brown's thorniest problem, even if it may have been his most frustrating one. The
thorniest problem was the cost.

Brown knew that a lot of voters will vote reflexively against any
bond issue, even one to hire police and build jails in the midst of a
crime wave. They would rather not pay taxes and buy guns, rather
not pay taxes and dig wells. This was especially true in southern California, the home turf of the John Birch Society and the Liberty Lobby.
Northern Californians were sure to be violently opposed, even if they
were promised some of the water. Northern Californians had always
resisted sending *their* water to L.A. Between metropolitan San Fran-

cisco, Sacramento, San Jose, Oakland, Stockton, and Contra Costa
County, there were four million people and at least a million voters
(out of three million who might vote statewide) who were certain to
go against him. Those votes had to be counterbalanced by "yes" votes
in southern California. But when those good Republican migrants from
the Middle West down there saw how much the project would cost,
they would blanch. How could he possibly win? There was only one
way, Brown decided. It was to lie.

"Lie" is a strong word, but in this case it is advised, because one
day Pat Brown would all but admit it himself.

It was, to begin with, hard to say how much the project would
cost, except that it would cost a bundle. Oroville would be not only
the world's tallest dam, but its fourth most massive. San Luis, in the
Coast Range foothills farther south, would be the *fifth* most massive
dam in the world, nearly two miles long. Two of the world's biggest
dams; the world's longest aqueduct; the world's highest pump lift,
surmounted by the world's most powerful pumps—five full batteries
of pumps; a chain of smaller dams and reservoirs strung out to receive
the water—all of this would be incredibly expensive. The Department
of Water Resource's feasibility report, known as Bulletin 78, offered
an estimate of $1,807,000,000, but an economist for the RAND Cor-
poration, Jack Hirschleifer, immediately tore it to shreds. Reading
between the lines, Hirschleifer noticed that though the report mentioned Oroville Dam at length, it *failed to include the expense of building
it*. It was an extraordinary omission, to say the least. The DWR explained that the dam wouldn't be needed right away and might be
built later. (It would be built right away.) The estimate also failed to
include the cost of branch aqueducts to San Luis Obispo and Santa
Barbara, although the DWR had promised those cities water and Pat
Brown was counting on their votes. And there was no "cross-Delta
facility," later known as the Peripheral Canal, on the price list, though
without it the project could never deliver its full annual yield of
4,230,000 acre-feet. In fact, it was unclear how the above-mentioned
facilities, immense as they were, and assuming all were built, could
deliver that much water every year. Even ignoring that, Hirschleifer
wrote in his report, "the correct figure, for capital costs only and
accepting official estimates, is certainly in excess of $3 billion." Three
billion dollars in 1959 was the equivalent of $13 billion in 1987. What
state would vote for a $13 billion bond issue today? Not one. Pat Brown
knew that very well. That was why he decided to say that the project
would cost $1.75 billion—just over half of what he knew, or should
have known, the estimate should have been.

Years later, a conversation with an unthreatening interviewer from Berkeley's Oral History Program finally brought out the truth. "We were questioning, could we even pass a bond act of $1.75 billion," Brown told his interviewer, Malca Chall. "We didn't know exactly the cost of the project. We hadn't priced it out to any exactitude. As a matter of fact, we thought it would cost *more* than the $1.75 billion, probably in the neighborhood of $2.5 billion. . . . We had to scrape and pull to put this project together. I mean don't kid yourself. [Laughs.] It was a close fit and $1.75 billion was about all that we felt we could get a bond issue [*sic*]. We were afraid to make it $2 billion. It was like $1.99 instead of $2. We thought that just sounded better to the people.

"I remember someone telling me how Huey Long operated in Louisiana where the legislature wouldn't give him money to build a road," Brown added. "He started at one end, built it to here, and left a great big gap."

$1.99 instead of $2. Like many of the New Deal politicians of his era, Brown had a habit of dropping the last few zeroes from his figures. These were *billions*, not pennies, that were being talked about, back when a billion was still real money. Brown's $750 million lie (and, if Hirschleifer was right, it was considerably more than that) was a $3 billion lie in modern money. And it would set the stage for a monumental predicament, one that the governor's son, ironically, would be the first to have to face. In order to embark on building the project, the DWR would have to have contracts in hand to sell water. That was the whole idea—demand before supply. Those contracts would ultimately demand that the state deliver 4,230,000 acre-feet of water. But if the initial bond issue failed to deliver the full amount, and the voters subsequently rejected bonds to expand the project, the state would expose itself to a torrent of crippling lawsuits from cities and farmers who had planned their growth and invested their money on the promise of water it could never deliver. The damage claims might cost more than the project itself. Back in 1959, however, all of that still seemed far in the future.

One of the reasons Pat Brown felt confident with his misleading cost estimate had to do with the tidelands oil contract between Long Beach and the state which, as attorney general, he had abrogated in 1954. Long Beach was understandably outraged, and immediately filed suit against the attorney general's office. To its amazement, the California supreme court sided with the state. It was a remarkable legal opinion. The attorney general had nullified a signed contract to let a city have some revenues, and the court had upheld him even though the state had no demonstrable need for the revenues. It didn't even have a plan to use them. What kind of court was this?

An answer—a speculative one—popped up in another part of the governor's long 1981 interview with Malca Chall. Actually, they were discussing something else—Brown's decision to try to use the old Central Valley Project bonds which the voters had authorized in 1933 to scrape together another $170 million in cash. That, in its own right, was a matter of peculiar legality: using a bond issue passed twenty-seven years earlier—a bond issue that was meant to finance the Central Valley Project—in order to construct an *entirely different water project*. But, mystifyingly, the California supreme court had okayed that, too. "That was Phil Gibson, the chief justice, with whom I worked very closely," Brown told his interviewer. Then, according to the transcript, he laughed. "He was a great chief justice and it was great to validate those bonds. . . . The chief justice worked very, very closely with me in all of those decisions. You see the supreme court didn't have to take original jurisdiction in those cases. But I would call the chief justice and say, 'Chief, this is very important. I want you to take it.' And invariably he did."

Phil Gibson died before he could be interviewed for this book, and Pat Brown, in a personal interview, hotly denied ever having tried to influence the court's decisions. But Gibson's obituary in the San Francisco *Chronicle* described him as perhaps the most powerful and influential chief justice in the history of the court, and he was, after all, Pat Brown's bosom friend. All of this leaves at least a *suggestion* that, in California, where an issue as important as water was concerned, strict legality, separation of powers, honesty, and other niceties of governmental conduct could easily be ground into mush.

In 1959, after intensive lobbying by Pat Brown, the California state legislature agreed to allot the tidelands oil money for the water project—an annual interest-free loan of $25 million, repayable . . . whenever. It was an open-ended deal; the Tidelands Oil Fund could keep feeding the project until the oil ran out, which might take a hundred years. Even Brown would admit in yet another startling little confession to Malca Chall that "it was another subsidy to the big farmers." But it was not just any old subsidy. It was a subsidy that had an architectural elegance, a wonderful symmetry to it. Several of the "big farmers" who would get much of the water from the Feather River Project were oil companies—the same oil companies that were paying into the Tidelands Oil Fund. In exchange for a modest extraction tax—quickly offset by the billions they would make on the easily accessible oil—they would have their barren, worthless acreages in the San Joaquin Valley turned opalescent green. *And* they would get the growth, and the cars, and the freeways, that would increase the demand for—and the cost of—the oil!

In the last days of the legislative session in 1959, the legislature gave final approval to the Burns-Porter Act, which authorized the Feather River Project—now rechristened the California Water Project—subject to a statewide referendum on the bond issue scheduled for November of the following year. Once again Pat Brown had shown what great Irish politician's instincts he possessed. One of the two sponsors, Hugh Burns, was a northern Californian who had made a reputation *opposing* water diversions from the north. Brown, among whose attributes modesty was notably absent, would later boast, "The fact that I selected Hugh Burns to carry the bill in the Senate . . . that was political *genius* if I do say so myself." Cyril Magnin, "Mr. San Francisco," was persuaded to serve as campaign chairman there. The supporters put on prominent display in southern California were fiscal conservatives and Republicans. Everyone would get a little water, too: Napa County, Alameda County, the Santa Clara Valley. But the Kern County Water Agency alone would get thirty times as much as all of California north of San Francisco.

Only in December, *after* the legislature had already authorized the project, did the Department of Water Resources make a stab at an economic justification, in a report called *An Investigation of Alternative Aqueduct Systems to Serve Southern California*. Instead of trying to justify the project by weighing costs against benefits—which is what the Bureau of Reclamation did, or went through the motions of doing—it compared the cost of the project to the most expensive alternative: desalinating seawater. On that basis, it concluded that the project made sense. But as Jack Hirschleifer disdainfully commented in his RAND Corporation report, you can justify *anything* if you compare it to a more expensive alternative.

The critics were too few and too late. On Friday, November 4, 1960, just four days before the referendum was scheduled, the Metropolitan Water District capitulated and signed the contracts that indicated its support. The Los Angeles *Times* was now in favor. The only widely read newspaper that adamantly opposed the plan was the San Francisco *Chronicle*. When the votes finally came in, forty-eight of the fifty-eight counties in the state had voted against the bonds. But the populous counties in the artificial paradise of southern California all went heavily for the project. It was, after all, early November, and they hadn't seen real rain since April. November—the last days before

the rainy season began. That was another little bit of subtlety from
Pat Brown. The bond issue passed by 174,000 votes.

The California Aqueduct begins at Oroville Dam, an inverted pyramid of such improbable dimensions—the height of the Pan Am
Building, the length of the Golden Gate Bridge—that it appears much
smaller than it actually is. In February of 1980, in the midst of a long
spell of wet Pacific fronts, Oroville Reservoir, despite its capacity of
something like a trillion gallons, was full, and the dam was spilling—
seventy thousand cubic feet per second, the Hudson River in full flood,
roaring down the spillway at forty miles per hour, sending a plume
of mist a thousand feet in the air.

Below the dam and the Thermalito Afterbay the Feather River joins
the Sacramento, which flows through the Delta out to San Francisco
Bay. In the winter of 1980, the Delta, a huge reclaimed marsh protected
by weakening dikes made of peat, was in danger of being reclaimed
by nature; the levees were being repeatedly breached by the flood, and
farmed tracts of three thousand acres were disappearing under twenty
feet of water. From a chartered Piper Cub, the odd vulnerability of
this Brobdingnagian contrivance was manifest: the levees keep in-
truding seawater from mingling with southern California's water as
it traverses the Delta on its way to the California Aqueduct, which
begins south of there. The Delta is the system's weakest link, and one
could see why from an airplane: below was the water on which a
million-plus acres and ten million people depend; a few miles west,
lapping hungrily at the first phalanx of levees, was the tongue of a
salty ocean that humbles all.

At the south end of the Delta, the Clifton Court Forebay appeared
below us—a receiving reservoir big as a Minnesota lake that rises and
falls like the Bay of Fundy in rhythm with southern California's thirst.
A wide canal leads out of the forebay toward a rectangular building
resembling the nonnuclear end of a very large nuclear power plant.
The building houses the delta pumps—a battery of ten-thousand-
horsepower machines that suck Feather River water thirty miles
across the Delta before it can escape to sea, then lift it the first three
hundred feet toward its ultimate thirty-four-hundred-foot rise over the
Tehachapi Mountains. The water disappears inside and reappears
thirty stories up the hill, at the beginning of the California Aqueduct.
From overhead one could see the water spurting out of the siphons,

each one wide enough to consume a freight car, as if shot from a water
cannon. The aqueduct wound southward through the pale foothills,
as level as a railroad grade, and disappeared in valley heat. It is 444
miles long, the longest river, if you can call it that, in California, and
it is entirely man-made.

Interstate 5 parallels the aqueduct for two hundred fifty miles
through the San Joaquin Valley. Not many years ago this was utterly
barren land: it sprouted some patches of green during the winter, then
lay dead during summer's drought. Now it is a wide swath of cotton
and orchards growing billions of new dollars in agricultural wealth.
A hundred miles south of the Clifton Court Forebay the water arrives
at San Luis Dam, now the ninth-largest dam in the world, a structure
almost as immense as Oroville. What is bizarre about San Luis is that
its basin, in the rain shadow of the Coast Range, is devoid of constant
streams. Nearly all the water in the huge reservoir, eight miles across,
is Feather River and Sacramento River water, pumped uphill. San
Luis adds stability and security in a state inclined toward unpredictable weather and tectonic upheaval; in such a theater of disaster, a
state utterly dependent on reservoirs needs to store its water in as
many places as possible. The penalty for this added security is the
giant jolt of electricity required to lift the water another three hundred
feet. It is a Sisyphean lift, for the water comes right back down again
when the San Joaquin Valley and Los Angeles call for more. You
recapture some of the expended energy in turbines when you release
it from San Luis, but the overall loss is around 33 percent. More than
anyplace else, California seems determined to prove that the Second
Law of Thermodynamics is a lie.

This whole hydrologic ballet, this acrobatic rise and fall of mega-
tonnages of water performed on a stage twice the length of Pennsylvania, is orchestrated by a silent choreographer in the Water Resources
building in Sacramento: a Univac Series 904 computer punched and
fed floppy disks by a team of programmers. At the south end of the
valley, the aqueduct arrives at its moment of truth. The Sierra escarpment curves westward and the Coast Range bends eastward and
they mate, producing a bastard offspring called the Transverse, or
Tehachapi, Range. The Tehachapis stand between the water and Los
Angeles, which sits in the ultramontane basin beyond.

The water is carried across the Tehachapis in five separate stages.
The final, cyclopean one, which occurs at the A. D. Edmonston Pumping Plant, raises the water 1,926 feet—the Eiffel Tower atop the Empire
State—in a single lift. To some engineers, the Edmonston pumps are
the ultimate triumph, the most splendid snub nature has ever received:

a sizable river of water running uphill. At their peak capacity, if it is ever reached, the Edmonston pumps will require six billion kilowatt-hours of electricity every year, the output of an eleven-hundred-megawatt power plant. Moving water in California requires more electrical energy than is used by several states.

Having surmounted the Tehachapis, the water charges downhill again through closed siphons and a battery of turbines that steal some of its energy back. Soon it is in an open aqueduct again, which ultimately forks like an interstate highway: the West Branch goes straight to Los Angeles, and the East Branch continues southward across the high Mojave Desert to the vicinity of Riverside, where it terminates in Lake Perris—a reservoir. Lake Perris is six hundred miles from Oroville Dam.

Walking along the East Branch Aqueduct, you see people strolling, bicycling, and fishing as if this were a river through a city park instead of a concrete highway of water under a blazing sun in a shadeless desert where it rains seven or eight times a year. The Department of Water Resources stocks the aqueduct with fish—that way it can write off a fraction of the project's cost to recreation—but fish seem to find their way in there anyway. In fact, sections of the aqueduct have respectable fishing for striped bass, which cannot easily tolerate the pollution of Chesapeake Bay or spawn in the freakish cross-Delta currents that the project pumps have caused, but which don't seem to mind a three-hundred-foot lift in a pressurized elevator of water. This turbid, computer-controlled, concrete-walled river is the unlikeliest habitat imaginable for striped bass—as fitting a symbol of wild, fecund nature as one could find. The water project seems as make-believe as California itself, in its relentless quest to deny its desert heart.

Aside from lying about the true cost of the State Water Project, Pat Brown and his water resources chief, Bill Warne, had been less than candid about another matter of supreme importance: how much water the initial bonds would actually buy. Most Californians, it seems, believed they were buying four million acre-feet or more. But, as early as October of 1960, Joe Jensen had predicted that the bonds would never suffice to develop that much, and he was right. The initial facilities, it turned out, could deliver around 2.5 million acre-feet, perhaps three to three and a third million in wetter years—at least a million acre-feet less than the various cities and irrigation districts had signed up to buy. Meanwhile, population projections for southern California continued their horrifying march; in 1961, the Los Angeles Department of Water and Power was estimating that forty million people would live in the South Coast area by 1990. By February of 1962, Alfred Golze, Bill Warne's chief engineer, was already calling for new reservoir construction on the North Coast as early as 1972; Warne himself said that "new reservoirs, dams, tunnels, and diversion projects must be undertaken somewhere in the North Coast area within the next twenty years."

As it turned out, a splendid opportunity to do just that arrived sooner than Warne dared hope. In December of 1964, California was hit by floods that were even wilder than the great floods of 1955. In three days, from December 21 to 24, Blue Canyon on the American River recorded twenty inches of rain. All the rivers were roaring, from Big Sur to the Oregon border and beyond. But the river that rampaged most was the Eel. The Eel rose seventy-two feet from its bed. It snapped bridges with surgical precision; it uprooted three-hundred-foot redwoods; it swept fifty million board feet of timber out to sea—driftwood which, for the most part, is still piled along California's beaches. At Scotia, near its mouth, the Eel was carrying the Mississippi River in a garment bag; 765,000 cubic feet of water were going by each second. Every town along the river was damaged—some were never seen again. The high-water mark can still be seen along the Avenue of the Giants, displayed on a number of redwood trees. It is about three stories above the road.

The Christmas flood—the second "hundred-year" flood in just nine years—had Brown, Warne, and the Army Corps of Engineers issuing statements expressing profound dismay while they privately rubbed their hands with glee. Within months, the Corps, the Bureau, and the Department of Water Resources had locked arms as the State-Federal Interagency Task Force, ready, once and for all, to choke California's untamed rivers into submission. Every river on the North Coast, except the Smith and the Klamath, was to get at least one big dam; the various forks of the Eel were to get eight. But the Bureau and the Corps kept getting into scraps over who was to build what first, and Pat Brown's term was running out, so, one by one, the dams fell into obscurity. By 1966, when Ronald Reagan became governor, the only dam in which strong interest was still being expressed was the largest, Dos Rios, on the Middle Fork of the Eel. With twice the storage capacity of Shasta Lake, Dos Rios was the ideal addition to the State Water project; it could deliver another 900,000 acre-feet, almost enough to bring the total yield, in normal years, up to the 4,230,000 acre-feet the state had promised to deliver. The site was reasonably close to the Central Valley; all one had to do was dig a twenty-one-mile tunnel through the

Yolla Bolly Mountains and dump the water into Stony Creek, a tributary of the Sacramento.

Dos Rios had three things going against it, though the Eel had
acquired such a black reputation that none seemed likely to prevent
its being built. One was the fact that it would do nothing to control
the Eel. During the Christmas flood, more than 500,000 cubic feet per
second had poured out of the South and North forks and the main Eel,
which would all remain undammed. What did it matter if one's house
was under twelve feet of water or eleven feet four inches? Those eight
inches at Scotia were the sum total of the flood crest that Dos Rios
would contain; a local rancher, Richard Wilson, who had a degree in
agricultural engineering from Dartmouth, proved it, and the Corps
could only wish him wrong.

Another drawback was that the reservoir would drown an Indian
reservation and the town of Covelo—population two thousand—but
that sort of thing had been done many times before. (The Corps had
included the flooding of the reservation in its benefit-cost analysis, but
had it down as a *benefit* because the Indians would get a "nicer" town
somewhere else.) The third drawback was that the new governor of
California, Ronald Reagan, wasn't particularly interested.

Reagan, as a westerner, should have been a friend of dams, but he
was growing more conservative by the hour, and true conservatives
tend to dislike great public works. He also distrusted the Corps of
Engineers—a feeling which the Corps, if anything, seemed to reinforce.
Reagan's resources secretary, Norman Livermore, remembers asking
the Corps to do two cost-benefit analyses—one using the 3¼ percent
interest rate which the Corps planned to use, the other using the 6¼
percent rate that reflected economic reality. "When they gave it to
me," remembers Livermore, "I looked at the two columns, and the
bottom line was exactly the same. I took it into a cabinet meeting and
really got a laugh."

For four and a half years, Reagan stalled on Dos Rios while the
water lobby was practically battering down his door. The head of
his Department of Water Resources, Bill Gianelli, a short, square man
with a Vince Lombardi temperament and an American flag perpetually
stuck in his lapel, was, according to Richard Wilson—who was the
leader of the ragtag opposition—an "absolute zealot" in favor of building the dam. So was Don Clausen, the Republican Congressman representing the North Coast. But Wilson was a friend of Norman
Livermore's, and Livermore had Reagan's ear. According to Wilson,
when the governor realized he finally had to say yes or no, he asked

Livermore to give him every argument he could think of against the dam. When Livermore was finished, he emerged from Reagan's office and almost fell into the arms of Don Clausen, who was waiting to give Reagan his arguments *for* the dam. Clausen was a voluble and persuasive man, but later he confided to his intimates what had really happened during the meeting. Halfway through it, Clausen said dispiritedly, the governor had fallen asleep.

Wilson insists he got the story from Livermore himself, though Livermore, still a Reagan loyalist in 1984, said he "couldn't remember" it. Whatever the case, in 1969, Reagan finally announced that he would not support Dos Rios Dam. In the press release explaining his reasoning, he talked about costs, poor economics, the frailty of the flood-control rationale. Privately, though, Reagan was upset about flooding the Round Valley reservation. "We've broken enough treaties with the Indians already," the old cowboy actor is reported to have said.

By the time Reagan left Sacramento, in 1974, the Department of Water Resources was predicting that the dreaded shortfall—demand for water greater than supply of water—might be as little as fifteen years away. To plan the final phase of the State Water Project, get it approved and funded, and build it would easily require fifteen years. Through an irony some found delicious, then, the person who took it upon himself to complete the project that Pat Brown had left unfinished was none other than the apostle of the "era of limits," the first politician to proclaim that "small is beautiful" and "less is more": Jerry Brown—Pat Brown's son.

"He did it for the old man" was how Jerry Brown's last loyalists explained the spectacle of the younger Brown promoting what seemed certain to become the most expensive water project in the history of the world. Depending on which of the Brown administration's estimates one believed—and a new one seemed to appear every six months or so—the cost of completing the project was either astonishing or flabbergasting. What Pat Brown hadn't foreseen, when he underfunded the bond issue to ensure that the voters would pass it, was inflation. Because of inflation, it would cost two to five times more to deliver the project's last 1,730,000 acre-feet than it had cost to deliver the first 2.5 million. The most detailed estimate, released by the DWR in 1980, pegged the cost at $11.6 billion. Interest on the bonds—based on a rate of 9 percent, which was then three points too low—would add another $12 billion. It was unheard-of. The only comparable schemes anywhere in the world were Canada's James Bay Project and

Itaípu Dam, which would end up costing $19 billion and help Brazil dig itself a bottomless financial hole. But Itaípu would at least generate 12,500 megawatts of electricity to help pay for itself. Brown's Phase Two water plan would *consume* an awesome amount of power, because the water, cubic miles of it, would be pumped not just uphill but over a mountain range.

Jerry Brown's dilemma—which was insoluble, but which he thought he could solve anyway—was trying to please the water lobby and his large environmental constituency at the same time. He wanted a project, but he wanted it to be "environmentally sound." To be environmentally sound, there could be no on-stream storage—no dams or reservoirs on any significant wild streams. The North Coast rivers, with 29 percent of the state's runoff, were therefore off limits. Instead, Brown wanted to skim high "surplus" flows from the Sacramento River during the winter and spring and store them. But all the natural storage basins were at elevations well *above* the river. His Department of Water Resources engineers, acting on orders some of them considered insane, finally settled on a basin in the foothills of the Yolla Bolly Mountains, near Red Bluff, which had a stream running through it and a couple of small preexisting flood-control dams. They would run a twenty-mile aqueduct up there, up a thousand-foot slope, and dump the Sacramento surplus flows in. The reservoir, to be called the Glenn "Complex," would be as large as San Francisco Bay. It would submerge both preexisting reservoirs and a couple of small towns. There would be some contribution from Stony Creek, but not much; a tremendous amount of energy would be required to pump water uphill. A second off-stream reservoir—smaller, but still a third the size of Shasta Lake—would be created farther south, in the foothills near Mount Diablo. Below there, water was already being pumped three hundred feet uphill for storage in San Luis Reservoir—another off-stream site— and farther south it was being lifted to improbable heights by the Edmonston Pumps. If it was all built, the California Water Project would require about as much electricity as both units of the $5.4 billion Diablo Canyon nuclear reactor could produce, and Brown *didn't* want that built. Where, then, would the energy come from? The DWR set loose a bewildering flurry of "soft path" proposals—geothermal plants, wind machines, solar-generating ponds. The meanest of the governor's critics, taking note of his interest in Buddhism, said it was all going to be powered by yaks.

Brown was so sympathetic to environmentalists in other ways that a lot of them were hesitant to oppose the plan. (The California Sierra Club's leadership first endorsed it, only to be overturned in a referendum taken to the members at large.) After all, his director of water resources was Ron Robie, a smart, elfin, fast-talking lawyer who had been instrumental, while on the Water Resources Control Board, in writing decision 1422, a decree requiring minimum fresh-water flows through the state's most important estuary, the Delta. Robie's assistant director was Gerald Meral, a former staff scientist for the Environmental Defense Fund. Meral, a gaunt, bearded zoologist, was a great fan of wild rivers, an expert whitewater kayaker—there was even a falls and pool on the Tuolumne River named for him. How could people like Jerry Brown, Ron Robie, and Gerry Meral propose anything really bad?

One answer came from Tom Graff, a lawyer for the Environmental Defense Fund and Meral's former colleague. The centerpiece of Brown's plan was called the Peripheral Canal, an outsize channel to be constructed around the collapsing Delta. The Peripheral Canal had been a top priority of the water interests for forty years. What Brown wanted to do to win the environmentalists' and northern California's support was guarantee minimum releases to the Delta from the canal—a big surge of water would be let out every few miles, turning the Brown Canal, in effect, into a giant sprinkler hose. Robie and Meral argued that their plan would mimic the primordial river inflows and eliminate the cross-flows caused by the Delta pumps; in so doing, it would help salmon and striped bass spawn and actually improve the fishery. In fact, if one listened to them long enough, the whole $11.6 billion scheme was mainly for the sake of the Delta fish. But Graff pointed out that the Peripheral Canal would remove another couple of million acre-feet of water from the Delta and San Francisco Bay, water that normally went through at high flows. Delta outflows had already been reduced from 35 million acre-feet to around 17 million, and the fresh water that still managed to escape the project pumps was needed to wash pollution out of the bay; besides, the whole bay ecosystem had grown dependent on large seasonal fresh-water flows over tens of thousands of years. Who was to say that the bay, having already seen its fresh-water outflow decline by 55 percent, wasn't on the brink of ecological ruin?

Besides, what if the legislature, dominated by southern California and the agricultural lobby, decided to overrule the Delta outflow guarantees? And what if it decided to dam the North Coast rivers? With the canal in place—it was, after all, to be four hundred feet wide, and would be capable of containing most of the Sacramento River—the water could finally be moved. The Glenn Reservoir site, curiously, was at the receiving end of the proposed Grindstone Tunnel, which was to have carried water from Dos Rios Reservoir through the Coast Range. The Peripheral Canal, according to Graff, was a "loaded gun pointed at the North Coast."

Brown's answer to that, in 1981, was yet another set of environmental guarantees. When his first canal legislation failed to pass the legislature, he supported a new package known as Senate Bill 200, which included an amendment to the state constitution keeping the North Coast rivers wild and scenic forever—which meant no dams. All of the larger ones had had such designation since 1972, but it was state, not federal, protection, and the legislature could annul it at will. Brown's constitutional amendment would have made it impossible to develop the Eel and the other rivers unless the state's voters, by a two-thirds majority, decided at some point to repeal it.

Jerry Brown was quite sure his proposal would mollify the environmentalists, but it had a totally different result. Until then, feeling about the Peripheral Canal—a term that became shorthand for everything else in the plan—had sloughed along traditional lines: northern Californians were mostly against it, the valley and the South Coast were mostly for it. But his decision to include constitutional protection for the North Coast rivers in S.B. 200 created a stranger alliance than Brown and the growers. It was, in the minds of some, the oddest alliance since the Hitler-Stalin Pact. All of a sudden, two of the mightiest, wealthiest growers in California were on the side of Friends of the Earth.

The two retrograde growers were the J. G. Boswell Corporation and the Salyer Land Company, which had long dominated affairs at the valley's southern end. Salyer and Boswell were two of the main beneficiaries of the Corp of Engineers' Kings River and Kern River dams, which gave them year-round irrigation water that was nearly free and tens of thousands of new acres in the old bed of Tulare Lake. They had figured prominently in the Feather River Project Association, which helped get the State Water Project authorized in the first place. In 1980, Boswell owned 206,021 acres in California, plus hundreds of thousands of acres elsewhere; it was the biggest grower in the state. Salyer's holdings were smaller, about 77,000 acres, more than the five boroughs of New York. In one year, Boswell's private political action committee, or PAC, ranked among the top ten in the nation in the amount of money it showered around. For all their power and money, however, Boswell and Salyer had a problem. They were located in the part of the valley with the severest groundwater overdraft. Someday, if pumping wasn't to become prohibitively expensive, more surface water would have to be brought in—a lot more water, since the valley's groundwater overdraft was projected to surpass the yield of the State Water Project by 1999. Boswell and Salyer felt there was only one place it could come from—the Middle Fork of the Eel. The idea of making the North Coast rivers wild and scenic seemed like a prescription for their economic demise; they were also incensed, as a Salyer spokesman put it, that "the Delta fish come before we do"—an allusion to the minimum Delta outflow guarantees in S.B. 200.

By the end of 1981, to everyone's amazement, Boswell and Salyer had poured $406,000 into the campaign against the Peripheral Canal, outspending the thirty-three largest contributors on the pro-canal side—who included Shell Oil, Getty Oil, Southern California Edison, Lockheed, the Fluor Corporation, and Walt Disney Enterprises—by $73,689. It helped, but not enough. Later that year, the legislature passed S.B. 200, subject to ratification by the voters in a special election to be held in June of 1982. The planning meetings among the canal's opponents, as they prepared for the referendum, must have been something to behold. Environmentalists and northern Californians were there because they thought S.B. 200 was too weak. Boswell and Salyer were there because they thought it was too strong. Delta interests didn't much care one way or the other; they just wanted to keep getting their irrigation water free. (As water on the way to the federal and state aqueducts flows between their levees, they simply slurp it out; they would have to pay to get it out of the canal.) After a series of ferocious catfights, the strategy that the canal opponents and Russo-Watts, the public relations firm handpicked by Salyer and Boswell, agreed on was to hammer away at the cost.

It wasn't a bad idea. The votes the canal would need were in southern California, and those voters would be saddled with most of the cost. About 70 percent of the original works of the State Water Project were being financed by the Metropolitan Water District's customers. Actually, they paid for the project twice: through daily water rates, and through an assessment of twelve cents on every $1,000 of property value in the service area, which they paid whether they got water or not. (The city of Los Angeles still got 93 percent of its water from the Owens Valley and Mono Basin, but paid the assessment like everyone else because it was subsumed under the MWD.) Using simple arithmetic, one could divide the number of Metropolitan customers into the $11.6 billion that Phase Two was supposed to cost, multiply that by .70 and come up with a figure of $3,000. That was the average cost of S.B. 200 to each household in southern California. If one added the $12 billion in interest that would have to be paid on the bonds, the figure doubled. As if that weren't bad enough, the California Energy

Commission was predicting that energy, in the year 2000, would cost thirty-three cents per kilowatt-hour, which was six times what it cost in 1981. At those rates, it would cost at least $50 just to pump one average family's share over the Tehachapis. And that share was only a fraction of the family's *annual use,* because the MWD's full entitlement to State Project water amounted to less than a third of all the water used in southern California. People would also be paying for water pumped sixteen hundred vertical feet from the Colorado River; they would be paying for water pumped from the ground. If one added it all together, the cost of water in southern California would be . . .

The estimates varied about as wildly as estimates can. State Senator Reuben Ayala, the chief sponsor of the Peripheral Canal bill, said it would cost the average family only $5 extra per year. The Met said $50 per year. Dorothy Green, the leader of the opposition in southern California and the founder of an organization whose acronym had somehow been tortured into spelling WATER, was saying that a year's worth of water would cost $1,400 in the year 2000 if the canal and everything else were built. The public remained utterly confused by all of this, which, as far as both sides were concerned, was fine. The campaign could then be run on fear. Magazine spreads began appearing in southern California showing a child's upturned tricycle at the edge of a dried-up reservoir. Northern California billboards were papered with huge letters (courtesy of Salyer and Boswell, who ended up spending $1 million on the campaign) that simply read, "It's Just *Too* Expensive." Everyone knew what "It" was, just as everyone knew what horrible fate the abandoned tricycle was supposed to represent. One leader of the stop-the-canal campaign, a businessman, talked off the record about how dirty a war over water in California can get:

"The business community in southern California has got the business community in northern California half out of its wits. Crown Zellerbach, the big San Francisco paper company, has been told it better not take an anticanal position if it wants to sell any more paper south of San Jose. They've stayed neutral. The San Francisco Chamber of Commerce is staying neutral, too, even though an informal plebiscite among its members showed 92 percent of them opposed to the Peripheral Canal. The chamber's board of directors has refused to share those results with the membership. *We're* going to have to tell them. The chairman of the board is opposed to the canal—he *hates* it—but he won't say so in public. These guys are representing the interests of their own corporations, not of northern California, or even the Chamber of Commerce. They're scared to death. It's hard for us to raise any money, because contributions are identifiable and everyone is scared they're going to be found out and blacklisted down south. It's like a banana-republic election where the houses of the opposition candidates all catch fire."

Nineteen seventy-six and 1977 were the third-driest and the driest years, respectively, in California history. Shasta Lake, the reservoir on which billions of dollars in farm income depend, was nearly dry, down to an eighth of its capacity; water rationing was imposed all over the state. But 1978, which looked as if it might herald the beginning of California's end, was, to everyone's surprise, a wet year; 1979 was even wetter. In 1980, Los Angeles was clobbered by a succession of subtropical Pacific storms that threatened to float it out to sea. By then, memories of the drought—which had panicked almost everyone in California, even environmentalists in Marin—were growing dimmer. 1981 was drier than normal, but not by much. 1982 marked the beginning of what climatologists called the "El Niño episode," when parts of the state got three times their normal rainfall and relentless storms caused $1 billion worth of property damage. It would be excessive to say that a string of five rain-laden years determined the outcome of the vote on the Peripheral Canal, but it would probably be true. Had the referendum been held in October of 1977, when most of the state had barely seen rain in a year and a half, Californians might have voted for anything, even dragging icebergs down from the North Pole. Memories of the drought had grown faint, but memories of inflation hitting 15 percent in 1980 were strong. Houses that had cost $35,000 in 1974 were being snapped up for $200,000. The referendum on the Peripheral Canal carried southern California by two to one. But in counties around San Francisco it lost nine to one. When the final tally was in, the Peripheral Canal had gotten less than 40 percent of the vote. It was trounced.

As it turned out, however, the big San Joaquin growers would have plenty of water—miraculously cheap water—for a long, long time.

Twenty-two years earlier, after Californians had voted in favor of building the State Water Project, the Department of Water Resources began to circulate water sales contracts in the San Joaquin Valley. Few of the farmers were willing to sign. The irrigation water would be relatively unsubsidized—the main subsidy being the $25 million annual contribution from the Tidelands Oil Fund, which was called a "loan" even though virtually none of it has been repaid—and it would be expensive. The development cost would be around $20 per acre-foot, plus the price of delivery, so most irrigation water would cost anywhere from $25 to $45 per acre-foot. And that was actually a discount price, held low by cheap power rates and a drawn-out repayment schedule, so that the farmers could afford to build laterals, headgates, and all the other appurtenances they would need to shift from groundwater to surface irrigation, or from no irrigation to irrigation. Eventually, the cost would shoot up dramatically to recover the initial discount. Farther north and east in the valley, farmers were buying water from the Bureau of Reclamation for $2.50 to $3.50 an acre-foot. The most anyone paid for Bureau water was $7.50. In a lot of places you could still pump groundwater for $15 an acre-foot. How could the State Project's customers compete? The difference between water at $3.50 an acre-foot and water at $30 an acre-foot—if you irrigated 320 acres and used four feet on your crops—was $33,920. That was more than the net income of a typical small farmer in a year.

There was, however, a way to make the water affordable. The phenomenal growth rate that California has sustained since the turn of the century was finally slowing down. (In 1969, for the first time ever, the state registered a net loss of a few thousand people.) The Metropolitan Water District wouldn't need its full entitlement for a good while—that was common knowledge—and now it looked as if it could do without it for even longer than expected. But water projects do not make more water available in small increments. Once Oroville Dam was completed, an immense amount of water would suddenly be available. The system was likely to have a big surplus sloshing around in it for years. What was wrong with letting the growers have that water for the energy cost of delivery?

The growers made their case to Bill Warne and found him sympathetic. Naturally, he said, there ought to be some restrictions. The surplus water should go only to lands that overlie the aquifer (the extreme southern part of the San Joaquin has no usable groundwater at all). Otherwise it would bring a lot of land into production that would be stranded when the surplus deliveries ended, creating even more pressure for new water development. The water would have to be sold on an interruptible basis, from one year to the next, and it ought to irrigate only pasture or alfalfa, not permanent crops such as orchards. Otherwise, when the surplus ran out, the farmers, having invested a lot of money in trees, would begin pumping groundwater like crazy to protect their investment, and demand still more dams, and the vicious cycle the State Water Project was intended to stop would begin all over again.

If Warne was amenable to the idea, Joe Jensen, the thin-lipped, mercurial Mormon chairman of the Metropolitan Water District, was not. The growers, he told Warne, were self-interested, avaricious cut-throats who wanted a free ride on the Met's customers. They—the urban users—would be paying capital and interest costs on each acre-foot developed, whether it was delivered to them or not. In fact, they would be paying higher development costs on the surplus water, without seeing a drop of it, than the growers would pay to have it delivered to them. Why, shouted Jensen, should *not* receiving water cost more than receiving water? The whole idea was an outrage. The Met, Jensen said, would never stand for it.

Jensen held his board of directors under "an almost absolute dictatorship"—those were Warne's words—so the prospect that the growers would get anywhere were slim. When Warne tried intervening on the growers' behalf with some friendlier members of the Met's board, they spurned him. One day, however, an old colleague of Warne's from his Interior days, who had since become chief counsel for the Kern County Water Agency—which owns the largest entitlement of all in the State Project, 788,409 acre-feet—called. The lawyer, Stanley Kronick, told him that the issue of cost-of-delivery surplus water was extremely important to the growers, and could jeopardize the whole future of the project, because without it the growers might not be able to pay their way, and the project could default. Kronick wondered whether he shouldn't go down to Los Angeles and speak with the board himself.

Warne remembers being faintly amused. "Sure, Stan," he said. "You're welcome to try. But you aren't going to get anywhere, you know. Joe Jensen is adamant, and the rest of them have got their heels dug in. I've been over it with them a dozen times already."

Nonetheless, Kronick said, he was going to try. A few days later, Warne received another call from him. When he picked up the telephone, he felt sure that he knew what he was going to hear.

"Well," said Kronick, "they agreed."

The fifty-two members of the board of directors of the Metropolitan Water District are protected, by charter, against conflict-of-interest disclosures. No one has to release information on stock ownership, business connections, or anything else that might provide a clearer picture of where their true interests lie. As a result, no one knew much about them—though many tried to find out—except what was obvious: in the 1960s most were white, male, middle-aged or older, wealthy, and passionately committed to water development.

Therefore, the most cynical interpretation of the Met's decision to let the San Joaquin growers have its customers' unused water cannot be proved. They did it, this argument goes, because when they realized what a bonanza it would be to the growers they all invested in valley agriculture themselves. After all, if they had their customers' interests truly at heart they should have held out for a higher price, forcing the growers to share at least some of the capital and interest costs. The people who make this argument usually take their listeners on a guided tour through California's verdant history of public graft to reinforce their point. Bill Warne, for his part, doesn't share such paranoid suspicions. In his mind, they did it because they knew they would need the valley's help when the time came to get more water from the north. "With the environmental opposition to new development, the Met realized it had to stop fighting with the valley and close ranks," he says. "Maybe this was their way of making peace."

Whatever the answer, the first of the surplus water was delivered to the San Joaquin Valley in 1973. Precipitation stayed near or above normal for the next ten years, except for the two freak years of the drought, and consumption in southern California remained well below predictions. As a result, there was a literal flood of surplus water in the valley, sold at an average price of $3.50 per acre-foot—the same incredibly low price the Bureau charged. Even in 1976, the beginning of the drought, the state inexplicably let go of 580,110 acre-feet of "surplus" water that it might well have husbanded for the near-catastrophe waiting around the corner. The Kern County Water Agency, whose clients include many of the biggest and wealthiest growers in the state, took 442,250 acre-feet of that amount, setting a pattern: since 1973, it has gotten between one-half and three-quarters of the share. By the end of 1981, it had received a total of 1.8 million acre-feet of surplus water. It got it for around $6 million—the alleged cost of delivery. Meanwhile, according to Richard Walker and Michael Storper, two analysts at the University of California, the Met's customers had been assessed about $170 million for the same water— water they never received. The growers had gotten a $164 million gift.

After peaking at 524,247 acre-feet in 1979, Kern County's surplus deliveries began to diminish as the Met called on more of its entitlement, but it was virtually guaranteed hundreds of thousands of acre-feet for years. Meanwhile, as the Peripheral Canal debate was raging on, and the Met was saying that without the canal southern California would perish, an internal study, not intended for release, predicted that "as much as 750,000 acre-feet [of unused water] in the MWD service area" would be available if the canal was built. In other words, the Peripheral Canal would not so much save Los Angeles as allow the growers to keep using hundreds of thousands of acre-feet of surplus water while metropolitan Los Angeles paid for it. The farmers, for their part, seemed to be counting on it, for they were using surplus water to expand their acreage well beyond a level sustainable with contract water alone. According to at least one agricultural economist at UC-Davis, they were also using it to irrigate permanent crops— exactly what Bill Warne had said they must not do. It would have been very foolish of them to do so unless they expected to have a lot of surplus water for a long time.

It was the same old story again. The big farmers had managed to get something (a lot of water) for next to nothing. People in Los Angeles, meanwhile, were being taught a different lesson: that you can get nothing for something.

All of this raises a further question: who, exactly, are the farmers getting most of the water?

In 1981, Les Melville had been growing olives for nearly fifteen years on his fifty-acre farm near Oroville, the town that grew up alongside the Water Project's monumental dam. He bought the farm in the 1960s, and through innovation, and a lot of lavish care, he raised the previous owner's average yields from around thirty tons a year to 250 tons a year. It was a remarkable effort. Then, having finally accomplished what he set out to do—prove you can make a good living on a fifty-acre farm—he began to go broke.

"When we started here in 1967," Melville told an interviewer, "we ended up with some $500 per ton of fruit. In 1980, we were down to $350 a ton. We're getting less for our fruit now than we were getting in 1946." Melville's costs, meanwhile, were constantly rising, and his disposition and his health were failing. The problem wasn't competition from imported olives; it was the California Water Project.

At the other end of the valley, the Prudential Insurance Corporation was farming more acres of olives than all four-hundred-odd olive growers in Tehama County, where Les Melville's farm is. Prudential had five thousand acres planted in olives on its McCarthy Joint Venture A ranch near Bakersfield, in which it owned a 75 percent interest, and those five thousand acres were only about a quarter of the entire ranch. Its olive trees were planted very close together, like hedgerows—not because the country wants more olives than anyone can produce, but because the fruit can then be harvested by machine. Machine harvesting wastes fathomless numbers of olives, but saves a substantial amount of labor. The olive-harvesting machinery was developed, in large part, by the taxpayers of California, who finance the agricultural experimentation programs of the University of California's extension service—which are largely devoted to inventing and perfecting labor-saving machinery. (One of its star creations, the tomato harvester, is said to have displaced twenty thousand agricultural workers.)

When Prudential's olive trees matured in 1978, they began producing all at once. California's production of olives increased by 46 percent in that year—a single year—and olive prices fell like overripe olives. Of all the state's growers, however, only one was relatively unaffected by the drastic drop in wholesale prices: the Prudential Insurance Corporation. The company was well aware that its prolific production would cause the collapse of the market, and therefore decided to write an unusual contract with Early California Industries, the state's largest independent olive processor. In exchange for an opportunity to defer the purchase price of $1 million, Early Cal agreed to buy Prudential's entire harvest. Previously, it had bought from many small growers around the state, like Les Melville, who now had to look elsewhere to sell their production. The deferred payment, Early Cal proudly remarked in its annual report, "bears no interest and is repayable only on termination of the contract." It was what labor unions like to call a "sweetheart" deal. With a single stroke, a New Jersey–based insurance corporation had, in its first year of competition, with a single gigantic orchard, pretty much captured the olive market of the United States.

Like a number of other corporations, holding companies, and investor cartels, Prudential got into farming in the 1960s, when Congress passed legislation allowing investors to deduct all expenses on a number of crops (chiefly orchard fruits and nuts) while the trees or vines are maturing and bearing no fruit. All of a sudden, a lot of land that wasn't worth very much was worth a great deal—in an inverted sense. According to economists at the University of California at Davis, the new tax provisions amounted to a tax break of $346 on an acre of land for persons in the 70 percent bracket. For corporations it was less of a break, but still a good one. With its 75 percent share of the McCarthy Ranch, Prudential could realize a tax saving of around $1 million per year, farming the government. Then, when the trees were mature, it could begin earning at least that much income every year. It was all made possible by the State Water Project.

The land on the far southwestern side of the San Joaquin Valley, where the McCarthy Ranch sits, is underlain by a brackish, boron-poisoned aquifer. The quality of the water ranges from execrable to unusable. The climate is so dry—around six inches of rainfall a year—that the few small freshets barely trickle during the rainy season. Until the late 1960s, when the first deliveries from the California Water

Project arrived, $50 an acre would have been a good price. Now it is worth at least $2,000 an acre.

In August of 1981, the California Institute for Rural Studies released a report on property ownership in five water districts within the service area of the State Project. Most of the districts are in Kern County; most of the farms are neighbors of the McCarthy Ranch. Together they accounted for two-thirds of all the entitlement water delivered to the San Joaquin Valley by the project. However, because the Kern County Water Agency, the region's main water broker, had been receiving a flood of surplus water (1.8 million acre-feet) as well, the five districts had actually received about half of *all* the water the State Project had delivered throughout the state.

The CIRS report corroborated what the Department of Water Resources had taken unusual pains to point out: that the majority of farmers receiving project water were small farmers. Of 479 identifiable owners in the five water districts, 291, more than half, had farm holdings of 160 acres or less. Nine out of ten worked farms smaller than 1,281 acres. But those farmers owned less than a third of the total acreage; the other two-thirds, which amounted to 227,545 acres, was owned by eight companies.

The largest of the farmers was Chevron USA, the main subsidiary of the Standard Oil Company of California. Chevron owned 37,793 acres in the immediate vicinity, in addition to 42,000 acres scattered elsewhere in the valley. In second place, with 35,897 acres, was the Tejon Ranch, one of the great land empires of California—272,516 acres all told. The principal stockholders of the Tejon Ranch are members of the Chandler family, which owns the Los Angeles *Times*—the strongest voice for water development in California for the past eighty years.

In third and fourth place were two more oil companies, Getty and Shell, which owned 35,384 and 31,995 acres, respectively. The presence of Getty (and Chevron USA) in the service area of the California Water Project again pointed up the architectural brilliance with which the project was conceived. They pay a severance tax to California on oil they pump off Long Beach, which is immediately put into a fund that makes annual interest-free "loans" of $25 million a year to the State Water Project, which delivers doubly subsidized irrigation water to their formerly worthless land.

Fifth place belonged to Prudential's McCarthy Ranch, whose total acreage was 25,105. (If these numbers are bewildering, it helps to know that a good-size Illinois farm consists of six hundred to a thousand acres.) In sixth place was the Blackwell Land Company, whose 24,663 acres are part of a burgeoning U.S. land empire being assembled by a company of foreign investors, among them S. Pearson and Sons of England, Mitsubishi of Japan, and Les Fils Dreyfus of Switzerland, an offshoot of Lazard Frères.

Tenneco, the huge chemicals and food conglomerate, was seventh among the eight largest owners, with 20,180 acres. A few years before, Tenneco executives had been making some unusually candid statements to the effect that small family farms are the most efficient food-producing units human beings could ever create, and said it might give up farming altogether. When the State Water Project became operational, the company began singing a different tune. In the early 1970s, it bought the old James Ben Ali Haggin–Lloyd Tevis estate, the Kern County Land Company—300,000 acres of prime valley land—and metamorphosed into one of the most ardently competitive agribusiness growers in the world.

In last place, with 16,528 acres—a plot of land that is still considerably larger than Manhattan Island—was the Southern Pacific Railroad, the largest private landowner in California. In 1981, besides owning 700,000 acres of California forest and range land, Southern Pacific owned a large portion of downtown San Francisco and 109,000 acres in the Westlands Water District, where, between the good graces of the Bureau of Reclamation and the dilatory expertise of its battery of lawyers, it was still receiving subsidized federal water for $7.50 an acre-foot.

In California, when the issue is water, the ironies seem to string out in seamless succession. Bill Warne, the man who built the California Water Project, was in government service nearly all his life, and never made a great deal of money. In his mid-seventies, Warne was still doing consulting work; he also owned a small almond orchard outside of Sacramento. The consulting work was lucrative, but unpredictable. The almonds, on the other hand, were a good, reliable source of income. Or they were until Tenneco, by far the largest almond grower in the state, made a bid in 1981 to control the market—the same kind of power play that Prudential made with olives. "The bastards really went for our throats," Warne admitted ruefully during an interview early in 1982. "They beat the hell out of the rest of us in the market, and that includes me." Of course, one could just as well have said that Warne beat the hell out of himself. It was *his* project that irrigated Tenneco's almond orchards; it was *his* aqueduct that flowed practically within view of his small almond ranch, destined for the huge factory farms in the desolate southern reaches of the valley. Because of the hot climate down there, the crops grown on irrigation water have always been, in large part, specialty crops: almonds, pistachios, grapes, olives, kiwis, melons, canning tomatoes. And because the national acreage given over to such crops is comparatively small (California accounts for most of it), a single big grower who doesn't mind being a little ruthless can whiphand the market pretty much as he pleases.

Bill Warne's project had become a Frankenstein's monster. But its maker still refused to turn against his creation. "The moment we began settling California, we overran our water supply," he said. "We've never gotten to the point where you could just stop. And we never will."

Whether or not that is true, it is hard to imagine, by 1985, how the State Water Project would ever be completed. The old war-horses, the Bill Warnes and Pat Browns, might still be talking about the "unconscionable waste" of water flooding down the Eel River each winter (as Warne did, to whoever would listen), or saying that "the Columbia doesn't need all that water that flows down there—it's ridiculous, between you and me" (as Pat Brown did during an interview in 1979), but those who followed them in public office and were faced with the nitty-gritty problem of diverting the Eel, or the Columbia, or any so-called "surplus" water that could be found, discovered that it was like uncovering a nest of killer bees. Jerry Brown's successor, George Deukmejian, was elected with large infusions of cash from the growers in the San Joaquin Valley, where he is from. As expected, Deukmejian, a deeply conservative Republican, proved himself ideologically double-jointed on the issue of water development; while wading through the state budget with a machete, he made a wide circle around the Peripheral Canal, which he wanted to build but call something else, and he spoke approvingly of plans to send a lot more water southward. The reaction from northern California politicians, who, in the meantime, had managed to seize control of the speaker's chair in the legislature, and, through Congressman George Miller (who represents the Delta) of a key committee in Congress that can probably thwart much of what Deukmejian hopes to build, was so intemperate that the governor, after a year in office, was hardly mentioning the canal anymore.

Deukmejian may merely have decided to lie low, but by 1985 the people who will feel the impending shortages most acutely—the growers and the cities of the South Coast—appeared to have given up on the idea; either that, or they were mollifying their opposition while they stealthily plotted some hydrologic equivalent of Pearl Harbor. In June, the State Water Contractors, an organization representing all

George Gillette, chairman of the Fort Berthold Indian Tribe Business Council, weeps as he watches Secretary of the Interior J. A. Krug sign a contract whereby the tribe sells 155,000 acres of its reservation's best land in North Dakota to the government for the Garrison Dam and Reservoir Project on May 20, 1948. Gillette said of the sale: "The members of the Tribal Council sign the contract with heavy hearts. . . . Right now, the future does not look good to us."

(AP–Wide World Photos)

For more than fifty years, the tiny man-made river in the foreground, the Granite Reef Aqueduct of the Central Arizona Project, has been viewed by Arizonans as the one thing that can save them from oblivion. In the next century, however, as seven states suck up their full share of the feckless and overappropriated Colorado River, the aqueduct may run as empty as the diversion canal on the right.

(Bureau of Reclamation)

Teton Dam, just as the flood abated. Hours earlier, the flow of four Mississippi Rivers was thundering through the breach. The big concrete structure on the left is the spillway, whose outlet works hadn't been completed and which couldn't be used to begin emptying the reservoir when the first signs of trouble appeared. The height from river level to the crest of the remnant of the dam is about thirty stories; at the spot from which the photo was taken, boiling waves were more than one hundred feet high. (*Bureau of Reclamation*)

The remains of Teton Dam, as seen from the air, hours after the flood. (*Bureau of Reclamation*)

The three main antagonists in the Narrows Dam controversy. To Colorado Governor Richard Lamm (AT LEFT) the dam was an offer he couldn't refuse. To water lawyer Glenn Saunders (BELOW, LEFT) the dam symbolized a spendthrift society clinging to obsolete hopes. Former Colorado State Engineer C. J. Kuiper (BELOW, RIGHT) still believes the dam could fail catastrophically, as Teton did

A section of the spillway at Glen Canyon Dam completely destroyed by raging floodwaters spilled during the very wet El Niño winter of 1982–83. Although the Glen Canyon spillways run directly beneath the dam through rock that is mainly sandstone, the Bureau of Reclamation insists that the structure itself was never threatened. *(Bureau of Reclamation)*

OPPOSITE, ABOVE: The desert blooms on the Gila Project near Yuma, Arizona. Not far from here, the Hohokam, one of the world's great irrigated civilizations, went extinct. *(Bureau of Reclamation)*

OPPOSITE, BELOW: The Control Room of the California Water Project, where the man-made flow of nearly a trillion gallons a year is orchestrated.

(© Peter Menzel, 1986)

The California Aqueduct winds through Lost Hills, turning nearby desert, once considered worthless, into a billion-dollar agricultural bonanza.

(© *Peter Menzel, 1986*)

The Wind Gap pumps, which send water from the Feather River over the 3,400-foot summit of the Tehachapi Range, consume the electrical output of a nuclear power plant and stand between Los Angeles and disaster. (© *Peter Menzel, 1986*)

Mono Lake, an inland sea in eastern California desert country, is slowly dying. Most of the water that used to flow into the lake is now being diverted and piped to Los Angeles, three hundred miles away. As the lake's depth has decreased, natural calcium formations called tufa towers have been exposed.

(© *Peter Menzel, 1986*)

Salt deposits cover ruined farmlands in the San Joaquin Valley. A million acres in California alone may ultimately be affected. (© *Peter Menzel, 1986*)

the customers of the State Water Project, issued a report predicting a shortfall, by the year 2010, of 4.9 million acre-feet state-wide—the domestic consumption of twenty million people. The deficit within the State Project service area alone would be about 1.9 million acre-feet. Without more construction, the San Joaquin Valley would receive 733,000 fewer acre-feet than it was counting on. The South Coast cities and irrigation districts, which signed contracts to buy 2,497,500 acre-feet from the State Water Project, could be guaranteed a firm yield of only 1,120,000 acre-feet. Only in wet years could each region hope for more; during extended droughts they would receive even less. Meanwhile, a state report on groundwater pumping was describing the overdraft as "potentially critical" in eleven subregions of the Central Valley, most of which were in the service area of the State Water Project. What made things worse was that the valley's ancient salt-water aquifer, lying below the fresh water, could eventually rise to take its place.

Those figures, if they were accurate, bespoke calamity from both regions' points of view. What was startling, therefore, was the fact that the report said virtually nothing about sending more water from northern California southward. Its solutions—which it admitted were only halfway solutions—were for the most part the same ones that had been proposed by the environmental lobby, and which the water lobby had scorned just a few years earlier. The Imperial Valley farmers, according to the report, could conserve about 250,000 acre-feet if they lined their earthen canals and improved their irrigation practices; the water could then be sold to Los Angeles. The occasional surplus Colorado River flows below Parker Dam, as long as they lasted, could be stored in groundwater basins near Los Angeles and San Diego. Reusing treated sewage water (the report didn't go so far as to advocate drinking it) could save a few tens of thousands of acre-feet. Delta channels could be widened and levees rebuilt to allow slightly greater flows. The state could buy the surplus water in the Central Valley Project, for as long as that lasted. It was nickel-and-dime stuff, no heroics; the water savings might amount to 1.6 million acre-feet, which would only make up a third of the projected statewide shortfall. Only two new reservoirs, both off-stream and judiciously located south of San Francisco, were even mentioned, and the report didn't even advocate that they be built; it merely called for "investigations." (Initial investigations by the Department of Water Resources suggested a per-acre-foot price range of $310 to $400 from one of the reservoirs, Los Banos Grande; since that was fifteen to twenty times the cost of Oroville water, it was hard to imagine who in his right mind would buy

it, at least as long as there was groundwater to overdraft.) Not a word
was said about the Peripheral Canal.

Ironically, the State Water Contractors' report was accompanied
by a rather lengthy history of the State Water Project, written by the
first head of the Department of Water Resources, Harvey Banks, which
called the project "a high water mark symbolizing the results of the
collective efforts of people of many points of view to resolve their ward
with a program of statewide benefit." Reviewing the history of the
project, it was hard for some to see how Banks managed to arrive at
such a conclusion. To begin with, Californians had been sold a pig in
a poke: a project whose cost was deliberately and extravagantly understated, and whose delivery capability was much less than they had
been led to believe. Completing just the first phase of construction had
required federal cost-sharing at San Luis Dam, nearly half a billion
dollars in tidelands oil subsidies, and several hundred million dollars
in scavenged new revenue bonds. Then, when spectacular agricultural
and urban growth had occurred on the promise of water the project
couldn't deliver, a new leadership of "new age" politicians had tried
to sell the voters an even bigger and more expensive pig, which they
had spurned. Los Angeles had fought with the growers, then formed
an alliance with them, then fought again, then formed another alliance; two of the biggest growers had been instrumental in launching
the project, then played an indispensable role in the defeat of the
Peripheral Canal; and, all the while, the state had remained bitterly
divided along the geographic and climatologic lines the project was
supposed to supersede. This was "cooperation"?

As for the "statewide benefit" Banks wrote about, the California
Water Project may have been necessary if the state was to continue
to grow at its historical, breathtaking rate. But that was the point.
The growth it created was not "orderly" growth, to use that buzzword
of which the water developers are so fond. It was giantism. It was
chaotic growth. In southern California, project water is allowing
hundreds of acres to be subdivided, malled, and paved over each week,
transforming what could have been a Mediterranean paradise into one
of the twentieth century's urban nightmares. In Kern County, it created, solidified, and enriched land monopolies that are waging economic war against the small farmers who are so important to the
state's economic stability, and who give its agricultural regions what
little charm they have. To drive from east to west across the San
Joaquin Valley, from a pretty little palm-colonnaded city such as
Chowchilla, made prosperous by the Central Valley Project and surrounding small farms, to a shabby town such as Huron, surrounded

by endless tracts of irrigated land farmed by distant corporate owners, is to fathom the sorry social impact agricultural monopoly can have.

And what is worse, the State Water Project fostered growth in the desert, willy-nilly, without a secure foundation of water. Twenty million people may live between Santa Barbara and San Diego in 2010; the current outlook, according to the State Water Contractors, is that five million of them won't have water unless some drastic conservation steps are taken and occasional surpluses are scavenged from every available source. Even if the groundwater overdraft in the San Joaquin Valley continues to increase—and the chairman of the California Water Commission said recently that it may become "intolerable" by the year 2000—a shortfall of nearly a million acre-feet looms ahead there. The likeliest "solution" to the shortages, as things now stand, will be a lot of land going out of production. The farmers who are apt to give up first are those who are wholly dependent on farming for their livelihood. The ones likely to continue are those to whom farming is a sideline to oil refining or banking or running a railroad, or a tax writeoff—a way to accumulate a little judicious financial loss.

When Pat Brown's two terms as governor were over, he opened a lucrative law practice in Beverly Hills. One of his firm's most important clients became the Berrenda Mesa Water District, where the lands of several of the biggest corporate growers are located. The Blackwell Land Company, for example, owns 16,000 acres within Berrenda Mesa and co-owns 4600 more; Getty and Shell both farm thousands of acres there; one company, Mendiburu Land and Cattle, controls some 250,000 acres statewide. Thanks to his beloved Oroville Dam and the Governor Edmund G. Brown California Aqueduct—it was finally given that name by his son, in 1982—he had an opportunity to build up a tidy nest egg for his retirement.

But in his later years Pat Brown remained unrepentant about his firm's client relationships, which some might have considered unseemly, and he was as proud of his project as ever. Another thing that hadn't changed about him, curiously, was his candor. During his interviews for the Bancroft Library's Oral History Program, he allowed himself some final thoughts about the meaning of the State Water Project in California's history. "This project was a godsend to the big landowners of the state of California," he confessed to Malca Chall. "It really increased the value of their property tremendously. . . . But also the ordinary citizen has been helped by it, too." When his interviewer asked if enriching the big landowners of the state at public expense was really the result he had in mind, Brown responded, "It was the extreme liberals who wanted to break up the big farms in the state of California. They felt that the device of the delivery of water would do it. I was never convinced that the small farmer could succeed or would be good for the economy of the state and I don't know today as I talk to you whether that's true or not."

Having said that, Brown suggested another motive that had made him, a northern Californian by birth, want so badly to build a project which would send a lot of northern California's water southward: "Some of my advisers came to me and said, 'Now governor, don't bring the water to the people, let the people go to the water. That's a desert down there. Ecologically, it can't sustain the number of people that will come if you bring the water project in there.'

"I weighed this very, very thoughtfully before I started going all out for the water project. Some of my advisers said to me, 'Yes, but people are going to come to southern California anyway.' Somebody said, 'Well, send them up to northern California.' I knew I wouldn't be governor forever. I didn't think I'd ever come down to southern California and I said to myself, 'I don't want all these people to go to northern California.' "

CHAPTER ELEVEN

Those Who Refuse to Learn . . .

Early in September of 1965, the Bureau of Reclamation's newest dam, Fontenelle, on the Green River in southwestern Wyoming, sprang a leak. A big leak.

Eighteen years later, Pat Dugan remembered it as vividly as if it had been yesterday. Dugan was then regional director in Denver; he was the person who held the keys to the Bureau's airplane. "Barney Bellport, the chief engineer, called me up at four A.M.," Dugan remembers. "He said, 'We've got to get that plane in the air quick. We've got a dam that's about to go.' Barney was a self-confident guy—a little bit of an arrogant bastard—so I figured if *he* was worried, we were in plenty of trouble. We were."

Fontenelle was an earthen dam of moderate size on a troublesome site; it stored water for the Seedskadie Project. That the site had geologic problems was apparent from the very beginning, but the Bureau, as it would in a number of cases, built it anyway, for the simple reason that it was running out of good places to erect its dams. "I think I was the first person who ever did up a detailed cross section of that site," Dugan remembers. "I didn't like it from the beginning. The left abutment was fine, but for some reason we had a lot of trouble with the right one. It was shaly and just generally lousy. I figured it would take a lot of grout." Asked what he thought of the Seedskadie Project itself, Dugan said, "It was one of the few lemons we could find in Wyoming that didn't make your mouth pucker completely shut."

Wyoming has had its share of powerful politicians in recent decades, from Senator Joseph O'Mahoney, who stopped FDR's plan to

pack the Supreme Court, to Senator Gale McGee, Lyndon Johnson's
most articulate ally on the subject of Vietnam. What the economy of
their high, harsh, hot, arid, and bitterly cold state could not produce
on its own, they could produce for it out of the national treasury. The
growing season in the region is extremely short: the altitude of most
agricultural land is between four thousand and seven thousand feet,
and there is frost nine months of the year, sometimes even in August.
The land is useless for growing anything but cattle browse. To build
an expensive dam, a spillway, an outlet works, and canals in order to
grow grass or alfalfa is not generally an economically rewarding proposition. It can, however, be a politically rewarding one. To paraphrase
what someone said about pleasure and pain, economics are an illusion,
while politics are real. Besides, as Wyoming's politicians never tired
of pointing out, their state had contributed substantial mineral royalties to the Reclamation Fund, and they were supposed to get some
projects in return. If they didn't, Wyoming's share of the Colorado
River—all of it contained in its biggest tributary, the Green—might
disappear down California's maw.

The leak began as a wet spot on the downstream face of the dam
which first appeared on the 3rd or 4th of September and grew steadily
larger. By the evening of the 6th it was a small waterspout. A waterspout is a signal that water is piping inside the dam—forming placer-
nozzle velocities and excavating channels which allow the dam to be
eaten from within. By the time Barney Bellport flew overhead, Fontenelle Dam was firehosing water from its downstream face. It appeared too late to save it.

"We left as soon as it was light enough to see," Bellport remembered. "Wyoming seems like a mighty big state when you're flying
across most of it to inspect a leaking dam. After we made a pass over
the dam, I didn't need to make another. I was really worried that we
were going to lose it." The Bureau plane landed at nearby Kemmerer,
the improbable site of the first J. C. Penney store. The chief engineer
then roared overland toward the Green River, wondering whether he
could get there in time to save his reputation.

It would have been one thing if the dam were newly completed
and the reservoir pool just forming behind it. But Fontenelle had,
oddly enough, held water for some weeks; filling the reservoir had
given no indication that some serious trouble lay inside the dam or
bedrock. The reservoir was therefore full, and had to be emptied fast.
"My project engineer hadn't begun emptying it because the contractor
was downstream fixing the apron of the power plant," Bellport recalled, sounding still disgusted with the man. "I asked him if he would

rather wash away the contractor's equipment or the town of Green River." With the dam hemorrhaging across a wide section of its face—huge burps of muddy water were gushing out of it, as if it were gagging on the reservoir and vomiting it up—Bellport ordered both outlet works opened full-bore. The water that was being stored to irrigate the surrounding high desert began flooding uselessly over it, reverting a large piece of Wyoming to something it had not been since the last Ice Age: a swamp. The outlet works carried off so much water so fast that the reservoir could be seen dropping visibly, like a bathtub. A crowd of tiny figures watched tensely from the canyon rim. Forty miles downriver sat the town of Green River, exposed and vulnerable, right on the riverbank. "You felt like you do when you're passing another car and suddenly there's an oncoming car coming right at you," Bellport recalled. "You've got to keep passing but your heart's fluttering and you wonder why you didn't buy a car with more pickup." Only in this case the almost unbearable tension was to last for hours instead of seconds. The outlet works could empty the reservoir only so fast; the dam was still belching out great surges of muddy water; its downstream face was steadily eroding under the force. Downriver, there were already reports that the rising Green was inundating the town golf course. Volunteers were furiously sandbagging the river's banks.

The Bureau was lucky. By early evening, the force of the huge leak finally began to expire. As the flow subsided, one could see the frightening gouges and gullies that the exit of superpressurized water had caused in the downstream face. The dam looked as if it had been pounded by artillery shells. But, miraculously, it had held.

In 1983, sitting in retirement in Rossmoor, California, Barney Bellport still echoes the attitude of the Bureau of Reclamation during the whole affair. When speaking of the crisis itself, he allows himself an excursion into melodrama. "It was damn serious," he says. "*We really thought we were going to lose it.*" But then, having talked himself through the incident, he jumps to his own and the Bureau's defense, like the sinner who avoided being caught and therefore believes he didn't sin. "We repaired it, and it held," he says. "It's been holding water ever since. The Bureau has built hundreds of dams, and they've all held beautifully, except Teton." That, it is suggested, was a pretty large exception. Bellport pauses, looks ironically at his wife, and lets his gaze drink in his surroundings. "Teton," he says firmly, "was either an act of God or human error. You do not blame an organization with a single blemish on its record for the mistakes—if they were mistakes—of a handful of employees who didn't live up to its reputation."

There is not now—there was not then—much evidence of soul-searching on the part of the Bureau's leadership, old or new. They did not seem to be asking themselves what they were doing building potentially dangerous dams like Fontenelle to serve demonstrably wasteful projects like Seedskadie. No one seemed to be wondering whether a bad project might not, through some Shakespearean inevitability, lead to a worse end.

Actually, that is not quite true. Pat Dugan was wondering, and so was Dave Crandall, the regional director in Salt Lake, whose office had to deal with the Fontenelle aftermath. Judging from the correspondence he carried on with his superiors in the wake of the near-disaster—correspondence that traveled the blue-envelope route—Crandall seemed to sense what the others did not: that the Bureau had committed the sin of pride. In a letter to Bellport, he mentioned a demand by some local citizens—people who would have to spend their lives immediately below a dam that had almost failed—asking that the Bureau convene a major investigation before rebuilding the dam. "I do not accede to threats," Crandall wrote, "but since there is this feeling in the local area, and also to preserve our position of impartiality and objectivity, I urge that you consider a Board of Review to appraise the repairs at Fontenelle." Such a board, Crandall pointedly added, should include "qualified non-Bureau non-federal professionals."

To this, Bellport's response was a peremptory harrumph. Ignoring Crandall, he took the matter directly to Commissioner Floyd Dominy. "As you know, the principal competence in earth dam design and construction lies within the Bureau," Bellport wrote to Dominy. "I strongly suspect that a review of the competent earth dam people in consulting firms throughout the country would reveal that a considerable portion of them have either Bureau or Corps background. I also take a very dim view," Bellport offered, "of a professor of geology from a university sitting in judgment on the Bureau."

However, what Bellport's "professor" might have told him, had he and the Bureau felt like listening, was that it had just about run out of good damsites. As Fontenelle was an inferior site compared with Flaming Gorge, as Glen Canyon was inferior to Hoover, as Auburn was vastly inferior to Shasta (but four times as expensive, even allowing for inflation), the Bureau was now being forced to build on sites it had rejected forty, fifty, or sixty years earlier. It was building on them because while the ideal damsites had rapidly disappeared, the demand for new projects had not. The demand for new projects had, if anything, increased, especially now that the Reclamation Act had been amended and re-amended to such a degree that federally supplied water was the closest thing left to a free good. The West and the Congress wanted more projects, and the Bureau wanted more work, but the good dam-sites were gone. The Bureau, of course, rationalized its decision to keep on building by claiming that advances in engineering were keeping up with the challenges. Even though it was now building dams on rotten foundation rock, between spongy sandstone abutments, in slide-prone canyons, and close to active earthquake faults, the dams held— for now.

"The country around Fontenelle is trona country," Barney Bellport says. "It's full of sodium carbonate—soda ash. The stuff speeds up the setting of concrete. We finally figured out that it had made the concrete we poured for the grout curtains set too fast. Somewhere it left a fissure where the water got through and entered the dam. After that we knew to mix and pour concrete in trona country that wouldn't set so fast."

Pat Dugan essentially agrees. "There hasn't been an ideal damsite since 1940," he says. "Every site we've built on since then would probably have scared hell out of a nineteenth-century engineer. But you wouldn't feel safe going a hundred miles per hour in a Model T, either, if you could get it going that fast down a hill. You might feel perfectly safe in a Porsche." That might be true, except that the dams built at less than ideal locations are usually larger than those built at the earlier, better sites, and with so many dams now in place one dam's failure could conceivably cause other dams to fail, resulting in a domino of disasters unlike anything the country has ever seen. The failure of one large, strategically placed dam (Glen Canyon, for ex-ample, which would surely take out Hoover as it went) could undo much of what the Bureau of Reclamation has built up over seventy years, leaving southern California a desert underwater and the econ-omy of the Southwest in ruins.

A modest version of that is what might have happened when Teton Dam collapsed. What actually happened was bad enough.

When Bob Curry got his first look at a cross section of the Teton damsite, his reaction was much like Pat Dugan's when he looked at his finished cross section of Fontenelle. "Holy Christ!" Curry, a geologist, remembers thinking to himself. "What a terrible site for a dam!" By then, however, the dam was already one-quarter completed.

The French colonizers of what is now Chad informally divided their hollow prize into two separate nations. The south was *Chad utile*—

"useful Chad"— and the north was *Chad inutile*. In the south, which delved into the fringes of the Central African rainbelt, you could raise a crop; there was wildlife. Northern Chad was deep in the Sahara, as barren as Antarctica. One's first impression of Idaho is much the same, only the polarity is reversed. Northern Idaho is green and welcoming; it is beautiful. Close enough to the Pacific to be influenced by its storehouse of winter warmth, mountainous enough to wring moisture out of passing weather, northern Idaho is the banana belt of the Rockies— warmer than the mountains of New Mexico a thousand miles to the south, wetter than eastern Oregon and Washington to the West. Wild rivers pour out of the mountains—the Salmon, the Clearwater, the Lochsa, the Boise, the Pend Oreille. Apple and cherry orchards thrive in the valleys. In the middle is a vast wilderness, the Salmon River breaks—the most expansive roadless area in the coterminous United States.

Northern Idaho, however, doesn't count much in the economic scheme of things. Real Idaho, serious-minded Idaho, is in the south, along the desolate reaches of the Snake River's old volcanic plain. Like barnacles on an anchor chain, Idaho's cities, its most productive farmland, and much of its wealth are strung along the Snake as it loops around the southern half of the state. Thanks to irrigation, it is a useless place made rich; nowhere except in Arizona and California's Central Valley has such an utter transformation been wrought in the West. Twenty miles from either side of the Snake there is little but desert, and more desert, and rockpiles of basaltic tuff. It was exactly this sort of landscape that appealed to early Mormons, who found a place attractive in exact proportion to its ability to repel anyone else. Drifting up from Salt Lake Basin, the Mormons glimpsed the Snake, incongruously big in the desert, and immediately saw a future. Diverting the few smaller streams, they made a tentative beachhead; then the Bureau of Reclamation arrived and built Jackson Lake and Minidoka and American Falls dams, and the beachhead became an invasion. Within the forty-mile corridor along the Snake River now exists an irrigation economy that has given Idaho a higher percentage of millionaires than any other state in the nation. The best-known crop is potatoes, which like their soil loose, friable, a little sandy, and well drained—the exact conditions of the Snake River Plain. One of the problems of Idaho irrigation farming, in fact, is that water, in places, tends to drain through the soil too quickly, requiring annual waterings in excess of ten feet. That, in part, is how Teton Dam came to be built.

The fountainhead of southern Idaho's agricultural wealth lies to the northeast, where the Yellowstone plateau and the Grand Teton

Mountains produce enough water to engorge the Snake to substantial size before it enters the state. On a bright day, the Grand Tetons are visible from the eastern reaches of the plain; a huge buttress wall facing north-south, ninety miles long and thirteen thousand feet high, the range wrings a lot of water out of passing Pacific storms. On the western side the runoff gathers into two rivers, the Henry's Fork and the Teton, which ultimately join the Snake above Idaho Falls. The Teton is, or was, the prettier river; for thirty miles, it whipsawed through a low, U-shaped canyon amid cottonwoods along the bottom and conifers that walked up the canyon's collapsed slopes. An oasis stream in a landscape that is at best austere, the Teton was coveted by the deer that wintered in its canyon, by the fat trout darting from pool to pool and by the humans who thought it could be put to better use. Since the 1920s and before, there was talk of a dam somewhere on the river, but the dam was never built. One reason can be seen in the granular rock of the canyon's steep slopes. The geology of the region is ultravolcanic: the rock is fissured, fractionated, cavitated, and criss-crossed by minor faults. The neighboring farmland, meanwhile, though productive enough, requires inordinate amounts of water. Those two drawbacks add up to poor economics, and though a Teton Dam was studied and restudied through the 1940s and 1950s, it was never built—until the 1960s.

The impetus, as in the case of many dams, was disaster, or what was called disaster—first a drought, then a flood. The drought occurred in 1961 and 1962, the flood the winter after. The flood caused some few hundred thousand dollars worth of damage, most of it because ice jams occurred at a couple of bridges during a sudden early melt. The drought was mainly a misnomer, nothing like the early thirties or the drastic rainless period in California in the mid-1970s; farm income remained high. In the West, however, a drought and a flood together set off a strong Pavlovian response. The first thing that enters anyone's mind is a dam.

For a project that had spent three or four decades in the pupa stage, Teton was authorized and built in a great hurry. The main reason was Willis Walker, a crotchety Mormon farmer and president of the Fremont-Madison Irrigation District, who managed to organize all of southwestern Idaho behind it. His task was not that difficult. This, after all, was the Mormon West. The closest thing to oppositoin was indifference. years later, speaking with a reporter, Walker remin-isced, "One of the arguments we used back there was that in '60 and '61 we had a lot of potatoes and a lot of sugar beets around here that didn't have enough water to finish them out. I figured I had better find that water or quit farming." The argument conjured images of crops wilting on the vine, of families ruined on the eve of harvest for want of water to bring their crops to ripeness. Everyone bought it, even though it was nonsense, for the most part. Years later, a graduate student writing a thesis discovered that production of some crops had actually *increased* during the drought. In Fremont and Madison counties, for example, the yield of potatoes in 1961, the worst year of the drought, was 212 hundredweight per acre. Between 1956 and 1959, a stretch of more or less normal years, the average yield was only 184 hundredweight per acre.

Even had the drought threatened ruin, there was a solution much simpler and cheaper than a dam, the same solution that California's farmers would fall back on during their far more apocalyptic drought of 1976 and 1977: groundwater. Idaho may have more groundwater in storage than any other state except Alaska. The Snake River Aquifer, lying directly beneath the Teton River, is still prodigious. During the 1960s, when the drought occurred, thousands of pumps were already operating, supplementing the diversion ditches. Pumping, of course, can be expensive, especially if one's crops require nine or ten feet of water a year. The answer then might be to grow something that requires less water, or to install more efficient irrigation systems. But the farmers of Fremont and Madison counties, good upstanding Mormon conservatives, wanted things their way—and they wanted the descendants of the people who had chased them out of Ohio and Illinois and Iowa to pay 90 percent of the cost. "Mormons get burned up when they read about someone buying a bottle of mouthwash with food stamps," says Russell Brown, one of the dam's most persistent critics. "But they love big water projects. They only object to nickel-and-dime welfare. They love it in great big gobs."

With the entire Congressional delegation from Idaho behind the dam, authorization was a snap, and in the later years the appropriations came fast and furious. However, the project had a little trouble getting going; it received only $3 million during the first six years following authorization, probably as a result of the Vietnam War. During that same period, Congress passed the National Environmental Policy Act of 1969, and the Bureau was forced for the first time to make a public assessment of the environmental effects of its new dams. Before it learned to flood its critics with a tide of ink, the Bureau merely went through the motions of writing an environmental-impact statement; in the case of Teton, it ran to fourteen pages and didn't say much of anything. The exercise, however, drew some attention to the project; both the *Idaho Statesman*, the state's preeminent newspaper, and the Idaho Environmental Council began to take a closer look at it, and liked little that they saw. Published in Boise, on the other side of the state, the *Statesman* could afford to be objective, but even had the project been next door, the paper's maverick young editor, Ken Robison, was not the sort who parrots the views of the local Chamber of Commerce. The Environmental Council, which included a number of scientists from the Department of Energy's Nuclear Testing Station at Idaho Falls, was unusually sophisticated for a tiny organization, and fed Robison a steady diet of statistics worked out on a federal computer.

The statistics, on their face, were quite damaging. The project benefits had been calculated by the Bureau on the basis of the worst drought on record, outrageously stacking the deck in the project's favor. The figures it used to calculate the annual value of flood prevention were about 200 percent higher than historical losses to floods. Of the thirty-seven thousand "new" acres to be opened to irrigation, twenty thousand acres were *already* being irrigated by groundwater pumping; the project would simply substitute surface water for sprinklers, which is a lot different from bringing new land into production.

No statistic, however, was as startling as one freely provided by the Bureau itself. According to its own report, on the 111,000 already cultivated acres that were to receive supplemental water from the Teton project, the average annual irrigation amounted to *132 inches*; the project would simply give the farmers, on the average, another five. One hundred and thirty-two inches is five times the annual rainfall of farmland in Iowa; it is ten times what prudent farmers in the Ogallala region of arid West Texas put on their crops. It is the precipitation of tropical forests. In fact, according to the Bureau, a common method of irrigating on the Rexford benchlands is subirrigation, which means literally what it implies: water is dumped on the ground in such prodigious quantities that the water table rises up into the root zones of the crops. In one of the driest zones of North America, the Bureau was going to sell dirt-cheap irrigation water to irrigators practicing the equivalent of hydroponic gardening.

The Teton project could be justified only by using an interest, or discount, rate of 3¼ percent. Even with that rate, which was unrealistic in the hyperinflationary 1970s, the best it could manage was a benefit-cost ration of 1.2 to 1. After getting rid of the phony flood-control figures, the phony "new" irrigated land, and the more implausible fish and wildlife and recreation benefits, the Idaho Environmental Council came up with a benefit-cost ratio of .73 to 1.00. Using a 6 percent discount rate, which was more realistic, the ratio dropped to .41 to 1.

Taking, for the sake of compromise, the midway point between the Environmental Council's more flattering figure and the Bureau's, the Teton project was exactly worthless as an investment of tax dollars: it would destroy a beautiful river for the sake of nothing in return.

Such arguments, persuasive though they might have been in an objective sense, seemed only to solidify the local support for Teton Dam. Since Willis Walker had won authorization for the project, the man who emerged as its chief propagandist was Ben Plastino, the political editor of the local newspaper, the Idaho Falls *Post-Register*. Plastino was the sort of small-town editor Twain or Mencken would have loved. It wasn't just his appearance, though that certainly helped. He was short, middle-aged, and pudgy, and his sartorial tastes ran to combat clashes of checks and plaids—vivid figurine shirts, loud polyester ties, acetate houndstooth-checked pants, multicolor Dacron-polyester jackets. Plastino felt a newspaper had two important roles. One was to bring as much federal money as possible into its region, especially in the form of a dam. The other was to rail against big government and creeping socialism. One senator's immortal words during the Watergate hearings—"Don't confuse me with the facts"— were words Ben Plastino had gratefully taken to heart. As recently as 1979, he insisted that Teton was primarily a flood-control project (it wasn't, or it would have been built by the Corps of Engineers), maintained that none of the farmers put anywhere near ten feet of water on their crops (some used up to thirteen), and insisted that every water project pays for itself, regardless of cost.

The *Post-Register* was magnanimous enough to publish an occasional letter opposing the dam, but in its news stories the opposition was usually referred to as "extreme environmentalists." Covering one meeting of dam supporters, Plastino wrote obsequiously about their efforts in behalf of Teton, describing the "warm thanks" and "warm applause" that greeted each self-congratulatory testimonial. The paper, however, was a lot more objective than some of its readers. "Those who would cramp and belittle America's dream and who labor to stalemate needed natural development," stated one letter to the newspaper, "have plans for a singularly small and feeble nation, a blueprint for weakening our nation in a time when enemy nations are straining to develop their resources and strengths." Another asked, "I for one would like to know who is the power behind these so-called environmentalists? Why are they so radical about condemning anything that would improve Idaho's irrigation?"

Jerry Jayne, who was then president of the Idaho Environmental Council, hardly looks like the communist many of his neighbors

seemed to think he was. Crew-cut, strong-jawed, erect as a cabinet,
he bears a strong resemblance to Mike Nomad, in the *Steve Roper* comic
strip, and one might expect to find him at the controls of a nuclear
power plant—which is exactly where one would find him, since he
works for the Department of Energy's nuclear testing facility at Idaho
Falls. "I don't know what it is about these Mormon irrigation farmers,"
Jayne said. "I can talk to the loggers, I can talk to the ranchers. I can
talk to the mining companies. I can say nothing to the irrigation farmers. They're not reasonable. They don't listen. They're true believers.
They're like communists—only in reverse."

Idaho has had one of the most convulsive recent geologic histories of
any state. Only a few million years ago, it was an almost continuous
cataclysm of volcanic eruptions, earthquakes, and lava flows. The Yellowstone plateau, two hundred miles off to the northwest, still exhibits
the remnants of such activity, as do the Cascade Ranges to the West.
(In the fall of 1983, one of the biggest earthquakes in recent U.S. history
struck a remote part of Idaho less than two hundred miles from the
Teton site.) The whole eastern Snake River Plain, including the Teton
site, is a vast bed of basaltic rock. The hazards of building a dam in
such terrain, however, became an issue almost entirely by accident.
In 1973, Robert Curry was teaching geology at the University of Montana; he did some occasional consulting work for the Sierra Club,
mostly on the effects of logging and mining operations. Though he was
quite familiar with the geological firmament of southern Idaho, and
knew it was anything but firm, he always assumed the Bureau knew
how to build a safe dam in such a locale. He also assumed it would
have the sense not to build one at an absolutely terrible site. "The first
time I heard anyone question the safety of Teton Dam," Curry remembers, "is when some people with the Idaho Environmental Council
called me up in 1973. They had been sitting around drinking beer with
some guys from the Geologic Survey and one of the Survey guys said—
I guess he didn't even mean to let it out—'Well, the Bureau's going
to have a hell of a time building Teton Dam.' An IEC member asked
him what he meant, and the Survey guy said, 'Well, it's really a
crummy spot to put a dam.' I was one of the few geologists around
who had much sympathy for the environmental side, so they called
me up and asked me what I knew. I didn't know anything. I figured,
well, they'd built American Falls Dam down there and some other
ones, so they must know what they're doing. But I asked the Survey
if I could see their cross section anyway. I looked at it and that's when
I said, 'Holy Christ!'

"The stuff they were going to build the dam on—all those ashflows and rhyolitic rock—may look solid to you, but it's really a veneer, sort of like the wood veneer on a cheap desk. It's brittle, it's cracked. It could peel off just like the veneer on the desk. They were going to scrape away the worst of it and then say that they were anchoring the dam in bedrock. But it isn't really what most geologists would call bedrock. The dam was not going to have a true bedrock foundation.

"It was such an obviously lousy site to a trained geologist," Curry added, "it makes you wonder what happens to human judgment inside a bureaucracy."

Accompanying the Geologic Survey's schematic of the Teton foundation was a report to the Bureau of Reclamation written by four geologists in its regional office, which—in its first version—raised "certain questions about the fundamental safety of the Teton Dam. . . . Despite the incompleteness of the data," the geologists cautioned, "we fell obliged to bring them to your attention now, while they may still be useful and on the chance that some factors may not have been adequately considered in design of the project."

From reading the memorandum, it was clear that the four geologists considered the possibility of an earthquake to be the greatest hazard associated with the dam. "Young ashflows and associated rhyolitic volcanics like those being used as buttresses for the dam," they wrote, "are cut by very young block faults." Often, they said, undetected faults with substantial destructive capability can exist in such terrain. "The Seismic Risk Map of the coterminous United States assigns southeastern Idaho to Zone 3," the code for highest seismic risk. Although the geologists—Steven Oriel, Hal Prostka, Ed Ruppel, and David Schleicher—stopped just short of urging the Bureau to abandon its plans to build on the Teton site, they asked that their observations "be given the serious consideration we believe they merit."

Actually, the tone of the memorandum was mild and rather conservative compared with an earlier internal draft prepared by Dave Schleicher, who had made the initial observations. In his draft, which was addressed to his colleagues instead of the Bureau itself and written in early December of 1972, Schleicher, besides mentioning all the risks that were included in the later memorandum, expressed amazement over the fact that the Bureau appeared oblivious to them. "Within the last five years five earthquakes less than 30 miles from the proposed Teton damsite have been detected," he wrote. "At least two of them had Richter magnitudes greater than 3.

"I find no recognition of this . . . in any of the documents for the project and no indication that the dam and reservoir would be designed to withstand seismic damage and prevent serious secondary damage. There is no recognition . . . that reservoirs have actually caused earthquakes.

"The[se] points appear to be significant enough," Schleicher warned, "that they should be presented to the Bureau as soon as possible—certainly within a month or two. I'd plead that we need a firm deadline on this: we've been aware that there's some need for concern for nearly three months, and we're being seriously delinquent if we don't pass this information on."

At the end of his memorandum, almost as an afterthought, Schleicher included a remark which, in retrospect, would take on a chillingly prophetic overtone. "A final point," he said, "is that flooding in response to seismic or other failure of the dam—probably most likely at the time of highest water—would make the flood of February 1962 look like small potatoes. *Since such a flood could be anticipated, we might consider a series of strategically-placed motion-picture cameras to document the process . . .*" (emphasis added).

Most, but not all, of the urgency in Schleicher's tone was gone by the time his three colleagues had redrafted his remarks. But even their toned-down version was never to be sent. The letter that finally arrived on the desk of the Bureau's Teton engineer, Robbie Robison, had the quality of weak tea. In place of Schleicher's remark about installing movie cameras at the site, the final paragraph of the delivered memorandum read, "We believe that the geologic and seismic observations, though preliminary, bear on the geologic setting of the Teton Basin Project. We are presenting them to you as promptly as possible for your consideration." The rest of the letter could have been lifted from a treatise on local geology—it did not warn of anything. Though Schleicher had made his initial remarks in December of 1972, the final version was dated April 3, 1973. By the time it had been routed through Boise and off to Denver, where any decision affecting the dam's fate would have to be made, it was already July. By then, the dam foundation was already being readied, and another $10.5 million had been appropriated for construction.

The metamorphosis of the report was mainly the work of the director of the Geologic Survey, Vincent McKelvey, but not all of the responsibility could be laid on him. It had just as much to do with the historic relationship between the Bureau and the Survey. Like an awkward older sibling who watches a younger one grow up to letter in four sports, the Survey held the Bureau in a certain awe. In 1902, when the Reclamation Service was newly fledged, the Survey, in a legal sense, became its parent. For the next couple of decades the Service and the Survey were more like sister agencies in pursuit of a common goal—the Survey mapping the West and its geology, the Reclamation Service taking the maps and transforming it. Since then, however, Reclamation had ridden a rising star; transformed from a mere Service into a *Bureau*, it had expanded its staff to as many as nineteen thousand, commanded half a billion dollars a year, and built half the wonders of the modern world. The Survey's great work, the mapping of North America, was essentially complete; it was now a rather small collegium of scratchers, samplers, and scientific scriveners. Who was *it* to tell the almighty *Bureau* what to do?

The Bureau, inflated by a sense of its own accomplishments, must have asked itself the same question. Steve Oriel, the most senior and diplomatic of the four USGS scientists, would later observe that "we got no feedback at all from the Bureau" after the Survey's letter was sent. The earliest evidence of a reaction—any reaction—from the Bureau was a confidential note by one of its geologists, J. D. Gilbert, concerning a telephone conversation he had with Oriel in October, seven months later. Regarding some continuing investigations at Teton by Hal Prostka, Oriel's colleague, Gilbert wrote, "Steve said that Prostka had found numerous recent faults on the Snake River Plain in the general Teton area, but Steve had no information on the right-abutment 'fault' at Teton. [Even though the Survey strongly suspected it had found a hidden fault right at the damsite, Gilbert was inclined not to believe it.] . . . Steve said that a 'Sierra Club' type individual [one of the Idaho Environmental Council people] involved in the Teton litigation had looked him up in the field to discuss the USGS work in the area."

What *really* had Gilbert worried, it seems, was the fact that "the Washington office [of the Survey] has published (or will publish shortly) the material contained in the USGS letter to the Bureau on Teton . . . in their 'Short Contributions.' Several other reports of a preliminary nature will also be published shortly on this portion of the Snake River Plain." Gilbert had gone back and underlined those last two sentences. Hand-scrawled next to them was a margin note which read, "We better develop our ideas on points in the GS 'prel.' rpt. and present some constructive criticism and make effort to get some hard data on 'rt. abutment' fault."

In the mind of a good Bureau man, the first priority was to attack—"constructively"—anyone who questioned his agency's judgment. The second priority was to see whether there was some truth in what he said.

In the opinion of Steven Oriel, the Bureau's response was "disappointing." The Bureau would not listen to the Survey, he was to tell a Congressional committee, "because they were already committed to the project politically." Bob Curry agrees. "You could have told them that they were building a dam on top of an active volcano," he says, "and they would have had a hundred guys out there trying to prove you wrong. I tried to get some more information out of them and eventually I gave up. All I got was Mickey Mouse. No one was listening."

It is irrelevant, but irresistible nonetheless, to point out that while Curry was getting what he called "Mickey Mouse" out of the Bureau, its acting director of dam design and construction was named Donald J. Duck.

Meanwhile, for an entirely different set of reasons, the Nixon White House was beginning to take a closer look at Teton Dam. It wasn't so much the cost—compared to, say, the Central Arizona Project, Teton was beer money—as it was panic over the OPEC-spawned inflation that had suddenly exacerbated the Vietnam-spawned inflation that already was. Also, an organization called Trout Unlimited, made up substantially of rich Republican fly-fishermen who had donated to Nixon's reelection campaign, was quite audibly upset about the loss of yet another blue-ribbon wild trout stream. Nixon's Council on Environmental Quality and the Environmental Protection Agency were similarly upset about the project, and their skepticism had partially infected the closest approximation of an environmentalist in the inner White House, Presidential adviser John Erlichman.

The strongest official opposition came from Nathaniel Reed, a wealthy Floridian whom Nixon had appointed Assistant Interior Secretary for Fish, Wildlife, and Parks. Reed, tall, intense, and witty, a blazered social lion from the Gold Coast, was to clash repeatedly with the prosaic engineers upstairs in the Interior building, and for a while rivaled Dave Brower as the Bureau's public enemy number one. "They took me on a tour of the engineering headquarters in Denver once," recalls Reed, "and I walked by some guy's office with a dartboard that had my smiling face on it. There was a dart stuck in each of my eyes. I didn't think anyone there even knew who I was."

Reed had the ear of Interior Secretary Rogers Morton, another wealthy southeasterner, and, together with Robert Cahn of the Council on Environmental Quality, slowly brought Morton around. The result was that on October 7, 1971, with contractors from across the country gathered in Idaho Falls to bid on the major construction contract for the dam, Morton suddenly gave instructions to postpone the opening for thirty days. His explanation was that he wanted to reevaluate the project one more time to see if its benefits would truly exceed its costs. Morton, of course, was already pretty well convinced that this was not the case. More likely, what he really wanted to do was gauge the reaction to something as moderately drastic as he had just done.

In the words of Nat Reed, "The shit hit the fan." The whole Idaho Congressional delegation was up in arms, and almost every Idaho newspaper carried an indignant editorial. In a matter of hours, an obscure project no one had heard of in a remote western state had become a main topic of discussion in the Nixon White House.

For a westerner and an ex-Congressman, Nixon himself had surprisingly little interest in water projects. It wasn't that he was a conservationist in his secret heart; he had almost no interest in nature, either. Nixon was interested almost exclusively in politics, and mainly in foreign affairs. Domestic policy bored him; public works were especially deadly. Nonetheless, Nixon was an outstanding politician, and he knew as well as Lyndon Johnson how to use the budget process to further his ends. "At the time, Nixon was about to open the gates to China," John Erlichman recalled in 1983. "Then there was the international monetary agreement, the SALT talks, détente with the Soviets. He couldn't get anywhere on those without Congressional support, and Congress knew that, and the Idahoans in Congress wanted that dam." Erlichman professed to remember little of the Teton Dam episode, though rumors at the time made him the principal point man at the White House. Whoever it was, someone in the White House turned Rogers Morton around very quickly. Eleven days after he postponed the contract opening, he announced that Teton was a sound project after all. Groundbreaking was to begin within weeks.

There was only one person who could have jerked a President and an Interior Secretary around so fast, and that was retiring Idaho Senator Len Jordan. When Nat Reed went out to Idaho soon thereafter to dedicate the Birds of Prey removal lands—a new national monument along the Snake River where hawks and golden eagles live in remarkable numbers—Jordan was with him, all smiles and camaraderie, posing for photographers. "As soon as the photogs went off," Reed remembers, "Jordan got crude and angry. He yanked me aside and said, 'Listen, Nathaniel Reed, we're going to building this fucking dam and you're going to come out to dedicate it. I've used every chip I've got on Teton Dam. What do you think I'm doing here dedicating this goddamned vulture site?' " At least, Reed added ruefully, Jordan was honest.

Without the support of Rogers Morton or Idaho's governor, Cecil

Andrus—who, if his later record on water projects as Interior Secretary is any clue, probably thought Teton was a bad project but didn't dare come out against it—the only hope left for the dam's opponents was the courts. There they went up not so much against the Bureau as against Fred Taylor, the presiding judge of the federal district court for Idaho—a man with deep local roots and a sense of religion about water development. Was *he* going to preside over the demise of the Teton project? Evidently not. Taylor refused to allow any discussion of economics, or of safety, during the trial, using as crabbed an interpretation of the National Environmental Policy Act as he could get away with without inviting a reversal by the court of appeals. The matter of safety did come up, once, as Sierra Club Legal Defense Fund attorney Tony Ruckel tried unsuccessfully to introduce some testimony to the effect that the dam might leak more that the Bureau admitted. Judge Taylor had a ready response. "Matter of fact," he told Ruckel, evidently thinking this was funny, "if the dam won't hold water, I don't think the fish and wildlife are going to be hurt." Then he disallowed Ruckel's testimony on grounds of irrelevance.

Ruckel had wanted to introduce testimony from Shirley Pytlak, a professional geologist who had worked briefly on the Teton project during the summer of 1973, drilling test holes at the damsite and injecting water into them. The idea was to see how fast the holes filled up, which would allow the Bureau to gauge—"guess" is a better word—the extent to which the surrounding rock was fissured and fractured and concomitantly leak-prone. For weeks, Pytlak said, the boreholes had been pumped with water at a rate of three hundred gallons per minute, which was like sticking a fire hose in them and turning it on full-blast. The holes never filled. If test holes leaked at such a rate, Pytlak asked her superiors, how much water would seep out of the reservoir and try to get around the dam?

Actually, none of this should have come as a surprise. Three years earlier, the Bureau had conducted a similar test-drilling program, and three deep holes—numbers 301, 302, and 303—turned out to be particularly thirsty. Injected with as much as 440 gallons of water per minute, all of them refused to fill. The three holes were all drilled in the right canyon wall. Number 303 was only 250 feet from what would be the dam's embankment. Clifford Okeson, the Bureau's regional geologist and the person supervising the drilling program, reported to his superiors: "The three deep drill holes which were completed on the right abutment of Teton Dam during 1970 encountered cracks capable of transmitting much more water than the cracks encountered in previous drill holes." This led Okeson to conclude that some reservoir leakage was inevitable. "Probably some of the reservoir water will leak around the ends of the dam, through cracks in the bedrock, and emerge from cracks at lower elevations of the bedrock surface downstream from the dam. The water would be under artesian pressure so it would gradually wet the thick cover of soil, thus turning [it] into a loblolly or quagmire. Loblolly conditions could also develop in places *within the impervious section of the dam* if one or more cracks is poorly grouted" (emphasis addedd).

Although he was loath to say so—using an adjective like "serious" is regarded by some engineers as unwarranted emotionalism—loblolly conditions inside the dam would be a serious occurrence, one under which the dam could conceivably be lost. The key to preventing them was proper grouting. Grouting, a commonly used technique in the dam builders' art, involves injecting liquid concrete under high pressure into drill holes in the abutment walls on either or both sides of a dam; the concrete moves like water, filling all the fissures, shear zones, and holes, and then hardens, leaving a supposedly impervious barrier against seepage. The plan at Teton was essentially the same as at Fontenelle—several grout curtains would be extended outward from the site, into the abutments, to block any flow of water trying to move around the dam. The grouting might be done improperly under three sets of conditions: if the engineers were inexperienced or otherwise incompetent; if the rock was so hopelessly fractured and fissured that a near-perfect job of grouting was impossible; or if the canyon wall surprised the engineers by taking so much more grout than expected that, at some point, they declared the job done and quit.

In 1969, the year before the water-injection tests, the Bureau had taken the rather unusual step of performing a test-grouting program, so unsure was it of the conditions at the Teton site. Holes were drilled in the rock, and grout was pumped in under high pressure; then the job was tested to see how well it worked. As far as the Bureau was concerned, it had worked fine. "Once we decided that the cracks in the abutments could be sealed with grout," Harold Arthur, then head of dam design and construction, told a reporter from the Los Angeles *Times*, "we never reconsidered the suitability of the Teton site, despite the difficulties we experienced later in construction."

Only one thing had been wrong with the Bureau's test-grouting program. There was a road leading to the damsite from Sugar City, a few miles to the southwest, but none from the north. All of the test grouting was performed on the south abutment of the dam. None whatsoever was performed on the north side, the right abutment of the dam—the side where three hundred gallons of water per minute injected into holes had simply disappeared, day after day after day.

With the defeat of the environmentalists in court, there was no way to stop the dam. From an appropriation of $1,575,000 in 1971, funding for Teton jumped to over $10 million in 1972 and went even higher for the next four years, reaching an apogee of $15,217,000 for fiscal year 1976, when the $85 million dam was completed. Or, to be more accurate, when the Bureau's engineers thought it was completed.

In his 1970 memorandum, Clifford Okeson, the Bureau geologist, had said that the largest cracks he could find after extending a miniature televison camera and light down the length of a thirty-five-hundred-foot borehole were about an inch and a half wide. That was a small crack, easily grouted—nothing to worry about. In February of 1974, however, as the Bureau's main contractors, Morrison-Knudsen of Boise and Peter Kiewit of Omaha, were excavating the huge keyway foundation trench—which would replace the worst of the fractured surface rock with a man-made concrete foundation—they came on the right abutment's great secret. It was a discovery that five years of boring, injecting, and test grouting had failed to reveal. What they found, Robbie Robison, the Bureau's project engineer, wrote his superiors in a memo, were "unusually large" fissures in the rock of the right canyon wall.

"Unusually large" was hardly apt. The fissures were gigantic. They were *caves*. One of them was eleven feet wide and a hundred feet long. Another was nine feet wide, in places, and 190 feet long. One by one, other fissures were discovered. The whole right canyon wall was full of them.

If Robison's description of what had to be considered an appalling discovery was understated in the extreme—even if the fissures weren't a safety problem, it was astonishing that they had been missed—his recommendation of a course of action displayed an arresting mental paralysis. "We do not recommend to grout these voids at this time." Robison wrote Harold Arthur in Denver. "The claims situation [by the contractor] . . . makes us hesitant to cause any delays. . . . Furthermore, grouting of these voids is not critical at this time as they are located outside the dam area and could be grouted at a later date if you should so desire."

Robbie Robison, barely thirty years old, was on his first big project. It had been a troublesome project from the beginning, racked by delay. Costs were up; schedules were behind. For four years the Teton project had been officially underway, and now, in 1974, there was still nothing to show for it but a huge amount of excavation at the bottom of the canyon and some trailer sheds and a lot full of earthmoving equipment. The two biggest voids alone would eat a trainload of grout. Who knew what others would be found? The important thing, Robison figured, was that they were *beyond* the keyway trench; they were *beyond* the point where the Bureau had arbitrarily decided no further grouting was required; they were, therefore, beyond the limits of reasonable concern. After all, if you wanted to be *really* secure, you could have extended the keyway trench all the way to Ashton, which was twelve miles out from the north abutment of the dam. That was what Robie Robison sarcastically told a reporter, later on. It might not have been a bad idea.

Though the airy caves in the rock were a shocking discovery, no one besides Robison, the contractors, Harold Arthur, and a small circle of Bureau officialdom knew about them. Gil Stamm, the commissioner, was probably never told. The people of Rexburg and Sugar City, the two towns lying directly in the Teton River floodpath, were entirely in the dark, as were the politicians who had so assiduously promoted the dam. Of course, had they known about the voids, it probably wouldn't have mattered to them anyway. After all, the Bureau of Reclamation had the best engineers in the world.

The dam was finished, more or less, on October 3, 1975, when the flow of the river was interrupted for the first time. Even with the biggest voids left unfilled, the job had taken 503,000 cubic feet of grout—more than twice as much as the Bureau predicted it would have to use. That winter, a series of Pacific storms bashed into the Teton Mountains, depositing a big snowpack. As spring was about to arrive, Robbie Robison had two worries: how he was going to settle with the contractors over the cost of the extra grouting, and how he was going to capture the snow that was about to melt out of the Grand Tetons without violating the Bureau's time-honored rule about filling reservoirs behind earthfill dams.

The rule is simple: the rate of fill is to be kept at or below one foot a day measured vertically along the reservoir walls. That way, if problems develop with the dam or the abutments, or back along the reservoir itself—where rising water sometimes loosens rock and causes landslides, or causes the bedrock to shift under its weight, producing the same result—they can be dealt with. At a slow rate of fill, such problems are less likely to develop in the first place. It was a sensible rule, and, like most sensible rules, it had already been violated on a number of occasions. Why not dispense with it again, with all that precious water coming down from the Teton Range? On March 3, 1976, Robison wrote Harold Arthur formally requesting permission for a two-foot-per-day filling rate. Ironically, one of the arguments he used in support of his request was that a faster rate of fill would permit the Bureau to observe how effective its grouting program had been. It was, in a way, like arguing for a hundred-mile-per-hour speed limit on the grounds that motorists would spend less time on dangerous highways if they drove twice as fast. But on March 23, Arthur readily acceded to Robison's request.

Actually, the whole business—formal request, formal permission granted—was a meaningless charade. The main outlet works—the tunnel and appurtenances that would carry water out of the reservoir and into the adjacent canals—were not yet finished. The auxiliary outlet works were, but they were designed to carry a maximum flow of 850 cubic feet per second. Engorged by a snowpack half again as deep as normal, the Teton River was about to peak at several thousand cfs. Without a functioning main outlet works, the reservoir would rise as fast as the Teton River felt like filling it. It was likely to rise a lot faster than two feet per day.

Harold Arthur was unconcerned about such a fast rate of fill because he had ordered a series of observation wells to be drilled around the dam, which would—in theory—inform the Bureau of any developing problems. The water table around a damsite will often show a rise as the reservoir fills, because a certain amount of seepage into outlying terrain is inevitable. If the water table rises precipitously, however, and if wells far from the reservoir are affected—especially wells downstream—it could mean that the reservoir is seeping excessively. The only other possibility is a pressure response, where the adjacent water table rises out of proportion to the actual rate of seepage because of hydrologic pressure, much as the constriction of a hose nozzle turns a placid gurgle into a sixty-foot jet.

From what Arthur had heard from Robbie Robison, the observation wells in the vicinity of the dam were showing what he termed a "predictable buildup"; that was the term he used in his March 23 memo. Obviously, he had not yet seen, or even been told about, a Bureau report written almost exactly at the same time, which disclosed the startling fact that "the rate of travel of the rising water table north of the reservoir is over 1,000 times that calculated for predicted movement of water." The memo came from Gordon Haskett, the Bureau geologist who had been monitoring the observation wells. To engineers who had spent their lives working with microtolerances, who considered almost any adjective hyperbolic, something as extreme as a thousandfold increase should have leaped off the printed page. Haskell's report, however, was routinely routed through the Boise regional office, from where it went to Denver, where it reached Harold Arthur's in-box on April 13, three weeks after he had already consented to the faster rate of filling.

It probably wouldn't have mattered if it had arrived the day after it was written. After looking Haskett's memo over, Arthur filed it away. In a way, he cannot be blamed. Having reported the bizarre thousandfold increase in the predicted rise of the groundwater table, Haskell had felt obliged to explain it. It was, he said, *too excessive* to be attributable to seepage. "Therefore," he concluded, "[it] must be a pressure response."

Actually, a relatively simple and inexpensive piece of gauging equipment, a piezometer, could probably have told the Bureau whether something drastic was going on or whether the inexplicably rapid rise of the adjacent groundwater table was merely a pressure response. Forty miles across the Rexford Bench, on Willow Creek, the Corps of Engineers had just erected Ririe Dam, and all forty-nine of its observations wells were equipped with piezometers. Their use had been routine practice for years. The closest thing to an official explanation as to why they weren't used at Teton came from Richard Saliman, the chief of the Bureau's design division. "We do use them on other dams," Saliman told a reporter, "but basically, we had such an excellent foundation we didn't feel it necessary. . . . With the rock types we had we just didn't see the need for it." For his part, Harold Arthur doesn't think the piezometers would have detected anything "unless one of them happened to be exactly where the leakage was occurring. It would have been a matter of luck." But even if luck had been on the Bureau's side, it might not have made a difference. The Bureau didn't believe in luck—it believed in itself. "Suppose we'd gotten a reading from a piezometer that there was massive seepage from the dam," Arthur told an interviewer in 1983. "We might not have believed it. We had a perfect record up to then. We might have thought the thing was giving us a wrong reading."

By mid-May of 1976, the Teton River was a frigid deluge. Square miles of snowfields were melting into it under a hot, high sun and the reservoir was rising much faster than it ought to have been, approaching four feet a day. As the reservoir filled, the emergency outlet works were the only real insurance against catastrophe. If the dam gave evidence that it was going to fail, the outlet works would permit a rapid but controlled drawdown of the reservoir. But the outlet works were still not operational; they were completely sealed off by a huge metal barrier and in the process of being painted. On May 14, Robison was finally concerned enough about the rapid filling to write his superiors. "Request your comments for flood control operations," he said in a terse memo. It was a pro forma exercise: the Bureau, by then, was completely in the river's hands.

On the 3rd of June, a Thursday, the first equipment operator arriving at the damsite early in the morning noticed a small leak pouring out of the canyon wall about a third of a mile below the dam. From the canyon rim, three hundred feet above the river, the leak looked like nothing; one could barely hear it bubbling above the quiet rush of what was left of the river flowing out of the auxiliary outlet works. The leak was coming out of the north abutment—the right canyon wall. The water was clear. Five hundred feet closer to the dam was another leak, even smaller, also clear. The next day there was still another. All three leaks were coming out of the right canyon wall.

Robbie Robison stood on the canyon rim watching the leaks for a while. Looking back at him impassively was his masterwork, Teton Dam: an average-size modern dam, but a monument that would have made a pharaoh reel. Although Robison was, as he later put it, "just a cog in a great big wheel," it was *his* monument. The reservoir was sitting quietly behind the dam, looking utterly serene. Suddenly set free, it would have a calculable energy release approximating that of a quarter-megaton bomb.

Robison returned to his office in the trailerlike project building. Then, restless, he went outside and watched the leaks again. Finally, he went down into the canyon and crossed the river by boat. The dam loomed above him, 305 feet high. Robison jumped over the rocky bed and climbed up the fifty-degree slope to the first leak and measured it. Sixty gallons a minute, about a seventh of a cubic foot per second. The second leak was flowing at about forty gallons a minute, the third—the one closest to the dam—at about twenty.

Robison went back across the river, climbed to the Bureau's trailers, and wrote a brief memo to Harold Arthur telling him about the leaks. At the end of the memo, he said, "I'll keep you advised."

Off and on during the day, Robison's men monitored the leaks through binoculars. By nine o'clock in the evening it became too dark to see, and they went home.

Saturday, June 5, dawned pellucid and bright, a warm and somnolent day. The first Morrison-Knudsen man arrived at the Teton site at seven in the morning. In the shadowy postdawn light, the downstream embankment, facing west, was still dark. He looked at it and saw nothing. Sometime around seven-thirty he looked again and saw something. There was a roiling creek of muddy water emerging from the right abutment adjacent to the dam.

The construction man immediately phoned Robison, who drove out at eighty miles an hour. By the time he arrived another leak had developed, almost exactly at the contact point of the dam with the abutment. Robison quickly ordered one of his men to try to divert the flow away from the powerhouse with a bulldozer. Then, at last, he decided to call his superiors in Washington, Denver, and Boise.

A Bureau report later said, "The project supervisors did not believe at this time that the safety of the dam was jeopardized."

At about nine-thirty, one of the men noticed an odd-looking shadow on the downstream face of the dam, twenty feet or so out from the right abutment. He looked at the sky. There was no cloud anywhere. The shadow was a wet spot. In a few more minutes it was a spring. Then it was a creek. Then it was a sizable torrent washing away the embankment of the dam. Robbie Robison called the sheriffs of Madison and Fremont counties and told them to prepare to evacuate twelve thousand people.

Watching the unprecedented spectacle beneath him, Robison was biting his lip until it almost bled. He thought of the main outlet works and did a quick mental calculation of how long it would take to open it. He decided hours, maybe a day, maybe two. He told his men to try anyway. Then he ordered a second bulldozer down to try to shove material into the widening hole. The two big Caterpillars crawled across the dam face like flies on a wall. As fast as they could plug the hole, the torrent swept away what they had filled in. The hole was now a crater, as large as a swimming pool. It was vomiting muddy water in rapid heaves.

At that same moment, a family of tourists was driving up the access road from Sugar City to take a look at the newly completed dam. It was just an unplanned side trip, prompted mainly by the sign at the junction of the access road with Highway 33 that proudly announced the existence of the dam. Through such a chance excursion, David Schleicher's wish was about to be fulfilled. On the seat of the car was a movie camera, loaded with film.

Nothing could plug the hole in the downstream face. After twenty minutes one of the Caterpillars fell halfway into it. Terrified as he was, the operator of the other dozer frantically tried to winch it out. Meanwhile, on the other side of the dam, a more ominous phenomenon was occurring. A whirlpool had begun to develop in the reservoir a few yards away from the face of the dam. Like the whirlpool over the outlet of an emptying bathtub, the vortex could only mean that water was leaving the reservoir in a hurry, and was sluicing directly through the dam. Two more dozer operators crawled down the canyon slope and onto the upstream side of the dam, shoving riprap from the embankment into the swirling hole. One of them was named Jay Calderwood. Jay Calderwood, like almost everyone else in the area, was a Mormon. "Every pass I made I wondered whether it would be my last," he recalled later. "I though, 'Well, Jay old boy, this is it. I'm going to go. Have I lived the righteous life my parents taught me?' I felt very close to the Lord at this time. I had Him on my mind all the time, when I was trying to stop the leak and save the dam. 'This is it, I can't do a bit of good at what I'm doing. But I'll go out fighting. I'll not be a coward.' "

Meanwhile, on the downstream side of the dam, the two bulldozers were still trying to plug the huge spring gushing out of the embankment. It was now regurgitating the dam's insides by the cubic yard. The audience on the canyon rim, which had grown to include a couple of local radio reporters, was helplessly spellbound. At almost exactly eleven-thirty, the sides of the hole suddenly collapsed some more, widening it by twenty feet. The Caterpillars began to drop as if through a trapdoor, two huge yellow machines in slow-motion aerial freefall. Both drivers launched themselves out of their seats and ran for safety along the dam's crest and up the canyon slope.

Now one could only watch. Robbie Robison, trembling and licking blood off his punctured lip, may still have been telling himself it couldn't happen. The dam was too big, too solid. It could not be moved. At eleven fifty-five, the crest of the dam fell into the reservoir as if a sword had whacked it off. Two minutes later, as the movie camera whirred in the hands of a speechless tourist, the second-largest flood in North America since the last Ice Age was heading out the Teton River Canyon.

The dam went almost noiselessly. It didn't so much break as melt. One second there was a dam, three hundred feet high and seventeen hundred feet wide at the base; the next minute it was gone. Actually, two-thirds of it was somehow left standing as the flood roared through the bombed-out hole on the right side. The reservoir spilled out in a great, fat, smooth, probing tongue; then, a couple of hundred yards downstream, it suddenly erupted into a boil about fifteen stories high. For a moment, the spectators on the canyon rim thought it might consume them; then it boomed off in a heart-stopping chaos of boils, whirlpools, and fifty-foot waves. The initial rapids resembled Lava Falls on the Colorado River, a Colorado River with two million cubic second-feet of water. The color was an awful brown.

Six miles beyond the dam, the Teton Canyon abruptly comes to an end; below there, flat as a slightly inclined board, lies the Snake River Plain. Two towns, Wilford and Teton, sat at the terminus of the canyon, four or five miles apart. Teton was south of the river and above it; it would be spared, barely. Wilford was just north of the river at bank elevation. A few miles beyond Wilford was Sugar City, and six miles farther down was Rexburg, a community of eight thousand people. Another sixty river miles beyond was Idaho Falls, population 35,776, the third-largest town in Idaho. All four towns were going to absorb a direct hit, but none would be hit like Wilford. When road atlases were republished a year later, Wilford would not be listed among Idaho's cities and towns.

The leading wave arrived twenty-five minutes after the dam broke. It was twenty feet high. The fastest egress to safety was the road north to St. Anthony, even though it went straight across the plain in sight of the river for three miles before it began to climb. As the last refugees from Wilford roared up the highway in their cars, they could see the flood approaching out of the east. It looked like a dust storm, until they saw the dust snapping huge cottonwoods in half. One of the first homes hit was Alice Birch's. The day before, she had celebrated living in the same house for fifty years. The twenty-foot wall crashed into it, tore it off its foundations, and lifted it onto a power line, which snapped in half. The shooting voltage ignited a ruptured propane tank and Alice Birch's house blew to smithereens.

Glen Bedford's aging parents-in-law, the Liedings, lived in Wilford. When the first radio announcements about the dam came around ten o'clock, he raced up to their house from Parker, on the Henry's Fork of the Snake, to help them get out. Roaring by his sister-in-law's home in St. Anthony, five miles before Wilford, Glen Bedford saw his mother-in-law already unloading a pickup with a few belongings. Her husband was nowhere in sight. Believing that he was still at home in Wilford, Bedford drove his foot into the accelerator pedal. His father-in-law, who had been behind the house and out of view, read Bedford's mind and roared off after him. When he got to Wilford he could already see the flood pouring out of the canyon. From a mile and a half away, he said, it looked fifty feet high. When Lieding caught up with his son-in-law at his house he screamed at him to turn back to St. Anthony. "I'll be there in four minutes!" Bedford yelled and ran upstairs to collect a last armload of valuables and mementos. They found him eleven days later, twisted almost beyond recognition amid a pile of trees and torn-up trailers.

Wilford went in an instant. The flood left only the two-story Mormon meetinghouse, and of that it left only the brick shell. The other 154 houses were intact or in pieces, riding the fifteen-mile-an-hour crest.

As the flood swept southwestward it spread to a width of two miles, but it had enough churning power to strip the topsoil off thousands of acres of first-class farmland. When it hit Sugar City the flood was no longer liquid, but semisolid.

There was a trailer park outside of Sugar City, and, according to witnesses in airplanes overhead, the flood hit town tumbling trailers like ice cubes, smashing houses off their foundations. Like Wilford, Sugar City was motionless one minute and moving fifteen miles an hour the next. Somehow, one of the victims there was killed by a shotgun blast.

In their desperation to flee Sugar City, Betty and Rodney Larson flooded their car's engine so badly that it wouldn't start. With the flood bearing down on them, it was too late to escape on foot. They ran upstairs with their three children and draped themselves over mattresses, hoping they would float. For three hours, their house felt as if a turbine generator were rattling itself loose in their basement. The house eventually came right off its foundation, but, miraculously, it did not move. Like a dud missile, it floated two feet off its pad and settled back down exactly where it had been. To pass the time, they counted dead cows.

Since eleven o'clock in the morning, the Rexburg police and civil defense had been herding people to higher ground. The Rexburg benchlands rise up from the eastern edge of the town, and on top of the first hill stood Mormon Ricks College, its dormitories recently emptied. Seven thousand people streamed up College Hill like the Hebrews during the Exodus, dragging whatever cars, wheelbarrows, and muscle could carry. By the time the flood hit Rexburg, the radio said, the crest would be only two to four feet deep. They saw the dust first, a four-mile-wide roiling cloud, then they saw the wall of water. It came just like a lava flow: five feet in front of it everything was dry, and then came the wave, seven feet high. Just before it hit town, the radio station went dead. The first thing the wave hit was the lumberyard outside of town. All the logs, thousands of them, were set loose. Dozens of them smashed against a bulk gasoline storage tank a few hundred yards away. The tank went off like a firebomb, setting flaming slicks adrift on the racing water. When the wave hit the front line of houses a hundred windows were instantaneously shattered. Witnesses said it sounded just like a rifle shot. Then the flaming gasoline poured into windows and set Rexburg on fire, like a floating-island dessert.

The throng on College Hill watched speechlessly as the wall of water washed their town away, burning it down as it went. A big white frame house floated over to the base of the hill below them and settled down in shallow water in the middle of a street. The water itself, moving only ten miles an hour now but engorged with a cubic quarter mile of topsoil, had force enough to separate homes from their foundations, but the real damage to Rexburg was done by Sugar City and Wilford. Reduced to giant pieces of flotsam—silos, walls, automobiles, telephone poles, pianos, trees—Wilford and Sugar City were a battering ram afloat, smashing Rexburg to pieces. When the flood passed after dusk, it had left six inches of silt on everything, as if it had snowed mud. A Greyhound bus sat on someone's lawn.

A hundred miles downriver on the Snake was American Falls Reservoir, holding four times as much water as Teton had held. American Falls was one of the Bureau's oldest dams. The dam was, in fact, unsafe—something the Bureau knew as early as 1966, but hadn't bothered to correct. (In 1967, chief engineer Barney Bellport wrote Floyd Dominy that "the need for replacement of American Falls Dam is largely governed by structural reasons, although the deterioration of the concrete due to alkali-aggregate reaction contributes to the poor condition of the structure. The lack of bond between constriction joints and the fact that the dam was not designed for ice pressures are of great significance." By 1976, however, the dam had been neither replaced nor fixed.)

If the dam was too weak to withstand the strain of the Teton flood coming on top of high flows in the Snake, the resulting calamity could only be guessed at. Instead of spreading out, the water would remain largely confined by the canyon of the Snake until it hit the Boise. Below, beyond Hells Canyon, the dams were lined up like dominoes: Ice Harbor, Little Goose, Lower Granite, Lower Monumental; then the Columbia River and McNary, The Dalles, John Day, and Bonneville dams. The bigger Columbia dams would have seen such a flood before, perhaps, but those on the Snake, unless their reservoirs could be emptied in time, might meet flows they were never designed to handle. There was only one course of action: empty American Falls. Over two days, the archaic dam would have to release more water than it ever had before, and its reservoir would receive more at one time than it ever got.

By nightfall on Saturday, Rexburg was a silhouette of wreckage, carnage, and flaring fires. The lower half of the town was a total loss. As Rexburg finally became a vast, slowly shrinking pool of standing water, the flood was washing up against the Menan Buttes, some low hills off to the west. Now six miles wide, it split suddenly into two streams. The one veering northward around the buttes struggled upward against the inclined plain and fell back into a channel it quickly dug down to bedrock. Within minutes, it was a replica of the chocolate-brown Colorado River at high water. Then, beyond the buttes, the two channels rejoined, and the flood went into Idaho Falls.

Two things saved Idaho Falls. One was the geologic bedrock and soil which had made Teton such a bad project, physically and economically. By the time the flood poured itself into the Snake River twenty miles above the town, a lot of it had drained off into the porous soil and deeply fractured bedrock beneath it. The other salvation was a night and a day spent by thousands of volunteers sandbagging the levees along the river, which goes through the dead center of town. The flood built toward a crest all day Sunday and finally peaked, at just over 100,000 cubic feet per second, at ten o'clock at night. As logs, fiberboard, and bales of hay crashed up against the Broadway Bridge, which retained only inches of freeboard, a reservoir began to form behind it. Nine pounds of dynamite and a sixty-foot dragline could not dislodge the debris. It was only after an escape channel was dug that officials decided they wouldn't have to blow up the bridge. It survived, looking as if it had been chewed by a hundred-foot shark. The town escaped with two hundred flooded homes.

At American Falls Dam, water was bursting furiously out of the outlet works. Ten thousand Bureau people and three million more downstream, all the way to the Pacific, held their collective breath as the reservoir began to fill early Monday morning. But the remains of the flood did not even overtop the spillway.

Eleven people died in the Teton flood, but the dam could just as easily have gone at two in the morning, in which case the toll could have risen into the thousands. Power and telephone lines between Sugar City and Rexburg were cut as soon as the flood struck, so the odds are there would have been no warning. The Bureau had installed no sensors below the dam to warn the towns if a flood was on the way.

Four thousand homes were damaged or destroyed; 350 businesses were lost. Damage estimates climbed to $2 billion, though settlements were to fall substantially short of that. Nothing, however, was as startling as what the flood had done to the land. The topsoil was gone from tens of thousands of acres—stripped off as if a plow a mile and a half wide had come along, scraping the earth down to bedrock.

According to one estimate, more land was destroyed—permanently, made incapable of ever growing anything again—than would have been opened to irrigation by the dam.

That was merely the first in a long string of ironies that followed in the wake of the tragedy. As it turned out, the farmers on the Rexburg bench, the rich irrigators for whose benefit the dam was mainly built, were entirely spared. Their riverbottom neighbors, whose means of livelihood vanished with the flood, would have to search the region, the state, even the country to find a decent farm they could afford with their settlement money. But the farmers on the Rexburg bench could relax; they might not even miss the water they would now never receive. "A lot of wells have been drilled up on the bench," explained Agriculture Commissioner Bill Kellogg, confirming what the dam's opponents had been saying all along, "and the dam was only intended for supplemental water." This same supplemental water—a life-or-death matter three days before—had suddenly become something they could do without. The dam's opponents had argued that, too. But even had the irrigators on the benchlands been ruined for want of water, there were only a handful of them. There were thousands of victims on the floodplain below.

The politicians who had fought hardest for Teton Dam, such as Frank Church, were the first to pounce on the Bureau after the dam failed, the first to search the disaster for whatever political refuge could be found. Church castigated the Bureau for being "a prisoner of stale engineering ideas"; he made no apology for the stalest idea of all, the Congressional pork barrel. "No one told me the dam was going to break," Church blustered when the local people, most of whom had wanted the dam as badly as he, tried to hold him responsible. Actually, Bob Curry had suggested just that to him three years earlier, when he wrote Church about the geologic defects of the site. Curry claims he never got a decent response.

As for the Bureau, it said as little as it could. Its reputation suddenly in shambles, it tried not to make a wretched situation worse. Its press releases after the catastrophe were a dry recitation of events. They were honest, but there was no hint of responsibility, not even sympathy for the flood's victims, and no suggestion that perhaps the dam shouldn't have been built.

None of Teton's principal designers and builders were fired. Harold Arthur voluntarily retired—he had reached retirement age anyway—and started up a lucrative consulting business in Denver. Though he never publicly entertained a doubt about the dam, though he approved every major decision during its construction, though he vetoed a plan to install three grout curtains instead of one, not once during the interviews in 1982 and 1984 did Arthur display a hint of remorse. "One minute I hear the dam is fine and the next minute it's failed," Arthur told me. "There wasn't anything *I* could do about it." Donald Duck was twice passed over for promotion, took early retirement, and moved to Chicago, where he became a vice-president of the Harza engineering firm, which builds dams. Robbie Robison drifted off and disappeared; in 1984, no one seemed to have any idea where he was. Commissioner Gil Stamm was, always, wooden as a cigar-store Indian. "I ran into Stamm in Washington after the dam went," his old friend Floyd Dominy said. "I said to him, 'Jesus Christ, haven't you committed suicide yet?' He just smiled," said Dominy. "He just smiled."

To this day, no one is exactly sure what caused the collapse of Teton Dam, though several million dollars were spent on four independent investigations to figure it out. It might have been a leaky joint between the foundation and the dam. It might have been a flaw in the impervious core of the dam itself. It might have been poor filler material. It might have been expansion and contraction caused by ice that formed during winter construction. The theory Harold Arthur maintained is "incredible, virtually impossible"—that water drifted around the grout curtain on the right side and immediately went back into the dam, turning it to mud—is the one that one former Bureau engineer, who would rather not be named, believes is the likeliest explanation. "With the other theories, you can blame it on the contractors," he says. "With the grout-curtain theory, you're saying it was a lousy design. But that's why it failed. All the other theories are so much b.s."

However, among all the ironies that piled up in the aftermath of the Teton tragedy, everything pales beside one: there are a lot of voices in Idaho calling for the dam to be rebuilt. When a plague of locusts struck Utah after the first Mormons arrived, huge flocks of migrating seagulls flew in and ate them up. When clouds of disease-ridden flies and mosquitos appeared in the wake of the Teton debacle, the same thing happened again, so the Mormon irrigation confederacy of southern Idaho has apparently decided that God, all evidence to the contrary notwithstanding, is still on its side. On December 10, 1976, only half a year after the disaster, the Idaho Water Users' Association issued a resolution calling for a "safe" Teton Dam to be rebuilt at or near the same site. Harold Arthur still believes he could design a dam at the Teton site that would not collapse, though no one seems inclined to let him try. His suggestion, offered in muted tones, came as close to an apology as anything he said.

The economics of the project are worse than ever, and with so much of the arid West screaming for more projects—projects that, for whole regions, are really a matter of life or death, at least if irrigation is to continue—it would be hard to justify an irrigation project for farmers still putting ten feet of water on their land. None of this is to suggest, however, that the tragedy of Teton Dam might not be repeated somewhere else. Colorado, for example.

Flowing through Denver, the South Platte River appears so insignificant it is hard to believe it is the city's main water supply, let along the sustenance of hundreds of thousands of irrigated acres downstream in Colorado and Nebraska. The South Platte is a mere fork of the main Platte, itself a tributary of the Missouri, itself a child of the Father of Waters. From a plane climbing up from Stapleton Airport, the South Platte is seen to meander forlornly out of town until it is quickly swallowed, as everything is, by the surreal endlessness of the Great Plains. Viewed from a low bluff two hundred miles downriver, it still appears of no consequence. The bottomlands are a tangle of shrubs and barbed wire interspersed with cottonwoods, and they are grazed bare by cows, which stare uncomprehendingly from the muck. It is a river without pretensions, haggard and used-looking, like a bag lady. In August, near the Nebraska border, the river dries up completely; all that reaches Nebraska is the underground flow. The Platte is one of the most hungrily used rivers in the entire world, surpassing even the Colorado. However, as far as the state of Colorado is concerned, it is not used enough.

The South Platte is one of two rivers left in Colorado that isn't utterly and irrevocably appropriated, now and forever. To a state which is second to California in the arid West in population, industry, and irrigated acreage—but which has at its disposal about one-tenth as much water—the fact that some 7 percent of its share of the river still escapes to Nebraska is a fact of overarching significance. That the Bureau of Reclamation has offered to build an enormous dam across it to attempt to correct that situation is another. This last glimmering promise, in the face of a hopeless, nonnegotiable finality, has been enough to lead the members of Colorado's political establishment into a world of fantasy, leaving both their senses and their principles behind.

Don Christenson's crew cut stands up about an inch and a half, like
a brush. A three-hundred-pound bear could nest down in that hair for
the night and in the morning, after the bear lumbered off, it would
spring right back up. The rest of Christenson fits the hairstyle: he is
lean, weathered, bronzed as a Comanche. His jaw is made of cast iron.
The one anomaly in his all-American countenance is a thick, volup-
tuous set of lips. In 1979, at the annual Conference on Rivers and
Water Policy in Washington, D.C.—better known as the Damfighters'
Conference—Christenson, surrounded by longhairs, environmental
lawyers, bureaucrats, and kayakers, stood out like the man from Mars.

Christenson's presence at the Damfighters' Conference was a signal
event. Unofficially, he was the first verifiable irrigation farmer who
had ever attended the conference. He was probably one of the first
who had ever opposed a dam, but he had a good reason. The Bureau
of Reclamation, having nearly run out of decent damsites, had finally
decided to turn the Reclamation Act inside out. It was going to flood
out a bunch of small farmers so it could give supplemental water to
a bunch of bigger farmers, several of whom would be in violation of
the Reclamation Act. Christenson was one of the small farmers, and
he was the one with the biggest mouth, so the Bureau wanted to drown
him with a vengeance.

When Don Christenson's father settled in the Weldon Valley in
1926, there was already talk of a big dam at the Narrows of the South
Platte, four or five miles downstream from his land. The first serious
proposal seems to have emerged in 1908. The farm, the whole town
of Weldona, and everything else from bluff to bluff for thirteen miles
was to go under, but the elder Christenson refused to let the prospect
faze him. "Dad would tell us, 'Maybe they are gonna build it. But
maybe they're not gonna build it. Maybe they're gonna build it but
they're not gonna build it for thirty years. I'm not going to sit here
and let the goddamned government worry me to death. We're gonna
farm our land and live a normal life and keep our property up, and
to hell with them.' " His prophecy was remarkable. Finally authorized
in 1944 as one of three-hundred-odd projects in the Pick-Sloan Act,
the Narrows Project had still not been built forty years later.

The senior Christenson's attitude managed to infect all three of his
sons, who raise their crops, paint their houses, fix their equipment,
and otherwise carry on as if the threat of a dam did not exist. The
same cannot be said, however, for the Weldon Valley as a whole.
Weldona has the look of a town losing hope: houses unpainted, shutters
askew, eerily quiet. "The people who've just decided to let their ol'

house decay may be the smart ones," Christenson says bitterly. "Why spend $15,000 to fix up your property when you know the Bureau of Reclamation"—he pronounces it *"Bee-yoor-o"*—"is gonna tell you your house is a slum anyway when they make you an offer?"

The Weldon Valley was settled in the 1870s, only forty years before the first proposal for a Narrows Dam; for most of its existence it has been threatened with extinction. Any day, any hour, someone might appear on one's front lawn to survey; someone might amble up one's walk with a sheaf of papers and an offer to sign or else. It is bad enough to live like this; it is worse to live under the shadow of a project as nonsensical as the Narrows Dam. And it becomes almost ludicrous if there is a distinct possibility that the dam, once built, may not hold water and could conceivably collapse, rendering seventy-five years of worry, agony, and divisiveness for naught. This has been Don Christenson's fate since the day he was born.

The dam will be immense—an earthen monster. Twenty-two thousand four hundred feet long, it would stretch, if laid across Washington, D.C.—which is where Christenson suggests it ought to be built—from outer Georgetown to the Capitol. In New York, it would stretch from the Empire State Building to the Staten Island Ferry. For all its length, it would be only 147 feet high, and the reservoir behind it would be drawn down much of the time, which has prompted its critics to rename the project "the Shallows." How anything this monumental— one of the largest dams on earth, longer even that the main dam of Itaípu, longer than Fort Peck—could be built for $226 million (the official cost estimate as of 1980) is anyone's guess. Actually, a great part of that expenditure—probably half—wouldn't even be used to build the dam. It would go to 844 landowners to pay compensation for the ninety-five farms, twenty-eight businesses, two churches, and elementary school that would be put underwater. It would also be used to relocate twenty-six miles of the Union Pacific's track and twenty miles of State Route 144. The remainder of the money would somehow erect a four-mile-long dam.

The Narrows Reservoir would submerge fourteen thousand to seventeen thousand acres of productive, privately irrigated farmland, none of which has ever received the kind of subsidy the beneficiaries of Narrows would automatically get. (This is some of the oldest continuously irrigated land in the West; the Weldon Valley ditch was dug by human and horse muscle in 1881.) Another forty thousand acres of unirrigated grazing land would also be drowned or affected. Some local waterfowl habitat would be harmed, but the real damage to nature would be downstream. Flows in the hugely depleted South

Platte, which are already critically low for the three-quarters of a million ducks and geese and the migrating whooping and sandhill cranes—the entire surviving U.S. population of whoopers—for which the river is a crucial feeding and resting spot, would be further reduced. Although the amount of water diverted would not be much—for a dam of such size and cost, it would be pathetic—its absence would be sorely felt by the waterfowl. On top of this, the concentration of fertilizers, pesticides, and sewage—Denver's and Fort Collins's—in the river would become worse.

The main benefit of building Narrows, on the other hand, would be supplemental irrigation water for 287,000 acres of land downstream from Don Christenson's farm, most of it between the towns of Brush and Sterling. There would only be enough water for inches per acre, and one could reasonably question whether, during the occasional severe drought whose ravages Narrows is supposed to make less severe, so little water would do any real good. In fact, the continued profitability of the farms making up those 287,000 acres through dry years and wet ones makes me wonder why the water is needed at all. The other benefits claimed for Narrows are recreation and flood control.

The high plains are home to some of the most freakish and violent weather in the world. Once, in Spearfish, South Dakota, thermometers leaped from two below to thirty-eight above zero in two minutes. A rutting ground for Canadian and Caribbean airflows, the plains are also known as tornado alley. Something like 90 percent of the world's tornados occur in North America, most of them between the Rockies and the Mississippi River. A much more frequent natural phenomenon, however, is the tornado's weaker sister, the hell-raising, rambunctious, exhilarating Great Plains thunderstorm.

One has to experience such a thunderstorm, preferably while lying scared to death in a ditch, to fathom the magnificent power of creation. In Texas, where the tropical flows are still saturated with moisture when they clash with colder air, a parade of thunderstorms dumped thirty inches of rain in twenty-four hours in the spring of 1978, far more rain than West Texas normally receives in a year. In Colorado, six- and seven-inch storms have been known on the plains, and since the natural groundcover is sparse, the flooding that results is spectacular. In 1964, such a flood occurred in the Bijou Creek watershed eighty miles east of Denver. Most of the time, Bijou Creek is less a creek than a dry wash; one has to search to find a puddle. During that storm, however, the Bijou became the second-largest river in the United States, carrying 465,000 cubic feet per second off the barren plains. Don Christenson was there. "It was the most unbelievable son of a bitch you ever saw," he says. The Bijou rose in a few hours and was almost dry a day later: a phantom monster. But the damage downstream was done.

Bijou Creek enters the South Platte from the south, exactly at the site of Narrows Dam. Upriver, the Platte is well controlled; the main untamed tributary below Fort Collins, and the main cause of damage downstream, is Bijou Creek. The damsite is flexible enough so that one can more or less choose to put the dam in front of the Bijou confluence or behind it. If the dam goes in upstream from the Bijou confluence, obviously, most of the flood-control benefits are lost.

Originally, the Bureau was intent on capturing the Bijou behind Narrows Dam because the economics of the project would automatically improve: greater flood-control benefits could be claimed, and a much larger proportion of the dam's cost would be nonreimbursable. Flood control, however, has always been the province of the Corps of Engineers, and the Corps, not the Bureau, would have to decide whether capturing the Bijou was worth it—or, for that matter, whether it was even safe to try to contain it.

The decision was to be made in 1965, shortly after Narrows was reauthorized by Congress and the Bureau began to push it seriously. On July 14, 1965, in a confidential letter to Commissioner Dominy, Pat Dugan, who had just left California to become the Bureau's regional director in Denver, described his efforts to ensure a decision that controlling the Bijou was worthwhile. Having just attended a meeting of the Colorado Water Conservation Board, Dugan reported, "I stressed the necessity for an early answer from the Corps of Engineers on the benefits to be provided for control of Bijou Creek by extension of the dam. The Board strongly expressed themselves as being in favor of this facet of the Narrows Project, and *I am confident that the Corps will be under continued pressure to provide the necessary answer*" (emphasis added). Unfortunately for the Bureau, the pressure to provide the "necessary" answer came to naught. The Bijou flooded mightily, but it flooded most infrequently, the Corps decided, so controlling it wasn't worth the extra cost. There was also some question as to whether another 450,000-cfs flood might not take out the dam.

With most of its flood-control rationale gone, Narrows went into eclipse during the remainder of the 1960s. In the early 1970s, however, the tide of fortune changed. Wayne Aspinall, the chairman of the House Interior Committee, was growing old and politically vulnerable, and Narrows, it seemed, was to be his swan song. The imperious old schoolteacher began pushing it so relentlessly that he even refused to let the project's opponents testify before his House Interior Committee. At the same time, the first OPEC oil crisis hit, and everyone began eyeing Colorado's huge reserves of oil shale. Some Coloradans seemed to want to turn the state into an energy colony and grow rich off it; others wanted to lock up as much water as possible so the oil, coal, and uranium industies would be forced to remain relatively small and the state's rural character, what was left of it, would remain fairly intact. One of the main adherents of the latter view was the new governor, Richard D. Lamm; an even stronger adherent was his commissioner of natural resources, Harris Sherman. The fact that Narrows was nowhere near the shale oil and uranium was somehow lost. What mattered was giving the state's unappropriated water to agriculture and locking it up, as best one could, now and forever.

As Midas turned everything he touched into gold, the Narrows Project had a miraculous ability to turn everyone it touched into someone else. It turned a crew-cut, rawboned young farmer like Don Christenson into an environmentalist. It turned a handsome young environmentalist like Senator Gary Hart into an avid water developer. Above all, it turned perhaps the three most powerful men in Colorado into bitter enemies.

One of the three was Glenn Saunders, the chief counsel for the Denver Water Board. A brilliant man with a silver tongue, Saunders had, for more than thirty years, been *the* water lawyer in a state where water lawyers wield power that makes them objects of profound respect. Under his tutelage, the Denver Water Board had become a kind of understudy of the Metropolitan Water District of Los Angeles: a well-oiled, well-funded suprapolitical machine trying to purloin water from every corner of the state, all in the interest of turning Denver into the Los Angeles of the Rockies—a goal which has been largely achieved. In a strictly legal sense, of course, the Water Board didn't steal water. But cross the Front Range and go into the mountains, where most of Colorado's water originates, and the response to a mention of the Denver Water Board is likely to be an oath.

Saunders was the perfect symbol of this rough-and-tumble political machine. With his Dickensian visage, in his checked suits and pastel shirts and vivid ties, he was the city sharpie making ruthless inroads into the virgin old West—terrifying witnesses in the docket, shouting down citizens at public hearings, and always scheming, pushing, plotting for more dams.

The second of the three men was Clarence J. Kuiper, who, through most of the 1970s, served as Colorado's state engineer. In a state such as Colorado, where both ground and surface water are regulated and everyone wants more than there is, the state engineer is a combination of judge, jury, and cop. He decides what is a reasonable diversion to each farm; he decides who can put in a well and how much he can pump; he decides when a diverter can no longer divert during a drought and when a pumper can no longer pump; he makes sure enough water reaches neighboring states to satisfy compact agreements; and, in the course of making such decisions, he wins the wrath and, if he does his job honestly and well—as Kuiper did—the grudging admiration of every water user in the state. Kuiper's whole life had been spent in water development: first as a young engineer in Turkey, for whose government the Bureau was building dams; later as a consulting engineer for the state of Wyoming, for which he drew up a water plan; and, finally, as Colorado's viceroy of water. A gigantic man whose ponderous gait and basso profundo voice bely a quick and encompassing intellect, Kuiper was light-years from being a conservationist. He was a water developer and an admirer of Ronald Reagan; he was enough of a westerner to call Jimmy Carter's water-projects hit list an "act of war," even if, in private, he referred to most of the projects in question as "dogs." Kuiper never stood in the path of a water project, unless it was a project in another state that threatened his own state's supply. But that would change.

The third man was the governor of Colorado, Richard Lamm. Young, humorless, thoughtful, intense, prematurely silver-haired, Lamm was a prototype of the New Age politician. As a state legislator he had made a name as an environmentalist, and a rather bold one— he was the leader of the successful effort to keep the lucrative Winter Olympics out of Denver. In 1978, the *Almanac of American Politics* described him as "far-out." He flew periodically to Chicago or New York to hobnob with people like Garrett Hardin, the ecologist, and Hazel Henderson, the "futurist," who served with him on the national board of the Council on Population and Environment. He staffed his administration with left-leaning people in their twenties and thirties— people like Harris Sherman, his resources secretary, who had served as counsel to the Environmental Defense Fund. Lamm was the sort of politician one could imagine drinking Red Zinger tea amid the whiskey-swillers in the smoke-filled rooms; he had backpacks and bicycles in his garage, and his wife, Dottie, was a well-known feminist. From every Chamber of Commerce in every mean little Colorado town there arose a collective groan. Dick Lamm—the *governor*?

But Lamm already had a reputation, in some circles, as a rather shameless opportunist. And even at the apogee of his alleged radicalism, he never was known as someone who didn't like water projects. In 1975, when Don Christenson and his Weldon Valley landowners'

417   Those Who Refuse to Learn . . .

group went shopping for a lawyer to represent them in what they were sure could culminate in a legal battle with the Bureau of Reclamation, they decided they had better choose well. "Everyone we talked to said, 'You want the best, go hire Glenn Saunders,' " Christenson remembers. "I said, 'Glenn Saunders, hell! Name one dam he's ever opposed. He isn't going to bother with a bunch of farmers like us.' Well, we went to see him anyway. At first he looked like he couldn't wait for us to go back out the door. But we served it up to him straight, and that man listened to us. You could watch his prejudices dissolve. I mean, he was a lawyer, first and foremost, and he knew we had a case."

"Here was this bunch of farmers marching in here saying they wanted to stop Narrows Dam," a raspy-voiced Saunders recounted. "I said to them, 'Stop Narrows Dam! We don't want that. We want to get everything we can built!' But they kept throwing facts at me, and they finally had me convinced Narrows is a boondoggle. When I took a closer look it was an even bigger boondoggle than they said."

"Old Saunders had sort of half agreed to represent us," Christenson recalls. "But I think he still wanted to hear what the Bureau had to say. So he ups and says, 'Get your coats! We're going out to see the Bureau.' Just like that! We drove out there to the Bureau's big box of a headquarters, Mr. Saunders and Marvin Etchison, our president, and me. Saunders knew just the man to see. We walked into the bureaucrat's office—I can't remember who he was—and sat down like we owned it. I was tickled—mad as I was at the Bureau, I never would have done something like that. And Mr. Saunders and this Bureau guy got into an argument right away. I don't even know about what, but the Bureau guy said, 'Well, Mr. Saunders, *you* of all people should know that.' " Christenson is given to explosions of laughter, and the recollection makes him almost giddy. " '*You* should know the answer to that!' Saunders doesn't say another word. He was *mad*! He gets up and kind of calmly says to Marvin and me, 'Come on, Marvin and Don. We can accomplish nothing further here.' And out we went, just like we came in. In the car, Saunders says, 'I want you to go back to the Weldon Valley and start raising a kitty of a hundred thousand dollars. That's what it's going to cost you to fight your government.'

"A hundred thousand dollars! You could have licked me if I thought we could raise that kind of money from a little old bunch of farmers."

In plotting their strategy, the Weldon Valley landowners' group had made one crucial mistake. They had always assumed that their main fight would be with the Bureau, the Colorado Water Conservation Board—a chamber of commerce for dams—and the Lower South Platte Conservancy District, which was scheduled to receive water from Narrows. The Lower South Platte Conservancy District was led by two brothers, Dave and Don Hamel, both influential in state and national politics; Dave Hamel had run unsuccessfully for governor and was a former administrator of the Rural Electrification Administration. (If Narrows was built, the Hamels would probably be the chief violators of the Reclamation Act in its service area, for they owned several thousand acres there.) But, as it turned out, the Bureau and the Lower South Platte people were merely a major and a minor irritant. The really tough opposition came from the person they had originally counted on for help: Colorado governor Dick Lamm.

What had happened to Lamm, the onetime radical environmental legislator? His former friend Alan Merson, who beat Wayne Aspinall in the Democratic primary in 1972, lost the general election, and ended up as regional administrator for the EPA, thought he had his finger on it. "Lamm got religion rather late in life," Merson told an interviewer. "Once a political aspirant gets elected, he finds he has this strange new dilemma: rather than worrying about what people want to *hear*, he has to worry about what they want to *have*. There's a big difference. People move out here because of the Rocky Mountains, but if some huge hand came down and swept away the Rocky Mountains a lot of them wouldn't even notice. They're too busy getting rich. Well, Dick Lamm was elected in the middle of the biggest boom in this state's history. He saw that the great big capitalist machine creating all the filth and ugliness and pollution was also making his constituents fat and sleek and happy. He came to feel that he had slighted the capitalist machine, which suddenly seemed to him to be working miracles. I mean, you look out from the capitol dome and all you see is brown inhospitable plains on the one side and ice-covered mountains on the other. It looks like a tough place. But the capitalist machine was scratching phenomenal wealth out of it. At some point Lamm realized that the whole damned machine runs on the impoundment of water. So he said, 'By God, we'd better impound some more water.'

"It isn't just Lamm," Merson went on disgustedly. "The whole Congressional delegation, except for Pat Schroeder"—a young Democratic Congresswoman from Denver—"is on the run from the irrigators—not even all the irrigators, but just those who are lucky enough to be sucking off the big federal teat. Gary Hart, Floyd Haskell, Tim Wirth—I like them all, they're my friends, but they're all scared to death of not liking water enough. This state is booming like crazy, and we're running out of water. So politicians tend to go blind in office. They're for *any* water project—they don't care how bad it is.

"At EPA, we tried to start a permit program for salinity discharges," Merson went on. "Some of these irrigators are poisoning rivers all the way to the ocean, returning water that's twenty times saltier than when they take it out. I explained it to Dick and he said, 'You're right. It's a good plan. But I can't support it. The legislature will kill me over it. Goddamn it, this could be another Interstate 470. I'll lose!' That was what really bothered him," Merson said, " 'I'll lose!' I took it to Harris Sherman and he said, 'It's unconstitutional, illegal, and immoral—*and* it will hurt agriculture.' "

Agriculture was key in Lamm's and Sherman's thinking, because what they wanted even more than growth was *stable* growth. In 125 years, Colorado's economy has boomed and busted more than that of any other state except, perhaps, Nevada. Nevada had introduced stable industries: gambling, prostitution, marriage, divorce. In Colorado, the only industry that had filled the fearful troughs between the boom cycles, when it looked as if the state might be virtually abandoned, was agriculture. It represented stability. Late in the twentieth century, it had also come to represent something else. Unlike eastern states, which can keep out development only by passing laws, western states have a natural means of halting industries they don't want at their gates: a scarcity of water. In the early 1970s, Colorado became the first western state that actually wanted to keep an industry out, or at least keep it from overwhelming its economy and way of life. The industry was energy—especially oil shale. And the means of holding back its growth was to try to put the remaining water in agriculture's hands and let the energy companies worry about wresting it away—or let them import water from somewhere else, as Exxon was proposing to run an aqueduct from Oahe Reservoir in South Dakota.

C. J. Kuiper, on the other hand, was charged with putting water to beneficial use, and it seemed silly to him to waste tens of thousands of acre-feet on crops with a low economic return—crops which were subsidized by the Reclamation program and, in the case of some, federal price supports—when half of America's oil was now coming out of the Middle East. Privately, Kuiper believed oil shale development was necessary: philosophically, he believed in the doctrine of highest use. Water had become so scarce in Colorado that whoever could pay the most should get what remained. Reclamation farmers paid the least of anyone.

Such thinking, however, was ultimately to have very little to do with the position Kuiper took on the Narrows Project. His position rested on his growing conviction that Narrows, if built, wouldn't even be able to *hold* water; that it would never be able to deliver the water it promised; and that there was a very real possibility the dam would collapse.

Never, since Narrows was first authorized in 1944, had anyone suggested that it might sit on an unsafe site. How much on-site testing the Bureau did prior to the 1970s is unknown; its main concern seemed to be drumming up enough local support to overwhelm the oppositon. But by 1976 it had its first sizable appropriation in hand, and finally decided it ought to learn something about the geology of the Narrows site.

One morning in the summer of that year, Corky Tomky, a neighbor of Don Christenson's and a leader in the battle against the dam, noticed that the Bureau had a man with a drilling rig down by the South Platte. Tomky wandered over to say hello. The man announced that he was drilling core samples to see what the foundation of the dam was like. Tomky asked him what he had found so far.

"Well, don't quote me," the driller answered, "but this site has big problems."

"Big problems?"

"*Big* problems. There's bedrock down there somewhere, but I can't find it. I've drilled two hundred and fifty feet down and still haven't hit it. All I get is gravel and loose rock, and sand."

"What do you suppose that means?" Tomky asked.

"It means," the Bureau man drawled, "that this dam is going to have a hell of a time holding water. The foundation is like a coffee filter. But don't tell 'em I told you that."

Tomky swore that he wouldn't, then he walked casually back to his truck and gunned it over the bumpy road toward Don Cristenson's place.

"As soon as Corky told me what he heard," Christenson recalled, "we called up our Congressman, Jim Johnson. He was one hundred percent for the dam, but we figured *this* was a piece of news. We got his assistant on the line—I can't even remember what his name is. Well, he sounded real concerned on the phone. He told me, 'I'll talk to the Congressman and get right back to you.' I wished I'd had a tape recorder on that damn line. He never got back to me. No, sir. And the next day, wouldn't you believe it, that well driller was *not* back on the job. They handcuffed him to a desk in Denver somewhere. He never came back again. It was about then," Christenson said, "that we decided to see the state engineer."

The point at which Christenson decided to pay a call on the state engineer coincided nicely with the collapse of Teton Dam. Teton, as Kuiper put it, "scared the living bleep out of Lamm and Harris Sherman." Both of them watched poor Cecil Andrus face the reporters on the news, and saw his hapless water-resources director, Keith Higginson, blamed for a tragedy he had had little to do with. Andrus had been lukewarm at most about the Teton Project. What if a dam Lamm and Sherman strongly backed wiped out a string of Colorado towns? After Teton Dam went, Sherman decided he had better review the safety questions surrounding any imminent project planned for Colorado.

"When Sherman called his meeting, I was just leaving on a trip," Kuiper recalls. "I had never paid much attention to the Narrows— I'm not required to in the case of a federal project. I knew the ancient Platte River left a great big alluvial bed and that the Bureau would have to get through a lot of alluvial wash to anchor the dam on anyting solid. But I figured they knew what to do. I could have walked into Sherman's meeting and said, 'Well, I know of a few problems with the site but I defer to the Bureau's expertise.' After *Teton*—good Lord, I didn't imagine that the Bureau was going to let something *that* stupid happen again." But, Kuiper figured, he was the state engineer; if a dam failed, and he had assayed the site, he would share in the blame no matter who deserved it. Besides, Sherman had asked for his opinion, and he might as well give an informed one. Therefore, as he left to go on his trip, he asked his assistants, in his absence, to prepare a schematic of the Narrows site, superimposing the dam over a big color diagram of what was known of the geologic conditions. When he got back he had only a few minutes to look over the schematic; a few minutes was all he needed. "I looked at that schematic," Kuiper said, "and in thirty seconds I saw why that test driller was right. The old alluvial bed of that ancient river is huge. There are about ten stories of gravel out there sitting on five stories of cobblestones. Way off on the south end of the site the alluvial bed is almost three hundred feet deep. Well, they can't clean all that stuff out—it would be much too expensive and God knows where they'd even put it. So they were just going to let the dam sit on top of the alluvium, not really anchored to rock except at the abutments. And the alluvium ran *under* the south abutment. To prevent seepage under the dam, they had a cutoff trench planned down to bedrock, sort of like the keyway trench they built at Teton. But basically they were just going to hang it under the dam like a curtain.

"Hell, that alluvium is so wide they've got to run that trench out on the south side, way beyond the dam, or water is going to creep around it—exactly the way it did at Teton. It looked to me, from the schematic, that they were going to have to extend it out a mile. Well, no way they were planning to do that—it would cost too much.

"I sat there staring at the schematic," Kuiper said, "and I said to myself, 'Here we go again. Doesn't the Bureau even know how to learn from a *disaster*?' "

Even if the seepage didn't reenter the dam immediately—which was what apparently happened at Teton—Kuiper guessed that the rate of water seepage would be so enormous that the reservoir would more or less disappear and emerge somewhere downriver, as a swamp. But where? The water would back up behind the dam, penetrate the porous reservoir bottom, and sneak around the cutoff trench, underground. Then, following the downslope of the plains, it would have to resurface at about the same elevation. That elevation coincided approximately with the town of Fort Morgan, population eight thousand, which lay fifteen miles downriver. "If they build the dam," Kuiper said sardonically, "those Fort Morganites had better learn how to swim."

When Kuiper walked into Harris Sherman's meeting, he was surprised to see his sometime nemesis Glenn Saunders smiling at him. Saunders had somehow caught wind of the gathering and had demanded admittance; Sherman, who could hardly have wanted him there, hadn't dared bar him. One did not invite the antipathy of the preeminent lawyer in Colorado.

Sherman opened the meeting by asking each of the assembled members to state flatly whether they had any misgivings about the Narrows site.

"The site's fine," said the Bureau geologist.

"The site's fine," said Felix Sparks, the head of the Colorado Water Conservation Board.

"The site's fine," said the state geologist.

Everyone else had the same answer, except Kuiper.

"Well," said Kuiper, "I might have agreed with you until ten minutes ago, when I saw the schematic my staff prepared for me. Maybe you should have a look at it, too."

Sherman looked pained. "What are you saying, Kupe?" he demanded.

"I'm saying that looking at that schematic gave me some serious reservations about the Narrows site," Kuiper said. "From the looks of it there could be major leakage right under the dam. If it were a nonfederal project, I'd never approve it."

Sherman, watching Saunders and Don Christenson, whom the lawyer had brought with him, cackling silently behind closed lips, was incensed. "On what basis do you say that? Why do you say that?"

Kuiper then laid out what the schematic had told him. Sherman acted as if he hadn't heard a word of it. "I don't care about your schematic," he finally interrupted. "I want to see a lengthy memo on all of this. You've made some very serious charges in the presence of two people who will obviously use them against this dam. You had better be right."

Kuiper stood up to his full six feet six and glowered at Sherman, who was at least twenty-five years younger. "Young man, you'll get your lengthy memorandum," he growled. "But don't you tell me what I'd 'better' be." Then he stalked out of the room.

Kuiper had hardly finished his memorandum later that day when he received calls from both Saunders and the *Rocky Mountain News*, which had obviously been put onto the story by Saunders, asking whether they could have a copy. The *News* reporter also wanted to take a look through his Narrows file. As a public servant, Kuiper had no other choice than to keep his files open, except on matters involving national security. He was also legally obligated to make public any document he wrote, including the Sherman memo. He invited both Saunders and the reporter to come over. The reporter from the *News* was just taking the file to an empty desk when Sherman stalked into Kuiper's office.

"What is *he* doing here?" Sherman demanded, pointing at the reporter.

Kuiper said he had given him permission to look through the file.

Sherman was aghast. "*I* haven't even had a chance to look at it," he protested.

"Well, he asked first," said Kuiper. Sherman looked as if he were ready to throw a punch. He walked over to the reporter and grabbed the sheaf of files. "I'm looking through these first," he said, plopping the stack on an empty desk as the reporter stood by dumbfounded.

In Kuiper, Sherman had a messenger whom he couldn't kill, and when he tried he seemed only to wound himself. After the incident with the reporter, the state attorney general removed Kuiper's Narrows file for safekeeping because of the lawsuit pending over the issue. Kuiper insists he did not ask him to do it, but Sherman evidently thought he had; the whole thing reflected badly on him, because it looked as if the attorney general thought someone might pilfer materials from the file, and the person who would have seemed to have the best motive—the person most ardently in favor of Narrows—was Sherman himself. Sherman was enraged. He immediately wrote Kuiper a long memorandum impugning, implicitly or explicitly, his integrity, his motives, his sense of judgment, and even his competence as an engineer. Because he was about to leave town, Sherman dictated the memo and asked his assistant, Jerry Sjaagstad, to sign it. After reading the memo, Kuiper sat down and wrote a blistering one of his own, which he walked downstairs and threw on Sjaagstad's desk. Ten minutes later, Sjaagstad rushed into his offfice and demanded that he retract what he had said. Kuiper refused. When Sherman returned and heard what had happened, *he* came storming into Kuiper's office.

"You are being insubordinate," he yelled at Kuiper. "I'm going to take disciplinary action against you. You are going to regret this."

Kuiper stood up and went chest to nose with Sherman, who was a full head shorter. "I'm civil service," he thundered. "You *can't* discipline me without cause. But I hope you try. I'll blow you right out of the water, young man."

It seemed that *nothing* could change Dick Lamm's and Harris Sherman's minds about Narrows: not the plight of the Weldon Valley; not the state engineer's misgivings about the safety of the damsite; not the Teton disaster; not even the fact—which became an issue again after Kuiper's skepticism was reported in the press—that there was an alternative to the Narrows site. It was an alternative that appeared to be safer, that would inundate a cow feedlot instead of homes, churches, and graves, and that made as little or as much economic sense as the Narrows Project.

Twenty-five miles upriver toward Greeley, the Hardin site had been under consideration for years as an alternative to Narrows. It was not authorized by Pick-Sloan mainly because it would have cost slightly more to build. In other, highly important respects, however, it was the superior site. The main "improvement" within the taking area was the Joseph Monfort feedlot, the largest cattle-feeding operation in the world. Qualifying "improvement" is especially advised here, because the Monfort feedlot—100,000 cows on a couple of thousand acres—was an insult to all five senses. Its downwind neighbors found themselves wishing wistfully that they could replace it with a paper mill. One of the largest sources of nonpoint pollution in the country, the feedlot would sooner or later run into the Clean Water Act, and might be shut down for good. Rumor had it that Joe Monfort would be happy to have someone pay him to take it off his hands.

But the Hardin site, if it was substituted for Narrows, would have to be authorized all over again. At its authorization hearings, it would run into cost-conscious members of Congress and the environmental movement, which hadn't existed when Narrows was first authorized.

Worse still—far worse—was the fact that it would have to be justified with a discount, or interest, rate twice as high. Since Narrows was the cheaper site, and *it* could barely pass muster at a 3¼ percent discount rate, it was hard to see how a Hardin dam could ever be authorized.

Now that the Hardin site had reemerged as an alternative, however, it could only be viewed as a threat. The Bureau of Reclamation, therefore, decided that there was only one course open to it. It had to break ground on the Narrows project quickly, and the first step was to move the people out of the way.

The history of "relocation"—removing people in the way of a project from their land and compensating them for what they lost—started early in the century with the Los Angeles Department of Water and Power, and was embellished a short while later by the New York City Water Department when it drowned the Catskill valleys to create a new water supply. These were the first times in our history—except, of course, for the indignities visited on the Indians—when thousands of people were dispossessed for the crime of impeding progress. What the TVA did in the 1930s, what the Corps of Engineers did along the Missouri in the later 1940s, and what the Bureau tried to do in the Weldon Valley in the 1970s followed the same script. They sniffed through the community, smelling out its most avaricious members, those most susceptible to an offer. They spread rumors; they spread lies. They offered extravagant settlements to the first few who bit, then grew less and less spendthrift with the holdouts, both to punish them and to balance the initial extravagance. They played on the social conscience of communities, accusing them of selfishness, of denying the greatest good to the greatest number. And in the final resort—judiciously at first, then more threateningly, then like a defensive line blitzing a quarterback—they invoked the prospect of eminent domain.

They did all this without a sense of shame, because they told themselves they were serving an ultimate good—they were preventing floods, feeding the hungry world, offering power and light to schools and heat and air conditioning to hospitals. They denied—to themselves as to their would-be victims—that the real reason they were doing it was that they couldn't bear the thought of no longer building dams. And the very majesty of their great works made it easier for them to do it. It may be easier to sweep hundreds of people out of the way than ten or twelve, as if a project important enough to call for the removal of so many must be worth building.

———

George Kyncl, an employee of the Colorado Department of Social Services, who witnessed firsthand the trauma the Bureau's relocation effort was causing in the Weldon Valley, was always struck by its indifference to its victims' fate. "They were like Jekyll and Hyde," Kyncl says. "When you met them on the street or in meetings or the coffee shop in Fort Morgan they'd smile and joke with the same people they were trying to throw off their land. They were in here for so long they almost felt like members of the local community. You had to keep reminding yourself of the real reason they were here."

When the Bureau's men approached Ben Schatz, a South Dakota farmer whose land it wanted for the Oahe Diversion Project, they said, "To us you're just a dot on the map. When you get in the way, we move you."

When you get in the way, we move you. Don Christenson was sitting at home one day in 1976 when the phone rang. Don's wife, Karen, picked it up. It was a neighbor, someone the Christensons did not see regularly. Don could see from Karen's expression that it was something bad. Karen kept saying, "No, no, no. No, it's ridiculous. It's crazy." When she hung up, she looked at Don with a pained expression—half laughing, half anguished. "The talk they hear is that we've sold out."

Don and Karen Christenson cannot prove that it was the Bureau that spread the rumor. Could it have been Felix Sparks? Harris Sherman? "The frustrating thing was we didn't know where the rumor came from," said Karen. "It was so evil, so nasty a thing to do. You feel so helpless, but you feel so mad. When I heard that rumor I just wanted to scream." It might have been the Lower South Platte Conservancy District. The district was not above some rather sneaky tactics. In the full-color brochure it was still using in 1981 as a propaganda piece for Narrows, it showed Bijou Creek coming in *behind* the dam, not in front of it, even though that plan had been dead since 1965. Called on this point, Gary Friehauff, its young executive director, offered what seemed like a lame explanation: "We're still going through the old stock of brochures."

It was not the first rumor, and by no means would it be the last. At the Damfighters' Conference in Washington, Christenson had been warned by someone who had watched the Corps of Engineers in action in the Middle West that divisive rumors spread innocuously in neighboring towns would be a prime tactic when the Bureau began trying to buy the land in the reservoir area. Not long afterward, Don got a call from a neighbor who had just gone to Fort Morgan to get a haircut. The man in the next chair had been talking about all the people who, according to scuttlebutt he had heard, were thinking of selling out early. He represented himself as a real estate broker from out of town. No one had ever seen him before.

The Bureau knew exactly whom to go after. Sandy Desmond (a pseudonym) was, for a time, one of the leaders of the Regional Landowners group. Everyone liked Sandy—he was amiable, a teddy bear, a sort of irrepressibly cheerful Mr. Micawber. His weakness was also Micawber's—Sandy loved money, and he liked to make a fast buck. Nonetheless, people didn't really worry too much about Sandy. He was, after all, a leader of the opposition. To hear him rave against the Bureau was almost an embarrassment—small children had to be kept out of earshot. "We're going to kick their goddamned butts out of here in six months," he said in 1975, after the Bureau set up its first project office since the 1940s.

One Weldon Valley resident remembers how she found out about Sandy. "My husband walked in the house one day," she says. "I think it was late in the afternoon. I was sitting right here at the kitchen table. I could tell from the look on my husband's face that something was really wrong. All the things that it could be flashed through my mind in a second and I just lighted on Sandy. I said, 'Sandy went over.' And he said, 'Yup. Sandy went over. They made him an offer he couldn't refuse.'

"They poisoned the atmosphere in this community something dreadful," said the woman. "They went after the people they thought were more likely to sell, but they also spread lies about the leaders of the Regional Landowners organization. We just heard the rumors. We didn't know who was spreading them, we didn't know if they were true. When I heard that rumor about Don Christenson selling out, I thought, 'Well, that's the end.' They created such an atmosphere of distrust it took years before we got over it. I'm not sure we have completely yet."

Another weapon in the Narrows lobby's repertoire was the old strategy of feint and dodge. "Every time we read one of their new reports the figures were different," says Don Christenson. "In one document they said they were going to have 100,000 acre-feet available from the reservoir each year. They they said 120,000 acre-feet. At one point they were up to 150,000. They never gave an explanation. They just changed the numbers on us all the time, so we had to get out the old calculator and prove them wrong again. All the while I'm trying to raise a thousand acres of corn and worry about a few hundred head of cattle. It was no picnic, I'll tell you. One thing about the Bureau, though," Christenson added grimly: "They sure know how to make a person mad."

Meanwhile, as the Bureau was doing battle with the Weldon Valley (or "poverty valley," as Gary Friehauff of the Lower South Platte organization described it to me) on the one hand and with the newly elected Carter administration on the other—one of Carter's first actions was to put Narrows on his initial water-projects hit list—the state engineer, C. J. Kuiper, thought he had discovered yet another fatal flaw in the scheme. It was one of those details that dwell in a special kind of obscurity reserved for the perfectly obvious. What if the water couldn't possibly get to where it was intended to go?

"At first I never thought much about the channel losses," the state engineer would remember later on. "But one of the biggest headaches of my job had always been getting water down to the senior irrigators along the South Platte. All the groundwater pumpers who came along during the fifties and sixties and seventies had been depauperizing the aquifer on both sides of the river. Some guy would call on fifty second-feet that were his rights and my river master would cut off the junior diverters and the pumpers upstream so he could get it. Nothing would arrive. He'd call on another fifty cfs and we'd send it to him and it still wouldn't arrive. I said to myself, 'What the hell's going on here?' Then we figured out that it was all being captured by the aquifers. The pumpers had emptied those aquifers so bad that *they* were acting like pumps. The water we sent down went right through the bottom of the Platte and migrated laterally and went into the aquifer. It was like it had a great big hole in it.

"So I went to the Bureau and told them their water was going to disappear on the way down from Narrows to the South Platte Conservancy District, and they said, 'Hogwash!' I said, 'Hogwash? We're cutting junior diverters up and down the river by four hundred cfs so the seniors at Julesburg can get twenty cfs and they're still not getting a goddamned drop!' The Bureau was saying that if they released a hundred thousand acre-feet out of Narrows Reservoir, maybe ninety thousand acre-feet would arrive at the headgates of the guys they'd contracted to sell water to. Well, they were full of baloney. They'd be lucky to get twenty-five thousand acre-feet.

"I kept telling the Bureau and Felix Sparks and Harris that I didn't care if they built their dam or not," Kuiper said. "But I'm the one who has to see to it that every irrigator gets the water he's entitled to. Well, if the Bureau promises them ninety thousand acre-feet, and I can only deliver twenty thousand, I'm the one who gets blamed. I can't give them ninety thousand unless I cut off others who have senior rights, and that's illegal. The Bureau was making a bunch of outlandish promises and I was the one who was supposed to keep them."

The Bureau, Dick Lamm, Felix Sparks, and Harris Sherman naturally refused to believe a word Kuiper said. Even so, his reputation was good enough, and his statements were colorful enough, that the newspapers listened to him, and soon the issue of channel losses—of the Bureau planning to build a $226 million project that might not be able to deliver water—was all over the local press. Sensing yet another impasse, the Bureau decided that it had better get someone else's opinion. "Someone else" turned out to be Woodward-Clyde.

A huge engineering firm of considerable reputation, Woodward-Clyde has enjoyed a comfortable relationship of long standing with the Bureau. The Bureau often relies on Woodward-Clyde to perform independent assessments of its plans; then it often rewards it with lucrative construction contracts. It was no surprise, therefore, when Woodeard-Clyde's estimate of channel losses in the Platte coincided nicely with the Bureau's. Kuiper, however, continued to insist that both of them were wrong. "Their whole calculation was based on average *annual* channel losses," he said. "Well, they may be right on an annualized basis, but an annualized basis doesn't mean a damn thing. Most of those channel losses occur in the summertime, when the Platte Valley aquifer has been pumped out. That's when it acts like it has a hole in it and the water going down the river just disappears into it. But summertime is when the Narrows customers are going to need their water. That's the irrigation season. They're going to call on it and it won't get there."

The Woodward-Clyde study was interesting in other respects, for it went on to examine the other questions that were being raised about the Narrows. It concluded that "a safe dam can be built at the Narrows site," but, as Kuiper pointedly noted, declined to say at what cost. In stark contrast to the conclusions of the U.S. Fish and Wildlife Service, it said that the project "would have no adverse effect on sandhill cranes." And even though, fifty miles away, the Badger and Beaver irrigation districts were pioneering an alternative to on-stream and off-stream reservoirs, groundwater storage, it concluded that groundwater storage was "not an economically feasible" alternative to the Narrows.

Nothing was more striking, however, than its conclusion that the Bureau's claimed benefits for recreation "remain valid"—even though "primary production biomass in the reservoir will exceed levels that are usually indicative of eutrophy." In that remarkable juxtaposition of irreconcilable conclusions, Woodward-Clyde was tacitly agreeing with the Environmental Protection Agency, which was convinced that Narrows Reservoir, touted by the Bureau as a fine new recreational "lake," would quickly turn into a fetid, grossly polluted agricultural sump. Betwen the partially treated sewage of nearly two million people, the untreated runoff of hundreds of thousands of cows (some of them defecating right in the river), and pesticides and fertilizers washing in from thousand and thousands of acres of intensively farmed land—between this, and the fact that the reservoir would be shallow and warm, with an evaporation rate of four feet a year, the water quality was going to be absolutely awful. The EPA was suggesting that it would not be fit for *contact*, which meant no swimming or water skiing without a waterproof covering over every inch of one's body. Woodward-Clyde seemed to agree—it said biomass would *exceed* the levels that produce eutrophication, which is when a body of water begins turning into an algae pool. But it also agreed, implicitly if not explicitly, with the Bureau and Dick Lamm when they said that this aspiring swamp was destined to become the most popular reservoir in the entire state.

If one believed the brochure of the Lower South Platte Conservancy District, as many people would be drawn to fetid, shallow, bath-warm Lake Narrows as are drawn to Yellowstone National Park.

Here was a dam that the state engineer said would deliver only a third of the water it promised and could conceivably collapse; a project whose official cost estimate—if what two officials of the Union Pacific had privately suggested was correct—would barely suffice to relocate twenty-six miles of railroad track; a project whose real cost, whatever it turned out to be, would therefore be written off, in substantial measure, to "recreation," though the water would be unsafe to touch; a project whose prevailing interest rate (crucial to justifying the whole scheme) was one-fifth the rates banks were charging in the late 1970s; a project many of whose beneficiaries owned more land than the law permitted in order to receive subsidized water (even *after* the acreage limit was stretched to 960 acres in 1982); a project that might, if the state engineer was correct, seep enough water to turn the town of Fort Morgan into a marsh; a project that would pile more debt onto the Bureau's Missouri Basin Account; a project that would generate not a single kilowatt of hydroelectric power and would be all but worthless for flood control.

And yet, on top of all this, there was to be still another development, one that ought to have finished off the Narrows Project once and for all: *Most of the farmers who were supposed to be the beneficiaries said they didn't want the water.*

The farmers, it turned out, were not as ingenuous as the Bureau wished them to be. During the years it had pushed for the Narrows Project, the Bureau had never quoted them a firm price for the water nor guaranteed them a fixed amount. Why, in that case, should they obligate themselves to buy it for the next forty years?

In a letter to the Denver *Post*, Jacob Korman, the president of the Irrigationists' Association, Water District Number One—the preexisting water district over which the Lower South Platte Conservancy District was trying to impose itself as a superagency—explained the farmers' position. "There are fifteen irrigation districts in our association from Kersey near Greeley to Balzac below Brush," Korman wrote. "If built, the dam would be in the heart of our district. These ditch companies provide irrigation water to 125,000 acres of land. Twelve of these ditch companies representing farmers irrigating 105,350 acres have taken positions *opposing* the Narrows. One, representing 1,100 acres, has indicated support. At the last report, two, involving 19,100 acres, have taken no position."

Although many of the ditch companies in his district were initially enthusiastic about the project, Korman said, "those which have been offered specific contracts by the Lower South Platte Conservation District have found these contracts to be unacceptable." The main reason was that the contracts demanded that the farmers pay for water released *at the dam*. The Bureau refused to guarantee delivery of the water at the farmers' headgates—as if to demonstrate that it knew all along that Kuiper's theory about channel losses was correct. The farmers might be a trusting lot, but they weren't dumb.

"The office of our state engineer," Korman concluded, "is the only office . . . which has both the data and the technical staff to make a professional assessment as to what the real impact of the Narrows would be on providing water for irrigated agriculture in northeastern Colorado . . . . Actually, most of the ditch companies in the association feel . . . that the unadjudicated water which we need to supplement our reservoirs and decreed waters *would in all probability be lost if Narrows were built*" (emphasis added).

That a majority of the farmers for whose benefit Narrows was to be constructed finally decided they would *lose* more water than they would gain from the dam was fascinating. The "unadjudicated" water of which Korman spoke were those high flows which, after every Colorado farmer had taken his water right and Nebraska had been guaranteed its share, the farmers could skim off for themselves. As of now, no one "owned" these occasional surpluses in the river; anyone could divert them for storage in offstream reservoirs or in the aquifer beneath his farm. But with Narrows Dam in place, all but the most extraordinary high flows—the fifty-year floods—would be captured and they would belong to the Bureau of Reclamation. The Bureau would charge money for them—charge even if it refused to guarantee that the water would ever arrive. The Bureau wasn't even offering the farmers a pig in a poke; it was offering them a poke without a pig.

In 1982, at the behest of Senator Gary Hart, Woodward-Clyde did yet another study of the Narrows problem—it was now a "problem" as often as it was a "project"—and reversed its earlier conclusions as innocently as if they had never been held. Its estimate of water available for annual delivery was now down from more than eighty thousand acre-feet to thirty-four thousand acre-feet, almost in line with Kuiper's estimate and fathoms below the Bureau's. The effects on the sandhill and whooping cranes and other migratory wildlife downstream were now regarded as "moderately negative" instead of insignificant. But the most startling reversal came when the firm recalculated the worthiness of the project in simple economic terms. Using an interest rate of 7½ percent, but retaining the doubtful flood-control benefit of $800,000 a year and a highly optimistic view of the recreational potential, Woodward-Clyde came up with a benefit-cost ratio of only .10 to 1.0—for every dollar invested, *ten cents* would be returned. Even with an interest rate of 3¾ percent, Narrows was a loser.

Like most water projects, though, Narrows refused to roll over and die. In 1983, Congress, at the urging of local Representative Hank Brown, voted it another $475,000 appropriation. It wasn't enough to build anything, but it was enough to keep it alive. The latest unofficial cost estimates, in 1984, were in the neighborhood of $500 million. If they are correct, each acre-foot (assuming 34,000 acre-feet is the annual yield) will cost $14,500 to develop. Few Colorado farmers can afford to pay more than $50 an acre-foot for water, and the Bureau has never charged any of its client farmers half that much (most get it for $7.50 or less). The taxpayers, presumably, will make up the difference, buying a couple of hundred farmers about the most expensive water on earth.

And yet, in early 1984, the politicians who had always been for the Narrows were still for it. Senator Gary Hart, a neoliberal, was for it; liberal Congressman Tim Wirth was not against it; Senator Bill Armstrong, a budget-conscious conservative Republican, supported it. But no one was for it as much as Dick Lamm—although Dick Lamm was the one politician honest enough to admit, discreetly, that it wasn't worth building. Once, at a Denver Broncos football game, Karen Christenson's sister and her husband found themselves sitting a few seats away from the governor, sporting their big, bright "Stop the Narrows" buttons. Lamm noticed the buttons, came over, and asked who they were. Then, in an odd small burst of candor, the intense young forward-thinking governor delivered himself of a private opinion about the project he had championed so relentlessly. "I know Narrows isn't the best project in the world. I'd much rather use the money to build up the state's economy in a more efficient way. But when Washington offers you that kind of economic impetus, a governor can't just turn it down."

Repeating the story, Don Christenson mused, "If that's the way they run a railroad, then this country hasn't got any hope."

Meanwhile, in Denver, Clarence Kuiper had taken early retirement. "The Narrows thing got so annoying to me I couldn't stand it, so I retired," Kuiper says. "I've lived too long to put up with that sort of nonsense." Early in 1984, he was no less convinced than ever that Narrows, if built, stood a respectable chance of collapsing like Teton—an issue that had become all but lost in the minutiae of the debate. "Unless they extend that grout curtain a hell of a lot farther than they plan to, they're going to get seepage, just like they did at Teton. Seepage is one of the worst things that can happen to an earthfill dam. I'd rather have water going over the top in a waterfall than chewing away at my abutments. That's still the number-one issue as far as I am concerned."

Neither the Lamm administration nor the Bureau was ready to listen to Kuiper. Lamm, however, had finally found a way to get even. The firm with which Kuiper now serves as a consultant, the Harza Engineering Company of Chicago, was the other contender, with Woodward-Clyde, for the lucrative South Platte Basin Alternatives study in 1982. Because of Kuiper's relationship with Harza, it didn't get it. Suave Bill McDonald, who relieved the intemperate Felix Sparks of his command of the Colorado Water Conservation Board, put it right into his letter. Unless Harza dumped Kuiper as a consultant, it stood no chance of getting the contract. As far as Kuiper is concerned, he is being blackballed throughout the state. "They've stolen a man's livelihood," he says. "My pension isn't enough to live on. I know this state as well as anyone, but they've made my name mud."

If one were to put an epitaph on this story, one might do no better than to quote Glenn Saunders, the man who championed water development for fifty years in Colorado and then, in the end, came up against a project he wanted to kill—and couldn't. As he readily admits, it changed his whole way of looking at things.

To Glenn Saunders, Narrows Dam was not so much a dam as a symbol of a senescent society seeking refuge in the past. "What that dam represents," he said, "is, first of all, the fact that there are very few honest people in the world. Ninety-eight percent of humanity cannot admit when it's made a mistake. This applies especially to politicians. A politician for some reason thinks it is political suicide to admit that he was wrong. Dick Lamm cannot bring himself to admit that he has been in error about Narrows. He has one of the finest minds in Colorado, his thinking on some subjects is some of the best thinking any politician in this age is capable of—but he cannot bring himself to say, 'I was wrong on the Narrows Dam.'

"The Bureau is the same way," Saunders went on. "It cannot admit when it has made a mistake. It has also run out of good projects. *And* on top of that it has all of these bizarre cash-register funds—the Missouri Basin Fund, which is behind the Narrows—that are supposed to make these projects self-financing. They do not, but no one understands that. The Bureau is like one of these crooks with money earning interest in twenty different banks—it has to spend the money on something. It is all *borrowed* money—it belongs to the people of the United States—but the people of the United States don't know that. The whole thing is a machine, a perpetual-motion machine that keeps churning out dams, which the politicians and most westerners are reflexively in favor of, and the whole business is running the country into the ground.

"The people who support these boondoggle projects are always talking about the vision and principles that made this country great. 'Our forefathers would have built these projects!' they say. 'They had vision!' That's pure nonsense. It wasn't the vision and principles of our forefathers that made this country great. It was the huge unused bonanza they found here. One wave of immigrants after another could occupy new land, new land, new land. There was topsoil, water—there was gold, silver, and iron ore lying right on top of the earth. We picked our way through a ripe orchard and made it bare. The new generations are going to go down, down, down. With projects like the Narrows, we're trying to pretend that things are as they always were. 'Let's just go out and find some money and build a dam and we'll all be richer and better off.' We've been so busy spending money and reaping the fruits that we're blind to the fact that there are no more fruits. By trying to make things better, we're making them worse and worse."

CHAPTER TWELVE

Things Fall Apart

O n a hydrographic map, the outline of the Ogallala Aquifer resembles the South American and African continents—broad and bulbous to the north, tapering to a narrow cape at the southern end. Driving its entire length—from southern South Dakota down into the heart of West Texas, where it feathers out just above the Pecos River—takes two long days and feels almost like a transcontinental trip, the more so because the landscape is relentlessly the same: the same flatness, the same treelessness, the same curveless thirty-mile stretches of road. All that changes is the crops: sorghum, then corn, then sorghum, then corn, then alfalfa, wheat, cotton—enough cotton, one would think, to clothe all humanity.

This was the country that Coronado traversed, looking for the gold cities of Cibola; it is the country that cost him half his men, his reputation, and nearly his life. In Coronado's time, it grew nothing but short grass, on which millions of buffalo feasted; feasting on them were grizzly bears, prairie wolves, vultures, and an unknown number of Sioux, Comanche, and Cheyenne. The tribes, widely regarded as ferocious, merely reflected the landscape itself. Even the Indians used the open plains mainly for seasonal hunting, retreating to river valleys when the weather became extreme—which was a good part of the time. The southern high plains, from Colorado south to the hill country of Texas, never knew a permanent civilization, as far as archaeologists can tell. There was a Llano culture as early as 10,000 B.C., followed by others that came and went like snow. Around 1300 B.C., Pueblo Indians occupied the region, but abandoned it less than a century later.

The Comanche, superb horsemen, may have shunned the open plains as much as possible because there was no tree where one could tie up one's mount. A place where one couldn't even secure a horse was no place to try to anchor a civilization.

White men were to learn that lesson, repeatedly, after the buffalo and the Indians were vanquished and gone. During the 1860s and 1870s the plains hosted great cattle drives from Texas to Kansas, but those ended in drought, overgrazing, and falling meat prices. Depauperized of much of its grass and invaded by mesquite and weeds, the region emptied out. But a decade of wet weather and demand for bread during and after the First World War sparked a repopulation, and the plains became a sea of wheat. Then came the Dust Bowl. After each calamity, a residual population managed to remain, surviving on a few cattle, some defiant wheat, the government, and finally, oil and natural gas. It was one of those survivors who sank a water well, hooked up a new invention—a diesel-driven centrifugal pump—and discovered the region's bounteous secret: underneath it, confined in a closed-basin aquifer, was enough fresh water to fill Lake Huron.

Everyone had always known there was water below. If you sank a well and erected a windmill-driven pump, you got enough for a family and a few head of stock. But windmills could bring up only a few gallons a minute and offered no clue as to how much water was actually down there. The centrifugal pump, which could raise eight hundred gallons a minute or more, did, and when geologists took a closer look they confirmed the evidence offered by the pumps. Under the plains was the trapped runoff of several Ice Ages, all nicely confined within gravel beds. The thickness of the aquifer varied; along the periphery it feathered to a few feet, but in the middle portions under Nebraska there were saturations of seven hundred feet. All in all, there were probably three billion acre-feet in confinement.

A flow of eight hundred gallons a minute will fill an Olympic swimming pool in just over an hour. It will also conveniently irrigate a hundred or more acres of crops. A hundred acres of irrigated land on the plains is worth five hundred acres unirrigated; actually, it is worth more, because a farmer need never again worry about going bankrupt during a drought. The water was free; all you needed in order to make money, real money—to watch your net income rise from $8,000 to $40,000 a year—was cheap fossil fuel or electricity, a big mobile sprinkler, and pumps.

The irrigation of the Ogallala region, which has occurred almost entirely since the Second World War, is, from a satellite's point of view, one of the most profound changes visited by man on North

America; only urbanization, deforestation, and the damming of rivers surpass it. In the space of twenty years, the high plains turned from brown to green, as if a tropical rainbelt had suddenly installed itself between the Rockies and the hundredth meridian. From an airplane, much of semiarid West Texas now appears as lush as Virginia. Where one saw virtually nothing out the window forty years ago, one now sees thousands and thousands of green circles. From thirty-eight thousand feet, each appears to be about the size of a nickel, though it is actually 133 acres—a dozen and a half baseball fields. The circles are created by self-propelled sprinklers referred to by some as "wheels of fortune." A quarter-mile-long pipeline with high-pressure nozzles, mounted on giant wheels which allow the whole apparatus to pass easily over a field of corn, a wheel of fortune is man-made rain; the machines even climb modest slopes which would ordinarily defeat a ditch irrigation system. Wheels of fortune are superefficient, but intolerant: they don't like trees, shrubs, or bogs. Therefore, the millions and millions of shelterbelt trees planted by the Civilian Conservation Corps have come down as fast as the region's fortunes have risen. All that now holds the soil in place is crops and water which cannot last.

In 1914, there were 139 irrigation wells in all of West Texas. In 1937, there were 1,166. In 1954, there were 27,983. In 1971, there were 66,144. Nebraska irrigated fewer than a million acres in 1959. In 1977, it irrigated nearly seven million acres; the difference was almost entirely pumping from the Ogallala. By that year, there were, depending on whose estimate one believed, somewhere around twelve million acres irrigated by the Ogallala Aquifer. One of the poorest farming regions in the United States had metamorphosed overnight into one of the wealthiest, raising 40 percent of the fresh beef cattle in America and growing a huge chunk of our agricultural exports. As West Texas sprouted corn, a water-demanding crop that had never been known there, Lubbock and Amarillo sprouted skyscrapers, most of them erected by the banks that ecstatically financed the farmer's road to wealth. On Fridays, the farmers cruise into town from eighty miles away, behind the wheels of their Cadillacs and big Buick Electras. After a conference with a deferential banker, they go off for drinks and a dinner of steak and lobster, then to watch a Texas Tech football game from fieldside seats. Since 1950, Lubbock's population has increased at about the same rate as Texas's irrigated land 7.5 percent a year. Anything growing at that rate doubles in size in a decade.

There is, however, a second set of statistics which offers a more meaningful depiction of what is going on. By 1975, Texas was withdrawing some eleven billion gallons of groundwater—per *day*. In Kansas, the figure was five billion; in Nebraska, 5.9 billion; in Colorado, 2.7 billion; in Oklahoma, 1.4 billion; in New Mexico, 1.6 billion. In places, farmers were withdrawing four to six feet of water a year, while nature was putting back half an inch. The overdraft from the Ogallala region in 1975 was about fourteen million acre-feet a year, the flow of the Colorado River; it represented half the groundwater overdraft in the entire United States. The Colorado is not a big river, but it would be big enough to empty Lake Huron in a reasonably short time.

The Ogallala region supports not so much a farming industry as a mining industry. If the pumping has been reckless, as some believe, it is an example of carefully planned recklessness, for all the states regulate the pumping of groundwater; their choice was to allow its exhaustion within roughly thirty to a hundred years after the pumping began in earnest back in the early 1960s. Except for petroleum and natural gas and coal, most mining industries affect a rather small area. This is one that affects an area larger than California. Actually, it affects the entire world, for the product of mining the Ogallala is a prodigious amount of food, much of it consumed overseas.

It is a dead certainty that the Ogallala will begin to give out relatively soon; the only question is when. Everything hinges on one constant—the weight of water—and two variables: the cost of energy and the price of food. As anyone knows who has ever carried a full pail up five flights of steps, water is one of the heaviest substances on earth; pumping it a hundred or two hundred feet out of the ground consumes a lot of energy. The Ogallala farmers do not benefit, as do many groundwater pumpers throughout the West, from hydroelectricity generated at Bureau of Reclamation dams and sold to them at discount rates. For the water table to drop fifteen or twenty feet during a period when the price of energy increases sevenfold is a catastrophe. This, however, is precisely what happened throughout much of Kansas, Oklahoma, and West Texas between 1972 and 1984. During the same period, the price of most farm commodities barely doubled, if that.

The odds are high, therefore, that long before all the water runs out, the farmers will no longer be able to afford to pump. In 1969, a study performed by Texas A and M University projected that the West Texas aquifer would decline to forty-four million acre-feet by the year 2015, down from 341 million acre-feet before the Second World War. Irrigated acreage would, by then, have fallen to 125,000 acres from a mid-sixties peak of 3.5 million acres. Sorghum yields would be down 90 percent; cotton would be down 65 percent; total agricultural value in the region would diminish by 80 percent. In those figures lay the

makings of a Dust Bowl-sized exodus, a social calamity, and a huge
rash of bankruptcies that could ripple through the nation's economy.
In 2015, the study predicted, there would be 300,000 fewer people in
the region than there were in 1969. A new set of figures compiled in
1979 by the Texas Department of Water Resources was somewhat more
optimistic, but the planning director of that same agency did not sound
as if he subscribed to the optimism himself. "It's pretty easy to conjure
up a disatrous series of events," said Herbert Grubb in 1980. "We're
sort of assuming that a lot of the farmers will stay in business raising
dryland cotton or wheat. But with interest rates high as they are, and
drylands yields down 70 or 80 percent from irrigation yields, I really
don't see how the farmers are going to carry their debts. The older
ones, maybe. But the younger ones, the newer ones, are up to their
ears in debt. So you could just as easily assume that millions of acres
suddenly go fallow. Then along comes a drought, some eighty-mile-
an-hour winds, and you've got another Dust Bowl. The shelterbellt
trees are gone. A lot of those farmers are milking every cent out of the
land while the water lasts. The conditions are ripe for something down-
right catastrophic."

The decision of the Ogallala states to treat the aquifer as if it were
a coal mine, thereby setting themselves up for a long, long fall, is
ironic in an extreme sense. Their economies—as the states recognized
and lamented long ago—are vulnerable to forces they can do little to
control. What supports Colorado besides irrigated agriculture? Mainly
minerals—coal, uranium, molybdenum, oil—and tourism, logging,
and ranching. Every one of those industries is subject to someone else's
whim: world supply and demand, international cartels, the price of
oil, the Federal Reserve Board, or, the ultimate caprice of all, nature.
Much the same applies to New Mexico. Oklahoma, Nebraska, and
Kansas are farm states whose prosperity or ruin, before irrigation,
depended on whether the isohyet of twenty-inch rainfall moved westward town the Rockies or eastward toward the Mississippi River. Once
they became dependent on a huge irrigation economy, all of the states
knew they would be in the same position as a junkie. Texas with its
vanishing oil and gas; Kansas astride the hundredth meridian; Colorado depending so much on tourists who, if oil prices doubled or there
was no snow, would stay home—each state knew that when the water
ran out, they would again face the same awful vulnerability that had
haunted them since the first settlers arrived.

Strictly regulated, the Ogallala could have been made to last
hundreds of years instead of decades. Irrigation farming could have
been slowly phased in, kept at a lower level, and gradually phased

out. In such a case, hundreds of thousands of people who became dependent on it overnight wouldn't face ruin, and the states' economies wouldn't go into sudden osmotic shock when the pumps began bringing up air. The states had begged the government to build them dams for irrigation, and they had lobbied to keep the price of water artificially low, arguing that agriculture was the only stability they had. the opportunity for economic stability offered by the world's largest aquifer, however, was squandered for immediate gain. The only inference one can draw is that the states felt confident that when they ran out of water, the rest of the country would be willing to rescue them.

A s deputy chief of planning for the Bureau of Reclamation in the mid-1960s, Jim Casey saw things from a fundamentally different perspective than the farmers, or, for that matter, most of his eleven thousand colleagues. The main concern of the typical Bureau engineer is building a bigger and grander project than the last one. It was Casey's job to think about what few of them did, or dared to: reservoirs silting up, river-basin funds drying up, salts building up—all the problems consigned through some unwritten conspiracy between politicians and bureaucrats to an amorphous, distant, and politically unrewarding future. Casey was, by definition, the Bureau's Cassandra. Peering into the future, he saw no place headed for deeper trouble than the Ogallala region. If surface water can be compared with interest income, and non-renewable groundwater with capital, then much of the West was living mainly on interest income. California was milking interest and capital in about equal proportion. The plains states, however, were devouring capital as a gang of spendthrift heirs might squander a great capitalist's fortune. To Casey's amazement, few of the farmers seemed to realize it. "They thought the water would last until the Second Coming," he says. Frustrated by the farmers' blindness, he finally decided that he had better address the one group that might listen to him: the region's bankers.

"I think it was about 1966 when I went out to give my speech," Casey says. His voice, after twenty years in Washington, is still thickly gravied with West Texas drawl. "You wouldn't believe how many bankers there are in Lubbock. I said to them, 'Look, you're all riding high out here and it's a great thing and we all like to pretend that great things are going to last. But no aquifer can sustain this rate of pumping. I don't know when you're going to run out of water, but I'd bet you're going to run out at about the same time you start running out of oil and gas. If that gets too expenseive the farmers won't be able to pump, anyway. There goes your whole economy. This corner of the world is going to be an Appalachia without trees unless you get off your fannies and try to save it. I don't know if you *can* save it; I frankly don't know if it makes any sense for the nation to invest billions of dollars in a rescue project to keep a few million acres irrigated and a few hundred thousand people employed out here. I don't know where the water would come from, but you'd better start thinking about it now, because it will take forty years to get a rescue project this big authorized and built, and you haven't got a lot of time left."

The effect of Casey's speech was remarkable. "I gave them religion. A few months hadn't gone by before I heard they were setting up a big new lobby to fight for a rescue project. They called it Water, Incorporated." Ambitious, perhaps even incredible, as its goal was—a project to rescue even a modest portion of the irrigated plains would be a project more grandiose than any yet built—Water, Inc., had a number of things going for it. California's voters had just approved the most ambitious and expenseive public-works project ever attempted by a single state in order to save its own agricultural industry. Would *Texans* countenance that upstart state building something they lacked the nerve to attempt themselves? In the mid-1960s, the age of limits had not yet dawned; a high-plains rescue project was seen by many people as the next logical step in "orderly" water development, something that might even capture the fancy of the nation at large. The generation of politicians then running the country had been suckled and reared on public works. And an astonishing number of them came from Texas.

The President of the United States, for example. As Robert Caro demonstrated in the first volume of his biography of Lyndon Johnson, *The Path to Power*, Johnson owed his political career largely to the Marshall Ford Dam. Begun under an emergency appropriation during the Depression—begun, just like Grand Coulee and Garrison dams, before it was even authorized, and built on land the government didn't even own—the dam was to make a reputation and ultimately a huge fortune for a couple of struggling small-time contractors named Herman and George Brown. At the time, however, it was just a big Bureau of Reclamation dam on Texas's Colorado River a few miles from Austin, a project which had run through its emergency appropriation before it was half built. To anyone else this didn't matter much—no one doubted that the dam would be completed someday—but to the brothers Brown it was a calamity. They had invested every nickel they owned and scraped together all the collateral they could in order to purchase one and a half million dollars' worth of construction equipment they needed and didn't have. (Until then, most of the Browns' contracts had been for road-paving jobs; what they owned in construction equipment didn't amount to more than a few fresno scrapers.) If more funds were not approved immediately, they would go bankrupt. But everyone was crying for relief funds, and an unauthorized project with a serious land-title problem in a remote corner of Texas was at a distinct disadvantage among its competition. In desperation, Herman Brown, the fiercely archconservative entrepreneur, pleaded for help from the district's most important politician, a newly elected twenty-nine-year-old liberal New Deal Congressman whose name was Lyndon Baines Johnson. Using his connections among the White House inner circle and his absolutely shameless flattery of FDR, Johnson managed to get Herman and George Brown a formal authorization, a resolution of the land-title dispute, and another $5 million to finish the dam as their lenders were about to smash down their barricaded door. Profoundly grateful, the brothers Brown poured enough money into Johnson's subsequent campaigns to catapult him into the Senate at a tender age. Their company, Brown and Root, was to grow into one of the largest construction firms in the world, mining the government just as Johnson mined the profits of their work—a symbiotic relationship that not only transcended ideology but subverted it, as public works are wont to do. And it all began with a dam.

Johnson was not the only local politician who had climbed to political power up the wall of a dam. There was Robert Kerr of Oklahoma, one of the princes of the United States Senate until he died in 1963. Besides unabashedly using his Senate seat to make himself rich—he was a cofounder of the Kerr-McGee Corporation—Kerr helped authorize a number of very large reservoirs in his native state which kept Oklahoma's construction industry perpetually busy, not only building new dams, but rerouting major highways around the ever-larger reservoirs that constantly formed in their path. Perusing a map of eastern Oklahoma, one would think that Kerr's ultimate goal was to put the state under water.

Then there was Jim Wright, who began representing Fort Worth in 1954 and was to become Majority Leader of the House in the late 1970s, a position he used to defy his own party's President in his attempt to knock off a few billion dollars' worth of water projects—including Wright's own favorite, the Trinity River Project, which was to turn Dallas and Fort Worth, sitting four hundred miles from the ocean, into seaports. Wright's dedication to water projects struck some of his colleagues as fanatical. He took time out in the late 1960s to write a book called *The Coming Water Famine*, in which he said, "The crisis of our diminishing water resources is just as severe (if less obviously immediate) as any wartime crisis we have ever faced. Our survival is just as much at stake as it was at the time of Pearl Harbor, or the Argonne, or Gettysburg, or Saratoga. . . . Pure water, when and where you need it, is worth whatever it costs to get it there."

There was also Ray Roberts, who represented Sam Rayburn's old district, and whose interest in water projects would elevate him to chairman of the House Public Works Committee. There was George Mahon, the chairman of the House Appropriations Committee in the 1960s and one of the five most powerful men in Congress, who happened to represent the district around Lubbock. There was John Connally, the governor of Texas, a Johnson protégé whose enthusiasm for grandiose undertakings, big-game hunting, and gigantic limousines made him into an unselfconscious parody of ambitious, superaffluent Texas.

With such men in power during an era of no limits, anything seemed possible—even a project to rescue the southern half of the Ogallala region by rerouting a substantial portion of the Mississippi River.

The origins of the project went back to 1958, when a U.S. commission—chaired by George Brown of Brown and Root—was appointed to come up with a systematic plan for developing the river basins of the state. The proposal called for eighty-three storage reservoirs and some water-conveyance works to be built by the year 2010, all of which, the commission modestly suggested, could be completed for around $4 billion. The great omission in the plan, however, was an aqueduct to West Texas. The reason for that appears self-evident: West Texas sits at an elevation more than three thousand feet higher than East Texas, where most of the state's water is, and nearly four thousand feet higher than Louisiana or Arkansas, the two states with enough of a water surplus to suggest themselves as the ultimate source. Pumping enough water to rescue several million acres that far uphill, over a distance of a thousand miles or more, would require a fantastic amount of energy. The commission did not say this in those exact terms, but its omission of any proposal to rescue the Ogallala overdraft region spoke volumes.

There followed, however, one of those peculiar metamorphoses in which a plan, as it evolves, conforms less and less to the constraints of nature, economics, and thermodynamics and more and more to the stridency of certain constituents and the desires of certain elected officials. John Connally saw in an Ogallala-region rescue project an opportunity to become a pharaoh in a pinstripe suit. George Mahon, subjected to merciless lobbying by Water, Inc., enjoyed the power of the purse by virtue of his being chairman of the Appropriations Committee; it was unthinkable that he would give East Texas and South Texas dozens of dams if West Texas got nothing in return. As a result, Connally, as governor, pointedly disregarded the Brown Commission's report and decided to draw up a proposal of his own. Its title was to be the Texas Water Plan.

The idea was for several million acre-feet of water to be diverted from the Mississippi River below New Orleans—a point from where, presumably, Louisiana wouldn't mind its being taken—and moved across the marshlands and swamp forests of the state in an aqueduct built to the dimensions of an airplane hangar. A river approaching the Colorado in size, running in reverse, the water would climb up to Dallas and Fort Worth, which sit at an elevation of 750 feet, by way of a series of stairstep reservoirs. A generous portion would head toward those two cities in a spur aqueduct; some of it would to to South Texas; but most of it would head toward Amarillo and Lubbock in the Trans-Texas Canal. There would be seventeen pumping stations en route lifting the water up the imperceptible slope of the plains; there would be nine terminal reservoirs waiting to receive it. Nearly a million acre-feet a year would be fed into the Pecos River; another half million would head toward Corpus Christi; 6,480,000 acre-feet would arrive on the Texas high plains, having climbed thirty-six hundred feet and traveled twelve hundred miles since New Orleans; 1.5 million acre-feet would perhaps go on to New Mexico. Two million acre-feet, the consumption of New York City and then some, would evaporate en route. It would take 6.9 million kilowatts of electricity to run it— about 40 percent of the electricity consumption of the entire state.

As a politician from a neighboring state put it after hearing the plan, "If those Texans can suck as hard as they can blow, they'll probably build it."

Without knowing anything but the vaguest outlines of the plan— without knowing whether the farmers could afford the water, whether its acid character was compatible with the plains' alkaline soils, whether Texas water law didn't exempt the farms from paying a dime once the water had percolated to the aquifer, whether the powerplants to move it could be financed and built, whether Louisiana had any intention of parting with one molecule of it—the voters of Texas suddenly found themselves, in August of 1969, being asked to appropriate $3.5 billion toward the Texas Water Plan's construction. Actually, the question was couched much more circumspectly than that. The proponents of the measure, which became known as Amendment Two, insisted that the voters were merely being asked to guarantee $3.5 billion in bonds to establish a "repayable loan fund" which any city or region in the state could tap in order to meet its water needs—an argument which was greeted by the referendum's opponents with catcalls. The fact was, they said, that the Texas Water Development Board, which could arbitrarily and peremptorily decide who got how much of the money, was deeply committted to a rescue project for West Texas. Governor Preston Smith, who was trumping Amendment Two up and down the state, was a native of West Texas. If the hidden agenda wasn't to build, or at least begin (since the $3.5 billion would never complete a project of such magnitude), the rescue project, why had the referendum been scheduled for August in an off-year election, when voter turnout was certain to be light, and organized elements behind the measure could affect the outcome much more dramatically than during a regular election year? The one place where turnout was likely to be heavy was in West Texas, because the farmers would be at home, busy with their crops, while a lot of East Texans would be off on vacation, escaping the humid heat. Why were the backers trying to distance themselves from the Texas Water Plan when that was the only plan that could absorb such a stupendous amount of money? This was, after all, 1969; in 1987, its equivalent would be more than $11 billion.

"We were being sold a bill of goods," recalls Ronnie Dugger, the publisher of the *Texas Observer*, virtually the only newspaper in the state that opposed Amendment Two. "It was actually $7 billion, not $3.5 billion, when you factored in the interest. Seven billion for what? No one was saying. No one knew. It was the biggest blank check in the history of the United States."

All such objections notwithstanding, the proponents of the measure had managed to amass as formidable a group of sponsors as Texans were ever likely to see. The backers included nearly everyone who was anyone in the state. Three former governors—John Connally, Allan Shivers, and Price Daniel—served as cochairmen. The editors or publishers of the San Antonio *Light*, the Austin *American-Stateman*, the Houston *Chronicle*, the Dallas *Times Herald*, the Fort Worth *Star-Telegram*, the Wichita Falls *Times-Record-News*, the Lubbock *Avalanche-Journal*, the Corpus Christi *Caller-Times*, the Beaumont *Enterprise-Journal*, the Port Arthur *News*, the El Paso *Times*, and the San Angelo *Standard-Times* were on it, not to mention dozens of smaller papers like the Bonhom *Favorite* and the Waxahachie *Times*.

The mayors of Midland, Dallas, Bay City, Corpus Christi, Austin, San Antonio, Laredo, Dallas, Lubbock, Fort Worth, and Arlington were on it. Presidents, chancellors, and regents of Texas universities were represented: Baylor, Texas Tech, the University of Texas, Texas A and M, Southern Methodist University. A hundred and forty-three of the 150 members of the Texas House of Representatives were on it. Twenty-eight of the thirty-one members of the Texas Senate were on it. The head of the Texas Parks and Wildlife Commission; lobbyists for railroads and manufacturers and municipalities; grocery-store magnates; retired Congressmen; Texas kingmakers such as Robert Strauss (later the head of the Democratic National Committee) and Leon Jaworski (later the Watergate special prosecutor)—the list read more like the sponsors of the United Way than a plan that appeared likely to end up dumping the Mississippi River on the expiring plains.

As an accident of geography made the rescue of West Texas so difficult and expensive, however, an accident of migration made passage of the referendum at least as difficult. In California, the conservative and reactionary factions of the state's electorate are concentrated mainly in the sprawl south of Los Angeles, in San Diego, and in the hard-bitten little cities of the San Joaquin Valley. Every one of those places is a desert or semidesert, haunted by extinction, and every one of them saw the State Water Project as salvation. An unknown number of people whose antipathy to government runs to things such as fluoridated water and Social Security voted enthusiastically for the most expensive public-works project in California's history. In Texas, the ultraconservative faction of the electorate tends to be spread more around the state. If it has a center, it is probably Houston, which stood to gain virtually nothing from the Texas Water Plan. The mayor of Houston, in fact, was conspicuously absent among those big-city mayors who had enlisted as members of the Committee of 500. Dallas, another conservative bastion, was to get water, but felt no sense of desperate need. Aside from that, Texas's population as a whole is skewed to the east, where the main problems with water tend to come in the form of thunderstorms, hurricanes, and floods. In California, two-thirds of the population resides in the drought-ridden south. "The opponents of Amendment Two were strange bedfellows," says Ronnie Dugger. "You had the Sierra Club voting with the little old ladies in tennis shoes." When the final count was in, the Texas Water Plan had lost, by sixty-six hundred votes.

About a year before the referendum on Amendment Two was held, Congressman George Mahon, the chairman of the House Appropriations Committee, had asked Jim Casey over for a chat between fellow West Texans. Mahon came from Lubbock and was old enough to remember it as a one-horse town. "He had a fear that things could return to that," Casey remembers. "It haunted him." As Casey expected, the conversation immediately came around to the Texas Water Plan. "He asked me what my gut reaction was," Casey remembers. "Did I think it would fly? I told him, 'I hate to tell you this frankly, but my gut reaction is that it's crazy.' I'll never forget the look on his face. It was like I was a doctor telling him his daughter was going to die." Casey's pessimism notwithstanding, Mahon was adamant about studying the plan. He even had an amendment ready for the appropriations bill giving the Bureau whatever money it needed to perform a feasibility report. "I told him that a feasibility report on a plan this big was going to consume a hell of a lot of man-hours and money. But I told him if he insisted we'd do it, because this plan made more sense to me than some of the other cockamamie ideas that were floating around. If it turned out to be infeasible, then we'd all better get ready to kiss irrigation on the high plains goodbye."

The figures Casey's staff began toting up over the next three years were appalling. Routing the aqueduct across wet Louisiana and southeastern Texas would be a costly nightmare, even if the Bureau didn't have to pay for bodyguards to protect its construction crews. "The Louisiana legislature told us to go ahead and study it, because the numbers would kill us anyway," Casey remembers. "But if we'd actually gone ahead and tried to build the thing God knows what might have happened. I felt like taking out a new life insurance policy before going into the bayous down there." The aqueduct would have to go underneath four major rivers by siphon; 142 minor streams would be siphoned under it. But what encumbered the Mississippi diversion most of all was its gluttonous appetite for energy. "Carry two buckets of water up the Washington Monument, take the elevator back down, and do it five more times. That was the lift we had to overcome to West Texas. We were talking billions of buckets. We were talking *trillions* of buckets," says Casey. There was not nearly enough surplus power in Texas, so the project would have to build its own generating plants. The Bureau decided to go the nuclear route, on the widely held belief that nuclear electricity would soon be dirt-cheap. "We took the most pie-eyed projections we could find from the Atomic Energy Commission. We figured the plants would cost $250 million apiece. The plan required about twelve of them. Twelve nuclear plants of a million kilowatts each. You couldn't build *one* nuclear plant in 1985 for the price we thought we were going to pay for twelve in 1971."

Notwithstanding power price estimates that were beyond the realm of fantasy, the Texas Water Plan—which, in the Bureau's version, was somewhat larger than Texas's own—would consume $325 million worth of power every year, in 1971 dollars. The West Texas farmers would end up with a water bill of $330 an acre-foot, all because of the relentless upslope of the plains. The most they could possibly pay, the Bureau decided, was $125 ("and that was hocus-pocus," says Casey). Taxpayers, therefore, would subsidize the rest. The benefit-cost ratio ultimately worked out to .27 to 1.00—for every dollar invested, there would be twenty-seven cents' worth of economic return. "The disparity between primary benefits and costs is so great that there is no reasonable prospect that any plan for transporting Mississippi River water to West Texas or eastern New Mexico would [become] favorable," the Bureau's report read. And it continued, "It is unlikely that the project described . . . could be completed in time to prevent virtual cessation of groundwater irrigation on the Texas High Plains and large-scale reduction of such irrigation in eastern New Mexico." If there was any justification for it at all, it was that "the project could contribute significantly to population dispersion in the 21st century, if this becomes a national objective."

In the Ogallala region, the Bureau's conclusions were met initially with discouragement, but not despair. Everyone knew that the Bureau and the Corps had built projects which made little better sense; they were merely smaller. The real issue, as far as Texas and Kansas and Oklahoma and Colorado and New Mexico were concerned, was that one couldn't simply abandon millions of acres of farmland to the desert from which it had so recently been saved. One couldn't let another Dust Bowl occur. The economics might look bad now, but who knew how they would look in thirty years? By the turn of the century, according to projections, there would be ten billion people, maybe more, on the planet. Who would feed them? Who still had land? The Russians did, but they couldn't feed themselves. Neither could Europe. Asia was thick with humanity; in Java, people would kill for enough land to raise a couple of cows. Australia was not only a desert, but, unlike the American West, a desert without rivers. Could anyone imagine Africa feeding the world? Canada was too cold to grow much of anything besides wheat and cattle. The only place left was South America, but when you chopped down rain forests and tried to grow crops the soil turned to laterite, hard as stone.

On the high plains, you still have five or ten feet of loamy topsoil. You had 1 percent of the farmers on 6 percent of the nation's agricultural land growing 15 percent of the wheat, corn, cotton, and grain sorghum. You had American technology, American know-how. You had the most productive region of the nation that was the food larder of the world. You had cities of 100,000, 200,000 people which depended utterly on irrigation farming and oil and gas. Could the nation just *abandon* them to fate, like the Leadvilles and Silver Cities and Bodies of a hundred years ago?

From the looks of things, it would. After the Bureau's report was released, one heard little about the Texas Water Plan for a number of years. In 1976, and again in 1981, Texans rejected water bonds that appeared likely to set the plan in motion. Arkansas and Louisiana began to talk of their water as if it were their daughters' chastity. The farmers, meanwhile, were still in business.

By the late 1970s, however, the Ogallala had dropped several more feet while energy prices had gone up sevenfold in a decade. The first farmers began going bankrupt—in Texas, in Colorado, in Kansas, in New Mexico, Tens of thousands of acres began reverting to dryland. The press, tantalized by the prospect of an imminent catastrophe, finally took some interest; newspaper and magazine stories appeared by the dozen. The result of all this was a predictable welter of federal studies, the most important of which was the 1982 Six-State High Plains-Ogallala Area Study, coordinated by the Economic Development Administration of the Department of Commerce. The study, as expected, predicted calamity, but decided it would not arrive as soon as most people thought. By the year 2020—which was as far ahead as it looked—Texas's share of the Ogalllala would be down to 87.2 million acre-feet from 283.7 million in 1977. New Mexico's would be all but used up. Colorado and Kansas would be somewhat better off. Irrigation in those states would increase over the near term, then begin to decline early in the twenty-first century. The real reckoning would come after 2020. Nebraska, however, would still overlie 1.9 billion acre-feet by then, and would be irrigating 11.5 million acres—far more than any state in the nation. Irrigation farming would simply move northward, leaving Lubbock, Clovis, and Limon behind. In Texas, according to the report, oil and gas production would be down to 7 percent of its 1977 level—a double blow that could make the fate of cities such as Buffalo appear benign. The economy of the southern plains would be a three-legged stool with two legs gone, unless some miraculous rise in agricultural prices, or some new source of cheap energy, or some revolution in DNA plant genetics came along, permitting corn to get by on fourteen inches of rain. The region would be, to use Jim Casey's phrase, an Appalachia without trees. The only commodities in abundance would be sun and wind.

Is it possible that the 1982 report's conclusions are overly optimistic? "It's possible," says Herbert Grubb, the planning director of the Texas Water Development Board. "When I saw the rate of increase they used for energy costs, I thought it was much too low. In the late seventies, I'd been hearing estimates of oil costing as much as $295 a barrel by the end of the century. It turns out now that they were pretty much on target, at least so far. A lot of us didn't expect an oil glut to materialize in the early 1980s. But no one can say how long it will last. If in ten years we get another series of price jolts like we did in the seventies, I don't see how the irrigators can keep pumping."

From a national perspective—forgetting about the farmers' plight—whether irrigation on the southern plains ends in thirty years, or in seven, or even in fifty years does not matter; the fact is, it will mostly end. The more important issue, from that same perspective, is what will happen then—not just to the farmers and the cost of food and the balance of payments deficits, but to the land.

When thousand of farmers on millions of irrigated acres can no longer afford to pump vanishing water, the dilemma they face will be universal: how to survive on a finite amount of acreage that has suddenly become one-fifth to one-eighth as productive as it was. The answer is foreordained: they cannot. Many of them, therefore, will sell out to more stubborn neighbors and head for the cities for work or relief. Those who remain on enough acreage to offer them a glimmer of hope will ponder their brief list of choices: they can try to raise dryland cotton or wheat or some desert crop—jojoba or guayule, perhaps—or they can try to revert their plowed fields to shortgrass prairie, and raise cattle.

Raising cattle, pehaps even buffalo—which outperform cattle in arid country—might seem the thing to do. However, it is hard to see how it will happen without billions of dollars' worth of federal support. To convert from, say, wheat to grassland, a farmer first needs to plant some fast-growing annual, such as rye, to develop a litter cover for the soil and build up its organic content; it will cost him perhaps $15 an acre and require a year. Then he has to seed gama grass; this costs him even more and takes another year. Finally, if the grass manages to take hold—a lot of it won't—he can begin grazing a few cattle and reseeding those areas that failed to propagate. If he owns a thousand acres, he will probably have spent $30,000 to $50,000 (valued circa 1984); it has taken him three years, and he hasn't earned a dime. He still has his living expenses to cover, and, unless he is a well-established farmer, a small mountain of unpaid debts. Once his grass is growing, he may still have to wait years for his cattle to mature. After seven years or so, he will finally begin to earn some income. But by then he will have fallen into a bottomless hole.

Farmers may therefore resist the temptation to raise cattle and do the economically sensible thing: raise a dryland crop. As Paul Sears wrote in *Deserts on the March*, "So long as there remains the most remote possibility that the drier grasslands, whose sod has been destroyed by the plow, can be made to yield crops under cultivation, we may count upon human stubbornness to return again and again to the attack. . . ." And in that effort lurks the likelihood of a recurrence of the catastrophe that inspired Sears's book: the Dust Bowl.

When a $1 million home perched on a fifty-degree slope above Malibu is clobbered by a mudslide after three weeks of rain—as thousands of houses throughout California were during the El Niño winters of 1982 and 1983—their owners tend to think of themselves as the victims of a "natural" disaster. The Dust Bowl of the 1930s is commonly regarded as such a "natural" disaster, because seven dry years in a row were accompanied by fierce winds, which scoured up the topmost layer of Oklahoma and blew it as far as Norway. The climate of the plains has remained relatively unchanged for hundreds of years, however, and there is no convincing evidence that such a disaster ever occurred before white men plowed up the sod and brought in cattle or, much worse, sheep to graze it down. Even after seven years of drought, the Dust Bowl would probably not have occurred had not man created the conditions for it. By 1932, in Texas alone, seventy million acres of land that had once been covered with a blanket of grass were growing mesquite and thorny weeds, which are poor at holding soil in place. The weeds had no business there; they were native to the ultramontane basins several hundred miles west. As Paul Sears wrote, "Weeds, like wild-eyed anarchists, are the symptoms, not the real cause, of a disturbed order. When the Russian thistle swept down across the western ranges, the general opinion was that it was a devouring plague, crowding in and consuming the native plants. It was no such thing. The native vegetation had already been destroyed by the plow and thronging herds—the ground was vacated and the thistles took it over."

The Dust Bowl occurred after a profitable wheat market had coincided for years with, by plains standards, a spell of abundant moisture. Prices were high enough to inspire greed; the farmers began plowing up everything in sight. Millions of acres of fragile, highly erodable land, from New Mexico all the way up to the Dakotas, had their sod pierced and replaced by wheat. The farmers actually began going bust before the drought even began; a glutted market, international competition, high tariffs, and the impoverished condition of postwar Europe conspired to do them in. The Dust Bowl was the *coup de grâce*.

The second Dust Bowl is apt to result from hardship rather than 1920s-style prosperity, though the pattern of land abuse will be pretty much the same. As the Ogallala aquifer steadily runs out and the surviving farmers watch their debts mount and their living standards decline, they will be forced by financial need to acquire and dry-farm as much new land as they can. Unless they can still afford to pump irrigation water on an emergency basis during droughts—if there is any water left to pump under their land—they will no longer be guaranteed a respectable harvest every year. Because of the high profits of irrigation, the plains farmers took a lot of marginal farmland *out* of production over the past few decades. They could afford to. Now it is likely to be returned to the plow. In the East, marginal land usually means rocks or swamps or steep hillsides. On the rockless, swampless, tabletop plains, it usually means fine sand. Most of western Nebraska is sand; so is a lot of eastern New Mexico and West Texas. In western Kiowa County, Colorado, 150,000 acres of sandy Class VI land (Class I is the best) are *already* in production, losing twenty tons of topsoil or more a year. There is also a lot of marginal land in production in the Portales region of eastern New Mexico.

The winds blow hardest on the southern plains in late winter and early spring—days of sixty-mile-per-hour gusts ripping across empty space, powered by convoluted airflows battling one another. On February 23, 1977, some of those winds blew into Portales country and began raising dust. A dust storm works on the principle of an avalanche: wind scours up some loose soil and forms a dense, stinging cloud of fine particles, which scours up more loose soil, and more, and more, until the horizon is filled by an advancing wave several thousand feet high, churning and swirling millions of tons of suspended matter. When these storms were first sighted in the 1930s, farmers ran inside their houses, fearing torrential rain. When they went back outside, their homes had lost their paint and their chickens were featherless. The Portales storm, which lasted only about a day and a half, removed forty tons of topsoil per acre from parts of Roosevelt and Kiowa counties—as much topsoil loss as rainfall causes in a year in the most erosion-prone parts of the East, and about three centuries' worth of topsoil formation on the arid plains. Early in 1984, the same thing happened in parts of West Texas, south of Lubbock. One reason the

storms did not grow out of control was that a lot of the surrounding
irrigated land was being prepared for planting, and was wet.

Wayne Wyatt, the manager of the Texas High Plains Underground
Water Conservation District in Lubbock—a man now presiding over
the most desperate water-conservation effort in the United States—
does not believe irrigation will end on the southern plains in a spectacular cloud of dust. "In the thirties," says Wyatt, "most of the farmers were still plowing with mules. They had power to dig down about
four or five inches. Now they have hundred-horsepower tractors, which
can easily bring up soil from two feet. It's either wet or it's clayey
enough to hold against the wind. The only way I can see another Dust
Bowl is if we have a real long drought. If it goes on for years in a row
and the farmers can't even manage one crop in between, and if it affects
this whole country and not just a piece of it, then maybe it could
happen again. But this region has never known a drought like that.
Even during the big one, there were a couple of years when you could
raise a dryland crop." I asked Wyatt how far back climatic records go
on the southern plains. "They go back to about the 1880s" was his
response.

Wyatt, a courtly ex-farmer ("I beat my brains out trying to make
a go of it") who speaks in an almost opaque drawl, is rather optimistic
about the future of the plains. "Half of the land around Lubbock is
still dry-farmed. Farmers have been getting crops for forty, fifty years.
Their costs are that much lower that they can make a profit, somehow.
And I'm not sure the aquifer is going to run out so fast. Conservation
is a religion around here now. We have farmers who've cut their water
use in half. Anyone who doesn't conserve tends to lose his friends fast.
We've begun experimenting with capillary water—the water that the
soil draws up from the aquifer and that saturates the layer above it.
You can't pump it, but by injecting compressed air into the soil there,
you squeeze it out like a sponge and it drains into the aquifer. Our
economist thinks capillary water could be available for $25 to $50 an
acre-foot. The farmers can lease air compressors from the oil industry
as their reserves give out. Then you still have to pump the water up,
but with enough conservation I think they can afford it. Capillary water
could prolong the life of the Ogallala by another twenty to forty years."

This, then, is the plains region today—a place that is reverting,
slowly and steadily, into an amphitheater of natural forces toying with
its inhabitants' fate. Besides the constant threat of drought and wind,
there are half a dozen other swords suspended over their heads. They
are as vulnerable to nuclear powerplant fiascos in Washington State

as they are to the debt crisis in Latin America. A couple of percentage-point increases in interest rates coupled with a collapse of the nuclear industry (which would put a premium on oil and gas), all of it occurring when rainfall drops from eighteen inches to twelve, could send them into a death spiral of debt, cost, and dust that might seal their fate. Meanwhile, the promise of water arriving from somewhere else when the aquifer begins running out is slipping almost out of view. Touring the region and speaking with farmers and politicians and bankers, one doesn't hear much of rescue anymore, though the subject is on everyone's mind. According to Steve Reynolds, the former state engineer of New Mexico, the odds against a rescue project being built "have gone from maybe fifty-fifty twenty years ago to eighty-twenty against today." Reynolds said he "frankly doesn't see how society will make this kind of investment in our behalf"—this despite his region's "tremendously important contribution to America's agricultural export production, the only thing that lets us pay for all we import." But then he spread his lanky frame out in his chair, scratched a plaster of mud off his boot (one of Steve Reynolds's leisure activities was walking along his state's meager rivers and pulling phreatophytes out by their roots), and began to veer toward one of his favorite subjects: microwave energy stations in space. "One of those microwave satellites could produce ten thousand megawatts of power. That's enough to power the whole project. I've never felt that we should give up on space. It's our last frontier, and we need one. One of these microwave satellites would be a way to make space exploration economically useful."

Even the economists who have looked into a water-importation project for the plains and pronounced it absurd seem unable to give up on the idea—such is our reluctance to let nature regain control, to suffer the fate of nearly all the irrigated civilizations of antiquity. In 1982, the High Plains–Ogallala Aquifer Region study projected an impossible cost of $300 to $800 per acre-foot for water imported into the region. But then it added: "The only long-term solution to declining groundwater supplies and maintaining a permanent irrigated agricultural economy in most of the High Plains region is the development of alternate water supplies. . . . Although emerging technologies for local water supply augmentation offer some potential for alleviating the overdraft of the aquifer, none can provide sustained and replenishable supplies to meet the region's needs. [Therefore] regional water transfer potentials . . . should be *continued and expanded to feasibility and planning levels*" (emphasis added).

Such investigations, the authors added in a cryptic sentence whose meaning will become clear later on, "should be international as well as national in scope."

The overdraft of groundwater on the high plains is the greatest in the nation, in the world, in all of human history—but it is merely an enormous manifestation of a common phenomenon throughout the West. On the east side of the San Joaquin Valley in California, enough groundwater disappears every year to supply Illinois. The overdraft is projected nearly to double in eighteen years. Tucson and El Paso have fewer than eighty years of water left even after raiding neighboring basins; they will have to get more from somewhere else. The overdraft in Arizona is rapidly forcing the state into an urban economy. There is a serious overdraft in parts of central and eastern Oregon, which pales so much beside the Ogallala, Arizona, and California overdrafts one hardly hears it mentioned outside the state. Groundwater overdraft is, moreover, a phenomenon not limited to the West. Long Island, sitting atop a closed-basin aquifer, is both depleting it and poisoning it with chemical wastes; where it could go for more water is an interesting question, since there isn't any available within four or five hundred miles that anyone seems willing to give up.

Of all these places, the only one that now appears likely to bring its use of water in balance with its supply is Arizona, mainly because it has little choice. The probable result, of course, is that irrigation farming will largely disappear unless Colorado River water, brought in through the Central Arizona Project, is sold to the farmers at incredibly subsidized rates. It was in Arizona, by ironic coincidence, that the only great desert civilization ever established in North America in earlier times disappeared—either for want of water, or, perhaps more likely, because of a surfeit of salt.

A few hundred million years ago, the waters of the oceans were still fresh enough to drink. It is the earth that contains the mineral salts one tastes in seawater. The salts are in all runoff, leached out of rock and soil. The runoff concentrates in rivers, which end up in the oceans—or, as in the case of Mono Lake and Great Salt Lake, in closed-basin sumps up to seven times saltier than the sea. Once in the ocean, the salts have no place to go; the seas are stuck with them. When the water is evaporated, the salts remain behind; when the water falls as rain and becomes runoff again, a fresh batch of salts washes in.

Like DDT in pelican egg shells, the salts in the oceans are testimony to the effects of concentration. As the evaporative cycle is repeated, day after day, year after year, millennium after millennium, eon after eon, the oceans grow saltier all the time. On March 24, 1992, the dissolved salt content in ocean water off San Francisco was about thirty-five thousand parts per million, perhaps a fraction of a ppm higher than it was ten thousand years ago. The process is so incredibly slow and immense that, for once, no act of man seems capable of affecting it by the tiniest measurable iota. What *is* changing—what has changed drastically in the very recent past—is the concentration of salts in some of the world's rivers, and in some of its preeminent agricultural land.

Explaining the collapse of ancient civilizations is a cottage industry within the anthropological and archaeological professions, like the riddle of the dinosaurs. The explanations vary considerably. Some blame their demise on chronic human failings: degeneracy, conflict, war. The decline of Rome, according to some, was the result of the Romans' use of lead in their eating and drinking utensils; since lead causes irreversible brain damage if eaten or ingested in fairly small amounts, the theory offers a tempting explanation for the obviously demented behavior of certain Roman leaders. (It does not, however, explain why most Romans were demonstrably sane, or why there was so much genius about.) Because most of the great civilizations rose in deserts or semideserts, a popular explanation has always been drought—a drought beyond any that modern mankind has known, perhaps caused by aberrant sunspot cycles or some huge volcanic eruptions that changed the climate.

The most fruitful of the ancient cultures grew up at the southeastern end of the Fertile Crescent, the broad valley formed by the Tigris and Euphrates rivers in what is now Iraq. From there civilization appears to have spread eastward into Persia, and on to Afghanistan, Pakistan, India, and China. Later, it spread to the west. Most of the Romans' fabled feats of hydrologic engineering were borrowed from the Assyrians, who borrowed them from their predecessors, the Sumerians. In the seventh century B.C., the Assyrians, under Sennacherib, built an inverted siphon into the Nineveh Aqueduct, a feat of hydrologic engineering which was not really improved upon until New York City built a pressurized siphon into its second Croton Aqueduct in the 1860s. For all its precocious brilliance and innovation, however, the southern part of the Fertile Crescent went into eclipse around the year 2000 B.C. When Babylon rose in the eighteenth century

B.C., many impressive Sumerian cities lay in ruins around it, as Babylon itself would lie desolated centuries later.

The story was repeated nearly everywhere, even in the New World, where a number of remarkable civilizations arose and prospered independently. One of the most impressive was the Hohokam civilization, in central Arizona, which left as its legacy some seven hundred miles of irrigation canals. Sometime around the fourteenth century, however, the Hohokam vanished—reason unknown. The Inca, Aztec, and Maya used irrigation, too, though they didn't rely on it as absolutely as the Hohokam. Their fate was sealed by European invaders, so it is perhaps idle speculation whether they would ultimately have gone the route of their predeceessors in Mesopotamia and elsewhere. Whatever the answer, it appears that only one civilization completely dependent on irrigation managed to survive uninterruptedly for thousands of years. That civilization was Egypt—but Egypt was fundamentally different from the others in one way.

The survival of a civilization depends mainly on sufficient food. But what makes a civilization great? Traveling across the United States, Lewis and Clark saw few fat Indians until they had arrived at the mouth of the Columbia River, where the Chinook were gorging themselves on salmon, oysters, and clams. With plenty of time for leisure, the Northwest Indians were making exquisite crafts and living in impressive lodges. Farther north, the Haida, similarly well fed, had ample time and energy to commit cruel depredations on fellow tribes in magnificent war canoes carved from whole trees. When we think of a great civilization, however, we think of great cities, of sublime architecture and monuments, of intricate governmental and social structures, of engineering ability which startles even the jaded modern observer. By that standard, neither the Chinook nor any of the other cultures in North America—except the Hohokam—was great. Individually, the Indians could be incredibly skillful—as horsemen, warriors, hunters, artisans—but their high achievement was just that: individual. Even where Indians shared a common language, they broke up into small, separate tribes that, for the most part, went their own way. In contrast to the collectivism of the great Mediterranean, Indus Valley, or Mayan city-states, North American Indian culture was fragmented, atomized, ephemeral.

Most of the great Mediterranean civilizations arose in a region notable for its benign weather. But the climate in California is very similar to that of southern Italy and Greece, and California was a gastronomical paradise on earth, with salmon in the rivers, acorns for the taking, whales grounding themselves on beaches, and enormous herds and flocks of game. But the Hurok and Miwok and Paiute tribes were living in caves and under trees when the Greeks and Romans were building aqueducts and the Parthenon.

An answer to this riddle begins to emerge when one considers that nearly all the great early civilizations were irrigated ones. That single act—irrigation—seems inextricably linked to their ascendance, as well as to their demise. Any people who, for the first time, managed to divert a river and seduce a crop out of wasted land had tweaked the majestic indifference of the universe. To bring off the feat demanded tremendous collective will: discipline, planning, a sense of shared goals. To sustain it required order, which led to the creation of powerful priesthoods, of bureaucracies. Irrigation invited large concentrations of people because of all the food; it probably demanded such concentrations because of all the work. Out of this, cities grew. Work became specialized. There had to be engineers, builders, architects, farmers—probably even lawyers, for the disputes over water rights among upstream and downstream irrigators could not have been much different from today's. The ample supply of food may have helped in the keeping of slaves; in California, during the mission days, some of the Indians signed themselves into absolute servitude with the padres in exchange for the certitude of being fed.

Once established, irrigated civilizations in the desert were incredibly well off. Before modern weapons, sheer numbers meant power, so they were formidable in war. Oases in hostile deserts, they would have been difficult to approach and attack. The desert was also a healthy place to live. There was no tsetse fly, no malarial swamp, no raging cold and chilling wind. Because everyone was out of doors much of the time, the spread of disease was much less of a risk than in colder climates. Famine was an almost forgotten nemesis. Food was also a wonderful commodity for trade. Mesopotamia had virtually no metals, but it produced enough food to trade not only for iron and bronze but for a phenomenal wealth of gold. Trade was also a way of exchanging ideas; it was through contact with the Assyrians and Greeks that the Romans learned to build aqueducts.

There were, of course, problems. Canals could silt up or wash out in floods. A rigid bureaucratic order could spawn revolution. Any disruption of the water supply—by an earthquake, a drought—would be catastrophic. But those were not the kinds of problems likely to crush civilizations as ingenious as these. They might take their toll, along with wars and plagues, but it seems unlikely they would have sent them into permanent eclipse or, as in the case of the Hohokam, cause a whole civilization simply to vanish off the face of the earth. There had to be another enemy—something subtle, unseen, subversive. It was likely to be something they could do little or nothing about, something which they may not even have understood, and thus might have been inclined to ascribe to vengeance from gods. Contemplating the list of enemies, natural and man-made, that might fit such a description, more and more anthropologists and archaeologists are concluding that the one that fits it best is salt.

Irrigation is a profoundly unnatural act. It hardly occurs in nature, and that which does occur is mostly along the rare desert rivers, like the Nile, that produce a reliable seasonal flood. In Africa and a few other places, there are natural depressions where runoff collects during rainy seasons, greening the land when it recedes. For every one of those, however, there are dozens of dead saline lakes or lake beds where the same thing used to happen and where, today, nothing can grow. They are common in Nevada—Groom Lake, Newark Lake, Goshute Lake, Winnemucca Lake, China Lake, Searles Lake, Cuddleback Lake—big saucers of salt left over from shallow Pleistocene seas, when the climate of Nevada was more like Szechuan. The waters that filled those lakes came down from ranges a short distance away, but in that brief intimacy with soil and rock had already accumulated enough salts to spell death for the basin below.

Man-made irrigation faces the same problem. In the West, many soils are classified as saline or alkaline. Irrigation water percolates through them, then returns to the river. It is diverted downstream, used again, and returned to the river. On rivers like the Colorado and the Platte, the same water may be used eighteen times over. It also spends a good deal of its time in reservoirs which, in desert country, may lose eight to twelve feet off their surface to the sun every year. The process continues—salts are picked up, fresh water evaporates, more salts are picked up, more fresh water evaporates. The hydrologist Arthur Pillsbury, writing in *Scientific American* in July of 1981, estimated that of the 120 million acre-feet of water applied to irrigated American crops the previous year, ninety million acre-feet were lost to evaporation and transpiration by plants. The remaining thirty million acre-feet contained virtually all of the salts.

Above a heavily irrigated strip of land along the Pecos River in New Mexico, water taken from the river has a measured salinity level of about 720 parts per million. Thirty miles beyond, salinity levels have shot up to 2,020 parts per million, almost entirely because of irrigation; 2,020 parts per million spells death for many crops. Near its headwaters in the Colorado Rockies, the Arkansas River shows only

a trace of salts. A hundred and twenty miles downriver, it contains
2,200 parts per million. The Colorado, a river whose importance is
absurdly disproportionate to its size, has the worst problem with salt
of any American river. There are small tributaries flowing out of the
salt-ridden Piceance Basin with measured concentrations of as much
as ninety thousand parts per million—three tablespoons in a cup—so
it is plagued by natural sources to begin with. In the Grand Valley of
Colorado, irrigation water runs through sedimentary salt formations
on its subterranean return to the river, reaching saline levels thirty
times higher than at the diversion point. Below there are two huge
reservoirs, Powell and Mead, evaporating a million and a half acre-
feet of pure water each year—at least a tenth of the river's flow. It
should come as no surprise, then, that by the time the Colorado River
has entered Mexico, its waters are almost illegal.

Behind Jan van Schilfgaarde's desk in his office at the Department of
Agriculture's Salinity Control Laboratory, in 1982, is a plaque proclaiming him a member of the Drainage Hall of Fame. Drainage seems
like a pedestrian business, and van Schilfgaarde is an uncommonly
sophisticated and witty man, so one wonders what odd fortune married him to this issue. As he explains it, however, drainage becomes
the most difficult aspect of irrigation—rather like fine-tuning a racing
car. In fact, on the face of things, drainage would appear a more
challenging problem than building dams. On the Columbia River,
Grand Coulee Dam is in place, impassive and content. Next door, in
the Columbia Basin Project, the battle against poor drainage and salts
is still going on.
    "When you apply irrigation water," says van Schilfgaarde, "it has
to go somewhere. If it drains back off into the river, quickly, then that's
fine. If it drains down to an underlying aquifer, fine—at least for a
while. If it doesn't drain or drains too slowly, then you have problems.
Salts build up in the root zones. The soil becomes waterlogged. Ultimately you can damage the structure of the soil, ruining it forever. So
you have to get rid of it. How? Where? These are tremendous problems
in places with lots of poorly drained land that apply tens of millions
of acre-feet of water per year, like the American West. Basically, you
can take the macro or the micro approach. You can build big drain
systems, desalination plants, and so on, but you are still left with saline
wastewater or pure salt to dispose of. Or you can tune your crop mix
and your irrigation system to the reality of poor drainage and saline
water and keep the problem at bay. That is what we have been doing

here, with considerable success. I keep telling people this but they don't want to listen to me."

"Here" is the Department of Agriculture's Salinity Control Laboratory, of which van Schilfgaarde was then director. It sits in the shadow of a hulking butte near the city of Riverside, California, surrounded by the very last agricultural land in the Los Angeles Basin. Sixty years ago, this was, acre for acre, the richest farming region in the world. Los Angeles County led the nation in farm income. Today, the main crop in the basin is tract housing. Displaced by twelve million people, agriculture moved eastward and northward into the San Joaquin Valley, which has one of the worst drainage problems in the world.

"Salinity is the monkey on irrigation's back," says van Schilfgaarde. "The good water goes up in the sky and the junk water goes down, so the problem gets worse and worse. Victor Kovda of the University of Moscow says the amount of land going out of production due to salinity now surpasses the amount being brought into production through new irrigation. In this country, we have lost a few tens of thousands of acres—actually a few hundreds of thousands if you include the Wellton-Mohawk Project in Arizona, on which we later spent a fortune in order to bring salted-out land back into production. But that figure is projected to increase drastically in the decades ahead. The problem is an abstraction to most people, like projections of declining oil reserves were back in the 1960s. If you want to see how bad it can get, go to Iraq."

Thousands of years before the birth of Christ, the Sumerians in the Fertile Crescent were already getting some experience with salinity firsthand. Counts of grain impressions in excavated pottery from sites in what is now southern Iraq—pottery that has been carbon-dated back to 3500 B.C.—suggest that at the time, the amount of wheat grown was roughly equal to the amount of barley. A thousand years later, wheat production had dropped by 83 percent. It wasn't that the Sumerians suddenly developed an insatiable craving for barley; they were forced to switch because wheat is one of the least salt-tolerant crops. Between 2400 B.C. and 1700 B.C., barley yields in Sumeria declined from twenty-five hundred per hectare (a highly respectable yield even today) to nine hundred liters per hectare. Not long afterward, massive crop failures began. "Sodium ions tend to be absorbed by colloidal clay particles, deflocculating them," reads an article in *Science* magazine from 1958—the first authoritative report linking the demise of Sumeria to salt. "[This] leaves . . . the resultant structureless soil almost impermeable to water. In general, high salt concentrations obstruct germination and impede the absorption of water and nutrients by plants. Salts accumulate steadily in the water table, which has only very limited lateral movement to carry them away. Hence the groundwater everywhere [in southern Iraq] has become extremely saline. . . . New waters added as excessive irrigation, rains, or floods can raise the level of the water table very considerably under the prevailing conditions of inadequate drainage. With a further capillary rise when the soil is wet, the dissolved salts and exchangeable sodium are brought into the root zone or even to the surface," killing the crops. As the authors—Thorkild Jacobsen and Robert Adams—suggested, Iraq is still struggling with its most ancient nemesis. It can feed itself mainly because it exports oil. At least 20 percent of its arable land (which doesn't amount to much) is permanently destroyed and can never be returned to cultivation. "Probably there is no single explanation," the authors wrote, "but that growing soil salinity played an important part in the breakup of Sumerian civilization seems beyond question."

Van Schilfgaarde's approach to the salinity problem is not the one favored by the farmers, the Bureau of Reclamation, and members of Congress in whose districts the problem lies. "The Bureau says we've analyzed the solutions I am talking about and they've been discredited, which is utter nonsense. Nobody has had the guts to implement them. I'm an outcast at every meeting I go to." The solutions favored by van Schilfgaarde belong to a kind of jujitsu style; the prevailing wisdom is to attack the problem with tanks and planes. "I have been saying for years that the solution to this problem is better management— very careful management," he says, his urbane Dutch manner giving way to rising exasperation. "Certain crops can take high salinity levels. At our experimental plots in the San Joaquin Valley, we have been growing cotton for six years with fifty-nine hundred parts per million water *and* getting 50-percent-higher yields. The salt stress seems to stunt the plants but doesn't affect their production of cotton flowers. The water also has boron in it—an average irrigator wouldn't touch it. This shows that you can use water on one crop, then on one that tolerates salt better, then bring it back and use it again on a still more salt-tolerant crop before letting it go. You use a lot less, which means that you have less to get rid of in poorly drained areas such as the San Joaquin Valley. The cost is low—about $10 an acre. The cost of the Yuma Desalination Plant is *officially* up to $300 million."

The Yuma Desalination Plant, its operation chronically delayed, is an example of the tanks-and-guns approach. In the 1940s, with the

Central Arizona Project deadlocked in Congress, the Bureau of Reclamation was anxious to build *something* in that state, not only to mollify
its citizenry and the increasingly powerful Carl Hayden but also give
its regional office, suffering existential malaise after the completion
of Hoover Dam, something new to do. Along the lower Gila River were
several tens of thousands of prime irrigable acres which had been
irrigated off and on by Spanish, Indians, and Americans for the
past three hundred years. Unfortunately, the region, named Wellton-
Mohawk after two desert hamlets located there, is plagued by poor
drainage. The Bureau revived the region by installing, at considerable
expense, an elaborate drainage system to carry the waste-water away.
Perforated tiles were laid several feet beneath the land, which led into
a master drain that emptied into the Colorado River above the Mexican
border. The project was completed in the early 1960s, just as the
Bureau was closing the gates of Glen Canyon Dam.

The effect of those two actions—a sudden surge of water containing
sixty-three hundred parts per million of salts accompanied by a drastic
reduction of fresh surplus flows from above—gave the Mexicans fits.
Below the border, the salinity of the Colorado River shot up from
around eight hundred to more than fifteen hundred parts per million.
The Mexicali region is the most productive in the entire country, which
suffers not only from frightening population growth but from a woe-
fully archaic, unbalanced, and inefficient agricultural sector. All the
irrigation around Mexicali is utterly dependent on the river. Only a
well-managed irrigation system, which the Mexicans did not have,
could tolerate such levels of salt, and even then under some duress.
Predictably, crop yields went into abject decline. The Mexicans were
all the more incensed because the United States seemed so unconcerned about their plight. We had promised them 1.5 million acre-feet
of water a year, which they were still getting. The Compact, U.S.
officials pointed out, contained no guarantees about water quality, as
long as there was enough. President Luis Echeverría campaigned heavily on the issue, and, after winning the election, threatened to keep
his promise to haul the United States before the World Court at The
Hague. In 1973, for reasons which are still obscure—but which might
conceivably have had something to do with the fact that Mexico
showed some promise of owning a great deal of oil—President Richard
Nixon appointed a former U.S. Attorney General, Herbert Brownell,
to work out a hasty solution. Signed six months later, in August of
1974, the agreement, known as Minute 242, calls for the United States
to deliver Mexico water whose salt content is not more than 115 parts
per million (plus or minus thirty ppm) higher than measured levels

at Imperial Dam in 1976—a level that turned out to be 879 parts per million. As a result, salinity levels at the border of a thousand ppm or above—and they have almost reached such levels—are a violation of international law.

The simplest and cheapest way to solve Mexico's salinity crisis would have been for the U.S. government to buy out the Wellton-Mohawk farmers and retire their lands. Even today, a generous settlement probably would not cost more than a couple of hundred million dollars, and a tremendous source of salts would be removed. Retiring some additional irrigated lands in the Grand Valley of Colorado, another prodigious source of salts, would be further insurance against the problem getting out of hand. None of this has, of course, happened. The solution of choice at Wellton-Mohawk has been the construction of a reverse-osmosis desalination plant—ten times larger than any in the world—which, while consuming enough electricity to satisfy a city of forty thousand people, will treat the wastewater running out the drain canal. The solution of choice in the Grand Valley is at least as expensive but more prosaic—lining irrigation canals to prevent seepage through subsurface salt zones is the main one. The legislation authorizing all of these works belongs in a class of Congressional sacred cows—whatever it costs to keep salinity levels down without retiring an acre of salt-ridden land is what Congress is willing to spend. The Yuma plant is now supposed to cost $293 million, a figure hardly anyone outside the Bureau believes, and the upper-basin works could cost another $600 million, perhaps much more. Energy costs could easily push the Yuma plant's cost to $1 billion or more over fifty years.

What Congress has chosen to do, in effect, is purify water at a cost exceeding $300 an acre-foot so that upriver irrigators can continue to grow surplus crops with federally subsidized water that costs them $3.50 an acre-foot.

"If the farmers at Wellton-Mohawk adopted efficient irrigation methods," says Jan van Schilfgaarde, "you could solve the problem without even retiring the lands. It would be quite possible to reduce their return flows from 220,000 acre-feet a year to 45,000 acre-feet. I'm not even talking about installing drip irrigation. I'm talking about laser-leveling fields and reusing water on salt-tolerant crops and not doing stupid things like irrigating at harvest time, which our neighboring farmer in the San Joaquin Valley did one year. A lot of these guys are actually absentee owners farming by telephone from their dentists' offices in Scottsdale. They hire some manager who may be competent or incompetent and they don't care. They're not in this business to farm crops, or even to make a profit. They're farming the government. They're growing tax shelters. But even if you *do* have a highly competent farmer who wouldn't mind reducing his wastewater flows, he has no incentive to conserve. Federal water is so cheap it might as well be free. What's the point of hiring a couple of additional irrigation managers to save free water? It's wrong to say the farmer is the culprit. He is being *forced* to consume water."

Van Schilfgaarde's outspokenness on this subject may well have had something to do with his departure from the USDA laboratory in 1984. Meanwhile, as his salinity-management approach is almost universally ignored and the Bureau's expensive solutions receive several hundred times more money than his laboratory does, salinity levels at Imperial Dam could reach 1,150 parts per million as early as the year 2000 and keep rising even if its desalination plant operates effectively—a prospect open to considerable doubt. New projects in the upper basin, oil shale development, the continued leaching of saline soil—all will contribute to salinity's inexorable march. This is bad news for the Mexicans, but it is bad news for Los Angeles, too. Each additional part per million of salts in the city's Colorado River supply is estimated to cause $300,000 worth of damage, basin-wide, to the things the water comes in contact with: pipes, fixtures, machinery, cars. A rise in salinity levels at Imperial Dam from 900 to 1,150 parts per million, then, will cost the citizens of southern California about $75 million a year.

The Bureau's answer to all of this appears on a chart which it has available for distribution. The answer is simply described as "further salinity-control projects under study." Adopting these unnamed solutions, at whatever cost, is supposed to hold salinity levels at about 1,030 parts per million at Imperial Dam, *still* too high to meet our Compact obligation to Mexico—which, since 1974, has become one of our three most important foreign suppliers of oil. The Bureau's answer to *that* appears on the graph as "future additional measures"—whatever those are.

In the Colorado Basin, the effects of wastefully irrigating saline lands are not, for the most part, being felt by those doing the irrigating. Thanks mainly to the taxpayers, the farmers who are contributing the lion's share of the salts to the river have had drainage facilities built which flush the problem down to someone else. In the San Joaquin Valley, it is a different story. The San Joaquin's problem is unique— an ingenious revenge by nature, in the minds of some, on a valley whose transformation into the richest agricultural region in the world was wrought at awesome cost to rivers, fish, and wildlife. Several times in the relatively recent geologic past—within the last couple of million years—the valley was a great inland sea, thick with diatomaceous life and tiny suspended sediments which settled near the middle of the gently sloping valley floor. Compressed and compacted, the stuff formed an almost impervious layer of clay that now underlies close to two million acres of fabulously productive irrigated land. In the middle of the valley, the clay membrane is quite shallow, sometimes just a few feet beneath the surface soil. When irrigation water percolates down, it collects on the clay like bathwater in a tub. In hydrologists' argot, it has become "perched" water. Since the perched water does not have a chance to mingle with the relatively pure aquifer beneath the clay, it may become highly saline, as in Iraq. The more the farmers irrigate, the higher it rises. In places, it has reached the surface, killing everything around. There are already thousands of acres near the southern end of the valley that look as if they had been dusted with snow; not even weeds can grow there. An identical fate will ultimately befall more than a million acres in the valley unless something is done.

For many years, the planners in the state and federal water bureaucracies talked about the need for a "master drain" to carry the perched water out of the San Joaquin Valley. It is more accurate to say that their *reports* have talked about it, while the officials, whose main concern was building more dams to satisfy the demands of the irrigators, ignored the need for drainage because neither they nor (they guessed) the public and the farmers could face the cost. "In the early and mid-1970's," says van Schilfgaarde, "the state's position was that no drainage problem exists. The early reports all said that the State Water Project makes no sense without a drain, because it would add inevitably to the perched water problem. But the public doesn't read reports, so no one mentioned them. Then, a few years ago, when the problem began threatening to become critical, there was suddenly an awful drainage problem that threatened the future of agriculture in California."

Today, three decades after the first reports spoke of the need for a huge, valley-wide drainage system, no such system exists. A modest-sized spur, called the San Luis Drain, was partially completed as a part of the Westlands Water District, which, by introducing a prodigious amount of new surface water into a relatively small area, threatened to waterlog the lands downslope. But the water carried off by the San Luis drain has nowhere to go until a master drain is built. For a while, it was dumped into a man-made swamp called Kesterson

Reservoir, near the town of Los Banos, which slowly filled and evaporated according to the intensity of the valley heat and the irrigation cycle. From the air, the reservoir, when it was full, was an attractive sight to migrating waterfowl, which descended on it by the tens of thousands as their ancestors once descended by the many millions on the valley's primordial marshes and shallow lakes. The presence of all of those coots, geese, and ducks at Kesterson Reservoir gave the Bureau an idea about how to solve one of the most daunting problems associated with the master drain: its enormous cost. By the time the San Luis Drain, a modest portion of the proposed master drain, is completed, its price tag will be more than $500 million. In 1984, Interior Secretary William Clark made an offhand projection that solving the drainage problem valley-wide could end up costing $4 to $5 billion. That comes to about $5,000 an acre to rescue the affected lands, which is more than any of the land is worth. The farmers, a number of whom are corporations or millionaires, are understandably loath to pay the bill. If one wrote off a third of the cost as a wildlife and recreational benefit, however, it would be easier to swallow. That is exactly what the Bureau and California's Department of Water Resources, in a 1979 interagency study entitled "Agricultural Drainage and Salt Management in the San Joaquin Valley," proposed to do in the case of the master drain, which, in that report, was projected to cost $1.26 billion in 1979 dollars. Ascribing annual benefits of $92 million to the master drain, the Bureau and the state's Department of Water Resources elected to write off about a third of that total, or $31.7 million, as a nonreimbursable benefit, payable by the taxpayers, for the creation of artificial marshes. If one were to divide the number of ducks which might be expected to use those man-made wetlands into $31.7 million, they would become very expensive ducks indeed. When the Bureau's dams went up, regulating the rivers and allowing the marshlands to be dried up—about 93 percent of the Central Valley's original wetlands are gone—it conveniently ignored the economic value of the millions of ducks whose habitat would be ruined. But later, when it became convenient to overvalue their worth, economic alchemy turned them into gold.

Due to a distressing twist of fate, however, the Bureau and California may consider themselves lucky if they succeed in writing off *any* part of the master drain to wildlife benefits. Sometime in 1982, hunters and biologists around Kesterson Reservoir began to observe that many overwintering birds seemed lethargic and sick—so ravaged by some strange malady that they could not even float on the water, and often drowned. At first, duck hunters and conservationists put

forth an explanation that the farm lobby had always pooh-poohed—
that pesticides and other chemical wastes in the sumpwater were
making the birds die. By 1984, however, biologists were quite certain
that the main cause of the ducks' awful fate was selenium, a rare
mineral, toxic in small doses, that occurs in high concentrations in
southern Coast Range soils—exactly those soils which, washing down
from the mountains over aeons, formed the Westlands Water District.
The San Francisco *Chronicle*, which has carried on a long, bitter battle
against water exports to the valley and southern California, has played
the story for all it is worth. But none of its news stories and editorials
had quite the impact of a poignant front-page photograph of a gorgeous
dying male pintail duck at Kesterson Reservoir, a duck about to sink
like a doomed boat to the bottom of the poisoned man-made marsh
its presence is to subsidize.

Since there can be only one ultimate destination for the wastewater carried by the master drain—San Francisco Bay—the spectacle
at Kesterson has infuriated many of the five million people who reside
in the Bay Area. They may pollute the bay badly enough themselves,
even if they do not admit it; but to have a bunch of farmers grown
wealthy on "their" water, and subsidized by their taxes, sending it
back to the bay full of toxic wastes, selenium, boron, and salt—that
is intolerable. The farmers—who have been stuck with much of their
toxic runoff since Kesterson was closed—might reject such reasoning
as simplistic and emotional. But the fact is that the people of the Bay
Area appear to have the political clout to prevent the drain from ever
reaching there, and they seem determined to use it. It matters little
that the salts in the wastewater (the selenium and boron and pesticides
are another matter) would hardly affect the salinity of a great bay into
which the ocean rushes every day. What matters is that the San Joaquin Valley farmers asked for water and got it, asked for subsidies
and got them, and now want to use the bay as a toilet. To their urban
brethren by the ocean, living a world apart, all of this smacks of a
system gone mad.

The one irrigated civilization of antiquity that remained intact for
thousands of years was Egypt, and we are now reasonably certain
why. Every year, the Nile, the world's most reliable river, would engorge itself in a spring flood and cover most of Egypt's agricultural
land. The floods would both carry off the salts and deposit a fresh
layer of silt. The farmers would then rush to plant their crops, which
grew lavishly on the residual moisture and the perfect soil. In the
1960s, however, the Egyptians, pumped up with a sense of grandiose

destiny by Gamal Abdel Nasser, decided to build a high dam on the
Nile at Aswan. The Soviet Union helped them do it against the United
States' advice. The result has been described as the worst ecological
mistake committed in one place by mankind. The spring floods are
gone; the nutrient-rich silts no longer come; the Nile sardine fishery
in the Mediterranean is going extinct; bilharzia, or schistosomiasis, a
gruesome disease borne by a snail that thrives in slack waters in Africa,
is rampant; the reservoir is silting up quite rapidly due to erosion from
primitive agriculture upriver; irrigation canals, meanwhile, are being
scoured by the silt-free water released by the dam; and the salts have
arrived. With their copious new supply of year-round irrigation water,
the Egyptian farmers have been irrigating madly, and the water table,
increasingly poisoned by salts, is rising dangerously. Recently, Egypt
hired a group of American engineers and agronomists, among whom
was former Reclamation Commissioner Floyd Dominy, to help them
figure out a solution. "Goddamned crazy Russians" was Dominy's
response when I asked him what things were like over there. "Anyone
should have seen that Egypt wouldn't be able to handle the effects of
that dam." The Egyptians now have no choice other than to install
drainage, which they can ill afford—partly because schistosomiasis
has become a national epidemic costing them some $600 million a
year. The hydrologic engineer Arthur F. Pillsbury, writing in *Scientific
American* in 1981, noted that Egypt, having avoided the fate of its
sister civilizations all these centuries, "is now faced with the universal problem of keeping salts from accumulating in the irrigated
fields."

In that same article, Pillsbury also wrote:

In order to maintain and ensure the long-term viability of irrigated agriculture and to provide enough water to carry the
salts to the ocean or some other natural sink, the development
of water resources *should be intensified.* . . . Before man began
harnessing the rivers, the seasonal floods were highly effective
in carrying salts to the ocean and keeping the river basin in
reasonably good salt balance. Today, with river flows being
regulated by storage systems, and with high consumptive use
of the released water, there is not enough waste flow left to
achieve anything approaching balance. The salt is being stored,
in one way or another, within the river basins. . . . Unless the
lower rivers are allowed to reassert their natural function as
exporters of salt to the ocean, today's productive land will eventually become salt-encrusted and barren.

In the end, Pillsbury concluded, there is only one answer. "Eventually, some grand-scale water diversion concept will be needed. . . ."

In 1946, after participating in a conference involving twenty-four eminent hydrologists and engineers, Dr. Charles P. Berkey had a moment of epiphany. Berkey was, at the time, one of the foremost hydrologists in the world. Newbury Professor Emeritus of Geology at Columbia University, he had been a consultant to the city of New York on its Catskill and Neversink water supply projects, and had a list of accomplishments and credentials four times as long as his arm—a list which had kept him so busy he never had a chance to contemplate the implications of his life's work until he was well into advanced age. Then it came to him—a sunburst of perception, a giant semantic leap.

What prompted Berkey's enlightenment was a talk delivered at the conference by J. C. Stevens, then the president of the American Society of Civil Engineers. Berkey was so dumbstruck by what Stevens had to say that he drafted a response as soon as he got back to his desk at Columbia—a response which reads more like a confession of blindness or an admission of personal failing than anything else. This is part of what he had to say:

> Although the principles involved in the paper by Mr. Stevens are well known, it is not certain that the implications are fully appreciated by many even in responsible relation to them. The Factual Data had been long known to the writer, but no statement before this one had brought so forcibly to mind their importance and bearing on long-range planning. . . . The United States has virtually set up an empire on impounded and redistributed water. The nation is encouraging development, on a scale never before attempted, of lands that are almost worthless except for the waters that can be delivered to them by the works of man. There is building up, through settlement and new populations, a line of industries foreign to the normal resources of the region . . .
>
> Effort to use water on desert lands is not a new adventure by any means; but a program involving development of a great region—inviting thereby a large new population under conditions that carry elements of certain future destructive encroachment in limited and computable time—that is new. Not only is it new, but in some of the implications it is fairly as-

tonishing. . . . The nearest thing in that respect was the settlement of the western high plains in earlier days by people who believed that these dust-bowl lands could be farmed in the same manner as those they came from in the Central Mississippi Valley, and no voice was raised to warn them. That was to be a vast and prosperous empire, too.

For the first time, after reading this paper, the long-range significance of the suffocating effect produced by accumulating silt in all these reservoirs was borne down on the writer. He had been so much taken with the fine things being done that he had not fully appreciated the fact that the program carried elements of destruction sure to bring some kind of ending. It was always evident, of course, that there were severe limitations, but it was too easy to overlook or belittle this element of damage from within.

The experience of founding, in difficult surroundings, settlements which finally grew into influence and power is not new; and neither is their decline, and even their ending. In the past, however, none of them carried, along with the agents that built them up, such relentless elements of destruction as in the present reclamation of arid lands. The astonishing thing is that the life of these relief works promises to be so short. One could forget it if the time vista were indefinite, or if there were promise of a thousand years. In that time most human subsistence and economic lines take new turns and become adjusted; but in some of these projects, typical of the average more or less, the beginnings of decline loom already and will certainly grow into a serious problem in three or four generations. One wonders how many settlers gathering around these projects appreciate what it means.

Of course, if one is able to divorce his interest from the future, there is nothing to worry about. In this generation, and the next and the next, an upgrade can be maintained. One can claim (and it is true) that much has been added to the world; but the longer-range view in this field, as in many others, is threatened by apparently incurable ailments and this one of slowly choking to death with silt is the most stubborn of all. There are no permanent cures.

The conference Berkey and Stevens had attended, "The Future of Lake Mead and Elephant Butte Reservoir," was, more precisely, a summit meeting on the subject of mud. Before Hoover and Elephant Butte dams were built, the Rio Grande and the Colorado River ran chocolate-brown in the spring and anytime a cloudburst occurred somewhere in the watershed. Now, the water emanating from the penstocks and spillways below the dams was an opalescent blue-green, colored only by the minerals and algae in it. Each year, millions of cubic yards of silt were coming to a dead halt behind both dams.

For all their breathtaking immensity, dams are oddly vulnerable things—a vulnerability that is shared and greatly intensified among the millions of people who depend on them. The engineers who have built them have gone to great lengths to make them safe from earthquakes, landslides, and floods. But their ultimate vulnerability, as Berkey wrote, is to silt. Every reservoir eventually silts up—it is only a matter of when. In hard-rock terrain with a lot of forest cover—the Sierra Nevada, the Catskill Mountains—a dam may have a useful life of a thousand years. In some overpopulated nations whose forests are nearly gone and whose farmlands are moving up mountains and whose rivers are therefore thick with silt, reservoirs built after the Second World War may be solid mud before the century is out. The Sanmexia Reservoir in China, an extreme case, was completed in 1960 and already decommissioned by 1964; it had silted up completely. The Tehri Dam in India, the sixth-highest in the world, recently saw its projected useful life reduced from one hundred to thirty years due to horrific deforestation in the Himalaya foothills. In the Dominican Republic, the eighty-thousand-kilowatt Tavera Hydroelectric Project, the country's largest, was completed in 1973; by 1984, silt behind the dam had reached a depth of eighteen meters and storage capacity had been reduced by 40 percent. In countries suffering from over-population, deforestation, which is the primary cause of reservoir siltation, can only be expected to grow worse.

As a matter of principle, any place where vegetation is relatively sparse, where soils are erodable, but where six inches of rain in a day or twenty inches in a month are not unknown is a less than ideal place to situate a dam. Those conditions, however, apply to a large part of the intermountain West—and, since the arrival of intensive agriculture, to a great portion of the Middle West as well. The Eel River in California is the most rapidly eroding watershed in North America— partly because the topography is ridden with erodable sediments, partly because of rampant clear-cutting earlier in the century from which the forests may never recover, partly because of stubble grazing by cattle and sheep that is still going on. There is no major dam on any branch of the Eel—at least not yet—but talk of building one there says a lot about what people are willing to ignore. Meanwhile, erosive forces are hard at work in the watersheds of the Missouri River, the Colorado, the Rio Grande, the Platte, the Arkansas, the Brazos, the Colorado of Texas, the Sevier, the Republican, the Pecos, the Willamette, the Gila—rivers on which there are dozens of dams.

Earlier in the century, it was thought by some that irrigation in those watersheds might actually slow the rate of erosion by creating more groundcover to hold the soil in place. In the 1920s, however, no one foresaw interest rates so high that farmers, pushed to the brink, would almost be forced to abandon careful husbandry of the soil for maximum profit. No one foresaw cheap fertilizers that allow land to be plowed year after year, never going fallow. No one foresaw six-ton tractors that tear up the soil and make it more apt to be carried off. No one foresaw a demand for U.S. agricultural exports that makes it profitable to farm Class VI land. As a result of all this—and because it was inevitable anyway—the dams are silting up.

Black Butte Reservoir, Stony Creek, California. Capacity in 1963: 160,009 acre-feet. Capacity in 1973: 147,754 feet.

Conchas Reservoir, Canadian River, New Mexico. Capacity in 1939: 601,112 acre-feet. Capacity in 1970: 528,951 acre-feet.

Alamagordo Reservoir, Pecos River, New Mexico. Capacity in 1936: 156,750 acre-feet. Capacity in 1964: 110,655 acre-feet.

Lake Waco, Brazos River, Texas. Capacity in 1930: 39,378 acre-feet. Capacity in 1964: 15,427 acre-feet.

Elephant Butte Reservoir, Rio Grande River, New Mexico. Capacity in 1915: 2,634,800 acre-feet. Capacity in 1969: 2,137,219 acre-feet.

Hoover Dam, Colorado River, Arizona-Nevada. Capacity in 1936: 32,471,000 acre-feet. Capacity in 1970: 30,755,000 acre-feet.

San Carlos Reservoir, Gila River, Arizona. Capacity in 1928: 1,266,837 acre-feet. Capacity in 1966: 1,170,000 acre-feet.

Howard Brothers Stock Dam, Driftwood Creek, McDonald, Kansas. Capacity in 1959: 26.58 acre-feet. Capacity in 1972: 14.18 acre-feet.

Ocoee Dam Number 3, Ocoee River, North Carolina. Capacity in 1942: 14,304 acre-feet. Capacity in 1972: 3,879 acre-feet.

Guernsey Reservoir, North Platte River, Wyoming. Capacity in 1929: 73,810 acre-feet. Capacity in 1957: 44,800 acre-feet.

Wilson Dam, Tennessee River, Tennessee. Capacity in 1928: 687,000 acre-feet. Capacity in 1961: 641,000 acre-feet.

Clouse Lake, Center Branch of Rush Creek, Ohio. Capacity in 1948: 234 acre-feet. Capacity in 1970: 142 acre-feet.

In thirty-five years, Lake Mead was filled with more acre-feet of silt than 98 percent of the reservoirs in the United States are filling with acre-feet of water. The rate has slowed considerably since 1963, because the silt is now building up behind Flaming Gorge, Blue Mesa, and Glen Canyon dams.

The Bureau of Reclamation has an Office of Sedimentation, which was being run in 1984 by a cheerful fellow named Bob Strand. One wonders whether his good cheer stems from the fact that sedimentation is the one problem the Bureau hasn't really been forced to deal with yet. "All of our bigger reservoirs were built with a sedimentation allowance," says Strand. "There's enough surplus capacity in them to permit most of the projects to operate according to plan over their payout lifetime. In most cases that's fifty to a hundred years. After that, silt will begin to cut into capacity. It hasn't happened yet to any significant degree." What will the Bureau do when it does happen? "We're working on it," says Strand.

"The dams are wasting assets," says Rafael Kazmann, a retired professor of hydrology from Louisiana State University and one of the world's foremost authorities on water. "When they silt up, that's it." Can't the mud be removed somehow? "Sure," says Kazmann, "but where are you going to put it? It will wash right back in unless you truck it out to sea. The cost of removing it is so prohibitive anyway that I can't imagine it being done. Do you understand how many coal trains it would take to haul away the Colorado River's annual production of silt? How would you get it out of the canyons? You can design dams to flush out the silts nearest to the dam, but all you get rid of is a narrow profile. You create a little short canyon in a vast plateau of mud. Most of the stuff stays no matter what you do."

The one place with some experience at desilting dams is Los Angeles, which has built a number of flood debris reservoirs around the basin whose capacity it can ill afford to lose. Between 1967 and 1977, the Metropolitan Water District and the Department of Water and Power removed 23.7 million cubic yards of mud from behind those dams. The cost was $29.1 million. At that rate, it would cost more than a billion and a half dollars, in modern money, to remove the silt that accumulated in Lake Mead over thirty years—if one could find any place to put it.

"The average politician," says Luna Leopold, another hydrologist who seems to have some appreciation of the magnitude of the problem, "has a time horizon of around four years. The agencies are tuned to Congress, so theirs is about the same. No one has begun to think about this yet. But keep in mind that thousands of big dams were built in

this country during a very brief period—between 1915 and 1975. Many
are going to be silting up at the same time. There already are some
small reservoirs in the East that are mud up to the gunwales. These
are little manageable reservoirs—nothing like the big canyon reservoirs we've built in the West. But I haven't heard of anything being
done about them."

The silt that is now accumulating behind the dams used to settle
near the mouths of the rivers. The Mississippi-Atchafalaya Delta,
which is bigger than New Jersey, is made up entirely of silt from the
West and Middle West. About half of the sediment that used to reach
it every year no longer does. Rafael Kazmann, who made a career of
studying the Delta and may understand it better than anyone else
alive, is convinced that a third to half of it will disappear within the
next few decades; a significant percentage already has. He also believes
the Mississippi will change course—probably by the year 2000—and
begin pouring down the Atchafalaya Basin, wiping out many miles of
interstate highway and several of the nation's largest gas pipelines.
"The river has been straitjacketed and robbed of its silt," says Kazmann. "It's a much more powerfully scouring river that it was. It's
just a matter of time before it eats away one of its bends and seeks
out a completely new course." Kazmann also believes that, in an economic sense, such an event could be the greatest peacetime disaster
in American history. The only thing that might eclipse it is the silting
up of the dams.

"The answer I have always heard from bureaucrats," says Kazmann, "is that scientific and technological progress has accelerated
at such a tremendous rate that some solution will come along. I don't
know that they think—that we're going to have fusion energy pumping
out the dams? The only answer I can see is to make the dams higher
or build new ones. Right now I can think of few places where it would
make economic sense to do that, even if it were feasible."

I n his book *Modern Hydrology*, Rafael Kazmann has written:

> [T]he reservoir construction program, objectively consid-
> ered, is really a program for the continued and endless ex-
> penditure of ever-increasing sums of public money to combat
> the effects of geologic forces, as these forces strive to reach
> positions of relative equilibrium in the regime of rivers and the
> flow of water. It may be that future research in the field of

modern hydrology will be primarily to find a method of extri-
cating ourselves from this unequal struggle with minimum loss
to the nation. . . . The forces involved . . . are comparable to
those met by a boy who builds a castle on the sandy ocean
beach, next to the water, at low tide. . . . [I]t is not pessimism,
merely an objective evaluation, to predict the destruction of
the castle. . . .

A Civilization,
if You Can Keep It

In May of 1958, while testifying at Senate hearings on the acreage provisions of the Reclamation Act, the then Associate Commissioner of the Bureau of Reclamation, Floyd Dominy, departed both from the issue at hand and from his prepared remarks to lecture some critical eastern Senators on what the federal irrigation program has meant to the American West.

"My people came here as farmers and settled in East Hampton, Long Island, in 1710," Dominy began. "As the generations progressed they moved westward as public lands were opened up and as the West was developed, until my grandfather, Lafayette Dominy, in 1845, was born on a farm in LaSalle County, Illinois, carved from the wilderness by his own father and grandfather. When Lafayette Dominy reached maturity and married and had his first child, who was my father, he wanted a farm of his own but discovered that within his means he could not acquire one in Illinois. . . . He borrowed $2,000 from a preacher in 1876 and migrated with his small family to Nebraska and took one of the 160-acre homesteads about which we have been speaking.

"Now as to the adequacy of that homestead I would like to have you know that they lived in a sod house. They lived out beyond medical attention, without any of the modern facilities that we feel are desirable for all Americans today. They lost all the girl children in the family to diphtheria. The three male children survived, or else I would not be here.

"I want you to know that on that 160-acre homestead it took that man from 1876 to 1919 to pay off the $2,000 that he borrowed. . . . [W]hen my father reached maturity he took a homestead in the same area, 160 acres. On that farm six of us children were born and six of us reached maturity on the substance of that 160-acre homestead. We had outside plumbing. We did not have deep freezers, automobiles, school buses coming by the door. We walked to school in the mud. We maybe had one decent set of clothes to wear to town on Saturday. . . .

"You take 160 acres that has to provide automobiles, modern school facilities, taxes for school buses, for good roads, to provide deep freezers, electric stoves, electric refrigerators, the modern conveniences that the farm housewife ought to have and deserves, it puts a much greater demand on the income of that land than was necessary to support us at a subsistence level, prevailing for my father or grandfather. . . .

"[When] I became a county agricultural agent . . . I saw the results of people who had decided 'this is the Utopia for which we seek,' and they had left Missouri and Iowa and other places where land was not available—they put their belongings in immigrant cars, and they went to Wyoming and Montana. They took out what was promised to them as an abundant chance for a great family living, 640 dryland acres. I want everyone in this room and I want this committee to know that most of those 640 acres could not sustain a family under any reasonable economic conditions that have prevailed then or now. I saw family after family, after devoting fifteen or twenty years of valiant effort . . . forced to sell out and start anew."

Considering all this, Dominy went on, how could you view the federal Reclamation program as anything less than the salvation of the West? The same 160 acres of flinty, stubbled, profoundly unwelcoming land that couldn't support a family, couldn't create a tax base, couldn't provide even dietary subsistence during drought years was magically transformed when water was led to it. Could one imagine what the West would be like if there *hadn't* been a Bureau of Reclamation? If the rivers hadn't been turned out of their beds and allowed to remake that pitless landscape?

It is a question worth thinking about. Nevada is the one western state without any mentionable rivers at all, and perhaps the closest approximation of how things could have remained if the landscape had suffered no improvement: its settlements a hundred miles apart, its economy rooted, for lack of a better alternative, in what used to be called sin, its ghost towns as numerous as those that managed to survive. Of course, in the states with rivers there was plenty of irrigation going on before the Bureau arrived on the scene, but an appalling number of those private ventures were destined to collapse. There were, as Dominy said, tens of thousands of heart-rending farm failures, and catastrophic overgrazing on the dryland ranches; irrigation helped put an end to both. There were all those rivers just *wasting* water to the Gulf and Pacific; there was the virgin Colorado, as Dominy liked to say, "useless to anyone." Did one prefer the tawdry mirage of Las Vegas to the palpable miracle of the Imperial Valley? Did one prefer a wild and feckless Colorado to one that measures out steady water and power to ten million people? Should we *not* have built Hoover Dam?

There are those who might say yes, who would argue that the West should have been left pretty much as it was. At the distant other end of the spectrum are the water developers and engineers who cannot rest while great rivers like the Yukon and the Fraser still run free, for whom life seems to hold little meaning except to subjugate nature, to improve it, to engage it in a contest of wills. For the rest of us, contemplating the modern West presents a dilemma. We mourn what has been lost since Lewis and Clark—the feast of wilderness, the mammoth herds of buffalo, the fifty thousand grizzly bears and the million antelope that roamed California, the coastal streams that one could cross on the backs of spawning salmon. On the other hand, to see a sudden unearthly swath of green amid the austere and mournful emptiness of the Mojave Desert or the Harney Basin is to watch one's prejudices against mankind's conquering instinct begin to dissolve. So we want to know, even if it seems an academic matter now, what it all amounts to that we have done out here in the West. How much was sensible? How much was right? Was it folly to allow places like Los Angeles and Phoenix to grow up? Were we insane or farsighted to build all the dams? And even if such questions seem academic, they lead to an emphatically practical one: What are we going to do next?

It isn't easy to get people to think along these lines, at least not yet, because the vulnerable aspect of our desert empire remains for most people, even most westerners, an abstraction, like the certainty of another giant earthquake along the San Andreas Fault. Drive through Los Angeles and see the millions of lawns and the water flowing everywhere and the transformation seems immutable: everything rolls along nonstop like the seamless ribbons of traffic; it all seems permanent. But then catch a flight to Salt Lake City and fly over Glen Canyon Dam at thirty thousand feet, a height from which even this magnificent bulwark becomes a frail thumbnail holding back a monstrous, deceptively placid, man-made sea, and think what one sudden convulsion of the earth or one crude atomic bomb or one five-hundred-year flood (which came close to occurring in 1983 and nearly destroyed a spillway under the dam) might do to that fragile plug in its sandstone gorge, and what the sudden emptying of Lake Powell, with its eight and a half trillion gallons of water, would do to Hoover Dam downstream, and what the instantaneous disappearance of those huge life-sustaining lakes would mean to the thirteen million people hunkered down in southern California and to the Imperial Valley—which would no longer exist. But the West's dependence on distant and easily disruptible dams and aqueducts is just the most palpable kind of vulnerability it now has to face. The more insidious forces—salt poisoning of the soil, groundwater mining, the inexorable transformation of the reservoirs from water to solid ground—are, in the long run, a worse threat. If Hoover and Glen Canyon dams were to collapse, they could be rebuilt; the cost would be only $15 billion or so. But to replace the groundwater being mined throughout the West would mean creating an entirely new Colorado River half again as large as the one that exists.

Like so many great and extravagant achievements, from the fountains of Rome to the federal deficit, the immense national dam-construction program that allowed civilization to flourish in the deserts of the West contains the seeds of disintegration; it is the old saw about an empire's rising higher and higher and having farther and farther to fall. Without the federal government there would have been no Central Valley Project, and without that project California would never have amassed the wealth and creditworthiness to build its own State Water Project, which loosed a huge expansion of farming and urban development on the false promise of water that may never arrive. Without Uncle Sam masquerading from the 1930s to the 1970s as a godfather of limitless ambition and means, the seven Ogallala states might never have chosen to exhaust their groundwater as precipitously as they have; they let themselves be convinced that the government would rescue them when the water ran out, just as the Colorado Basin states foolishly persuaded themselves that Uncle Sam would "augment" their overappropriated river when it ran day. The government—the Bureau and the Corps of Engineers—first created a miraculous abundance of water, then sold it so cheaply that the mirage filled the horizon. Everywhere one turned, one saw water, cheap water, inexhaustible water, and when there were more virgin rivers and aquifers to tap, the illusion was temporarily real. But now the desert is encroaching on the islands of green that have risen within it, and the once mighty Bureau seems helpless to keep its advance at bay; the government is broke, the cost of rescue is mind-boggling, and the rest of the country, its infrastructure in varying stages of collapse, thinks the West has already had too much of a good thing. So the West is finally being forced back onto solutions it should have tried decades ago: the cities are beginning to buy water from farmers; groundwater regulation is no longer equated with heavy-handed bolshevism. But to say that a new era has dawned is premature. Poll the rugged-individualist members of the Sacramento Rotary Club and a majority will say that their bankrupt government should by all means build them a $2.5 billion Auburn Dam.

There were excesses of both degree and style. For thousands of years Egyptian farmers irrigated by simple diversions from the Nile and nothing went badly wrong; then Egypt built the Aswan High Dam and got waterlogged land, salinity, schistosomiasis, nutrient-starved fields, a dying Mediterranean fishery, and a bill for all of the above that will easily eclipse the value of the irrigation "miracle" wrought by the dam. In the American West, the Bureau and the Corps fostered a similar style of water development that, though amazingly fruitful in the short run, leaves everyone and everything more vulnerable in the end. Only the federal government had the money to build the big mainstem reservoirs, which will end up being choked by silt or, at the very least, will require billions of dollars' worth of silt-retention dams to keep the main reservoirs alive (these smaller reservoirs will, of course, silt up fairly quickly themselves, even assuming it makes economic sense to build them). It was through the federal government that millions of acres of poorly drained land not only were opened to farming but were sold dirt-cheap water; the farmers flooded their fields with their cheap water and made the waterlogging and salt problems even worse; now that the lands are beginning to succumb to salt it looks as if the farmers will, in many cases, have to solve things on their own, and a lot of land that cost a fortune to bring into production is going to be left to die.

We didn't *have* to build main-stem dams on rivers carrying vast loads of silt; we could have built more primitive offstream reservoirs, which is what many private irrigation districts did—and successfully—but the federal engineers were enthralled by dams. We didn't *have* to mine ten thousand years' worth of groundwater in a scant half century, any more than we had to keep building 5,000-pound cars with 450-cubic-inch V8's. We didn't have to dump eight tons of dissolved salts on an acre of land in a year; we could have foresworn development on the most poorly drained lands or demanded that, in exchange for water, the farmers conserve as much as possible. But the Bureau still sells them water so cheaply they can't *afford* to conserve; to install an efficient irrigation system costs a lot more. The Israelis, who have far too little water to waste any of it, are stunned when they see the consumption of a typical western farm. And it is no coincidence that most of the water-saving innovations of the past years, such as drip irrigation, originated in Israel instead of here.

But the tragic and ludicrous aspect of the whole situation is that cheap water keeps the machine running: the water lobby cannot have enough of it, just as the engineers cannot build enough dams; and how convenient that cheap water encourages waste, which results in more dams. No one loses except, of course, the taxpayers at large.

Recently, the magnitude of these losses has finally begun to come to light. In August of 1985, the Natural Resources Defense Council (NRDC) released a report on the Central Valley Project that it commissioned from a team of economists supported by a Ford Foundation grant. Through that report, a window was thrown open for the first time on the kinds of liberties the Bureau has been taking with public funds and the law in order to perpetuate the myth of abundance and keep up the demand for more dams.

According to the report, the Bureau not only has been giving its California clients—the nation's richest farmers—cheap water; it has been inventing a whole new realm of subsidies, which are quite possibly illegal, in order to keep the price from going up. For one thing, it adopted, years ago, a completely unwarranted interpretation of the principle of "ability to pay," which is one of the main instruments by which water prices are set. Originally, adjusting water rates according to the farmers' "ability to pay" meant that the price of water could vary from good years to bad ones, as long as the momentum of the fifty-year replayment schedules was maintained. But the Bureau undercharged its client farmers so regularly that the CVP repayment schedule had fallen drastically into arrears by 1985. By that year—some three decades after the project was essentially completed—the farmers had repaid a mere $50 million of the $931 million in capital costs that they are obligated to pay back. (Remember that the farmers are exempted from paying interest on this amount, a subsidy worth at least a couple of billion dollars in its own right.) What is worse, since 1982, payments for water and power have not been sufficient even to cover the operation and maintenance costs of the project, and the Bureau has been cannibalizing the capital-cost fund to keep it from running out of operating funds. This, of course, is robbing Peter to pay Paul, and according to the NRDC it is perfectly illegal. It would have been perfectly *legal* for the Bureau to raise its water rates—it may even have been required by law—but that was never done.

A multibillion-dollar interest exemption, a repayment schedule allowed to slip drastically toward default, an amazingly magnanimous interpretation of "ability to pay"—that would seem to be subsidy enough; but the Bureau wouldn't even stop there. A substantial chunk of the project's cost has been written off to fish and wildlife "benefits," even though the main impact on fish and wildlife has been a drastic reduction in salmon and waterfowl populations. In addition, the NRDC report disclosed, the Bureau has for years been selling power to the farmers for considerably less than it pays to wheel it down from the dams in the Pacific Northwest.

The effect of everything, according to the economists, is that a few thousand farmers will, over the course of fifty years, receive a billion and a half dollars' worth of taxpayer generosity that was never supposed to be theirs. (The value of the interest exemption isn't included in this figure; that was their right.) And the result, according to the NRDC, is that "the repayment of [capital] costs of the CVP is likely to be *zero* by the time most of the water contracts expire in the 1990s." The farmers, who were entitled to incredibly cheap water, have ended up getting it nearly free.

Who are the beneficiaries of this vast unintended largess? The report found that the biggest subsidies, on a farm-by-farm basis, are going to the Westlands Water District, which is where the biggest farmers in the CVP service area happen to reside. (The Westlands, in fact, consumes about 25 percent of the water the project has for sale, enough to supply all of New York City.) By the economists' calculations, the true cost of delivering water to Westlands has now reached $97 per acre-foot; the farmers are being charged between $7.50 and $11.80. Taking the average farm size in the district, this translates into a subsidy of around $500,000 per farm—per year.

That sounds bad enough, but it is even worse than it sounds. Spread across the district, the subsidy to Westlands amounts to something like $217 per acre per year; the average annual *revenue* produced by an acre of Westlands land is only $290. This means that 70 percent of the profit on what is supposed to be some of the richest farmland in the world comes solely through taxpayer subsidization—not crop production. Not only that, but the main Westlands crop was then cotton, which in the 1980s had become very much a surplus crop. So the same subsidies that were helping to enrich some of the wealthiest farmers in the nation were at the same time depressing crop prices elsewhere and undoubtedly driving unsubsidized cotton farmers in Texas and Louisiana and Mississippi out of business.

It was these same Westlands farmers, incidentally, who, with the help of their good friends Senator Alan Cranston and Representative Tony Coelho, led the successful effort to expand the acreage limitation from 160 to 960 acres in 1982. Even so, when their ten-year "grace" period expires in 1992, many will still be in violation of the law unless they sell off their excess lands; farms of 2,000 and 3,000 acres are commonplace; "farms" of 30,000 acres are not unknown; not a *single* 160-acre farm exists within its borders. (Why such a group of farmers should have received subsidized water in the first place is a good question.) After saying all this, it hardly seems worth mentioning that the Westlands Water District's irrigation return flows are the main source of the valley's high levels of selenium, which have been poisoning tens of thousands of waterfowl in the valley wildlife refuges and, from the available evidence, all the way into San Francisco Bay.

There, in a nutshell, is how one of the nation's preeminent examples of reform legislation is stood completely on its head: illegal subsidies enrich big farmers, whose excess production depresses crop prices nationwide and whose waste of cheap water creates an environmental calamity that could cost billions to solve. And what was the response of the Bureau to the NRDC report? It quibbled about the actual size of the subsidies but, strikingly enough, didn't deny that they are occurring or even that they are illegal, and it didn't deny that the Central Valley Project is at least hundreds of millions, if not billions, of dollars in debt. Its response was a strange, calm, qualified agreement, as if to say, "Of course this is what has been going on. But it isn't really our fault."

In a sense, the Bureau is right. If blame is laid anywhere, it ought to be laid at Congress's door. Congress authorized the Central Valley Project; Congress approved the Westlands contract; Congress persistently refused to reform the Reclamation Act in any way except to enlarge the subsidies and to permit subsidized water to be sold to bigger farms; Congress, instead of offering incentives to conserve water, issued a multibillion-dollar license to waste it in the form of more and more dams. What cynic can blame it? To Congress, the federal water bureaucracy has been the closest thing to a schmoo, the little creature out of "Lil'l Abner" that reproduced mightily and lived only to be eaten by us. The dams created jobs (how efficiently is another matter) and made the unions happy; they enriched the engineering and contracting firms, from giants like Bechtel and Parsons to small-time cement pourers in Sioux Falls, and made them happy; they subsidized the irrigation farmers and made them happy; they offered enough water to the cities to make them happy; they gave free flood protection to the real estate developers who ran the booming cities of the West out of their pockets and made them happy; and as a result of all this, the politicians were reelected, which made them happy. No one lost except the nation at large.

What federal water development has amounted to, in the end, is a uniquely productive, creative vandalism. Agricultural paradises were formed out of seas of sand and humps of rock. Sprawling cities sprouted out of nowhere, grew at mad rates, and ended up as Frank Lloyd Wright's sanitary slums; while they were being rescued from the tyranny of the desert they gave themselves over as slaves to the automobile. Millions of people and green acres took over a region that, from appearances, is unforgivingly hostile to life. It was a spectacular achievement, and its most implacable critics have to acknowledge its positive side. The economy was, no doubt, enriched. Population dispersion was achieved. Land that had been dry-farmed and overgrazed and horribly abused was stabilized and saved from the drought winds. "Wasting" resources—the rivers and aquifers—were put to productive use.

The cost of all this, however, was a vandalization of both our natural heritage and our economic future, and the reckoning has not even begun. Thus far, nature has paid the highest price. Glen Canyon is gone. The Colorado Delta is dead. The Missouri bottomlands have disappeared. Nine out of ten acres of wetlands in California have vanished, and with them millions of migratory birds. The great salmon runs in the Columbia, the Sacramento, the San Joaquin, and dozens of tributaries are diminished or extinct. The prairie is civilized and is dull; its last wild features, the pothole marshes in the Dakotas, could all but disappear at the hands of the Garrison Diversion and Cendak projects, if they are ever built. And it didn't happen only in the West. Much the same thing happened in the East, especially in the South, where an incredible diversity and history and beauty in the old river valleys lies submerged under hundreds of featureless reservoirs. The vast oak and cypress swamps of the old South have been dried up, courtesy mainly of the Corps of Engineers, and converted to soybean fields (another crop of which we have an enormous glut). In fact, the Corps of Engineers is responsible for creating a lot more artificial farmland, wisely or unwisely, than the Bureau of Reclamation; by its own estimate, it has converted some 26 million acres of marshy or flood-threatened land, most of it in the East, into permanent crops. Depending on one's point of view, this achievement has been a monstrous travesty against nature, a boon to the local economies, or—the viewpoint most likely held by the Corps of Engineers—a fine opportunity to keep building more drainage projects and dams in order to protect what is only a precarious foothold against the forces of nature.

As we discover afresh each day, those forces can only be held at bay, never vanquished, and that is where the real vandalism—the financial vandalism of the future—comes in. Who is going to pay to rescue the salt-poisoned land? To dredge trillions of tons of silt out of the expiring reservoirs? To bring more water to whole regions, whole states, dependent on aquifers that have been recklessly mined? To restore wetlands and wild rivers and other natural features of the landscape that have been obliterated, now that more and more people are discovering that life is impoverished without them?

We won't have to. Our children probably won't have to. But somewhere down the line our descendants are going to inherit a bill for all this vaunted success, and between a $4 trillion national debt (a good bit of it incurred financing the dams) and the inevitability of expensive energy, it will be a miracle if they can pay it.

None of this is to say that we shouldn't have gone out and tried to civilize the arid West by building water projects and dams. It is merely to suggest that we overreached ourselves. What we achieved may be spectacular; in another sense, though, we achieved the obverse of our goals. The Bureau of Reclamation set out to help the small farmers of the West but ended up making a lot of rich farmers even wealthier at the small farmers' expense. Through water development, the federal government set out to rescue farmers from natural hardships—droughts and floods—but created a new kind of hardship in the form of a chronic, seemingly permanent condition of agricultural glut. We set out to tame the rivers and ended up killing them. We set out to make the future of the American West secure; what we really did was make ourselves rich and our descendants insecure. Few of them are apt to regret that we built Hoover Dam; on balance, however, they may find themselves wishing that we had left things pretty much as they were.

S uppose, though, that it were possible to solve at one stroke all the West's problems with water. Suppose you could import into the American West enough water to allow irrigation to continue, even to expand, for another three or four hundred years—to continue even after the great dams built during this century have largely silted up. Suppose you had enough surplus water to flush all the accumulated salts out to sea, thereby avoiding the hoary fate of almost every irrigated civilization. Suppose that, in the process of storing all this water behind great dams, you could create between 50,000 and 80,000 megawatts of surplus power—power that would be available for general consumption even after all of the irrigation water had been moved to where it was needed. (In 1985, the total installed electrical generating capacity of the United States was 600,000 megawatts, so if we take the higher figure we are talking about increasing the U.S. electrical output by nearly one-seventh.) This would be clean hydroelectric power—no pollution, no CO_2, no acid rain. The cost would be stupendous, but perhaps not much greater than the $300 billion the Pentagon has managed to dispose of annually since 1984.

Physically, such a solution appears within the realm of possibility. In a $6-trillion economy, it may even be affordable, disregarding the question of whether it makes economic sense. In the West, many of the irrigation farmers who are threatened by one catastrophe or another regard it as a matter of life or death, and it has long been an obsession to no small number of engineers and hardhat politicians. Its main drawbacks are that it would largely destroy what is left of the natural West and it might require taking Canada by force.

Larger than California and Oregon and Washington stitched together, flooded by up to two hundred inches of rain annually, bisected by big rivers whose names few people know, British Columbia is to water what Russia is to land. Within its boundaries are, in whole or in part, the third-, the fourth-, the seventh-, the eighth-, and the nineteenth-largest rivers in North America. It is debatable how much of the world's accessible and renewable fresh water the province holds, but the usual estimates are between 4 and 10 percent. The Fraser River alone gathers nearly twice the runoff of California; the Skeena's flow approaches the runoff of Texas; both run to sea all but unused. The Talchako River, the main branch of the Bella Coola, which empties into the Pacific halfway between Vancouver and Prince Rupert, is fed by ice fields the size of eastern counties, and in the early summer the river runs like the Mistral, a riverine expressway in a Yosemite canyon that would make a dam-builder gasp. Among the larger rivers of British Columbia it barely rates a passing mention.

The relative proximity of so much water to so much arid land has been a source of compulsive longing in the American West for years. It wasn't until the late 1950s, however, that anyone began thinking seriously about moving some of that water south. It is undoubtedly the grandest scheme ever concocted by man, and it was conceived, rightfully enough, in an engineering office in Los Angeles.

NAWAPA—like the mouth of the Amazon River or Itaípu Dam, it

is a thing one has to see to comprehend, and since it hasn't been built, even its architects may undervalue its brutal magnificence. Visualize, then, a series of towering dams in the deep river canyons of British Columbia—dams that are 800, 1,500, even 1,700 feet high. Visualize reservoirs backing up behind them for hundreds of miles—reservoirs among which Lake Mead would be merely regulation-size. Visualize the flow of the Susitna River, the Copper, the Tanana, and the upper Yukon running in reverse, pushed through the Saint Elias Mountains by million-horsepower pumps, then dumped into nature's second-largest natural reservoir, the Rocky Mountain Trench. Humbled only by the Great Rift Valley of Africa, the trench would serve as the continent's hydrologic switching yard, storing 400 million acre-feet of water in a reservoir 500 miles long. The upper Columbia and Fraser, which flow in opposite directions in the Rocky Mountain Trench, would disappear under it. Some of the water would travel east, down the Peace River—which would be remade and renamed the Canadian–Great Lakes Waterway—all the way to the Great Lakes and the Mississippi. It would be enough to raise the level of all five lakes, double the power production at Niagara Falls and down the St. Lawrence (New York, after all, has a large Congressional delegation), and allow some spillover into the Illinois River and the Mississippi, permitting ocean freighters to reach St. Louis and providing a fresher drinking supply for the cities now withdrawing carcinogenic wastes from the river. The rest of the water would go south.

Imagine the Sawtooth Lifts, a battery of airplane-hangar siphons shooting 30,000 cubic feet per second through tunnels in the Sawtooth Range of Idaho and on to California, Nevada, Arizona, and Mexico. Imagine Lake Nevada. Imagine the Columbia-Fraser Interchange, by which the West's two largest rivers would be merged; a Pecos River Reservoir the size of Connecticut (the feckless Pecos having received a huge jolt of water from the north); another giant reservoir in Arizona which, through some probably unintended irony, would be called Lake Geneva. Imagine 19 million acre-feet of new irrigation water for Saskatchewan and Alberta. Imagine 2.3 million acre-feet for Idaho, 11.7 million acre-feet for the Texas high plains, 4.6 million for Montana, 13.9 million for California (under the NAWAPA plan, water would, as usual, flow uphill toward political power and money). Imagine the Mojave Desert green. Imagine, on the other end of the continent, a phalanx of hydroelectric dams across the bigger rivers pouring into James Bay, the lower appendage of Hudson's Bay. Actually, those dams are the one part of the NAWAPA plan one needn't imagine. Over the past fifteen years, at a cost of $16 billion, Canada has gone ahead and built the James Bay Project itself.

NAWAPA—the North American Water and Power Alliance—was conceived in the early 1950s by Donald McCord Baker, a planning engineer for the Los Angeles Department of Water and Power. Baker took the idea to Ralph M. Parsons, the head of the Pasadena-based firm bearing his name, who instantly fell in love with it, as, he would later insist, "everyone who has worked on it has fallen in love with it." Before his death, Parsons created the NAWAPA Foundation, a tax-exempt receptacle for surplus profits from his company—which had fed on dams and aqueducts until it became the third- or fourth-largest engineering firm in the world—and dedicated it to enlightening the ignorant and converting the unappreciative about the project that was to become the obsession of his twilight years. In the 1960s, when anything big and brutish got at least a passing nod of attention, the NAWAPA scheme excited a considerable spasm of interest. Stewart Udall was able to declare, as Interior Secretary, "I'm for this kind of thinking." Some exploratory discussions were apparently held between Canada and Secretary of State Dean Rusk. Groups of dignitaries began making excursions into Canada under the auspices of the NAWAPA Foundation and the Wenatchee, Washington, *Daily World*, whose publisher, Wilfred Woods, was as enchanted by NAWAPA as Parsons.

In the 1970s, however, as the environmental movement and Canadian nationalism waxed, NAWAPA's fortunes waned. Udall, having become a conservationist in office, began ridiculing the idea. Even the Bureau of Reclamation, which had been secretly assisting the NAWAPA lobby along with the Corps of Engineers, began to hold it at arm's length. (In April of 1965, Commissioner Floyd Dominy went so far as to deliver a mild reprimand to an overenthusiastic Bureau engineer who had spoken too loudly and fondly of NAWAPA. "While I agree that . . . potential interregional water transportation . . . is a subject in which the Bureau is intensely interested and with which, I hope, the future will find us closely identified," Dominy wrote his subordinate, whose name was Lewis Smith, "I do not believe the time is ripe for us . . . we should, however, be prepared to move quickly should we have the opportunity.") But the idea was kept alive by diehard believers: former Utah Democratic Senator Frank Moss (who in 1985 was still being kept on retainer by the Parsons company as a NAWAPA lobbyist), Hawaii Senator Hiram Fong, the late Governor Tom McCall of Oregon (proving that one could be a conservationist and a NAWAPA booster, too). "This is a plan that will not roll over and die," Moss lectured anyone who would listen. "It may be fifty years or it may be a hundred years, but something like it will be built."

By the late 1970s, Frank Moss was beginning to feel vindicated. People were gunning each other down in gas lines. California had just come through the worst drought in its history by a gnat's eyelash. Nuclear power seemed on the verge of collapse. The Islamic revolution was the latest threat to America's imported oil. Thousands of lakes and whole forests were dying from acid rain, a consequence of sulfur and nitrogen emissions from fossil-fuel power plants. Suddenly, the monster project that had been all but given up for dead began to twitch again. In October of 1980, at a California conference on "A High-Technology Policy for U.S. Reindustrialization" sponsored by the Fusion Energy Foundation—an offshoot of the U.S. Labor Party, which despises the Soviet Union but envies its inveterate commitment to gargantuan public works—Dr. Nathan W. Snyder of the Parsons Company reintroduced NAWAPA to a large and enthusiastic audience. "Ultimately, the decision to build NAWAPA—or a project similar to it—will determine, in some part, the future economic well-being in North America," said Snyder. "Water is the most basic of all resources. Civilizations grew or withered depending on its availability."

The Canadians, for their part, have viewed all of this with a mixture of horror, amusement, and avarice. Few seem to believe that NAWAPA will ever be built, but anyone important who mentions it on either side of the border usually rates several column-inches in the Vancouver *Sun*. A number of times in the past several years, Canadian television crews have trooped into the United States to film the sputtering irrigation pumps in West Texas, the salt-encrusted lands in the San Joaquin Valley, and the ghostly abandoned orchards in central Arizona. In western Canada, at least, paranoia about NAWAPA seems to be the reigning state of mind. A few years ago, a British Columbia television journalist named Richard Bocking wrote a blistering book entitled *Canada's Water—For Sale?* which attacked not only NAWAPA but the huge and, as far as Bocking is concerned, pointless dams and reservoirs being built and planned by the provincial utility, B.C. Hydro—reservoirs that, as Bocking pointedly noted, could serve someday as off-the-shelf storage basins for a water-exportation scheme. The more conspiratorially minded in Canada's environmental community are convinced that an intimate confederacy exists among water developers—a kind of freemasonry of engineers—which makes them willing, even eager, to aid one another's grandoise ambitions at the expense of their own nation's interests. It happens to be true that in Canada most of those favorably disposed toward NAWAPA belong to the water-development fraternity. A Canadian professor of hydrologic engineering, Roy Tinney, has even proposed a somewhat less stupefying version of the plan, nicknamed CeNAWP, that would divert the Peace and Athabasca rivers and some of the water in Great Slave Lake to southern Alberta and the American high plains. Every now and then a British Columbia politician has dropped a coy hint that his province (which is, politically speaking, far more independent of Ottawa than an American state is of Washington) might be open to some mutually profitable continental water scheme—someday. Moira Farrow, a reporter for the Vancouver *Sun* who has covered water policy for years, says that some of the province's leading political figures are privately awed by the NAWAPA plan—as if they wished they had thought of it themselves.

There is, in fact, a great deal in the plan for Canada, as there is for Mexico, which has a surplus of oil but a chronic, and grim, and worsening shortage of food. Canada would get more hydroelectric power than the United States—some 38 million kilowatts under one version of the plan; Mexico would get 20 million acre-feet of water, enough to triple its irrigated acreage. Canada would also get a great deal of irrigation water, and, if the contemplated navigation canals are built, a shipping route between its mineral-rich northland and the Mississippi and Great Lakes.

It is Canada, however, that would have to suffer the worst of the environmental consequences, and they would be phenomenal. Luna Leopold, a professor of hydrology at the University of California at Berkeley, says of NAWAPA, "The environmental damage that would be caused by that damned thing can't even be described. It could cause as much harm as all of the dam-building we have done in a hundred years."

Every significant river between Anchorage and Vancouver would be dammed for power or water, or both—the Tanana, the Yukon, the Copper, the Taku, the Skeena, the Stikine, the Liard, the Bella Coola, the Dean, the Chilcotin, and the Fraser. All of these have prolific salmon fisheries, which would be largely, if not wholly, destroyed. (Since the extirpation of around 90 percent of the Columbia's salmon run, the Fraser, the Stikine, and the Skeena have become the most important salmon rivers on earth.) In the western United States, the plan would drown or dry up just about any section of wild river still left: the Flathead, the Big Hole, the Selway, the Salmon; the Middle Fork of the Salmon, the Yellowstone, the Madison, the Lochsa, and the Clearwater would largely disappear. In Canada and the U.S. alike, not just rivers but an astounding amount of wilderness and wildlife habitat would be put under water, tens of millions of acres of it. Surface aqueducts and siphons—not to say hundred-mile reservoirs—would cut off migratory routes. Hundreds of thousands of people would have to be relocated; Prince George, B.C., population 150,000, would vanish from the face of the earth. In general, though, the project's proponents display a peculiar blindness to the horrifying dislocation and natural destruction it would cause. They are far more comfortable talking about how NAWAPA is our only hope of averting worldwide famine.

Because of its unprecedented destructiveness, and due to a natural reluctance on the part of Canadians to let go of so much water for the sake of their paternalistic and overambitious neighbors, the tours organized by the Wenatchee *Daily World* in the 1960s encountered pickets at every airstrip in the bush carrying signs that read WATER THIEVES BEWARE. By 1981, anti-NAWAPA sentiment in British Columbia had, if anything, intensified. Everyone seemed to have heard of it, and nearly everyone was against it—"nearly," because here and there one finds someone who is for it, at least for some smaller version of it. Declining emphatically to be identified, a fairly well known professor at a major university said, "The thing is too big and destructive as is, but a smaller version is worth considering. Compared to the damage the loggers are inflicting on the coast, a few big new reservoirs and canals might appear harmless. The water is worth a lot of money to us, potentially. We wouldn't have to go out and fell whole forests for income. Besides that, I think Canadians are being very narrow-minded about the whole thing. We depend on you for food, and why *shouldn't* we help our neighbor when she is running out of water if we have far more than we can ever use?"

The logging of which the professor spoke is by far the largest source of income in the province of British Columbia, and is being conducted with a careless abandon that might make even the U.S. Forest Service wince. Logging is also a cyclical industry, expanding and contracting in rhythm with such imponderable forces as U.S. deficits and housing starts. Agriculture is more stable, and water could be sold through forty-year contracts like those of the Bureau of Reclamation, ensuring a steady, predictable income every year.

Derreck Sewall, who teaches at the University of Victoria and is widely acknowledged as the foremost authority on water in Canada, says that Canada has its own water shortages looming, particularly in the Okanagan region of southern British Columbia—western Canada's fruitbasket—and on irrigated parts of the Alberta plains, where the farmers are overdrafting groundwater as determinedly as their American counterparts. For the foreseeable future, he sees no possibility of NAWAPA's being built unless Canada itself broaches the idea. "There's a xenophobic, *dirigiste* mood in this country today," Sewall says. "Canadians feel like a colony of the U.S., which is in a certain sense justified. You own 95 percent of our oil industry, for example. So the mood is against exporting our most vital natural resource. But eventually Canada will approach the United States and say, 'You want some of our water? O.K. Here's the price to be paid. We'll deal with you in realistic terms. Water will be part of an overall program of resource development and protection. You want our water, then don't build the Garrison Diversion Project, or keep the return flows out of Lake Winnipeg. We'll give you a certain amount of water for each certain percent reduction in acid rain.' Canadians will eventually come to realize that, as far as the U.S. is concerned, water has a value far beyond that which prevails today. You could almost say that we've got you over a tub."

So what, all things considered, are the odds that NAWAPA will be built?

"We're going to solve the water problem through conservation," says one venerable U.S. hydrologic engineer. "We're not going to build any NAWAPA projects, even if the Canadians invite us in. The Bureau of Reclamation is going to have to start charging realistic rates for water and the farmers are going to live with them by saving a lot of water. We're going to solve the energy problem with coal. I don't know what we're going to do about salinity—put it off into the future, probably. I don't know if we're even going to build any more big water projects in this country. The economics went sour forty years ago. A lot of irrigated land will go out of production and we'll just watch it go out."

"NAWAPA is the kind of thing you think about when you're smoking pot," says another. "People who say it will be built are crazy. Ralph Parsons himself told me he wasn't really serious about it. He just needed the foundation as a tax dodge."

"We won't build the big NAWAPA," says a third. "But I'd bet we'll build a baby NAWAPA. No one knows how much money water will be worth in the future, but it's going to be worth a lot. When we see we're about to lose millions of acres of the most productive farmland in the country and thousands of towns are going to go bust, it will just be a tremendous shock. If we stop talking about water importation for a while, the Canadians will bring it up themselves."

Recently the Soviet Union decided, after many years of planning, to shelve a scheme that would divert the Ob River, three-quarters the size of the Mississippi, from its northerly course into the Arctic Sea and send it fifteen hundred miles or so deep into the steppes of central Asia. A second diversion, which would shunt the Sukhona River into the Volga, has not yet been shelved, but remains in doubt. Together, the two projects are about as ambitious as a NAWAPA scheme built to two-fifths scale. As a result of the decision, the Aral Sea will continue to decline indefinitely at its current rate of eleven and a half feet per year, due to irrigation withdrawals. "Central Asia will simply have to get along with more rational use of its own resources," said a group of Soviet water planners in an official statement. Then they added, "At least until the 21st century."

On April 21, 1981, the premier of British Columbia, Bill Bennett, on a tour of California, gave a speech at San Francisco's Commonwealth Club. Castigating those who wanted to stop building dams, Bennett told his audience that a way must be found to harness and preserve the fresh water pouring out of British Columbia to the ocean. "Dams are more than hydro," he explained. "They preserve our greatest resource and control wild runoff." A questioner then asked whether, since British Columbia at the moment had no plans to use the water Bennett wanted to "conserve" for anything other than hydroelectric power, his call for more dams meant that his government was considering the exportation of water to the United States. The answer was no, Bennett said firmly. Then he added, "But come and see me in twenty years."

Shortly after Bennett's speech, Canada was smacked particularly hard by the worldwide recession that followed in the wake of the Reagan economic policy. In British Columbia, the timber industry went moribund, and plans for several huge hydroelectric dams on the Peace, Liard, and Stikine rivers were indefinitely shelved. The provincial utility, B.C. Hydro, cut its staff force from 11,000 to 6,000, and unemployment went into double digits throughout the country. As a severely chastened Canada began crawling, slowly and unsteadily, out of the deepest economic morass it had seen since the Depression, one could detect a strikingly different attitude on the part of some of its prominent politicians toward a NAWAPA-style water-diversion scheme. Early in 1985, the leader of Quebec's Liberal Party, Robert Bourassa, began to push an eastern Canadian version of NAWAPA, the GRAND Canal (for Great Replenishment and Northern Development Canal Concept), which would turn James Bay into a freshwater lake by constructing a tremendous dike across its northern side. The big rivers feeding the bay would pool below the dike, forming a freshwater

A Civilization, if You Can Keep It

reservoir nearly the size of Lake Ontario. The water would then be led by aqueduct into the Great Lakes, and from there, according to engineers from the Bechtel Corporation—which was spending a million dollars to study the plan—to the American high plains. The estimated cost would be $100 billion.

"On the whole I find more interest in the idea than opposition," said Robert Bourassa.

"I view the prospect with enthusiasm," said Brian Mulroney, Canada's new Prime Minister.

Meanwhile, the people of northern Quebec—mainly Cree Indians—are seeing their culture disintegrate under the waters of monstrous reservoirs being erected by the $35 billion James Bay Hydroelectric Project, which is selling power—though not yet water—to the United States.

"I don't think the people of the province would stand for it," said Frank Miller, the premier of neighboring Ontario.

Afterword to the Revised Edition

In 1978, the year I moved to San Francisco to begin writing this book, the Fifth Horseman of the Apocalypse rode off into a bowl of heat and dust and the Sixth flashed in on a flood. The previous water year—which, in California, runs from October to the following September—had been the driest since recordkeeping began; water year 1976 had been the third driest. But late in 1977 the skies miraculously opened, and water year 1978 ended up as one of the wettest on record. It was a first act. By February of 1979, spillways were roaring at dams whose reservoirs had almost gone dry two years before. In 1980, the third year in a row categorized as "very wet," the jet stream, carrying storms like aircraft in a landing pattern at O'Hare, took aim at southern California, and for weeks the Los Angeles River was so swollen with runoff there was talk of building an aqueduct to send it north.

Then came the really big water years, the *El Niño* winters of 1982 and 1983. No one fully understands why the ocean warms during *El Niño* episodes—vast climatic oscillations are involved—and you can't safely predict the result, but strong *El Niño*s tend to coincide with heavy precipitation years. The early Eighties *El Niño* was the sharpest warming on record. The first huge storm hit the California coast just after Christmas in 1982. Winds over Mount Tamalpais, north of the Golden Gate Bridge, blew a hundred and ten miles an hour, and, after a truck tumbled onto its side, the bridge itself was closed for only the second time since it was built. The thousands of gouges, slumps, and landslide tracks that you see in the hills surrounding San Francisco

Bay were mostly caused by that storm, which dumped more rain in an hour than parts of California ordinarily see in a year. During the following winter, superstorms such as this were routine. In the Sierra Nevada, the standing snowfall record of 750 inches, set in 1906, was eclipsed by fifteen feet. Yosemite Valley was underwater. The storms, bloated with subtropical moisture that seemed to be flash-evaporating from the ocean, were not wrung out, as they usually are, by the Sierra-Cascade blockade. Mirages in Nevada and Utah filled with real water; the Great Salt Lake flooded highways miles from its fleeing shore. The Colorado River at spring melt was unofficially gauged at 350,000 cubic feet per second; that was the flood that damaged the spillway directly under Glen Canyon Dam and—by washing in millions of cubic yards of silt—hastened Lake Powell's ongoing metamorphosis from reservoir to farmland.

The *El Niño* episode played itself out by 1985, and the weather returned to normal for a year or two, until, on Valentine's Day in 1986—just as this book first went to press—one of the three biggest California storms since the turn of the century decided to make landfall.

I was in a Santa Monica hotel room when the frontal system approached the coast. I awakened to a radio weatherman in midsentence, saying something about an electronic buoy a few dozen miles offshore that was sending in low-pressure readings such as you measure inside the eye of a hurricane. I scrapped my plans and decided to flee for home. The ocean below my window was all whitecaps and tremendous gunmetal-gray swells. Ocean Avenue was already a litter of palm fronds torn off the trees by flailing winds. My flight was one of the last to leave before the LAX tower radioed pilots to reroute or wait out the worst of it. Forty tons of flying machine felt like a hummingbird in a gale; a flight attendant tumbled across three rows of passengers when the aircraft fell down an elevator shaft. As we landed in San Francisco in horizontal sheets of rain, screams and prayers turned to tears of relief. A few of us went straight to an airport bar and, at two in the afternoon, got stone-drunk.

The storm series lasted, almost without interruption, for ten days, lending credibility to Noah's flood. Central and northern California, where most of the big reservoirs are, were the hardest hit. I had always had a mordant wish to watch a dam collapse, and this seemed like the best opportunity I might get in my life. I arrived at Oroville Dam just as the storm was beginning to break up. (It took me hours longer than usual to get there, because shallow lakes had formed across Interstate 680, creating instant new refuges for mallards and pintails.)

In the previous week and a half, the Feather River watershed at five thousand feet had unofficially recorded fifty-five inches of precipitation, most of it as rain, which melted several feet of snow lying on the ground. Tampa gets that much rain in an average year. The spillway at Oroville is a big concrete channel that loops around the right abutment of the immense earthen dam. It was dumping a hundred and fifty thousand cubic feet of water per second, a couple of rivers the size of the Tennessee. That much water in that confined a space—the spillway is about as wide as a basketball court—is in a hurry-up mood. My guess is that it was moving thirty or forty miles per hour. Small trees and shrubs lining the spillway fence were bent double under the force of vortex winds created by so much mass in a rush. A crow, sailing arrogantly a few feet overhead, suddenly executed some frantic maneuvers to avoid being sucked in himself; he too had never seen anything like this before. Where the spillway poured the river back into the river below the dam—it didn't so much pour in as fly in—a dense plume of mist mushroomed eighty stories high, split by three arching rainbows.

A dam did actually burst during the flood, though I didn't see it happen. It was a temporary cofferdam built at the prospective site of Auburn Dam, whose construction had been mired in lawsuits and debate for years. The cofferdam held back about a hundred thousand acre-feet of water—thirty-two billion gallons—that merged, almost instantaneously, with a river already swollen to ten times its normal size. The flood-on-a-flood headed into Folsom Lake, which sits twenty miles above Sacramento and has a capacity of about a million acre-feet. Folsom Dam would have to spill the whole reservoir, 320 billion gallons of water, in three or four days in order to absorb the mythic flood pouring in. If it did not, the dam itself would be jeopardized, and if Folsom ended up like Teton Dam then a lot of Sacramento would float under the Golden Gate Bridge. When I arrived, a whole crowd of disaster buffs was already there, held at bay by dozens of highway patrol. I managed to sneak briefly onto the dam crest anyway; it trembled as a bank might tremble during a hurricane. The spillway at Folsom, a concrete and rock dam, was built into its center; it's really a man-made, two-hundred-foot waterfall. At the time, it was dumping much more water than Niagara Falls. You couldn't have heard a jet taking off five hundred feet away; that's the kind of noise a million pounds of water makes—a million pounds a *second*—as it tumbles a couple of hundred feet and crashes into a canyon riverbed. (If Folsom was going to be destroyed, it would probably be a consequence of the falling river chewing out the bedrock on which the dam was built.)

The waterfall reversed direction about eighty yards downriver and rose up in a towering, backfalling hydraulic wave that raced back and crashed into the dam's downstream face, as if it wanted a second chance to knock it to smithereens. Rapids with big reversal waves are the kind that kayakers fear most, because you can be trapped forever in the churning backwash. In a reversal of such monstrous size, a kayaker would have had the free will of a toothpick. A group of boaters was standing near me, screaming at one another over the river's roar; they were debating how long it would take before a trapped boater was ground down to individual molecules.

The Department of Water Resources later estimated that ten million acre-feet of runoff—enough for the city of San Francisco for forty years—had poured out the Golden Gate in two weeks. The crew of a freighter miles out to sea that was plowing through huge waves off the Gate said the wash coming across the bow tasted almost like Evian.

Californians didn't know it yet, but they were riding a meteorological roller coaster, and the great '86 storm was the crest before the giant drop. During all but one of the five subsequent water years, the *annual* runoff of all the rivers emptying into San Francisco Bay was less than the runoff measured from February 14 to February 28 in 1986. By 1992, nearly all of the state had suffered through six dry or critically dry years in a row—the fiercest drought since the Dust Bowl, when California had seven million people instead of the thirty-one million who officially live there today.

Unlike the drought of the mid-Seventies, which held the state in a vise grip for a couple of years and suddenly let go, this drought was like a lobster headed for the pot—it clamped down savagely, held on relentlessly, and then really began to squeeze. By 1990, one of Santa Barbara's two water-supply reservoirs was a plain of sun-cracked mud. The other, bigger one was about a quarter full. A few years earlier, therapists in southern California reported that they were seeing lots of people showing clinical signs of depression because the sun had disappeared for weeks. Now some of the same people were spray-painting their lawns green and hiring Indian rain dancers to try to coax in a cloud. Santa Barbara, a pretty city situated on a sliver of plain between hulking mountains and the sea, used to cherish its geographic isolation and its minimal water supply because both helped to constrain growth, which most people there abhor; it is the one major city in southern California that decided not to hook into the State Water Project. By 1991, however, panicked Santa Barbarans had voted to build a spur to the California Aqueduct through ranges of mountains, at a cost of hundreds of millions of dollars, *and* to construct one of the world's largest desalination plants, which will cost them many millions more.

Had it not been for a series of storms that came onshore in March, when the rainy season is usually about to end, the 1991 water year would have been the driest in California history. Until those storms arrived, precipitation in some parts of the state was less than 20 percent of normal, and measured runoff was as low as 5 percent of normal. Even when it did rain, hardly any runoff made it into the reservoirs—the famished landscape soaked it all up. Nineteen ninety-two—the year in which I am now writing—has been much the same. December and January, which are usually the wettest months, were numbingly dry, but toward the close of the rainy season, for two or three weeks, southern and central California were battered by storms. Not much of the runoff could be captured, because from Monterey south California has few reservoirs of real size—it doesn't rain enough in the south to make building them worthwhile, and when it does rain it often rains violently, so the rivers carry great volumes of sediment and debris. (A small reservoir built on Malibu Creek in the 1920s had utterly silted up by the mid-1940s.) Meanwhile, northern California, where the real reservoirs are, was again bypassed by the biggest storms, and so, as I write this, the state is entering the dry season—and its sixth consecutive year of drought—with less than half its usual water supply on tap.

As it happened, the drought was just a backdrop against which a patently Californian *sturm und drang* was being acted out. In 1989, northern California was hit by an earthquake that, though not exactly colossal—it released about 3 percent as much energy as the San Francisco earthquake of 1906—killed dozens of people and caused seven billion dollars' worth of damage to homes, buildings, and public infrastructure. Two years later, an enormous wildfire swept the Oakland Hills, destroying twenty-five hundred homes, taking more lives, and inflicting at least two billion dollars' worth of damage. Only a few weeks afterward, on Interstate 5, the worst mass highway collision in U.S. history occurred, involving 151 cars. About a year later, a pair of walloping earthquakes jolted the Mojave Desert, which has become suburban Los Angeles. In the midst of this litany was a hard winter freeze that wiped out a $1.5 billion citrus crop and yet *another* earthquake, which reduced much of the lovely town of Ferndale, far up on the north coast, to rubble.

Joan Didion once described the state as an "amphitheater of natural disaster," and all these events bore her out—life in California was imitating a heavy metal cartoon. Only none of these was a *natural* disaster in any true sense. Earthquakes are quite harmless until you decide to put millions of people and two trillion dollars in real estate atop scissile fault zones. California is not Brazil, and it is far north of Florida—orchard growers are always gambling with frost. The mass collision, a macabre excitement on the world's most boring stretch of interstate, was caused by a huge cloud of dust blowing off a cotton field that had been plowed bare and then fallowed due to the drought. Everything about California that is contrived and man-made and therefore vulnerable came together for the Oakland Hills fire: It began with a match or a cigarette dropped in a field of Turkish grass gone to straw (the native bunchgrasses, which can tolerate drought, have been all but usurped by invasive varieties); the grass fire spread into a grove of Australian eucalyptus trees, which can stand a drought but not a hard freeze; the resin-rich eucalyptus, which burn fiercely when frost-killed, went off like Roman candles, showering embers from roof to wood-shingle roof.

The drought itself, which may end up a more costly disaster than all of these combined, qualifies best as punishment meted out to an impudent culture by an indignant God. But the worst damage—ecological and economic—could have been averted, even after six dry years, had it not been for acts of man precipitated by the usual combination of wilfulness and avarice. It wasn't a man-made drought, but man made it very much worse.

Before the Gold Rush, the streams that drain into the Central Valley from the Sierra Nevada and the northern Coast Range represented so many miles of salmon-spawning habitat that you could have stitched it all together and run it across the continent and back again. By the 1960s, 97 percent of it was gone. Friant Dam single-handedly wiped out a spawning run of a hundred and fifty thousand fish by blocking and dewatering the entire San Joaquin River. Salmon could live with the small hydroelectric dams built high in the mountains decades ago; they cannot live with giant, impassable multipurpose dams built low in the foothills, whose main purpose is usually to capture as much water as possible that can then be taken out of the rivers.

Despite the worst disruption of salmon habitat that you can find anywhere on earth, the Sacramento River and a few tributaries, in the late 1960s, still supported a surprisingly robust salmon fishery—the most productive south of the Columbia. There were four distinct subspecies: a fall run, reared mainly in hatcheries, that was the bread and butter of the commercial salmon fleet; a distinct late-fall run; a large winter run; and a rapidly declining spring run, a superfish that goes over forty pounds and blasts through Class Five rapids on its way to spawning reaches nearly a mile above sea level in the Sierra Nevada. (The Sacramento River is unique in the world for its four runs of chinook salmon.) In good years, after the war, the Sacramento fishery could sustain a harvest of several hundred thousand fish, and in great years a million or more fish.

The tenacity of the Sacramento River salmon was remarkable because of the deadly obstacle course the fish, juveniles and adults, have to run from the beginning to the end of their lives. Shasta Dam blocked off enormously productive spawning beds in the watershed; other dams on important tributaries, especially the Yuba and the American, did the same. The Red Bluff diversion dam, at the gateway to the last mainstem spawning reach, frustrates many thousands of upriver-migrating adults despite a fish ladder that goes around it. The intake at the Glenn-Colusa Irrigation District, capable of diverting 3000 cubic feet per second, swallows millions of downriver-migrating juveniles each year. In drier years, when Shasta Lake swelters for months in hundred-degree heat, the warm water emanating into the lower river cooks vast numbers of eggs and juveniles, which usually cannot tolerate water warmer than 60°. An abandoned mint near Shasta leaches ghostly wastes when it rains, and agriculture adds pesticides and herbicides.

But the worst hazard to the fishery is the battery of pumps at the south end of the Delta, which feed the aqueducts that sustain southern California. When the State Water Project began operating in the late Sixties, joining the Central Valley Project, another couple of million acre-feet of water that used to pour out to sea was sucked across the Delta by the pumps, confusing the upriver-migrating adults and entraining tens of millions of hapless juveniles, which go wherever the river currents, natural or artificial, want them to go. In wet years, in the Sixties and Seventies, when the Delta pumps diverted only 20 percent of the Sacramento outflow, the escapement ratio was high and millions of young fish made it to sea, where they could fatten in ocean pastures and return in great numbers to spawn. But in drier years, when as much as 50 percent of the Sacramento River outflow was sucked toward southern California, escapement was low, salmon mortality was high, and the commercial fleet—still comprised of many hundreds of boats—braced itself for poor seasons in the years immediately to come.

As it happened the 1986 floods coincided perfectly with a heavy outmigration of young fish, so the escapement ratio was better than great. It was fabulous. The offshore catch in two or three years, when fish of the 1986 class returned to spawn, was going to be the best in decades.

I first encountered that prediction a few weeks after the floods in an obscure publication called *Fridays*, the biweekly house organ of the Pacific Coast Federation of Fishermen's Associations, which is put out by the PCFFA's only paid staff member, a fish processor's son with a law degree named Zeke Grader. He is one of a handful of people in the world who are paid to think exactly as a salmon would think, which means that his thinking tends to be the opposite of most everyone else's.

In the dry months and years following the 1986 floods, Grader's optimism about the 1986-class fish was counterweighted by a deepening pessimism over the fishery's long-term prognosis. His reasoning was simple and not arguable: Salmon have to confront a drought right away. Everyone else, cushioned by years' worth of reservoir storage, does not. It might not be obvious to people, but it was already obvious to the fish: California, in 1987, had entered a year of severe drought, and because droughts tend to come in cycles, there was apt to be another dry year—and then, conceivably, several more. No big floods ("surplus flows" in water-buffalo argot) were going to flush tens of millions of newly hatched salmon and steelhead past the insistent pull of 300,000 horsepower Delta pumps—not to mention the 160-odd diversion intakes, most lacking fish screens, between the Delta and Shasta Dam. In July of 1987, Grader observed in *Fridays* that 85 percent of the spring flow of the Sacramento River had been either diverted or held in storage that year, with unknown but potentially devastating consequences for the fishery. He quoted Dr. Michael Rozengurt, an expatriate Russian fisheries biologist, who compared California's situation to what the Russians had done to the Sea of Azov, a spectacular fishery turned into a biological desert by Stalin's directive to irrigate a limitless acreage of cotton.

During the next several years (I know this because I recently read five years of *Fridays* over a weekend), Zeke Grader sounded more and more like John the Baptist, although he must have felt more like Sisyphus. *Fridays* has only a few thousand readers, most of whom are West Coast fishermen or fisheries biologists—who needed no convincing that the drought could mean disaster for the salmon if extraordinary measures weren't taken to protect them. That is the sometimes fatal weakness of anadromous fish: By insisting on spawning in rivers and estuaries, they are like an army trapped in a mountain cul-de-sac, easy pickings for forces, natural or unnatural (which is to say, human) that are far beyond their control. But after years of intense drought, as Grader noted again and again, the Bureau of Reclamation and Department of Water Resources—which essentially run the Sacramento River watershed—were still allocating water as if these were normal times. They had taken nearly all of the salmon habitat; now they were taking most of the water—and the fish with it.

The most significant statistics from the drought—which Zeke Grader, to my knowledge, was the first to elucidate—really had nothing to do with precipitation and everything to do with what happened to the precipitation after it fell.

In 1987, which was categorized as a "critically dry" year—the driest of five classifications—the Central Valley Project and State Water Project gave their agricultural customers (who consume 95 percent of the CVP supply and around 65 percent of the SWP's average yield) every acre-foot of their water entitlements, based on the "carryover" they held in storage. The water managers could have argued, in 1987, that they were blindsided by the suddenness of the drought, but in 1988, another critically dry year, agriculture got full entitlements again. In 1989, a year classified as "dry," nearly all CVP and SWP customers received full water deliveries *again*. It wasn't until 1990, a desolately dry year despite some late rains in May, that the two huge water agencies began cutting back their agricultural customers. But even in that year a big block of users with water rights predating the Central Valley Project received normal-year water supplies.

One consequence of this policy (or lack of a policy) was that carryover storage in Shasta Lake dropped so low that, in February of 1991, the Bureau predicted that the reservoir—by far the largest in California—would be the world's biggest mudflat by fall of that year, down to 2 or 3 percent of its capacity of 4,500,000 acre-feet. The Bureau was rescued, barely, by another late spell of wet weather in March, but had those storms not come through there would have been no CVP water for anyone—fish, fowl, humans, or crops—by summer's end.

From the fisheries point of view, though, the most devastating consequence was that most of the runoff that reached the California Delta in those years never reached the Bay; it was immediately diverted across the Delta by the projects' huge batteries of pumps. In fact, from 1987 through 1989, Delta exports *increased* every year as river flows and reservoir storage dropped abysmally. In those three years, runoff to the Delta averaged nine or ten million acre-feet, while Delta diversions climbed from 5.2 million acre-feet in 1987 to 6.1 million acre-feet in 1989—a level barely surpassed in the wettest years.

On the other hand, the four runs of salmon, whose young rode out to sea on twenty to thirty million acre-feet of runoff before the great projects were built, had had *their* water supply reduced by almost 90 percent. Young salmon tend to go where most of the water flows, and most of it was now flowing into the deadly maws of the south Delta pumps.

No one could even guess how many tens of millions, or hundreds of millions, of juvenile salmon perished at the pumps' vast graveyard during the first several years of the drought. But the perverse irony was that, as the *future* California salmon fishery was being decimated as never before, the fishermen in 1988 hauled in the biggest harvest since 1945. As Zeke Grader had predicted two years earlier, the numbers of returning salmon that year—mostly fall run from the 1986 class that zoomed out to sea on the February flood tide—were greater than all but the oldest commercial fishermen could remember. The offshore catch that year totalled 1,400,000 fish, weighing more than fifteen million pounds—a bonanza worth about a hundred and fifty million dollars. Sport fishermen hauled in hundreds of thousands more, and another couple of hundred thousand spawners—about as many as the depleted rivers could handle—swam to upriver redds. As newspapers published photographs of salmon boats listing into port with huge piles of salmon on board, Zeke Grader was devoting whole issues of *Fridays* to a new, antithetical prognosis: that the salmon industry would suffer catastrophically in the years ahead. It's possible his own constituency wasn't listening by then.

But he was right.

In the 1960s, about a hundred and thirty thousand winter-run salmon returned to the Sacramento River to spawn—the remnants of a run that probably numbered in the half-million range before the state and federal projects were built. By the early seventies, the winter run was down to about twenty thousand fish. By 1987, it was down to two thousand. By 1991, the biologists counting the fish may have come close to outnumbering the fish; 191 spawners made it to the Red Bluff Diversion Dam. The spring run, much harder to count, was probably down to two thousand survivors—mainly due to depleted rivers, which were partly the fault of the drought, and unnatural Delta flows, which were not. By then, the fall (hatchery) run, which made up most of the huge 1988 catch, had crashed too. In 1992, the Pacific Fisheries Management Council imposed the most stringent quotas in history on the commercial fleet, and they applied, to varying degrees, from central California to the Canadian border because California salmon tend to head north once at sea. The offshore California harvest in 1992 was about 150,000 fish. A lot of boats never bothered to go out; if they had, the whole season would have yielded a few dozen fish per boat, worth less than the fuel required to catch them. But even boats in Washington State were forced to languish at dockside for weeks because farmers in California, twelve hundred miles away, were granted normal deliveries of subsidized water during the first several years of the worst drought in that state's history.

As it turned out, however, the hand of justice could be as perverse as the kiss of irony. In 1991 and again in 1992, the CVP and SWP water contractors finally experienced the same sort of water rationing—and worse—that salmon and fishermen had endured since the drought's first week. The State Water Project made no deliveries to agriculture in 1991—none. Most of the Bureau's customers saw their water supply reduced by 75 percent. In 1992, an election year, they got a little more water through the direct intervention of someone who had received millions of dollars in San Joaquin Valley PAC money, the president of the United States. Many growers shifted from surface water to groundwater, but they paid a price (groundwater can be several times more expensive); meanwhile, hundreds of thousands of acres were taken out of production. Tens of thousands of people—mostly farmworkers—lost their jobs, welfare caseloads rose astronomically, and in some agricultural counties unemployment rates brushed 30 percent.

Because the reservoirs had been so drastically depleted during the first four years of the drought, the Department of Water Resources and the Bureau had no choice but to cut the growers off. In 1991 and again in 1992, the CVP had just over five million acre-feet in storage in May (when most runoff has entered the reservoirs), and the growers—irrigating millions of acres—could have used it all up by July. But now there was an entirely new reason why they couldn't let much of the water go. By 1992, the winter-run chinook was listed as a threatened species by the federal government and as an endangered species by the state of California. The spring-run salmon was not yet listed because, as part of the recovery plan, almost all salmon fishing off California and Oregon might have had to be banned. (By the fall of 1992, however, the spring run, now represented by fewer than a thousand survivors, looked as if it might be listed too.) The late-fall-run chinook was regarded by fisheries biologists as a species of special concern, which meant that it might have to be listed too. It was not inconceivable, if the drought went on, that almost every salmon in California might eventually join the endangered species list.

The San Joaquin Valley growers, of course, were inclined to blame the whole situation on everything and everyone but themselves: if not exclusively on nature's drought, then on high-seas drift-net fishing, on ocean warming, on overfishing by the West Coast salmon fleet (the most drastically policed fishing fleet in the world), on dredge spoils dumped into San Francisco Bay, on seals and sea lions, on logging in the watersheds, on polluted runoff from abandoned mines—on any cause with a quarter-gram of plausibility. All of these horrors resulted in the loss of some fish; all of them combined are less responsible than the combination of empty rivers, intolerably warm rivers, and rivers flowing in reverse toward power and wealth.

So the fate of California agriculture is now helplessly entwined— because of its insatiable thirst for water—with the fate of the California salmon fisheries. In October of 1992, Congressman George Miller of California, the new chairman of the House Interior Committee, and Senator Bill Bradley of New Jersey saw their Central Valley Project Reform Act blown through the House and Senate and onto the president's desk. Members of Congress from the Northwest voted for the bill in order to protect their own salmon fleets; members from urban California voted for the bill because their constituents had endured severe water rationing while agriculture had not; members from nearly every other state voted for the bill because, in their opinion, agribusiness in California has gotten everything it wanted for far too long, often at the expense of farmers in their own states. Among other things, the Miller-Bradley legislation takes 800,000 acre-feet of water from agriculture and dedicates it to wetlands and fisheries—the first such reallocation since the Central Valley Project Act was passed in 1933. The only question is whether it isn't already too late. In the fall of 1992, more than 300 of the 350-odd salmon boats that comprise the fleet at Fort Bragg, California, were for sale, and winter-run salmon from the class of 1991, tatters of evolution, were being reared for tanks at San Francisco's Steinhart Aquarium, like the condors at the San Diego Zoo.

"You can replant an orchard and have it back in ten years," Zeke Grader told me one morning in the summer of 1992. "You lose a salmon that took twenty thousand years to evolve and you never get it back. The fishermen know that closing the season is their only choice. They know it's their only *hope*—if they have to starve for a year, or two, or a decade, it's the only way to save their industry. We're chucking a whole heritage. Fishing is the oldest industry in California. You have to go up the coast to appreciate the despair. Even then you really can't. You just can't. Everyone's broke. Everyone's living off relatives or on welfare. This was pure plunder. It's basically like the bison and the

Indians: The settlers and the hide hunters killed all the buffalo, so they didn't *have* to kill the Indians. The Indians couldn't survive without the buffalo. Now the cotton and alfalfa farmers killed most of the salmon, with some help from everyone else. I don't know if they consciously wanted to get us out of the way. As long as we have salmon, we'll have fishermen, and as long there're fishermen they're going to be a pain in the ass. But a destitute fishing industry isn't a lobby. It's no one's constituency—it's just a sentimentality. All we have now, besides Miller-Bradley, is the Endangered Species Act. I don't know how long it's going to last. If the growers had the political power to get all the water they wanted when California was drying up and blowing away, they might have figured that overturning the act—or seeing that it didn't affect *their* water supply—would be a piece of cake."

On May 20, 1979, an enormously tall, charismatic, and obsessed young man named Mark Dubois hiked into the canyon of the Stanislaus River, concealed himself near the river's edge, threw a length of chain around an undercut boulder, padlocked the ends of the chain together, tossed the key into the river, and leaned back against the boulder, waiting to drown.

The flood that was going to submerge Mark Dubois within a day or two wasn't moving downriver from the thick snowfields melting rapidly in the Sierra Nevada. This was a flood moving in reverse, up the river. A few months earlier, the U.S. Army Corps of Engineers had closed the gates of New Melones Dam, its most recent snub of nature, a mammoth rockpile wedged in Iron Canyon a few miles downriver. The reservoir had already submerged the older, much smaller Melones Dam and its reservoir, and now its tentacles of turbid water were creeping up the side creeks and the main river itself. Dubois had concealed himself somewhere in Camp Nine Gorge, nine miles of superlative Class Three whitewater that could have been conceived by Disneyworld engineers on amphetamines; after the Youghgighenny River in Pennsylvania, it was the most popular rafting and kayaking run in the United States. Dubois, an expert boater and evangelical environmentalist, was the sort of fixture on this river that old Harry Truman was on the slopes of Mount St. Helens before it buried him in volcanic ash—you could hardly think of the Stanislaus River without thinking of Mark Dubois. He had invested ten years of his life battling New Melones Dam, and for a while it almost looked as if he might win. But in the Seventies, in a contest with the Corps, the Bureau of Reclamation, and California's unquenchable irrigation lobby, he and his minions really had no chance. They were the cavalry; he was the Sioux; the chain and padlock were his Wounded Knee.

By then the Corps's regional hierarchy knew Dubois almost intimately and chose not to undervalue his inhuman will. If he said he was prepared to die, he probably was. Within thirty hours, the spill gates of the dam were opened, and a posse of searchers combed the river canyon on foot, by helicopter, and in rafts, trying to find his hiding spot. Even though some of them must have passed within a few yards of it, they did not. Meanwhile, the whole story had blown around the world—Dubois was being compared to the monks who incinerated themselves in Vietnam—and reporters and people from all over the place were roaring toward the Stanislaus to see what the fuss was all about.

I was one of the first of them, and, probably for the only time in my life, I saw a river born again. A short distance below the old Parrott's Ferry Bridge, where eighty thousand boaters had hauled out in the river's final year, was a small bouncy rapids, an effervescence of frothy, jumping haystack waves. On the morning of May 21, the reservoir was beginning to eat through them. I sat on the bank and watched. One after another, the big waves flattened out, their booming stilled, their splashing stopped . . . then they disappeared under gurgling little whirlpools, and where there had been rapids minutes earlier the river went dead calm. Late that day, however, the Corps began spilling the reservoir, and as it receded, the rapids began to reappear. First there was still water, then the water began to move, then it grew riffly and agitated, and then the rapid waves began rising up, gaining height, gaining force, splashing and spraying and churning as they had for thousands of years—suddenly, from one minute to the next, there was a river again.

But not for long.

Jerry Brown, who was governor at the time, decided to intercede personally with Mark Dubois, promising to try to hold the reservoir below the Parrot's Ferry Bridge, and Dubois, who had told a single emissary where he was and given him a padlock key, walked out of his hiding place. Between its clenched teeth, the Corps mumbled something about respecting the will of the governor of a sovereign state, which was its way of saying it would just wait everyone out. During 1982, the heavy rains and snows of the late 1970s returned. The Corps's and the Bureau's constituency—mostly conservative farmers and Republican towns with a God-given right to subsidized water and power and free flood control—staged demonstrations in Sacramento after releases from New Melones Dam overtopped the river levees and began

flooding their fields. Jerry Brown, possessing one of the shortest attention spans of any politician who ever lived, soon lost interest in the
whole mess. The Bureau of Reclamation, which was supposed to market the water in the reservoir the Corps got to build, complained about
all the waste—even though it hadn't signed a single contract to sell
any of the water and had no means of getting it to any of the growers
who allegedly wanted it. But this only meant that, if the reservoir was
filled, southern California, by default, had a new water supply. What
did a bunch of *rafters* matter, stacked against this? New Melones Lake
had filled all of Camp Nine Gorge by the following spring. Another
river that had flowed wild for hundreds of thousands of years was a
memory.

Coincidentally or not, however, the filling of New Melones Lake
brought the first Age of Dams to a close—at least in the American
West. In California, virtually nothing has been built since. It has been
the same everywhere else. The Narrows Dam in Colorado, Orme Dam
in Arizona, the Garrison Project in North Dakota, O'Neill Dam in
Nebraska, Auburn Dam, the North Coast dams—none of the projects
whose construction seemed likely when I began writing this book exists. There has been no NAWAPA-scale apotheosis; it's hardly mentioned anymore. The dam-building machine didn't even coast down
like a turbine going off-peak. It just suddenly fell apart.

So many factors have played a role that it's hard to judge which
mattered most. You have to give some credit to Mark Dubois: Like
Rosa Parks climbing defiantly aboard her segregated bus, he started
something that couldn't be quelled. Millions of people who had never
seen the Stanislaus River found themselves feeling upset, if not
infuriated, over its loss. Among environmentalists, "Remember the
Stanislaus" is what "Stay the Course" was to the Reagan faithful.
Meanwhile, river recreation—rafting, kayaking, fishing, just watching
the river go—boomed all through the Eighties, in a way that hauling
a sinister, gas-guzzling fighter jet of a motorboat to the local mudflat
did not. (Wallace Stegner estimates that about five thousand Americans who were alive in the 1930s had ever floated a whitewater river;
by the early 1990s, thirty-five million had.) Rafting is fairly big business in states like Colorado, where whitewater companies advertise
on billboards that once promoted agricultural chemicals, shale oil
development, or Wayne Aspinall. Having a captive audience helps: A
couple of days spent floating a beautiful, threatened river can turn
whole families into environmental radicals where the fate of that river
is concerned.

But the water lobby itself deserves most of the credit for its sudden drought of opportunities. Back in the days when most members of Congress cheerfully voted for each other's dams, the best sites disappeared as fast as the rivers on which the dams were built. By the eighties, you were left with ludicrous projects like the Narrows Dam, where you had to build a subsurface dam twice as large as the one aboveground in order to stop the river from seeping out underneath. A full-size Lake Auburn, which could hold 2,400,000 acre-feet of water—but would deliver only two or three hundred thousand acre-feet a year, because most of the American River is already captured and appropriated; Auburn Dam would need awesome runoff in order to fill up and remain full—is projected to cost about two billion dollars, which means it will cost twice as much. Hoover Dam, which captures thirty million acre-feet of water (and routinely delivers nine or ten million acre-feet a year) was completed in 1936 for forty-eight million dollars—*million*—and change. If you are the Bureau of Reclamation, you are left trying to justify a dam that would yield 3 percent of Hoover's water, and perhaps 8 percent of its power, and cost ten times more in *uninflated* dollars. You also have to explain why you are building a gigantic dam next door to a presumably active earthquake fault.

Finding the money to erect pyramids such as this was no problem for the pharaohs who ran Congress thirty or forty years ago, when the whole federal budget was smaller than the portion that pays interest every year on a $4 trillion national debt. But today, when a clutch of visionaries representing Utah water districts troops into the U.S. Capitol to lobby for some new taxpayer-financed dam, they get the same response the departing bunch from Texas just received: It's conceivable—*conceivable*—that Congress might find a little money for the project, if the local sponsors agree to pay, let us say, one-half of the cost—up front. That is how water projects that are a matter of life or death become projects a region can live without.

But the thorniest desert in which today's water lobby finds itself wandering is the ecological legacy of its predecessors. By erecting thirty thousand dams of significant size across the American West, they dewatered countless rivers, wiped out millions of acres of riparian habitat, shut off many thousands of river miles of salmon habitat, silted over spawning beds, poisoned return flows with agricultural chemicals, set the plague of livestock loose on the arid land—in a nutshell, they made it close to impossible for numerous native species to survive. So today, if you want to erect a dam on any tributary of the Colorado River, you have to worry about its effects on the squawfish, a federally listed endangered species. If you want to siphon more fresh water out of the Sacramento-San Joaquin River Delta, you have to ponder the effect on the spring- and winter-run chinook salmon, on the nearly vanished striped bass (an introduced species, but one with a big and tough sport fishing lobby), on the Delta smelt (a serious candidate for listing under the Endangered Species Act), and on two dozen, three dozen, who knows how many land-based species whose precarious hold on existence might be lost through the conversion of remnant deserts or marshes or grasslands to crops, or of fecund estuaries into sterile saltwater sumps.

The fiercest environmental battles of the 1990s are likely to be fought in the American West, and many of them—most of them—may, to one degree or another, involve the Endangered Species Act. But some would be fought even if that act were written out of law. The battles over salmon in California will probably seem as nothing compared to those in the Northwest, because there salmon are a *real* industry; the Columbia River's commercial and sport fishery is valued in the many hundreds of millions of dollars a year. The Columbia was once the greatest salmon river in the world: Fifteen million fish returned every year to spawn; today there are fewer than two million, and half of the watershed's salmon runs (dozens in all) are in fairly imminent danger of going extinct.

What it all boils down to is undoing the wrongs caused by earlier generations doing what they thought was right. The Bureau of Reclamation and the Corps of Engineers knew that their dams would ruin the Columbia River fishery, or most of it, as the years and decades went by. But they convinced themselves, and the Congress—and, for that matter, most people living in the Pacific Northwest—that all the new power and water was worth the price. It was simply how everyone thought—then. In 1967, in order to be ready for Vice President Hubert Humphrey, who was coming out to dedicate John Day Dam and who wanted to feel the thrum of its turbines, the Corps closed the dam gates before the fish ladders were operational, condemning a migration of *hundreds of thousands* of salmon and steelhead to death. The vice president's schedule couldn't be changed. The Corps, a perfect representative of its era, never bothered to ask whether the same might be true of the fish.

By the seventies, however, America's values were utterly different, because everyone's experiences had changed. People who came through the Depression didn't just eat salmon, they survived on it, and they were sick of it; it was known as poverty steak, because it sold for ten cents a pound. Those who were born later could only listen to stories of rivers you could cross on the backs of salmon, of creeks where they crowded themselves out of the water and flopped into the woods. Suddenly there was plenty of cotton and fruit grown on irrigation water; there was plenty of cheap steak, because subsidized water was raising millions of cattle on irrigated alfalfa and grass. There was plenty of cheap hydroelectricity, just two or three generations after the Depression, when many rural towns in the West had no electricity at all. All things man-made had become plentiful, but a great menu of things once abundant in nature had become scarce.

And now people were demanding some of it back.

It didn't seem possible when I began writing this book, but by now it is beginning to seem plausible after all. After damming the canyons and dewatering the rivers in order to spill wealth on the land, we are going to take some of the water back, and put it where, one could argue—as more and more Westerners now do—it really belongs. Law has been the ignition, but a great, almost epochal shift in values has worked as the engine of change. In the mid-eighties, after being hammered by a landmark public trust decision, the city of Los Angeles reduced its diversions from the streams feeding Mono Lake by 60,000 acre-feet a year. The level of the lake, a vast salty haven for migratory waterfowl, began to stabilize after dropping dramatically over forty years. A few years later, the city actually returned some water to the Owens River, which began to flow again for the first time in almost half a century. It didn't flow as it once did, but at least you could call it a river again. It flowed out of new history. William Mulholland was dead. The board of his Department of Water and Power had been all but taken over by environmentalists. The mayor of Los Angeles, Tom Bradley, said with genuine contrition that he wanted to repair some of the damage his city had done.

It was the same everywhere. In 1992, the newly appointed Commissioner of Reclamation, Dennis Underwood, hailed not from Bountiful, Utah, or Orchard City, Colorado, but from Santa Monica. His new regional director in California, Roger Patterson, had just decided to dedicate outflows from Folsom Lake to the California Delta instead of cotton farmers and was holding hundreds of thousands of acre-feet in Shasta Lake for the sake of fish instead of alfalfa. Patterson said he *looked forward* to implementing the just-passed Central Valley Project Reform Act—legislation that might have prompted Floyd Dominy to resign in disgust. After all, he had acquired a much more important constituency—a public that was beginning to wonder why such an agency even exists—and a loaded gun called the Endangered Species Act was aimed at his head.

Even in the Northwest, where the sheer size of the dams, and the sheer value of hydroelectricity, make change terrifically difficult, it almost has to occur. You can perhaps imagine California salmon going extinct, but you can imagine no such thing in the Pacific Northwest, a region the salmon very nearly symbolizes. The great mainstem dams will never be torn down, but smaller dams may be. The federal government already has plans to purchase a high dam on the Elwha River, which drains the north side of the Olympic Range and hosts all five species of Pacific salmon, in order to tear it down. And the mainstem dams, at a cost of hundreds of millions of dollars, will be re-engineered in order to block fewer adult salmon and pass more juveniles through. Holes may be punched through their immense, solid insides and then sealed with ponderous metal gates; when the fish are running downriver, the gates may be opened to let them pass without becoming chopped liver in the turbines. The river may be "managed" (for better or worse, it is in human hands) in a completely different way: the reservoirs rapidly drawn down to quicken the current, the gates opened for the fish, the whole process repeated, again and again, water tumbling down a ladder, until each successive run is safely at sea.

Forty years ago, only a handful of heretics, howling at wilderness, challenged the notion that the West needed hundreds of new dams. Today they are almost vindicated. There is more talk of deconstruction than of construction: of minor dams demolished, of big dams made "environmentally sound," of marginal acreage retired and water returned to its source, of flows bypassing turbines to flush salmon and steelhead out to sea. How can this happen? The region's population is growing and, in places, exploding. (California has added seven million people since New Melones Dam.) More people need more water and power and food. Asia sends its surplus population to California and the Northwest; the Mexican border is porous as a sieve.

It's only recently—mainly in the years since this book first appeared—that Westerners have begun to ask where their water goes, what it costs, and what it earns. That inquiry may produce the most revolutionary results since the Reclamation Act.

In California, for example, enough water for greater Los Angeles was still being used, in 1986, to raise irrigated pasture for livestock. A roughly equal amount—enough for twenty million people at home, at play, and at work—was used that year to raise alfalfa, also for horses, sheep, and (mainly) cows.

The more one tries to make sense of this, the less success one has. Feeding irrigated grass to cows is as wasteful a use of water as you can conceive. Pasture is hydrologically inefficient in the extreme, and, metabolically speaking, so are cows: You need seven or eight feet of water in the hot deserts to keep grass alive, which means that you need almost *fifty thousand* pounds of water to raise *one* pound of cow. (Feeding alfalfa to cows requires even more water, but at least alfalfa fixes nitrogen in the soil.)

If the livestock industry earned California real money, and if cows (unlike avocadoes or artichokes) couldn't be raised on rainfall in thirty-five other states, then giving more water to cows than to humans in the nation's richest and most populous state—a semidesert state at the mercy of a precarious water supply—might make a grain or two of sense. In 1985, however, the pasture crop was worth about $100 million, while southern California's economy was worth $300 *billion*, but irrigated pasture used more water than Los Angeles and San Diego combined. When you added cotton (a price-supported crop worth about $900 million that year) to alfalfa and pasture, you had a livestock industry and a cotton industry consuming much more water than everyone in urban California—and producing as much wealth in a year as the urban economy rings up in three or four days. (Rice, another crop that needs lots of water, consumed more than the entire Bay Area, but the state's rice acreage supports much of the Pacific Flyway on waste grain and an enormous winter production of invertebrate food, so I am leaving the rice acreage alone.)

It isn't much different in any other western state. In Colorado, the alfalfa crop is worth a couple of hundred million dollars a year, while tourism is worth about five billion dollars a year. To raise alfalfa, you have to dam, dewater, and otherwise destroy the rivers that many of the tourists come to fish, to raft, or simply to see. The hydroelectricity that could be generated down river by water used to raise alfalfa is potentially worth more than the crop. In Idaho, the money crop is potatoes, but the crops that use most of the state's water are alfalfa and grass. Each cow raised in the Columbia River watershed—where millions of cows are raised—indirectly consumes water for several salmon. Then the cow pollutes the rivers, overgrazes the hillsides, erodes the streambanks, and conspires, beyond the workings of its feeble brain, to ruin the fish and their habitat in other ways (for example, by sending forth acres of methane-rich flatulence that hasten the greenhouse effect).

In an arid or semi-arid region, you can irrigate low-value, thirsty crops such as alfalfa and pasture grass only if you have cheap water—if your fields are riparian, or if your dams and aqueducts were built decades ago, or if you get your water subsidized by the taxpayers, as one of every three of the far West's full-time irrigation farmers does. If you need forty or fifty thousand pounds of water in places like

California and Colorado to irrigate enough fodder to raise two dollars worth of cow, you can't even consider it if forty thousand pounds of water costs seven or eight dollars (as it would if you bought it from the California Water Project). But it makes perfectly good sense if the government sells you the same quantity for thirty or forty cents—as it does if the Central Valley Project is your source.

If free-market mechanisms—which much of western agriculture publicly applauds and privately abhors—were actually allowed to work, the West's water "shortage" would be exposed for what it is: the sort of shortage you expect when inexhaustible demand chases an almost free good. (If someone were selling Porsches for three thousand dollars apiece, there would be a shortage of those, too.) California has a shortage of water because it has a surfeit of cows—it's really almost as simple as that.

The urban areas in the West have been slow to recognize all this, but lately they have begun to recognize it with a vengeance. The Metropolitan Water District is flooding its millions of customers with literature that shows how a thousand acre-feet of water used in high-tech industry can create sixteen thousand jobs, and how the same thousand acre-feet of water used on pasture farms creates eight jobs. Eight. This kind of stuff infuriates the San Joaquin Valley, its erstwhile ally in the water wars, so valley mouthpieces respond in a manner that inspires the Met not just to anger but to retribution. All the old alliances are falling apart. Southern California wants nothing more to do with the San Joaquin; its water barons would rather scheme over sushi with environmentalists, because they represent the new nexus of power. Even the rice growers in the Sacramento Valley want little to do with the San Joaquin Valley; they raise lots of waterfowl food on acreage that the birds of the Pacific Flyway have come to depend on, and most conservationists now acknowledge that fact, and some have even begun to *like* rice—so why should the rice industry, which gets little subsidized water, carry the San Joaquin Valley's hod?

Meanwhile, all kinds of new alliances are beginning to form. The Sacramento Valley has its own water lobby, which has begun to hold meetings with the salmon fishermen, searching for solutions to *their* water shortage—which is devastatingly real. Las Vegas and Reno, which represent 95 percent of Nevada's economy but use 10 percent of its water (alfalfa growers use most of the rest), may fight like hyenas over monstrous gambling palaces that Japanese companies want to build, but they are in sweet accord on water policy. The new chief of the Las Vegas Valley Water District, a forceful woman named Patricia Mulwray (the murdered hero in *Chinatown* had the same last name), also happens to be chairman of a new Washington lobby group representing most of the urban water agencies in the western states. Its agenda is simple: more water for cities, more for the environment, and less for agriculture—especially water-gorging, low-value agriculture, which usually means cows. "It's not really the irrigators' water," says an urban water agency lobbyist, still too cautious to let me use his name. "It belongs to the people of the states. They have allowed the growers to put all that water to a reasonable and beneficial use. But those words could mean something entirely different in the future. What's so reasonable and beneficial about ruining salmon rivers to raise subsidized surplus crops while industries that employ lots of people decide to relocate to wetter states?"

The irrigation lobby still has a few things going for it, mainly sentimentality, tradition, and law. In many western states, it's the irrigation districts that set water policy: They can forbid sales of water rights from farms to cities beyond the district boundaries, and many of them do. And the irrigation lobby still has a few people convinced that, if it doesn't get almost all the region's water, then the whole world will starve. But the growers and their allies (anyone who wants to build more dams) are fighting a rearguard battle, and they know it. A number of states have legitimized water transfers, and a number of others—notably California—are going to soon. With George Miller now presiding over the House Interior Committee, the growers may be lucky to get any more subsidized federal water at all.

The West's real crisis is one of inertia, of will, and of myth. As Wallace Stegner wrote, somehow the cow and the cowboy and the irrigated field came to symbolize the region, instead of the bison and the salmon and the antelope that once abounded here. Stegner said that he spent much of his writing career breaking lances against windmills turned by the cowboy mystique. You needn't even get rid of the cowboys, who add color and relief to a culture that is becoming depressingly urbanized and, worse, suburbanized. But they might be driving bison, in reasonable numbers, instead of cows, and raising them, for the most part, on unirrigated land—which bison tolerate far better than cows. In a West that once and for all made sense, you might import a lot more meat and dairy products from states where they are raised on rain, rather than dream of importing those states' rain.

You would have a West where most people live in contained cancers called cities (as they already do, anyway), and where more rural people would provide the opportunities for people from the cities— for people from all over the world—to enjoy the region's splendors as they once were. A region where people begin to recognize that water left in rivers can be worth a lot more—in revenues, in jobs—than water taken out of the rivers. Maybe even a region where a lot of people really don't give a damn how much money a river can produce.

At some point, perhaps within my lifetime, the American West will go back to the future rather than forward to the past.

M.R.
October 1992

ACKNOWLEDGMENTS

It would have been absolutely impossible for me to write this book without the love, support, and indulgence of my wife, Dorothy Lawrence Mott.

Second only to hers was the faith and support of my agent and dear friend F. Joseph Spieler, who talked me out of quitting several times, and not for selfish reasons; and of my parents, Konrad and Else Reisner, who rescued me from insolvency more than once.

I must also acknowledge and thank my brother-in-law, Roald Bostrom, who convinced me that I should try to write for a living in the first place.

This book managed to consume three editors in the process of being written. Alan Williams liked the idea, bought the book, and provided much encouragement at the beginning. William Strachan offered moral support and advice along the way. I am most grateful, however, to Dan Frank, who replaced Bill Strachan when the book was nearly completed but treated it as if he had been with it from the beginning. His aesthetic sensibility, resonant judgment, and clear thinking rescued many parts of the book that had managed to beach themselves on the shoals of muddleheadedness, and he wouldn't have tolerated a hackneyed metaphor like that if I'd showed it to him first.

I owe a tremendous debt to the Alicia Patterson Foundation, which got me going; to E. Philip LeVeen and Robert Wolcott of Public Interest Economics, who helped keep me going; to Robert Rodale and the Rodale Foundation, who helped keep me going a while longer; and to the now-defunct American Edition of *Geo* magazine, whose generous expense policy helped fund a good bit of the research.

I can't imagine how the book could have been written had it not been for a handful of people who were extraordinarily generous with their time, candid in their observations, and forthcoming with memoranda, anecdotes, documents, and private letters. I would especially like to thank C. J. Kuiper for many hours of his time and a superb memory and storytelling flair. I am much in debt to Floyd Dominy, another great storyteller, who believes in open files and is as fearless of consequences as his reputation suggests. H. P. Dugan, Daniel Dreyfus, and Jim Casey, all former high officials of the Bureau of Reclamation, were also exceptionally candid and helpful.

Peter Carlson of the Environmental Policy Institute is as knowledgeable as anyone alive on the subject of water projects, and answered countless questions over the telephone. John Leshy of the Arizona State University Law School and Tom Graff of the Environmental Defense Fund were also especially helpful, not only in answering questions but in reviewing portions of the manuscript. A lot of thanks are also owed to James Flannery, Jim Free, Robert Edgar, Alan Merson, Patrick Porgans, Robert Smythe, David Shuster, Jim Cook, and Jan van Schilfgaarde.

Among the hundreds of others I interviewed, I want to single out a few dozen for special thanks. They are Philip Bowles, Helen Ingram, Frank Welsh, Robert Witzeman, Don and Karen Christenson, Richard Wilson, James Watt, Tom Barlow, John Gottschalk, Gilbert White, Bill Martin, Sam Steiger, Stewart Udall, David Brower, Dorothy Green, Phil Nalder, Steven Reynolds, Herbert Grubb, Arleigh West, former Governor Edmund G. Brown, Sr., John Erlichman, Nathaniel Reed, Pete van Gytenbeek, Derrick Sewell, Wayne Wyatt, William Gookin, Mohammed El-Ashry, Richard Madson, the late Horace Albright, Jack Burby, Willoughby Houk, George Baker, Jeffrey Ingram, Ronald Robie, Oliver Houck, Lynn Ludlow, Joe Moore, Barney Bellport, Kendall Manock, John Lawrence, George Ballis, Michael Catino, Keith Higginson, Peter Skinner, Edwin Weinberg, Ben Yellen, Samuel Hayes, Myron Holburt, Don Maughan, Moira Farrow, Bob Weaver, Sandy White, Felix Sparks, Russell Brown, Terry Thoem, Glenn Saunders, Robert Curry, Gus Norwood, Mason Gaffney, John Bryson, Bill Dubois, Mark Dubois, Alex Pesonen, the late Paul Taylor, Gilbert Stamm, Daniel Beard, Irving Fox, Lorelle Long, Stanford P. McCasland, John Newsom, Mary Ellen Morbeck, Brant Calkin, Carolina Butler, and W. R. Collier.

The American Heritage Center at the University of Wyoming is a hospitable, if not luxurious, place to work and contains a monstrous trove of archives relating to the settlement of the West and water development; I would like to extend special thanks to Gene Gressley and his staff. The Lyndon Baines Johnson Library at the University of Texas, Austin, the Bancroft Library at the University of California, Berkeley, the main library at the University of California, Los Angeles, and the Interior Department Library in Washington, D.C., were also most helpful in providing source material.

For many favors and services rendered I am grateful to Tom Turner, the staff of *Not Man Apart,* and the now-defunct San Francisco office of Friends of the Earth. Thanks also to Donna Wilcox and the Washington office of the Natural Resources Defense Council; to John Adams for many favors; to Elyse Axell and Janice Cornwell for indenturing themselves as underpaid and underemployed typists; to Jerold Ordansky for the index; to Joe Kane; and to Il Fornaio, Edible Delights, and the Howard Johnson's in Mill Valley for providing thousands of coffee refills and a pleasant place to go to write.

NOTES AND BIBLIOGRAPHY

CHAPTER ONE: **A Country of Illusion**

Wallace Stegner's *Beyond the Hundredth Meridian*, the preeminent source for this chapter, remains one of the finest biographies in print. It covers not only the life of John Wesley Powell but the lives of those in his circle— some of the most interesting Americans of the nineteenth century; how such things as laws and climatic aberrations influenced the settlement of the West in the nineteenth century; and the ideas that formed much of our present policy regarding natural resources. There are several Powell biographies, but Stegner's is the best.

Hamlin Garland's *A Son of the Middle Border* is as good a portrayal of life on the plains and the imperative that drove people there as has been written. See also O. E. Rolvaag's *Giants in the Earth* and Fred Shannon's *The Farmer's Last Frontier*.

Bernard De Voto, along with Stegner, is probably the finest of the modern western historians. *The Course of Empire* and *Across the Wide Missouri* were both a great help.

Walter Prescott Webb's *The Great Plains* is scholarly, prickly, readable, and as clean a dissection of the huge body of myth that has been built up around this region as anyone ever wrote. Fascinating visual imagery of the virgin West is contained in *Artists and Illustrators of the Old West*, edited by Robert Taft.

An interesting biography—really a hagiography, which makes it all the more interesting—of Henry Miller, the most acquisitive land baron in California history, is Edward Treadwell's *The Cattle King*. Though he is remembered mainly for his 1,090,000 acres, much of it acquired through a dubious legality, Miller's real contribution to history is *Lux v. Haggin*, a legal case which, to a considerable degree, formed the doctrine of western water law. The lawsuit pitted Miller and his lifelong partner, Charles Lux, against Lloyd Tevis and James Ben Ali Haggin, two rival land barons with a fiefdom of their own near the Kern River, who were prevented from irrigating when Miller tried to invoke his riparian water rights. Haggin and Tevis argued, unsuccessfully, that riparian doctrine would doom most of California's best land to dryland ranching, and that land-owners with river frontage should not be allowed to hog all the water. Public reaction against Miller and Lux's victory was so strong that most western states who hadn't already opted strongly for the "appropriative"-rights doctrine soon did. (This doctrine awarded water rights to anyone who used them first, even if his acreage did not border water.) California, for its part, has modified its legal code to allow a complex coexistence of riparian- and appropriative-rights doctrine.

The Education of Henry Adams, one of the most peculiar books ever written by an American, is interesting, in the context of this chapter, for its depiction of the mood of empire that swept the nation in the late nineteenth century.

Powell's journal—actually an embellished and edited version published for public consumption—is a very lively account of his Colorado River adventure and is worth reading, as is his original *Report on the Arid Lands*. Few, if any, bureaucrats since Powell have written as well.

A. B. Guthrie's *The Big Sky*, though a work of fiction, is the most compelling and realistic portrait of the mountain men I have seen. It is one of the few great American novels. Harrison Clifford Dale's account of the Ashley-Smith expeditions is a fairly rich account of some astonishing exploratory feats.

BOOKS

Adams, Henry. *The Education of Henry Adams*. Boston: Houghton Mifflin, 1918.

Athearn, Robert G. *High Country Empire*. Lincoln: University of Nebraska Press, 1971.

Boulton, Herbert Eugene. *Coronado: Knight of Pueblos and Plains*. New York: Whittlesey House, 1949.

Dale, Harrison Clifford. *The Ashley-Smith Expeditions and the Discovery of a Central Route to the Pacific, 1822–1829*. Glendale, Calif.: Arthur H. Clark, 1941.

De Voto, Bernard. *Across the Wide Missouri*. Boston: Houghton Mifflin, 1947.

———. *The Course of Empire*. Boston: Houghton Mifflin, 1952.

Dodge, Richard. *The Plains of the Great West.* New York: Archer House, 1959.

Dunne, John Gregory. "Eureka! A Celebration of California." In Jonathan Eisen and David Fine, eds., *Unknown California.* New York: Macmillan, 1985.

Garland, Hamlin. *A Son of the Middle Border.* New York: Macmillan, 1917.

Guthrie, A. B. *The Big Sky.* New York: Sloane, 1947.

Hafen, Leroy. *Mountain Men and the Fur Trade.* Glendale, Calif.: Arthur H. Clark, 1969.

Hoffman, Wilbur. *Sagas of Western Travel and Transport.* San Diego: Howell North Books, 1980.

Hollon, W. Eugene. *The Great American Desert, Then and Now.* New York: Oxford University Press, 1966.

Ise, John. *Sod and Stubble.* Lincoln: University of Nebraska Press, 1936.

Lewis, Meriwether, and William Clark. *The Journals of Lewis and Clark.* Edited by Bernard De Voto. Boston: Houghton Mifflin, 1953.

———. *The Journals of the Lewis and Clark Expedition.* New York: Dodd, Mead, 1906.

Lilley, William, and Lewis Gould. "The Western Irrigation Movement 1878–1902: A Reappraisal." In Gene Gressley, ed., *The American West: A Reorientation.* Laramie: University of Wyoming Publications, 1966.

Robinson, Michael. *Water for the West.* Chicago: Public Works Historical Society, 1979.

Rolvaag, O. E. *Giants in the Earth.*

Roosevelt, Theodore. *The Winning of the West.* Reprint. Fawcett House, 1963.

Shannon, Fred A. *The Farmer's Last Frontier.* New York: Farrar and Rinehart, 1945.

Smith, Henry Nash. *Virgin Land.* Cambridge, Mass.: Harvard University Press, 1950, 1970.

Stegner, Wallace. *Beyond the Hundredth Meridian.* Boston: Houghton Mifflin, 1953.

Taft, Robert. *Artists and Illustrators of the Old West.* New York: Scribner's, 1953.

Treadwell, Edward. *The Cattle King.* Fresno, Calif.: Valley Publishers, 1931.

Webb, Walter Prescott. *The Great Plains.* New York: Ginn, 1931.

Winship, George Parker. "The Coronado Expedition, 1540–1542." Washington, D.C.: U.S. Bureau of American Ethnology, Fourteenth Annual Report, 1892–93.

Winther, Oscar Osburn. *The Transportation Frontier: Trans-Mississippi West, 1865–1890.* New York: Holt, Rinehart and Winston, 1964.

CHAPTER TWO: **The Red Queen**

The story of how Los Angeles went to the Owens Valley for water has been told now and then, though not too accurately. The movie *Chinatown*, which came out in the mid-1970s, is a great film that may be responsible for misinforming a lot of people who consider it completely factual. (Oddly, Mulwray, the character whose name is a play on "Mulholland," comes across as a hero in the movie—and is murdered for his honesty—so the film may actually have polished Mulholland's reputation, which it probably did not intend to do.)

The most thorough and believable account, by far, of the whole Owens Valley–Los Angeles episode is William Kahrl's *Water and Power*, which was not published until 1982. Kahrl's prodigious research shows in the text. Remi Nadeau's *The Water Seekers* is considerably less exhaustive than Kahrl's book and is biased fairly heavily, in the end, in favor of Los Angeles. Nonetheless, it does contain some good anecdotal material, which I used in the chapter.

For a critical appraisal of Harrison Gray Otis, Harry Chandler, and the Los Angeles *Times* (the old *Times*, not the unrecognizably superior newspaper published by the third-generation Chandler, Otis), William Bonelli's *Billion Dollar Blackjack* is recommended. David Halberstam's *The Powers That Be* is also very good, though it deals more with the post-Otis newspaper. Anyone really interested in the mentality of the Los Angeles power structure at the turn of the century should peruse some old issues of the paper on microfilm; though more temperamental than most of his peers, Otis was no aberration.

Robert Matson's *William Mulholland: A Forgotten Forefather* provides some interesting personal detail about a very complicated man. Originally written as a thesis, the monograph is not easy to find in libraries.

Carey McWilliams's *California: The Great Exception* has to be considered required reading for anyone seriously interested in how California came to be the state and culture that it is. In fiction, James M. Cain may have captured southern California best, especially in *Mildred Pierce*; his essay "Paradise" is singular.

Important interviews for this chapter: Horace Albright, Jack Burby, Dorothy Green, David Kennedy, William Warne, Samuel P. Hays, and William Kahrl.

BOOKS

Bain, Joe S., et al. *Northern California's Water Industry*. Baltimore: Johns Hopkins University Press, 1966.

Beck, Warren A., and David A. Williams. *California: A History of the Golden State*. Garden City, N.Y.: Doubleday, 1972.

Bonelli, William G. *Billion Dollar Blackjack*. Beverly Hills, Calif.: Civic Research Press, 1954.

Carr, Harry. *Los Angeles: City of Dreams*. New York: Appleton-Century, 1935.

Chalfant, Willie Arthur. *The Story of Inyo*. Privately printed. Chicago, 1922.

Cooper, Erwin. *Aqueduct Empire*. Glendale, Calif.: Arthur H. Clark, 1968.

Dunne, John Gregory. "A Celebration of California." In Jonathan Eisen and David Fine, eds., *Unknown California*. New York: Macmillan, 1985.

Fogelson, Robert M. *The Fragmented Metropolis: Los Angeles 1850–1930*. Cambridge, Mass: Harvard University Press, 1967.

Gottlieb, Robert, and Irene Wolt. *Thinking Big: The Story of the Los Angeles Times, Its Publishers, and Their Influence on Southern California*. New York: Putnam, 1977.

Halberstam, David. *The Powers That Be*. New York: Knopf, 1979.

Hays, Samuel P. *Conservation and the Gospel of Efficiency*. New York: Atheneum, 1975.

Jorgenson, Lawrence C. *The San Fernando Valley, Past and Present*. Los Angeles: Pacific Rim Research, 1982.

Kahrl, William. *Water and Power*. Berkeley: University of California Press, 1982.

Kahrl, William, ed. *The California Water Atlas*. Sacramento: Department of Water Resources, 1979.

Keffer, Frank. *History of the San Fernando Valley*. Glendale, Calif.: Stillman, 1982.

Longstreet, Stephen. *All Star Cast: An Anecdotal History of Los Angeles*. New York: Thomas Y. Crowell, 1977.

McWilliams, Carey. *California: The Great Exception*. Santa Barbara: Peregrine Smith, 1949, 1976.

Matson, Robert W. *William Mulholland: A Forgotten Forefather*. Stockton, Calif.: University of the Pacific, Pacific Center for Western Studies, 1976.

Nadeau, Remy. *The Water Seekers*. Santa Barbara: Peregrine Smith, 1974.

Outland, Charles F. *Man-made Disaster: The Story of Saint Francis Dam*. Glendale, Calif.: Arthur H. Clark, 1963, 1977.

Watkins, T. H. *California: An Illustrated History*. Palo Alto: American West Publishing, 1973.

ARTICLES

Amaral, Anthony. "A Struggle in the Owens Valley." *American Forests*, August 1964.

Hayden, Frederick. "Los Angeles Aqueduct." *Building and Engineering News*, August 15, 1915.

Hoffman, Abraham. "Joseph B. Lippincott and the Owens Valley Controversy: Time for Revision." *Southern California Quarterly*, Fall 1972.

———. "Origins of a Controversy: The United States Reclamation Service and the Owens Valley–Los Angeles Water Dispute." *Arizona and the West*, Winter 1977.

Lippincott, Joseph B. "William Mulholland: Engineer, Pioneer, Raconteur." *Civil Engineering*, February/March 1941.

Los Angeles Times, 1898–1928. (Author's note: So many issues of the newspaper were reviewed for this chapter that it seems pointless to list them all here. Most of the citations in the chapter are dated. The newspaper is well indexed for anyone who wishes to review its coverage of the water issue during the period.)

"Mulholland Retires after 50-Year Service at Los Angeles." *Engineering News-Record*, November 22, 1928.

Wood, R. Coke. "Owens Valley as I Knew It." *Pacific Historian*, Summer 1972.

Yonay, Ehud. "How Green Was My Valley." *New West*, March 28, 1977.

CHAPTER THREE: First Causes

The chronicle of the political events leading to the passage of the Reclamation Act is based largely on William Lilley and Lewis Gould's "The Western Irrigation Movement 1878–1902: A Reappraisal," in Gene Gressley, ed., *The American West: A Reorientation*. The essay is revisionist history at its best—provocative yet sturdy—and few people seem to know of it. The chronicle of natural events that helped lead to passage of the Act is taken largely from Wallace Stegner's *Beyond the Hundredth Meridian*.

Samuel Hays's *Conservation and the Gospel of Efficiency* is a very good account of the early conservation movement and its utilitarian tenets.

Michael Robinson's *Water for the West* contains some good material on the failures of private irrigation ventures and contrasts vividly with William Smythe's supremely glorified view in *The Conquest of Arid America* (which was written much earlier). Eugene Hollon's *The Great American Desert, Then and Now* provided outstanding general background for this chapter, as did National Land for People's "Reclamation History" (three-part series).

BOOKS

Delano, Alonzo. *Life on the Plains and Among the Diggings*. Ann Arbor, Mich.: University Microfilms, 1966.

Gaffney, Mason. *Diseconomies Inherent in Western Water Laws*. Riverside, Calif.: January 1961 (unpublished monograph).

Hawgood, John A. *America's Western Frontiers*. New York: Knopf, 1967.

Hays, Samuel P. *Conservation and the Gospel of Efficiency*. New York: Atheneum, 1975.

Hollon, W. Eugene. *The Great American Desert, Then and Now*. New York: Oxford University Press, 1966.

Ise, John. *Sod and Stubble*. Lincoln: University of Nebraska Press, 1936.

Lilley, William, and Lewis L. Gould. "The Western Irrigation Movement 1878–1902: A Reappraisal." In Gene M. Gressley, ed., *The American West: A Reorientation*. Laramie: University of Wyoming Publications, 1966.

Robinson, Michael. *Water for the West*. Chicago: Public Works Historical Society, 1979.

Smythe, William E. *The Conquest of Arid America*. New York: Macmillan, 1905.

Warne, William. *The Bureau of Reclamation*. New York: Praeger, 1973.

CHAPTERS FOUR AND EIGHT: **An American Nile (I) and (II)**

These chapters (and the subsequent ones in the book) are drawn mostly from interviews, hitherto unseen files from the Bureau of Reclamation, and articles and reports. Anyone wishing to consult a single source for more background on the Colorado River and the conflicts over its use should read Philip Fradkin's *A River No More*.

Empires in the Sun, by Robert Gottlieb and Peter Wiley, contains an interesting account of how Kaiser, Bechtel, Morrison-Knudsen, and other firms that built Hoover Dam became instant giants through its construction. A detailed account of the actual construction work is in the Bureau of Reclamation's "Hoover Dam."

Helen Ingram's book, *Patterns of Politics in Water Resource Development*, is the best account I have seen of the political jockeying and compromising that led to passage of the Colorado River Basin Project Act. Dean Mann's *The Politics of Water in Arizona* is also helpful.

The Congressional debates over the Colorado River Storage Project (the 1956 act), especially those involving the late Senator Paul Douglas, one of the brainiest, wittiest, and most eloquent Senators we have ever had, are well worth reading. Economists were some of the earliest critics of water projects, but Douglas was even ahead of most economists.

Anyone who wishes to see how desperately Arizona wanted the Central Arizona Project built should review articles and editorials in the *Arizona Republic* and other state newspapers, particularly from the mid-1960s (prior to passage of the CAP legislation) and the late 1970s (the dread Carter years). Frank Welsh's *How to Create a Water Crisis* is a slightly dry but devastating dissection of the CAP and Arizona's perceived shortage of water, written by a former engineer with the Corps of Engineers and past president of the Phoenix chapter of the American Society of Civil Engineers.

David Brower's interviews for the Bancroft Library's Oral History Program (University of California, Berkeley) contain a lot of interesting anecdotal material about the battles over Echo Park, Glen Canyon, Marble Gorge, and Bridge Canyon dams. The Dominy archives at the University of Wyoming reveal what a pest Brower was to the water developers and make for an interesting dig.

In the 1980s it is striking to read the matter-of-fact tone with which the Pacific Southwest Water Plan and United Western Investigation propose monumental engineering works with staggering environmental consequences, and for what reasons. Both are in the author's files; they have become extremely difficult to find, though the Interior Department Library in Washington, D.C., ought to have them.

George Sibley's "The Desert Empire" is the best magazine article on the Southwest since Bernard De Voto's earlier essays in *Harper's*.

Important interviews for this chapter: Helen Ingram, John Leshy, Wesley Steiner, Daniel Dreyfus, David Brower, Jeffrey Ingram, Robert Young, William Martin, C. J. Kuiper, Stanford P. McCasland, William Warne, Myron Holburt, William Gookin, Daniel Beard, Nancy Laney, Robert Witzeman, Frank Welsh, Sam Steiger, Floyd Dominy, Tom Graff, Steven Reynolds, Patrick Dugan, Donald Maughan, Stewart Udall, Wayne Aspinall, Arleigh West.

Brower, David. *David R. Brower—Environmental Activist, Publicist, and Prophet*. Berkeley: Bancroft Library Oral History Program, University of California, 1980.

Fradkin, Philip L. *A River No More*. New York: Knopf, 1981.

Gottlieb, Robert, and Peter Wiley, *Empires in the Sun*. New York: Putnam, 1982.

Hollon, W. Eugene. *The Great American Desert, Then and Now*. Lincoln: University of Nebraska Press, 1985.

Holmes, Beatrice Hort. *History of Federal Water Resources Programs and Policies, 1961–70*. Washington, D.C.: U.S. Department of Agriculture Publication 1379, 1979.

Howe, Charles W., and K. W. Easter. *Interbasin Transfers of Water*. Baltimore: Johns Hopkins University Press, 1971.

Ingram, Helen M. *Patterns of Politics in Water Resource Development*. Tucson: University of New Mexico Press, 1969.

Mann, Dean. *The Politics of Water in Arizona*. Tucson: University of Arizona Press, 1963.

Ten Rivers in America's Future. Washington, D.C.: The President's Water Resources Policy Commission, 1950.

Trimble, M. *Arizona: A Panoramic History of a Frontier State.* Garden City, N.Y.: Doubleday, 1977.

U.S. Bureau of Reclamation. *Critical Water Problems Facing the Eleven Western States.* Washington, D.C.: U.S. Department of the Interior, 1975.

U.S. Water Resources Council. *The Nation's Water Resources.* Washington, D.C., 1968.

Welsh, Frank. *How to Create a Water Crisis.* Boulder, Colo.: Johnson, 1985.

ARTICLES

"Agency Mismanagement Responsible for Colorado River Flooding." Friends of the Earth, June 18, 1983.

"Alarm Over Deep 'Cracks' in Arizona." *San Francisco Chronicle,* July 4, 1982.

"All You Ever Wanted to Know About the CAP." Citizens Concerned About the Project (undated).

"Alternative to Orme Could Save Millions." *Phoenix Gazette,* June 2, 1979.

Animas-La Plata Project (feasibility data). U.S. Bureau of Reclamation, 1961.

"Arizonans Pushed Sierra Club Probe." *Phoenix Gazette,* July 1, 1966.

"Aspinall Raps Opposition to Project" and "Solon Blasts Detractors." *Albuquerque Journal,* November 11, 1966.

"Babbitt Appoints Water 'Czar.'" *Arizona Republic,* September 16, 1980.

"Battle Against Central Arizona Project Grows." *Rocky Mountain News,* March 30, 1966.

Boslough, John. "Rationing a River," *Science 81,* June 1981.

Bradley, Richard C. "Attack on Grand Canyon." *The Living Wilderness,* Winter 1964–65.

Brooks, Donald. "Testimony of Donald Brooks, Director of Planning, Metropolitan Water District of Southern California" (undated).

Brown, Howard. Memorandum to Senator Paul Fannin, "Wellton-Mohawk," May 5, 1975.

———. "The Central Arizona Project." Congressional Research Service, April 20, 1976.

"Build Orme Dam!" *Arizona Republic,* February 28, 1980.

"CAP Allocation Plan Criticized from All Sides." *Scottsdale Daily Progress,* October 28, 1980.

"Captured Flood Water Seen Aiding City Supply." *Phoeniz Gazette,* December 16, 1978.

Casserly, J. J. "Andrus Is Maneuvering to Cement His Anti-West Water Policy." *Arizona Republic* (undated).

Central Arizona Project, Environmental Statement (Final). U.S. Bureau of Reclamation, Washington, D.C. (undated).

"Colorado River, Vital to Southwest, Travels Ever Rockier Course." *Wall Street Journal* (undated).

Congressional Record, April 18, 1955. Senate debate on Colorado River Storage Project Act.

Dallas Creek Project (feasibility data). U.S. Bureau of Reclamation (undated).

Dallas Creek Project. U.S. Department of the Interior, Water Projects Review, April 1977.

Dams in Grand Canyon—A Necessary Evil? Sierra Club, August 1965.

"Debate Roils over Utah's Troubled Waters." *High Country News*, April 4, 1980.

"Dr. Strangelove Builds in the Desert." Maricopa Audubon Society, 1976.

"Dolores Project." Private papers of Roger Morrison.

Dolores Project (feasibility data). U.S. Bureau of Reclamation (undated).

"The Echo of Echo Park." Colorado River Water Conservation District, Glenwood Springs, Colorado, July 1981.

Etter, Alfred B. "Reservoir of the Unknown." *Defenders of Wildlife News*, April 1965.

———. "The Reclamation Machine." *Defenders of Wildlife News*, 1967.

"Excerpts from Memoirs of E. W. McFarland." Arizona Water Commission hearings, February 22, 1977.

Facts About the Proposed Grand Canyon Dams and the Threat to Grand Canyon. Colorado Open Space Coordinating Council, March 15, 1967.

"Farm Interests Lose Battle over Ground Water." *Arizona Republic*, May 22, 1979.

"Farmers Face Dilemma on Water Usage." *Desert News* (undated).

"Farmers See Little Help in CAP." *Phoenix Gazette*, October 28, 1980.

"Farms Called Big Losers in Water Battle." *Arizona Republic*, October 19, 1980.

"Fed Up? Here's a Way to Tell Washington." *Arizona Republic*, February 27, 1980.

"Flow Figures Suggest Orme as Inadequate." *Phoenix Gazette*, September 23, 1980.

"GAO: Colorado Basin in Trouble." *Rocky Mountain News*, May 22, 1979.

"GAO Glum on River Yield." *Denver Post*, May 22, 1979.

"Groundwater Balance Sought Early." *Arizona Daily Star*, March 20, 1982.

Hanson, Dennis. "Pumping Billions into the Desert." *Audubon* (undated).

Holburt, Myron. "California's Stake in the Colorado River." Colorado River Board of California, August 1979.

Ingram, Helen, et al. "Central Arizona Project: Politics and Economics in Irrigated Agriculture." John Muir Institute for Environmental Studies, August 1980.

Ingram, Helen, et al. "Water Scarcity and the Politics of Plenty in the Four Corners States." *The Western Political Quarterly*, September 1979.
"Interior Studying CUP Survival." *Desert News*, February 23, 1977.
"Irrigation Costs in Upper Colorado Basin." Private papers of Roger Morrison.
Kennedy, John F. "Special Message to Congress on Natural Resources," February 23, 1961. *Public Papers of the Presidents*.
Khera, Sigrid. "The Yavapia: Who They Are and from Where They Come."
Lichtenstein, Grace. "The Battle over the Mighty Colorado." *New York Times Magazine*, July 31, 1977.
McCasland, S. P. *United Western Investigation: Interim Report on Reconnaissance.* Bureau of Reclamation, Salt Lake City, January 1951.
———. "Water from the Pacific Northwest for Deserts of Southwest." *Civil Engineering*, February 1952.
McCaull, Julian. "Wringing Out the West." *Environment*, September 1974.
"New Colorado Water Fight." *Arizona Republic*, March 31, 1966.
"North Water Plan Detailed." *Sacramento Union*, March 17, 1965.
Ognibene, Peter. "Water Wasteland." *Washington Post*, May 3, 1978.
"The 160-Acre Limit and the 1902 Reclamation Act." Citizens Concerned About the Project, November 1977.
"Orme Still Needed." *Arizona Republic*, March 4, 1978.
"Panel Urges Land to Be Set Aside to Conserve Water." *Arizona Gazette*, December 14, 1978.
"Proposed Arizona Water Commission Staff Allocations of CAP Water." Arizona Water Commission (undated).
"Rio Salado, Not Orme." *Scottsdale Daily Progress*, March 8, 1978.
Riter, J. R. "Colorado River Basin Project," Memorandum to Chief Engineer, Bureau of Reclamation, February 13, 1967.
"River Plan Introduced by Kuchel." *Denver Post*, April 23, 1964.
Salisbury, David. "Managing Arid Lands." *Christian Science Monitor*, February 28, 1979.
San Miguel Project (feasibility data). U.S. Bureau of Reclamation, January 1966.
"Senate OKs Funds for CAP." *Arizona Republic*, September 11, 1980.
"Sen. Jackson Seeks More Congressional Authority over Bureau of Reclamation." *Medford Mail Tribune*, July 1, 1965.
Sibley, George. "The Desert Empire." *Harper's*, October 1977.
"Suit Seeks to Block CAP Dam." *Phoenix Gazette*, June 11, 1975.
"Thirsty Tucsonans Soaking the City Dry." *Arizona Daily Star*, February 7, 1982.
"Two Solons Decry Signing Reclamation Opponent." *Salt Lake Tribune*, October 23, 1966.
Udall, Morris. "Arizona and Water," February 1982.
"Udall Infuriated at Meet." *Arizona Republic*, March 31, 1966.

"Udall's Water Plan Shaky." *Arizona Republic*, January 24, 1964.

U.S. Department of the Interior. "Colorado River Storage Project." Washington, D.C.: U.S. Government Printing Office, 1964.

U.S. Department of the Interior. *Pacific Southwest Water Plan Report*, January 1964.

"Water Allocations: Central Arizona Project." William Gookin and Associates, March 3, 1975.

"Water and Politics." *Arizona Republic*, August 29, 1980.

"Water Code to Slow Growth." *Scottsdale Daily Progress*, June 6, 1980.

"Water from the Colorado Drawn into Arizona as Big Project Opens." *New York Times*, November 16, 1985.

Welsh, Frank. "Arizona Loses Water with CAP." *Arizona Reviews and News*, December 23, 1976.

———. "Sell Water to California." *Arizona Reviews and News*, March 3, 1977.

"Who, if Anybody, Is to Blame for Floods Along the Colorado?" *Wall Street Journal*, July 12, 1983.

Witzeman, Robert. "Consumers Ripped Off." *Arizona Republic*, October 19, 1978.

LETTERS, MEMORANDA, MISCELLANEOUS

Aspinall, Wayne. Letter to David Brower, November 22, 1966.

Beaty, Orren. Memorandum for S. Douglass Cater, Jr., The White House, February 10, 1967.

Bellport, Barney. Blue envelope letter to Commissioner Dominy, "Augmentation of critical water needs of major river basins," January 30, 1968.

———. Blue envelope memorandum to Commissioner of Reclamation, "Colored Dams," June 2, 1966.

———. Blue envelope memorandum to Commissioner Dominy, "Sierra Club," April 1, 1966.

Brower, David. Letter to Congressman Wayne Aspinall, November 19, 1966.

———. Letter to Congressman Wayne Aspinall, November 28, 1966.

———. Testimony, Colorado River Storage Project Hearings (undated, in files).

———. Letter to Interior Secretary James Watt, July 25, 1983.

Bureau of Reclamation, "Responses to Questions on Colorado River Water and Potential Raids on Pacific Northwest Water," internal memorandum, December 1975.

Clinton, Frank. Blue envelope memorandum to Floyd Dominy, Commissioner of Reclamation, "Eden Project, Wyoming," August 12, 1963.

Crandall, David. Blue envelope letter to Commissioner of Reclamation, "Incidental Scuttlebutt," April 4, 1968.

———. Blue envelope memorandum to Commissioner of Reclamation, "Summary of Upper Basin discussions concerning H.R. 4671 held in Denver January 18 and 19, 1966," January 20, 1966.

———. Blue envelope memorandum to Commissioner, "Wyoming items for possible inclusion in Colorado River project legislation," January 27, 1967.

Dick, James. "Water Pricing Policy of the Bureau of Reclamation." Memorandum to John Leshy of Natural Resources Defense Council (undated, in files).

Dickenson, Philip. Letter to Stanford P. McCasland, July 21, 1972.

———. Memorandum to Congressman Clair Engle, April 9, 1964.

Director, Resources Program Staff. Memorandum to Secretary of the Interior, "Pacific Southwest Water Plan—briefing material for meeting with President," January 21, 1964.

Dominy, Floyd. Letter to Ellis Armstrong, January 6, 1967.

———. Speech before the Southern California Water Conference, Los Angeles, December 14, 1964.

Gordon, Kermit. Memorandum to President Lyndon Johnson, April 22, 1968.

Graff, Tom. Letter to Gray Davis, Governor's Executive Secretary (California), January 8, 1980.

Harvey, Dorothy. Letter to Nelson Plummer, Regional Director, Bureau of Reclamation, July 19, 1978.

Hearings on Colorado River Storage Project Act, House of Representatives, *Congressional Record*, July 14, 1955.

Hilliard, E. H. Open Letter to Wayne Aspinall, May 1966.

Hogan, Harry. Handwritten memo from associate solicitor, Department of the Interior, to "Floyd," "Columbia River Compact Will Block Diversion of Col. R. Water to the Southwest," February 20, 1964.

Jensen, Joseph. Letter to Stewart Udall, December 31, 1964.

Jukes, Thomas H. Letter to George Marshall, President, Sierra Club, September 2, 1966 (and several others by same author).

Kuchel, Thomas. Remarks on Senate floor, "Protecting the Present Users of Lower Colorado River Water," *Congressional Record*, April 25, 1964.

McCasland, Stanford P. Letter to Philip Dickenson, August 10, 1972.

Manatos, Mike. Memorandum to Lawrence O'Brien, The White House, May 11, 1964.

Mitchell, A. L. Memorandum to Chief Engineer, Bureau of Reclamation, "Attendance at meeting on the Colorado River sponsored by the Colorado Mountain Club and the Sierra Club—Phipps Auditorium, March 22, 1966," March 23, 1966.

Palmer, William. Memorandum to Commissioner of Reclamation, "Water

resource development in the Lower Colorado River Basin," August 3,
1962.

Peterson, Ottis. Blue envelope letter to Commissioner Floyd Dominy,
March 29, 1964.

Pugh, C. A. Blue envelope letter to Commissioner Dominy, "Reply to Aspinall letter to Colorado River Basin Governors regarding Central Arizona Project water supply," June 3, 1965.

———. Blue envelope memorandum to Regional Director, Bureau of Reclamation, "Report on debate with Mr. David Brower of the Sierra
Club relative to Bridge Canyon and Marble Canyon Dams, February
10, 1965," March 20, 1965.

Riggins, Ted. Memorandum to F. Dominy, G. Stamm, N. B. Bennett, "Status Report: Central Arizona Project legislation," October 5, 1967.

Riter, J. R. Memorandum to Chief Engineer, Bureau of Reclamation, "Special Meeting of Colorado Water Conservation Board," August 18, 1965.

Robins, J. W. Memorandum to Project Manager, U.S. Bureau of Reclamation, Grand Junction, Colorado, "Colorado Water Conservation
Board meetings of September 7 and 8, 1967," September 13, 1967.

Rorke, H. B. Note to Ottis Peterson, Director of Public Information, Bureau
of Reclamation, "IntRevService clamp on the Sierra Club," June 21,
1966.

Schultz, Charles. Memorandum to President Lyndon Johnson, "1968 new
construction starts for the Corps of Engineers," December 31, 1966.

Straus, Michael. Letter to William Warne, December 30, 1952.

Udall, Stewart. Memorandum to President Lyndon Johnson, "Senator
Hayden and the Lower Colorado Project," August 9, 1967.

Warne, William E. Letter to Harry Bashore, November 12, 1963.

———. Letter to Senator Carl Hayden, March 4, 1964.

West, Arleigh. Blue envelope letter to Commissioner of Reclamation,
"Randy Riter's latest 'ciphering' on the Lower Colorado River Basin
water supply," June 17, 1965.

Witzeman, Dr. Robert. Letter to Cecil Andrus, Secretary of Interior, May
23, 1980.

———. Letter to Governor Jack Williams of Arizona, August 29, 1977.

———. Memorandum to Lee Thompson, January 24, 1980.

CHAPTER FIVE: **The Go-Go Years**

William Manchester's *The Glory and the Dream* is anecdotal history at its
best, and contains much fascinating stuff on the Roosevelt years and the
New Deal's glorification of public works.

Donald Worster's and Paul Bonnifield's books give strikingly different
impressions of the Dust Bowl. To Bonnifield, it was a natural event that

would have happened even if the plains hadn't been overgrazed and plowed up; to Worster, it was almost entirely a man-made disaster. Paul Sears's *Deserts on the March* is still the classic book on the subject, and Sears's conclusions land much closer to Worster than to Bonnifield (who, interestingly, is an Oklahoman).

George Sundborg's *Hail Columbia* is the story of the damming of the river from the viewpoint of an ardent New Deal water developer (he was administrative assistant to the late Senator Ernest Gruening, who wanted to dam the Yukon, too, and became exasperated that the Soviet Union was building bigger dams than ours). Albert Williams's book is more balanced, but not as detailed.

Daniel Jack Chasan's *The Water Link* and Anthony Netboy's *The Columbia River Salmon and Steelhead Trout: Their Fight for Survival* both contain mournful accounts of the fabulous fisheries destroyed by dams and logging in the Northwest.

Considerable information on the WPPSS fiasco, an indirect result of the huge dam-construction program in the Northwest (and something which I passed over rather lightly in the chapter), is in the files of the Natural Resources Defense Council in San Francisco.

Important interviews for this chapter: Phil Nalder, Frank Weil, Larry Meinert, Floyd Dominy, Ralph Cavanagh, Jim Casey, Horace Albright, Samuel Hays, Gilbert Stamm, C. J. Kuiper, Gus Norwood, A. J. Voy, Daniel Dreyfus.

BOOKS

Bonnifield, Paul. *The Dust Bowl*. Albuquerque: University of New Mexico Press, 1979.

Chasan, Daniel Jack. *The Water Link*. Seattle: University of Washington Press, 1981.

Columbia Basin Irrigation Project Report. Washington, D.C.: House Committee on Irrigation and Reclamation, 1928.

The Columbia River: A Comprehensive Report. Washington, D.C.: U.S. Bureau of Reclamation, 1947.

Holbrook, Stewart. *The Columbia River*. New York: Holt, Rinehart, and Winston, 1965.

Ickes, Harold. *The Secret Diary of Harold Ickes*. New York: Simon and Schuster, 1953.

Johanson, Dorothy O., and Charles M. Gates. *Empire of the Columbia*. New York: Harper and Row, 1957.

Jones, Fred O. *Grand Coulee from "Hell To Breakfast."* Portland, Ore.: Binfords and Mort, 1947.

Lavender, David. *Land of Giants*. Garden City, N.Y.: Doubleday, 1958.

Lowi, Theodore. *The End of Liberalism*. New York: Norton, 1969.

———. *Legislative Politics U.S.A.* (esp. "How the Farmers Get What They Want"). Boston: Little, Brown, 1962.

Manchester, William. *The Glory and the Dream.* New York: Bantam, 1975.

Netboy, Anthony. *The Columbia River Salmon and Steelhead Trout: Their Fight for Survival.* Seattle: University of Washington Press, 1980.

Neuberger, Richard. *Our Promised Land.* New York: Macmillan, 1938.

Schad, Theodore, and John Kerr Rose. *Reclamation: Accomplishments and Contributions.* Washington, D.C.: Library of Congress Legislative Reference Service, 1958.

Sears, Paul. *Deserts on the March.* Norman: University of Oklahoma Press, 1935.

Sheridan, David. *Desertification of the United States.* Washington, D.C.: Council on Environmental Quality, 1981.

Sundborg, George. *Hail Columbia: The Thirty-Year Struggle for Grand Coulee Dam.* New York: Macmillan, 1954.

Ten Rivers in America's Future. Report of the President's Water Resources Policy Commission. Washington, D.C., 1950.

Warne, William E. *The Bureau of Reclamation.* New York: Praeger, 1973.

Williams, Albert N. *The Water and the Power.* New York: Duell, Sloan and Pearce, 1951.

Worster, Donald. *Dust Bowl.* New York: Oxford University Press, 1979.

ARTICLES AND REPORTS

Case, Robert Ormond. "Eighth World Wonder." *Saturday Evening Post,* July 13, 1935.

Davenport, Walter. "Power in the Wilderness: Grand Coulee and Bonneville." *Collier's,* September 21, 1935.

George W. Goethals and Company. *Columbia Basin Irrigation Project, State of Washington: A Report.* Washington State Department of Conservation and Development, 1922.

"Grand Coulee Dam Is Again the Biggest." *Seattle Post-Intelligencer,* August 17, 1980.

Marshall, Jim. "Dam of Doubt." *Collier's,* June 19, 1937.

"Salmon Shortage Cited as Eagles Shun Stream." *New York Times,* December 5, 1982.

Taylor, Frank. "The White Elephant Comes into Its Own." *Saturday Evening Post,* June 5, 1943.

Tucker, Ray. "Interior's Adventures in the Missouri Basin." *Public Utilities Fortnightly,* July 17, 1952.

"Washington's Power Problem." *New York Times,* February 15, 1983.

"What's Happening on the Ainsworth Project?" *Nebraska Farmer,* January 1964.

"Where's the Limit on Reclamation Projects?" *Nebraska Farmer*, March 1964.
"WPPSS Preparing to Sell What Remains of Two Huge Nuclear Units, Piece by Piece." *Wall Street Journal*, August 4, 1983.

LETTERS, MEMORANDA, MISCELLANEOUS

Bellport, Barney. Blue envelope memorandum to Commissioner Dominy, "Delegation of design and specifications work and contracting authority for drains—Columbia Basin Project," May 27, 1966.
Dominy, Floyd. Letter to Claire and Donald Hanna, April, 15, 1955.
———. Memorandum to Chief of Allocation and Repayment Division, "Delay in amendatory repayment contract material review," November 2, 1949.
Dugan, H.P., et al. Blue envelope memorandum to Commissioner Dominy, "OBE-ERS Presentation," March 30, 1965.
Dugan, Patrick. Blue envelope letter to Commissioner Dominy, April 22, 1966.
Lineweaver, Goodrich. Letter to Floyd Dominy, September 2, 1949.
———. Memorandum to E. D. Eaton, September 2, 1949.
Nelson, Harold. Memorandum to Commissioner Dominy, "Extension of Columbia River Basin Account Benefits to Older Projects," February 19, 1968.
Pafford, Robert. Letter to Brigadier General Arthur H. Frye, Jr., November 8, 1963.
Peterson, E. L. Letter to Secretary of the Interior, "Garrison Diversion Project," November 20, 1957.
Saylor, John. "Is Power Really Reclamation's Paying Partner? Or Hominy Dominy Sat on the Wall." Extended remarks in *Congressional Record*, February 11, 1965.
Straus, Michael. Memorandum to regional director, Billings, Montana, "Proposed Repayment Contracts, Milk River Project," July 12, 1949.
Straus, Michael, Lewis Pick, J. A. Krug, and Kenneth Royall. Letter to the President, April 11, 1949.

CHAPTER SIX: **Rivals in Crime**

The account of the Corps of Engineers' coup on the Tulare Basin rivers in California is taken mainly from Arthur Maass's *Muddy Waters*. For the story of Garrison Dam and the drowning of the Three Tribes, I have relied largely on Arthur Morgan's *Dams and Other Disasters*.

The competition between the Corps and the Bureau is something of which I was completely unaware (as most conservationists are, too) until

I came across the Bureau's secret "blue envelope" files. Spokesmen for the Corps of Engineers were of no help in corroborating this information. The Marysville Dam episode, however, was largely corroborated in interviews with Robert Pafford, one of the chief actors. The self-defeating competition on California's North Coast rivers was similarly corroborated by David Shuster, formerly operations manager of the Central Valley Project, and to a lesser degree by William Warne.

Some of the Rampart Dam story is based on interviews with Floyd Dominy and John Gottschalk.

Other important interviews for this chapter: David Weiman, Richard Madson, George Piper, Ed Green, General John Woodland Morris (ret.), H. P. Dugan, Peter Carlson, John Marlin, Tom Barlow, Jim Cook, Norman Livermore, Richard Wilson, Jim Casey, Edmund G. Brown, Sr., Ronald B. Robie, Gerald Meral, James Flannery, Brent Blackwelder, Anthony Wayne Smith, Raphael Kazmann, Guy Martin.

BOOKS

Frank, Bernard, and Anthony Netboy. *Water, Land and People.* New York: Knopf, 1950.

Hart, Henry C. *The Dark Missouri.* Madison: University of Wisconsin Press, 1957.

Maass, Arthur. *Muddy Waters: The Army Engineers and the Nation's Rivers.* Cambridge, Mass.: Harvard University Press, 1951.

Morgan, Arthur. *Dams and Other Disasters.* Boston: Porter Sargent, 1971.

Schad, Theodore, and John Kerr Rose. *Reclamation: Accomplishments and Contributions.* Washington, D.C.: Library of Congress Legislative Reference Service, 1958.

Terral, Rufus. *The Missouri Valley.* New Haven: Yale University Press, 1947.

Williams, Albert N. *The Water and the Power.* New York: Duell, Sloan and Pearce, 1951.

ARTICLES AND DOCUMENTS

"Audit reveals Pick-Sloan poorly run, loss of funds." *Lincoln Sunday Journal and Star,* 1977.

"Audits Show Unbusinesslike Management of Western Basin Accounts." Environmental Policy Center, Washington, D.C. (undated).

"Budget includes funds for CENDAK planning." *Huron* (S.D.) *Daily Plainsman,* February 1, 1983.

Brooks, Paul. "The Plot to Drown Alaska." *The Atlantic Monthly,* May 1965.

"CENDAK benefits might not offset costs, new study says." *Huron* (S.D.) *Daily Plainsman,* December 9, 1981.

"Conflicts Among Agencies Peril Water Development." *Willows* (Calif.) *Daily Journal,* July 27, 1965.

De Roos, Robert, and Arthur Maass. "The Lobby That Can't Be Licked: Congress and the Army Engineers." *Harper's,* May 1949.

"Friction Periling Vast Water Plan." *Willows* (Calif.) *Daily Journal,* July 27, 1965.

"Garrison: The Canadian Concern." Manitoba Department of Natural Resources, Winnipeg, Canada (undated).

Garrison Diversion: Opposing Views. A compendium published by the Red River Valley Historical Society, March 1981.

Gruening, Ernest. "The Plot to Strangle Alaska." *The Atlantic Monthly,* July 1965.

Jacobs, Mike. "The Garrison Diversion Project is North Dakota's history, and destiny." *High Country News,* September 17, 1984.

Oakes, John. "Pork—U.S. Prime." *New York Times,* July 29, 1981.

A Review of the Environmental, Economic, and International Aspects of the Garrison Diversion Unit, North Dakota. Committee on Government Operations, U.S. Congress 1976.

U.S. Department of the Interior. "Alaska Natural Resources and the Rampart Project," June 1967.

LETTERS, MEMORANDA, MISCELLANEOUS

Bellport, Barney. Blue envelope letter to Commissioner, "Determination of optimum size of power plant—Auburn Dam," March 21, 1968.

Bureau of Reclamation. "Need for an Accelerated Water Resources Development Program, North Coast Project, California" (undated).

Cassidy, General W. F. Speech at Topical Conference on Hydraulics, Davis, California, August 16, 1962.

Chief, Division of Project Development, Bureau of Reclamation. Memorandum to Commissioner, "GAO Report on Central and Southern Florida Project," Corps of Engineers, December 24, 1964.

Dickinson, Phil. Letter to J. A. Krug, Secretary of the Interior-designate, March 28, 1946.

Dominy, Floyd E. Blue envelope letter to Pat Dugan, April 13, 1960.

———. Letter to Claire and Donald Hanna, April 15, 1955.

———. Memorandum to Kenneth Holum, "Understandings with the Corps—Columbia River Basin," December 20, 1961.

Holum, Kenneth. Letter to Elmer Staats, "Coordination of Bureau of Reclamation and Corps of Engineers investigations programs," November 14, 1961 (includes report on subject).

Jennings, Robert. Blue envelope letter to Floyd Dominy, Associate Commissioner of Reclamation, March 5, 1959.

Johnson, Bruce. Blue envelope memorandum to Commissioner of Reclamation, "Joint Study Missouri River, Fort Peck to Great Falls," May 9, 1963.

———. Blue envelope memorandum to Floyd Dominy, Commissioner of Reclamation, "Memorandum of Agreement with the Corps of Engineers," May 19, 1962.

———. Blue envelope letter to Floyd Dominy, Commissioner of Reclamation, "Oahe Unit Feasibility Report," April 18, 1960.

Mangan, John. Memorandum to Regional Director, Bureau of Reclamation, Boise, Idaho, "Conversation with Mr. Don Lane, Oregon State Water Resources Board," January 22, 1965.

Nelson, H. T. Blue envelope memorandum to Commissioner of Reclamation, "Relationships with the Corps of Engineers—Umatilla River Basin, Oregon," April 24, 1962.

Nelson, Harold. Blue envelope memorandum to Commissioner, "Army Bureau relationships in the Northwest states," November 30, 1961.

———. Blue envelope memorandum to Commissioner, Bureau of Reclamation, "Interior and Insular Affairs Committee vs. Public Works Committee approach to water resource development authorizations," February 9, 1965.

———. Blue envelope memorandum to Floyd Dominy, Commissioner of Reclamation, "So-called 'Information Bulletin' issued by the Corps of Engineers re: Rogue River Basin, Oregon," February 20, 1962.

———. Blue envelope memorandum to Floyd Dominy, Commissioner of Reclamation, "Subcommittee on Public Works visitation to flood disaster areas—California–Oregon," January 22, 1965.

Pafford, Robert. Blue envelope letter to Floyd Dominy, Commissioner of Reclamation, February 4, 1966.

———. Blue envelope letter to Floyd Dominy, Commissioner of Reclamation, August 4, 1966.

Project Manager, Bureau of Reclamation, Bismarck, North Dakota. Blue envelope memorandum to Commissioner, "Meeting with Local Interests in Minot," July 10, 1959.

A [Missouri] *River Basin Management Post-Audit and Analysis.* Arthur O. Little, Inc., Cambridge, Massachusetts, May 1973.

Roberts, Daryl L. Blue envelope letter to Floyd E. Dominy, April 7, 1961.

Roosevelt, Franklin Delano. Letter to Harold Ickes, May 29, 1940.

———. Letter to Harold D. Smith, June 1, 1940.

Sanders, Barefoot. Memorandum to Jim Jones, the White House, July 29, 1968.

Spencer, John. Blue envelope letter to Floyd Dominy, February 19, 1960.

Stamm, Gilbert. Blue envelope memorandum to Floyd Dominy, Commissioner of Reclamation, "Establishment of a wild river on Yellowstone River above Emigrant, Montana," February 3, 1965.

White, Lee C. Memorandum to President Lyndon Johnson, "Meeting with Alaska Congressional Delegation," March 31, 1964.

CHAPTER SEVEN: **Dominy**

The most important source for this chapter was Floyd Dominy himself. He regaled me with exploits and achievements that made for irresistible listening. Anyone who has ever worked with Dominy has a story or tale to relate—and I've tried to select the best.

John McPhee's *Encounters with the Archdruid*, in which Floyd Dominy and David Brower raft the Colorado River together, arguing nearly all the way, is some of the best journalism published in years.

BOOKS

McPhee, John. *Encounters with the Archdruid*. New York: Farrar, Straus and Giroux, 1970.
Robinson, Michael. *Water for the West*. Chicago: Public Works Historical Society, 1979.
Warne, William. *The Bureau of Reclamation*. New York: Praeger, 1973.

LETTERS, MEMORANDA, AND ARTICLES

Anonymous. "Ode to Domine Dominy," April 18, 1955.
Bellport, Barney. Blue envelope memorandum to Commissioner, Bureau of Reclamation, "Criticism for development of irrigation projects," February 18, 1965.
———. Blue envelope memorandum to Commissioner, Bureau of Reclamation, "Replacement of American Falls Dam," May 24, 1967.
Buckman, H. H., President, National Rivers and Harbors Congress. Letter to Floyd Dominy, April 19, 1962.
"The Crisis in Water: Its Sources, Pollution and Depletion," *Saturday Review*, October 23, 1965.
Crandall, David. Blue envelope letter to Commissioner, Bureau of Reclamation, "Wyoming items for possible inclusion in Colorado River Project legislation," January 27, 1967.
———. Memorandum to Commissioner, Bureau of Reclamation, June 23, 1967.
Director, Columbia Basin Project. Blue envelope letter to Commissioner Floyd E. Dominy, January 28, 1969.
Dominy, Floyd. Blue envelope letter to B. P. Bellport, Bureau of Reclamation, August 13, 1965.

———. Blue envelope letter to Clyde Spenser, Regional Director, Bureau of Reclamation, Sacramento, July 12, 1954.

———. Blue envelope letter to Gilbert Stamm, November 7, 1952.

———. Blue envelope memorandum to Assistant Commissioner and Chief Engineer, "Escalation in Construction Contracts," March 15, 1960.

———. Blue envelope memorandum to Assistant Secretary for Water and Power, Department of the Interior, "Proposed employment of Robert J. Pafford, Jr., as Regional Director, Region II," December 11, 1962.

———. Blue envelope memorandum to Chief Engineer, "Bureau of Reclamation Flag," April 25, 1966.

———. Blue envelope memorandum to Regional Director, Bureau of Reclamation, Sacramento, "Audubon Society Convention," August 4, 1966.

———. Blue envelope memorandum to Secretary of the Interior, "Planned engagements and related travel for Commissioner of Reclamation," May 8, 1962.

———. Letter to Alfred Etter, November 23, 1965.

———. Letter to Senator Carl Hayden, September 6, 1966 (with attachments).

———. Personal memorandum to files, "Meeting with Commissioner Dexheimer et al.—Columbia Basin anti-speculation and excess land problems—February 28, 1956," March 12, 1956.

———. Professional Diary (miscellaneous items, 1954–55).

"Dominy Foresees Water Sharing Need After Northwest Meets Requirements," *Idaho Daily Statesman*, January 22, 1965.

"Dominy's Appointment Tipoff to Repayment," *Columbia Basin Herald*, January 18, 1961.

Dugan, H. P. Blue envelope memorandum to Commissioner, Bureau of Reclamation, "January 18 meeting of Nebraska Mid-State Reclamation District Board." January 20, 1965.

Etter, Alfred G. "How Reclamation Can Kill the West," *Defenders of Wildlife News*, April/May/June 1966.

Frantz, Joe B. *Floyd E. Dominy Oral History*. Lyndon Baines Johnson Library Oral History Program, University of Texas, Austin, November 14, 1968.

Holum, Kenneth. Blue envelope memorandum to Commissioner of Reclamation, "Briefing Session—House Interior Committee—January 28," January 28, 1965.

Nelson, H. T. Blue envelope memorandum to Commissioner, Bureau of Reclamation, "Interest of the Idaho Water Resources Board in Middle Snake River Development Program," October 20, 1967.

Nelson, Harold. Blue envelope memorandum to Commissioner, Bureau of Reclamation, "Relationships with the Corps of Engineers—Umatilla River Basin," April 24, 1962.

Peterson, Ottis. Memorandum to Mr. Dominy, March 30, 1967.

———. Blue envelope letter to George N. Pierce, District Manager, Bureau of Reclamation, Juneau, Alaska, June 1, 1965.

Pettingill, Olin. "Convention Country," *Audubon*, 1966.

"Reclamation Boss Chides Utah Chief," *Arizona Daily Star*, October 20, 1962.

Regional Directors, Bureau of Reclamation, Boise and Denver. Blue envelope letter to Commissioner, "Meeting with the Governor of the State of Wyoming," April 3, 1962.

Stamm, Gilbert. Blue envelope letter to Floyd Dominy, Director of Irrigation, Bureau of Reclamation, April 26, 1954.

Stuart, Russell. Letter to Stewart Udall, February 18, 1966.

Udall, Stewart. Personal memorandum to Floyd Dominy, Commissioner of Reclamation, February 26, 1966.

"Udall Effects Troubled Truce in Two-Year Carr-Dominy Feud," *Pueblo* (Colo.) *Chieftain*, February 25, 1963.

"Warning to Interior," *Pueblo* (Colo.) *Chieftain*, September 3, 1962.

West, Arleigh. Blue envelope memorandum to Commissioner, Bureau of Reclamation. "Excess Lands—Imperial Irrigation District—Boulder Canyon Project, California" (undated).

———. Blue envelope memorandum to Commissioner Dominy. "CONFIDENTIAL: Excess land survey in areas served by Metropolitan Water District, Southern California," December 30, 1964.

CHAPTER NINE: **The Peanut Farmer and the Pork Barrel**

This chapter is based mainly on interviews and newspaper reporting. Sources who should be mentioned are Robert Smythe, Richard Ayres, J. Gustave Speth, Jane Yarn, Claude Terry, James Flannery, Peter Carlson, David Conrad, Jim Free, Guy Martin, John Leshy, Laurence Rockefeller, Tom Barlow, David Weiman, Ronald Robie, Congressman Robert Edgar, Brent Blackwelder, former Congressman Robert Eckhardt, Congressman Tom Bevill, John Lawrence, Congressman John Myers, Ruth Fleischer, William Dubois, Daniel Beard, and Steven Lanich.

Congressman Jim Wright's *The Coming Water Famine* makes for interesting reading if one wishes to understand how thoroughly a basically self-interested politician can delude himself into thinking he is serving the commonweal.

The Tellico story is drawn partly from Fred Powledge's *Water*. A good critical appraisal of the TVA's record in Appalachia is William Chandler's *The Myth of TVA*.

BOOKS

Chandler, William U. *The Myth of TVA*. Cambridge, Mass.: Ballinger, 1984.
Powledge, Fred. *Water*. New York: Farrar, Straus and Giroux, 1982.
Reid, T. R. *Congressional Odyssey*. San Francisco: W. H. Freeman, 1980.
Wright, Jim. *The Coming Water Famine*. New York: Coward-McCann, 1966.

ARTICLES AND REPORTS

"Accord Reached in Westlands Pact." *Sacramento Bee*, 1979.
American Rivers, December 1977 (entire issue).
"Andrus, Governors Weigh Drought." *Denver Post*, February 21, 1977.
"Andrus Sees No Major Shifts under Successor." *New York Times*, November 18, 1980.
"Andrus's Popularity Washcs Away in West." *New York Times*, February 20, 1978.
"Belly Up to the Trough, Boys!" *New Republic*, October 14, 1978.
Broder, David S. "A Most Puzzling Maneuver." *Washington Post*, March 16, 1977.
"California Farmers' Clout Preserves Federal Water Subsidy." *Washington Post*, August 19, 1979.
Carter, President Jimmy. "To the Congress of the United States," February 21, 1977.
———. "To the Congress of the United States," June 6, 1978.
"Carter in Full Retreat in 'War on West.' " *Washington Post*. January 1978.
"Carter Opts to Sidestep Fight on 160-Acre Limit." *Sacramento Bee*, June 22, 1979.
"Carter Water Policy Hurt." *Washington Post*, March 31, 1978.
"Carter Will Ask Hill to Halt Aid for 18 Major Water Projects." *Los Angeles Times*, April 18, 1977.
"Carter Won't Seek Cut in Big Projects." *New York Times*, January 14, 1978.
"Carter Yields on Water Projects." *Philadelphia Inquirer*, January 15, 1978.
"Carter's Water Policy: Furor in an Election Year." *Washington Post*, June 11, 1980.
"CEQ Releases Summary of Water Resource Project Deletions." Council on Environmental Quality, February 23, 1977.
"Congress' Going Away Gifts." *Washington Post*, December 10, 1980.
"Congress Makes Waves over Carter's Water Policy." *National Journal*, July 1, 1978.
"Devastating Blow Dealt Water Projects Pork Barrel." *Science*, October 27, 1978.

"Energy and Public Works Appropriations Bill." *Congressional Record,* October 5, 1978.

"Energy and Water Development Appropriations for 1981." Hearings, House Appropriations Committee, 1981.

"Environmentalists Slam Carter." *Rocky Mountain News,* March 31, 1978.

"Executive Summary: Water Resources Option Paper." Carter Domestic Policy Staff (internal document, undated).

Gardner, Don. "The Trinity River: Water and Politics." *Texas Observer,* May 20, 1977.

"Governors Assured by Andrus on Water." *New York Times,* June 24, 1979.

"Hart, Haskell Demand Data on Water Projects." *Denver Post,* 1977.

"House Sustains Veto of Public Works Bill." *Wall Street Journal,* October 6, 1978.

"Issue Paper: Federal Water Resources Policy." Carter Domestic Policy Staff (internal document), January 28, 1977.

Lamm, Richard D., and Scott M. Matheson. "Deficits: A Noose." *New York Times* (undated).

"Louisiana Girding to Save Waterway from the 'Hit List.'" *Washington Post,* March 28, 1977.

"Plan to Share Water Project Costs Is Gaining in West Under Reagan." *New York Times,* September 12, 1982.

"President Is Warned by House Democrats." *New York Times,* May 23, 1977.

"Roll Out the Barrels, We'll Have a Barrel of Funds, Folks Say." *Wall Street Journal* (undated).

"Senate Vote Defies Carter." *Washington Post,* March 11, 1977.

"Senators, White House Wrangle over Powers." *Washington Post,* March 1977.

"A Threat to Block Valley Water Pact." *San Francisco Chronicle,* October 4, 1979.

"Top Western State Officials Blast Water Project Cuts." *The Missoulian,* February 21, 1977.

"Turning Off the Water." *Newsweek,* April 4, 1977.

U.S. Department of the Interior, Water Projects Review. "Auburn-Folsom-South Project, California," April 1977.

"Water Policy: Battle over Benefits." *Congressional Quarterly,* March 4, 1978.

"Water Policy Reforms Going down the Drain?" Environmental Policy Center, *Resource Report,* May 1978.

"Water Project Budget Remains Virtually Untouched." American Rivers Conservation Council, March 1980.

"Water Projects Dispute: Carter and Congress Near a Showdown." *Science,* June 17, 1977.

"Watt Studies Sharing of Costs for Western Water Projects." *New York Times*, June 19, 1983.

"Watt Wading into Water Policy." *Washington Post*, April 18, 1981.

"Watt Would Lift Irrigation Limit, Reduce Subsidy." *Washington Post*, December 10, 1981.

"Westlands Hearings Not Likely to Bring Immediate Decisions." *San Francisco Examiner*, March 20, 1980.

LETTERS, MEMORANDA, MISCELLANEOUS

Dugan, Patrick. Blue envelope letter to Floyd Dominy, Commissioner of Reclamation, "Folsom South Unit," November 23, 1962.

Gordon, Kermit. Memorandum for the President, "Policies for Handling Navigation Projects," March 8, 1965.

Green, John A., Environmental Protection Agency. Letter to David Crandall, Regional Director, Bureau of Reclamation, Salt Lake City, November 28, 1976.

Kirwan, Michael, et al. Letter to the President, April 26, 1966.

Udall, Morris, et al. Letter to the Honorable Jimmy Carter, February 14, 1977.

Watson, Marvin. Memorandum to the President, March 24, 1965.

Wright, Jim. Letter to the President, April 7, 1966.

CHAPTER TEN: **Chinatown**

All the quotations from former Governor Pat Brown are in *California Water Issues, 1950–1966*, a bound volume of interviews conducted by Malca Chall of the University of California's Bancroft Library Oral History Program. The Bancroft Library has also conducted interviews with William Warne, Ralph Brody, and some of the other important participants in California's recent water-development history that are well worth reading.

Lynn Ludlow of the *San Francisco Examiner* has done an excellent job of chronicling abuses of the Reclamation Act in California. So has George Baker of the *Sacramento Bee*, whose coverage of the Peripheral Canal wars was also the best in the state.

Patrick Porgans of Red Tape Abatement, Inc., a private research and consulting firm, provided considerable assistance in understanding the financial aspects of the State Water Project. E. Philip LeVeen and Rob Stavens of Public Interest Economics have also published much useful material, as has Dorothy Green of WATER and the Contra Costa County Water Agency. Anyone trying to fully understand the project should also consult the annual reports of the Department of Water Resources.

Carey McWilliams's *California: The Great Exception* is highly recommended for its portrayal of how agribusiness, banking, food processing, the university extension system, cheap imported labor, and publicly subsidized water have created a huge economic juggernaut in the state. It may be the best general book written about California. The best essayist rooting around where California culture and politics meet, in my opinion, is not Joan Didion, but her husband, John Gregory Dunne. His "Eureka! A Celebration of California" is especially fine, though Didion's more famous essay, "Holy Water," is not to be missed.

A sense of the concentration of agricultural wealth in California can be gained from "Getting Bigger," by the California Institute for Rural Studies, which profiles the 211 largest farming companies in the state (the smallest of the 211 is a 5,000–acre operation). The study, a superb piece of research, reveals a good deal about interlocking directorates, holding companies, vertical integration in the food market, parent companies, hidden partnerships, market penetration, and so on. Most of the information on the big growers benefiting from the State Water Project comes from CIRS.

It is almost impossible to understand water and California history without consulting the *California Water Atlas*, a huge (in dimension), beautifully produced work that really does deserve to be called unique. To anyone with a keen interest in the subject, the LANDSAT photos and graphs (depicting river flows, rainfall records, floods, droughts, irrigation deliveries, pumping energy consumed, etc.) will be fascinating. The text is persistently neutral when discussing the political wars.

For thirty or forty years, a Berkeley professor named Paul Taylor kept up a largely futile but unflagging effort to reform the enforcement of the Reclamation Act (rather than "reform" the Act). His essays on the subject are meticulous and readable, especially when they delve into the social effects of agricultural giantism. Much useful information on the acreage limitation, and violations thereof, has also been published by National Land for People; though it is portrayed by the growers as a "radical" organization, its only real goal is enforcement of one of the most poorly enforced laws in the nation.

Important interviews for this chapter: Ronald Robie, Dorothy Green, Lorelle Long, Tom Graff, George Ballis, Kendall Manock, Edmund G. Brown, Sr., Gerald Meral, Ellen Stern Harris, Lawrence Swenson, Patrick Porgans, E. Philip LeVeen, Myron Holburt, Jack Burby, Willoughby Houk, Paul Taylor, David Weiman, H. P. Dugan, Robert Pafford, Michael Catino, David Shuster, Jim Cook, Kenneth Turner, Richard Wilson, Philip Bowles, David Kennedy, James Flannery, John Bryson, John Leshy, Ben Yellen.

BOOKS

Bakker, Edna. *An Island Called California*. Berkeley: University of California Press, 1971.

Berkman, Richard, and W. Kip Viscusi. *Damming the West*. New York: Grossman, 1973.

Caughey, John, and Laree Caughey. *California Heritage*. Itasca, Ill.: F. E. Peacock, 1971.

Chall, Malca. *California Water Issues, 1950–1966*. Berkeley: Bancroft Library Oral History Program, 1981.

Eisen, Jonathan, and David Fine, eds. *Unknown California*. New York: Macmillan, 1985.

Fogelson, Robert. *Fragmented Metropolis*. Cambridge, Mass.: Harvard University Press, 1967.

Haslam, Gerald, and James Houston. *California Heartland*. Santa Barbara: Capra, 1978.

Hawgood, John A. *America's Western Frontiers*. New York: Knopf, 1967.

Kahrl, William, ed. *California Water Atlas*. Sacramento: California Department of Water Resources, 1979.

Lavender, David. *California*. Norton, 1976.

McWilliams, Carey. *California: The Great Exception*. Santa Barbara: Peregrine Smith, 1976.

Nadeau, Remi. *The Water Seekers*. Santa Barbara: Peregrine Smith, 1974.

Rogers, G. L. *A History of the Canal System of the Central California Irrigation District Prior to 1940*. Privately published, 1920; archives of G. M. Bowles.

Taylor, Paul. *Essays on Land, Water, and the Law in California*. New York: Arno, 1979.

Treadwell, Edward. *The Cattle King*. Fresno, Calif.: Valley Publishers, 1931.

ARTICLES AND REPORTS

Acreage Limitation, Interim Report. U.S. Department of the Interior, Washington, D.C., March 1980.

Acreage Limitation Review. Hearings Before the Subcommittee on Irrigation and Reclamation, Committee on Interior and Insular Affairs, United States Senate, April and May 1958.

Allman, T. D. "Jerry Brown: Nothing to Everyone." *Harper's*, July 1979.

"Association Notes These Are Difficult Times for Water Development." *Sacramento Bee*, September 23, 1979.

Baker, George. "An icy reception ahead for the state's new water plan." *California Journal*, August 1977.

Barnum, J. D. "What Shall I Say More Than I Have Inferred?" Unpublished monograph, 1969.

"Big Money for Delta Canal Votes." *Oakland Tribune*, July 13, 1980.

"Bill OK Stops Flow of Funds for Mid-Valley Canal Study." *Sacramento Bee*, June 20, 1979.

"Billions Urged to Tap North Coast Rivers." *Los Angeles Times*, March 11, 1965.

"Brown Turnaround." *Fresno Bee*, November 10, 1977.

"California Canal Plan to Divert River South Stirs a Flood of Protest." *Wall Street Journal*, February 12, 1981.

The California State Water Project—1977 Activities and Future Management Plans. Department of Water Resources, Sacramento, November 1981.

The California State Water Project—1978 Activities and Future Management Plans. Department of Water Resources, Sacramento, November 1979.

The California Water Project in 1964. Department of Water Resources, Sacramento, June 1964.

"California's Coming Water Famine." California Water Resources Association, Burbank, California, September 17, 1979.

Cannon, Lou. "High Dam in the Valley of the Tall Grass." *Cry California*, Summer 1968.

"City Wants MWD to Stop Overcharging." *Los Angeles Times*, February 10, 1982.

"Cost of Water Will Eliminate Need for Peripheral Canal." Working Alliance to Equalize Rates, Los Angeles, April 3, 1980.

Delta Water Facilities. Department of Water Resources, Sacramento, July 1978.

"Early Obstacle to Water Plan." *San Francisco Chronicle*, February 15, 1984.

"Environmentalists Split over Water Projects." *San Francisco Chronicle*, January 23, 1978.

Fact Sheet: Delta Alternatives. Department of Water Resources, Sacramento (undated internal draft).

"Farmers, Ecologists Argue 160-Acre Limitation." *San Francisco Examiner*, November 8, 1977.

"Federal Water Limitation Issue Revived in Westlands." *Sacramento Bee* (undated).

Flaxman, Bruce. *The Metropolitan Water District of Southern California: California's Billion Dollar Hidden Empire*, June 1978.

———. *The Price of Water: Who Pays and Who Benefits?* Public Policies Studies, Claremont Graduate School, Claremont, California, May 1976.

Hagan, Robert, and Edwin Roberts. *Energy Requirements of Alternatives in Water Supply, Use, and Conservation: A Preliminary Report*. California Water Resources Center, Davis, December 1975.

How "Firm" Is the "Risk" of a Water Shortage in California—An Important

Aspect of the California Water Debate. Meyer-Sangri Associates, Davis, February, 1982.

"Is the Canal a Disaster for Everybody?" *San Francisco Examiner,* February 10, 1980.

Is There Water for California? Bank of America Economics Department, San Francisco, September 1955.

"Kern Farms Bloom with State Water." *Sacramento Bee,* March 20, 1980.

Kirsch, Jonathan. "Politics and Water." *New West,* September 10, 1979.

Koch, Kathy. "Senate Water-Use Bill Pits Big Firms Against Small Farms." *Congressional Quarterly,* September 29, 1979.

"Liberalization of Federal Water Subsidy Law Sought." *Sacramento Bee,* July 3, 1980.

McCabe, Charles. "The 160-Acre Law." *San Francisco Chronicle.* November 21, 1977.

————. "Reactive Politics." *San Francisco Chronicle,* November 23, 1977.

————. "Small Is Ugly?" *San Francisco Chronicle,* December 19, 1977.

————. "That Damned Canal." *San Francisco Chronicle,* 1982 (series of four articles).

"MWD Consistently Exaggerates Water Needs." Contra Costa County Water Agency, California, January 9, 1980.

"MWD Subsidizes Agriculture at Expense of Urban Taxpayer." Contra Costa County Water Agency, California, December 5, 1980.

"MWD Water Charges Are Highly Inequitable." Contra Costa County Water Agency, California, December 11, 1980.

"North Coast Program Is Bared." *Sacramento Bee,* March 11, 1965.

"Offshore Oil May Bring State $1 Billion." *Los Angeles Times,* March 2, 1962.

"Oil, Water, and Boom: Will They Mix?" *New York Times,* October 28, 1980.

"Overview of Future Water Supply Available to Metropolitan." Metropolitan Water District of Southern California, Los Angeles, January 4, 1979.

People, Land, Food. National Land for People, Fresno, November 1977.

"Peripheral Canal Proponents' Solid Front Starts Crumbling." *Sacramento Bee,* May 25, 1981.

Phase II: Alternative Courses of Action. Department of Water Resources, Sacramento, March 1976.

Porgans, Patrick. *The State of the State Water Project.* Red Tape Abatement, Inc., Chico, California.

"Powerful Groups Battle over Peripheral Canal." *Sacramento Bee,* March 18, 1980.

"Projected Electric Power Costs for the California State Water Project." Department of Water Resources, March 20, 1980.

"Reclamation History" (three-part series). National Land for People, Fresno, California, 1979.

"Report: Westlands Violates 160-Acre Law." *Fresno Bee*, November 5, 1977.

Report of Activities of the Department of Water Resources. Department of Water Resources, Sacramento, February 2, 1980.

Responses to Analysts' Questions and Comments Prepared by the Staff of the Metropolitan Water District of Southern California. Metropolitan Water District, Los Angeles, January 27, 1981.

Review of the Central Valley Project. U.S. Department of the Interior, Department of Audit and Investigation, Washington, D.C., January 1978.

Robie, Ronald. "Statement on SB 200." Department of Water Resources, Sacramento, February 27, 1980.

"Santa Barbara Defeats $102 Million Water Issue." *San Francisco Chronicle*, March 8, 1979.

"Scare Tactics Issue." *Waterlog*, October 31, 1981.

"Senate OKs $3.4 Billion Water Projects Measure." *Sacramento Bee*, June 24, 1977.

"Senators Agree to Increase Historic Farm Acreage Limit." *Sacramento Bee*, June 21, 1979.

"South State Keeps Up Search for Water." *Sacramento Bee*, April 16, 1980.

State Water Project—Status of Water Conservation and Water Supply Augmentation Plans. Department of Water Resources, Sacramento, November 1981.

Stead, Frank M. "California's Cloaca Maxima." *Cry California*, Spring 1969.

Storper, Michael, and Richard Walker. *Subsidy and Uncertainty in Financing the State Water Project* (undated).

Taylor, Paul. "Evasion of Federal Acre Limit Laws," April 1967.

"Ten Farmers Who Astound Experts." *San Francisco Examiner*, May 6, 1979.

Thomes-Newville and Glenn Reservoir Plans: Engineering Feasibility, Department of Water Resources, Sacramento, November 1980.

"Top Cash Farm Products in California—1980." *San Francisco Examiner*, October 4, 1981.

Turner, Kenneth, and Steven Kasower. *Drought in Northern California: Implications for Water Supply Management*, December 1978.

"$23 Billion Tag for Water Project." *Sacramento Bee*, February 14, 1980.

"U.S. Aides Tread Water over Peripheral Canal." *Sacramento Bee*, March 16, 1980.

"Valley May Not Need the Water." *San Francisco Examiner*, January 22, 1978.

Villarejo, Don. *New Lands for Agriculture: The California State Water Project*. California Institute for Rural Studies, Davis, California, 1981.

Vizzard, James L. "The Water Poachers." *America*, February 13, 1965.
"The War over the Peripheral Canal." *San Francisco Chronicle*, June 26, 1980.
"Water Crisis in Year 2000?" *Sacramento Bee*, June 9, 1977.
"The Water Miners" (three-part series). *San Francisco Examiner*, March 26–28, 1979.
"Water Plan Leaks." *Sacramento Bee*, June 29, 1977.
"Westlands Hit in UC Report." *Fresno Bee*, June 12, 1980.
Westlands Water District, A Study of the Proposed Contract with the Bureau of Reclamation. Department of Water Resources, Sacramento, September 17, 1975.
Will the Family Farm Survive in America? (Federal Reclamation Policy: Westlands Water District). Joint Hearings Before the Select Committee on Small Business and the Committee on Interior and Insular Affairs, United States Senate, July 17 and July 22, 1975.
Willey, W. R. Z. *Economic and Environmental Aspects of Alternative Investments in California's Water System.* Environmental Defense Fund, Berkeley, California (undated).

LETTERS, MEMORANDA, MISCELLANEOUS

Dugan, H. P. Blue envelope letter to Commissioner of Reclamation, November 23, 1960.
———. Blue envelope letter to Commissioner of Reclamation, January 30, 1962.
———. Blue envelope memorandum to Commissioner of Reclamation, "Discussion of Region 2 Activities under Secretary Carr," May 4, 1961.
Frandsen, L., of Southern Pacific Company. Letter to Harvey O. Banks, California Department of Water Resources, and Clyde Spencer, Regional Director of Bureau of Reclamation, October 1, 1956.
Harris, Ellen Stern. Memorandum to Carl Kymla, Metropolitan Water District, "MWD's Proposed 1979–80 Budget," June 4, 1979.
Interview between Lawrence W. Swenson and Donald Sandison, Department of Water Resources, and Patrick Porgans, Red Tape Abatement, Inc., August 4, 1981.
Lineweaver, Goodrich W., Senate Committee on Interior and Insular Affairs. Letter to William E. Warne, Commissioner of Agriculture, State of California, January 20, 1960.
McDiarmid, John M. Letter to Ronald Robie, May 30, 1980.
Pafford, Robert. Blue envelope memorandum to Commissioner of Reclamation, "Eel River Basin Planning and Interagency Coordination on Route Selection," January 19, 1968.
———. Blue envelope memorandum to Commissioner of Reclamation,

"Proposed Alternative to Kellogg Unit, Central Valley Project," December 24, 1968.

————. Blue envelope memorandum to Commissioner of Reclamation, "Subsidence in the Westlands Water District" (undated).

Regional Director, Sacramento. Memorandum to Commissioner of Reclamation, "Westlands Water District—Ground water pumping and excess lands." March 16, 1964.

Robie, Ronald, Department of Water Resources. Letter to Thomas Graff, Environmental Defense Fund, July 19, 1977.

Robie, Ronald B. Letter to Michael Storper (undated; circa May 1979).

Staats, Elmer. Letter to Lee C. White, The White House, "The Westlands Distribution System," September 19, 1964.

Stamm, Gilbert. Memorandum to Commissioner of Reclamation, "Excess Land Problem of the Imperial Irrigation District," October 1, 1965.

West, Arleigh. Blue envelope memorandum to Commissioner of Reclamation, "Excess Lands—Imperial Irrigation District—Boulder Canyon Project, California," April 15, 1965.

————. CONFIDENTIAL. Blue envelope memorandum to Commissioner Dominy, Bureau of Reclamation, "Excess land survey in areas served by Metropolitan Water District, Southern California," December 30, 1964.

CHAPTER ELEVEN: **Those Who Refuse to Learn . . .**

The main sources for the Fontenelle story were H. P. Dugan, Barney Bellport, Floyd Dominy, and a series of blue envelope memoranda discussing the near-disaster.

The chronicle of the Teton disaster was put together largely from excellent reporting by the *Idaho Statesman*. Coverage in the *Los Angeles Times* was also exceptionally good. John Erlichman, Pete van Gytenbeek, and Nathaniel Reed provided much of the political background.

Major sources for the Colorado and Narrows Dam section of the chapter were C. J. Kuiper, Don and Karen Christenson, Glenn Saunders, Felix Sparks, Guy Martin, Robert Weaver, Gary Friehauff, Daniel Beard, Alan Merson, Sandy White, and Wayne Aspinall.

BOOKS

Athearn, Robert G. *High Country Empire*. Lincoln: University of Nebraska Press, 1960.

Gottlieb, Robert, and Peter Wiley. *Empires in the Sun*. New York: Putnam, 1982.

McPhee, John. *Basin and Range.* New York: Farrar, Straus, and Giroux, 1981.

Thomas, Janet, et al., eds. *That Day in June.* Rexburg, Idaho: Ricks College Press, 1977.

ARTICLES AND REPORTS

Actions Needed to Increase the Safety of Dams Built by the Bureau of Reclamation and the Corps of Engineers. Government Accounting Office, Washington, D.C., June 3, 1977.

"Analyst: Narrows benefit distorted." *Rocky Mountain News* (undated).

Arthur, Harold. *Preliminary Report on Failure of Teton Dam,* Denver, U.S. Bureau of Reclamation, June 7, 1976.

"Avid Environmentalists Castigated for Tactics." *Idaho Falls Post-Register,* June 4, 1972.

"B of R Eyed, Rejected Acrial Survey of Leaks." *Idaho Statesman,* July 29, 1976.

"B of R Left Fissure Unfilled to Avoid Delay of Dam Work." *Idaho Statesman* (undated).

Brown, Russell. Testimony, Teton Dam Hearing, Senate Interior Committee, Idaho Falls, Idaho, February 21, 1977.

"Bruised Teton Dam Wins Another Political Fight." *Idaho Falls Post-Register,* October 20, 1971.

Bureau of Reclamation. *The Fryingpan-Arkansas Project,* June 1977.

"Bureau of Reclamation Won't Fight Teton Report Criticizing Agency." *Idaho Statesman,* January 7, 1977.

"Bureaucratic Gamble Ended in Disaster." *Idaho Statesman,* September 6, 1976.

"Charge of Scapegoating." *Idaho Statesman,* May 20, 1977.

"Colorado, Carter, and the Dams." *Rocky Mountain News,* February 23, 1977.

"Colorado Water Projects—Impacts and Alternatives." *Denver Post,* April 17, 1977.

Committee on Government Operations. *Teton Dam Disaster,* Washington, D.C., 1976.

"Currents of Argument Swirl over Poudre River." *Denver Post,* May 24, 1981.

"Dam Opposition Said Obstructionism." *Idaho Falls Post-Register,* October 10, 1971.

"Dam Safety: No National Answer." *ENR News,* May 8, 1980.

"Devastation in Eastern Idaho Boggles Mind." *Idaho Statesman,* June 8, 1976.

"Environment Aims Cloud Idaho Industrial Role." *Idaho Falls Post-Register,* October 10, 1971.

"Fissures Made Teton Difficult." *Idaho Statesman*, July 9, 1976.
"Groundwater Rise Preceded Collapse." *Idaho Statesman*, July 9, 1976.
"Grout Curtain Failure May Have Triggered Teton Dam Collapse." *Engineering News Review*, June 10, 1978.
"He Protests Too Loudly." *Rexburg Standard*, September 14, 1971.
"Idaho Dam Disaster—New Blame." *San Francisco Chronicle*, December 27, 1979.
Idaho Environmental Council. *The Fraud of Teton Dam*, 1972.
Idaho Environmental Council. *The Teton Dam Symposium*, June 4, 1972.
"Impossible Dam Failure Spurs Study of Liability," *Idaho Statesman*, June 10, 1976.
"Lamm Reaffirms Narrows Support." *Denver Post*, August 4, 1976.
"Looters Hit Chaotic Rexburg," *Idaho Statesman*, June 9, 1976.
"Memos Tell of Teton Project Deficit." *Idaho Statesman*, July 13, 1976.
"More state water projects likely." *Rocky Mountain News*, July 23, 1979.
"Narrows Dam Gets 'Clean Bill of Health.' " *Denver Post*, August 17, 1968.
The Narrows Unit, Colorado. Lower South Platte Water Conservancy District, Sterling, Colorado.
Narrows Unit, Pick-Sloan Missouri Basin Project. U.S. Department of the Interior, Water Projects Review, April 1977.
"New Dangers confront victims of Idaho flood." *Deseret News*, June 8, 1976.
Odell, Rice. "Silt, Cracks, Floods, and Other Dam Foolishness." *Audubon*, September 1975.
"Officials Dispute Time of First BOR Warning." *Idaho Statesman*, June 9, 1976.
"Opposition to Teton Dam Hurts Environmentalists," *Idaho Falls Post-Register*, June 6, 1971.
Regional Landowners Group. *Narrows Fact Sheet*, March 1978.
Regional Landowners Group. *"The Narrows and Its Alternatives—A Benefit-Cost Comparison,"* March 1978.
Regional Landowners Group. *Response to the Bureau of Reclamation's "Special Report—Narrows Unit,"* February 1978.
Reed, Scott. Untitled monograph for Snake River Regional Studies Center Conference, April 1, 1977.
"Safety questions scarcely raised by opponents of ill-fated dam." *Denver Post*, June 8, 1976.
Saunders, Glenn. *Colorado Water Rights—A Briefing Paper* (undated).
"State Engineer on Narrows" (three-part series). *Fort Morgan Times*, April 5, 6, and 7, 1978.
"Sunset for the Narrows." *Empire Magazine (Denver Post)*, January 11, 1976.
"Team Says Teton Dam Eaten Away from Within." *Rocky Mountain News*, July 16, 1976.

"Teton: Background of the Dispute." *Intermountain Observer*, June 10, 1972.

"Teton Dam: The Sorry Lesson." *High Country News*, June 18, 1976.

"Teton Dam Collapse: How Disaster Struck." *Los Angeles Times*, July 18, 1976.

"Teton Dam Designers Called 'Stale.' " *Idaho Statesman*, February 22, 1977.

"Teton Meets Economic Test, Asserts Walker." *Idaho Falls Post-Register*, October 17, 1971.

"Teton Project Impresses Andrus, Safeguards Eyed." *Idaho Falls Post-Register*, May 17, 1971.

"Teton River Dam Stirs Controversy." *High Country News*, June 11, 1971.

"Teton Served as Battleground." *Idaho Statesman*, September 8, 1976.

"Teton Dam Termed Non-Political Issue." *Idaho Falls Post-Register*, October 20, 1971.

U.S. Bureau of Reclamation. Statement Before the Subcommittee on Conservation, Energy, and Natural Resources of the Committee on Government Operations, August 6, 1976.

U.S. Bureau of Reclamation. "Teton Dam Failure," press release, June 9, 1976.

U.S. Department of the Interior, Water Projects Review. *Fruitland Mesa Project, CRSP*, April 1977.

U.S. Fish and Wildlife Service. *Comments of the Fish and Wildlife Service on the Bureau of Reclamation's Special Report, Narrows Unit, February 1978* (undated).

U.S. Water Resources Council. *Manual of Procedures for Evaluating Benefits and Costs of Federal Water Resources Projects*, Washington, D.C., February 9, 1979.

"USGS Response to Queries About Teton Dam." U.S. Geologic Survey, Reston, Virginia, June 15, 1976.

"Warm, Lazy day in Rexburg, then crash." *Deseret News*, June 7, 1976.

"Water Project 'Hit List': A View from the West." *Denver Post*, March 27, 1977.

"Water Projects' Need Debated." *Denver Post* (undated).

"Water Users Support Rebuilding Teton." *Idaho Statesman*, December 11, 1976.

"Water Won't Stretch for Western Cities' Growth." *High Country News*, October 7, 1977.

"Welcome to Wrecksburg." *Sundowner*, Winter 1977.

Williams, Philip. "Dam Design: Is the Technology Faulty?" *New Scientist*, February 2, 1978.

LETTERS, MEMORANDA, MISCELLANEOUS

Acting Chief Geologist, Bureau of Reclamation. Memorandum to D. J. Duck, "Seismic Monitoring Program—Teton Dam and Reservoir," June 20, 1973.

Bellport, Barney. Blue envelope memorandum to Commissioner of Reclamation, "Fontenelle Dam and Reservoir—Seedskadee Project, Wyoming," May 11, 1965.

————. Blue envelope memorandum to Commissioner, Bureau of Reclamation, "Replacement of American Falls Dam," May 24, 1967.

Crandall, David. Blue envelope letter to Chief Engineer, Bureau of Reclamation, "Fontenelle Dam," September 30, 1965 (with response undated).

Curry, Robert. Letter to V. E. McKelvey, U.S. Geologic Survey, July 6, 1976.

Green, John, Regional Administrator, Environmental Protection Agency. Letter to Keith Higginson, Commissioner of Reclamation, July 13, 1977.

Kosman, Jacob, Irrigationists' Association of Water District No. 1. Letter to President Jimmy Carter, et al., April 1, 1978.

Kuiper, C. J. Personal correspondence, November 21, 1979.

Kyncl, George. "A New Look at the Narrows," January 15, 1977.

McCabe, Joseph, Environmental Protection Agency. Letter to Daniel Beard, U.S. Department of the Interior, February 23, 1978.

McDonald, William, Colorado Conservation Board. Letter to Frank M. Scott, Harza Engineering Company, May 15, 1980.

McKelvey, V. E. Letter to Robert Curry, Department of Geology, University of Montana, July 21, 1976.

————. Letter to Senator Henry Jackson, June 11, 1976.

Nelson, Harold. Blue envelope letter to Commissioner of Reclamation, "Expedited processing of interim reports—Fremont Dam, Lower Teton Division: Ririe Dam; and raising Blackfoot Reservoir," February 24, 1962.

Oriel, Steven, et al. Unaddressed memorandum, "Preliminary Report on Geologic Investigations, Eastern Snake River Plain and Adjoining Mountains," June 1973.

Parenteau, Patrick A., National Wildlife Federation. Letter to Cecil D. Andrus, February 23, 1977.

Phipps, E. Personal correspondence, September 25, 1979.

Schleicher, David. Unaddressed memorandum, "Some geologic concerns about the Teton Basin Project," December 26, 1972.

Sparks, Felix. Memorandum to Colorado Water Conservation Board, "Consideration of FY 1980 Funding Requirements for Colorado Reclamation Projects," March 6, 1979.

CHAPTER TWELVE: **Things Fall Apart**

Most of the background on the Texas Water Plan comes from coverage in the *Texas Observer* and from "You Ain't Seen Nothing Yet" in *The Water Hustlers*.

The Ogallala situation is well described in the Economic Development Administration's report and in an excellent series of articles that ran in the *Denver Post* in 1979 (see bibliography). Desertification and its potential consequences are thoroughly covered in David Sheridan's *Desertification of the United States* and in Paul Sears's *Deserts on the March*. Sheridan's book, though not as eloquent, is considerably more up-to-date and crammed with information.

The Department of Agriculture's Salinity Control Laboratory in Riverside, California, is a great source of information on salinity, its consequences, and its avoidance. A good compendium on irrigation in general is Cantor's *World Geography of Irrigation*.

Information, much of it not so up-to-date, on reservoir siltation is available from both the Corps of Engineers and the Bureau of Reclamation. Most libraries are almost devoid of literature on this gigantic problem. This section of the chapter draws heavily on interviews with Raphael Kazmann and Luna Leopold, and on Kazmann's book *Modern Hydrology*, one of the few exceptions to the above statement.

Other important interviews for this chapter: Jan van Schilfgaarde, Jim Casey, Daniel Dreyfus, C. J. Kuiper, Joe Moore, Steven Reynolds, Herbert Grubb, Ronnie Dugger, Mary Ellen Morbeck, Bob Strand, Wayne Wyatt, Floyd Dominy, Jay Lehr, Philip Williams, Mohammed El-Ashry, George Pring, W. R. Collier.

BOOKS

Cantor, Leonard Martin. *A World Geography of Irrigation.* New York: Praeger, 1970.

Goldsmith, Edward, and Nicholas Hildyard. *The Social and Environmental Effects of Large Dams.* Cornwall, England: Wadebridge Ecological Center, 1985.

Graves, John. "You Ain't Seen Nothing Yet." In Robert Boyle, et al., eds., *The Water Hustlers.* San Francisco: Sierra Club, 1971.

Kazmann, Raphael. *Modern Hydrology.* 2nd ed. New York: Harper and Row, 1972.

Peterson, Dean F., and A. Berry Crawford, eds. *Values and Choices in the Development of the Colorado River Basin.* Tucson: University of Arizona Press, 1978.

Peterson, Elmer T. *Big Dam Foolishness.* New York: Devin-Adair, 1954.

Sheridan, David. *Desertification of the United States*. Washington, D.C.: Council on Environmental Quality, 1981.

ARTICLES, REPORTS

Adams, Daniel B. "Last-Ditch Archaeology." *Science 83*, December 1983.

Agricultural Drainage and Salt Management in the San Joaquin Valley. San Joaquin Valley Interagency Drainage Program, Fresno, California, June 1979.

Ambroggi, Robert P. "Underground Reservoirs Control the Water Cycle," *Scientific American*, May 1977.

"Arid West Is Trying Drip Irrigation." *New York Times*, June 28, 1983.

"Back to Basics: Mining Water Deep Below Heart of Texas." *New York Times*, July 27, 1980.

Began, Ann. "National Public Water Policy: The Colorado River." Environmental Policy Center, Washington, D.C., August 1977.

Briggs, Jean A. "There's no synwater industry to bail us out." *Forbes*, March 16, 1981.

Brown, Howard. "Wellton-Mohawk." Memorandum to Senator Paul Fannin, Library of Congress Congressional Research Service, May 5, 1975.

"Critics Call Billion-Dollar Water Project in Arizona Wasteful." *Washington Post* (undated).

"Department of State Sells U.S. Taxpayers and Mexico Down the Drain," Environmental Policy Institute (undated).

"Desalting Needs Reduced by Water-Savings Program." *Denver Post*, November 20, 1978.

"The Drying Out of America." *Discover*, April 1981.

Elephant Butte Reservoir, 1980 Sedimentation Survey, U.S. Department of the Interior, July 1983.

"Ground water levels dip in 10 Oregon farming areas." *Oregonian*, April 10, 1983.

"The High Plains: Depleting the Ogallala" (series of six articles). *Denver Post*, December 16–21, 1979.

Holburt, Myron. "The 1973 Agreement on Colorado River Salinity Between the United States and Mexico." Presented at the National Conference on Irrigation Return Flow Quality Management, Fort Collins, Colorado, May 1977.

The Hydrologic History of the San Carlos Reservoir, Arizona, 1929–1971. Geological Survey Professional Paper 665-N, 1977.

Jacobsen, Thorkind, and Robert M. Adams. "Salt and Silt in Ancient Mesopotamian Agriculture." *Science*, November 1958.

Kuiper, C. J. "The Ground Water Rush of 1978." Office of the Colorado State Engineer, February 5, 1979.

Kuiper, C. J., Floyd Dominy, et al. *Nile River Irrigation System Rehabilitation and Improvement Program.* Report of the Scope Team.

"Lamm Calls 'Water Grab' Time Bomb in Constitution." *Denver Post*, April 3, 1979.

Lehr, Jay H. "Groundwater—in the Eighties." *WATER Engineering and Management*, March 1981.

LeVeen, E. Philip. *A Political-Economic Analysis of the Prospects for Irrigated Agriculture in California*, Berkeley, California, November 1984.

———. "Toward a New Food Policy." Public Interest Economics—West, Berkeley, California.

"Losers all around in irrigation battleground." *Fresno Bee*, November 25, 1984.

"Mexico Won't Pay for Texas Oil Spill Mess." *San Francisco Chronicle* (undated).

The 1964 Sedimentation Survey of Boysen Reservoir, Wyoming. U.S. Bureau of Reclamation, November 1965.

"Ogallala Overdraft Could Start Big Battle over Rescue Strategies." *World Water*, September 1982.

Pillsbury, Arthur F. "The Salinity of Rivers." *Scientific American*, July 1981.

"Plans to Hold Down Colorado River's Salt Content and Avoid Irking Mexico Are Hit by Rising Costs." *Wall Street Journal*, June 21, 1979.

"The Price of a Sweet River." *Boston Globe*, May 27, 1979.

Rates of Sediment Production in Midwestern United States. U.S. Department of Agriculture, Soil Conservation Service, Washington, D.C., December 1948.

"Representative Brown Calls Colorado River Desalting Plant 'Boondoggle,' Fights to Halt Project." *Los Angeles Times*, May 8, 1979.

Rhodes, J. D., and R. D. LeMert. "Use of San Joaquin Valley Saline Drainage Waters for Irrigation of Cotton." USDA Salinity Control Laboratory, Riverside, 1981.

Risser, James. "Worse Than the Dust Bowl." *The New Farm*, February 1983.

"Running Dry: Huge Area in Midwest Relying on Irrigation Is Depleting Its Water." *Wall Street Journal*, August 6, 1980.

Russell, Dick. "Ogallala—Half Full or Half Empty?" *The Amicus Journal*, Fall 1985.

S. 496, "A Bill to increase the appropriation ceiling for title I of the Colorado River Basin Salinity Control Act." U.S. Senate, February 22, 1979.

Sedimentation Report of Buffalo Bill Reservoir, Shoshone Project." U.S. Bureau of Reclamation, May 24, 1949.

Sedimentation Study, Glen Canyon Dam, Colorado River Storage Project, U.S. Bureau of Reclamation, May 1962.

"Seventy-six Tons of Tomatoes per Acre with Subsurface Trickle Irrigation." *USDA Agriculture Research Service News*, November 4, 1982.

Shaw, Gaylord. "The search for dangerous dams: a program to head off disaster." *Natural History*.

Shoji, Kobe. "Drip Irrigation." *Scientific American*.

Skogerboe, Gaylord V. "Agricultural Impact on Water Quality in Western Rivers." Department of Agricultural Engineering, Colorado State University, Fort Collins.

"SPLAT! Clements belly-flops over Arkansas water." *Texas Monthly*, May 1981.

"State's Bold Plan to Create Power, Drinking Water." *San Francisco Chronicle*, August 14, 1984.

Sudman, Rita Schmidt. "Salt in the San Joaquin." *Western Water*, May/June 1982.

Texas Observer, 1966-1981. (Author's note: Most of the material relating to the Texas Water Plan was gleaned from excellent reporting in this publication over some fifteen years. Unfortunately, this file was lost or misplaced. However, the *Texas Observer* maintains a good index for those who wish to read more about water and Texas. Reprints of many issues are available.)

Theodore Roosevelt Lake—1981 Sedimentation Survey. U.S. Department of the Interior, July 1983.

"Upper Colorado River Basin Energy Development Project Water Needs and Water Supplies." Unpublished Interior Department Discussion Draft, May 23, 1974.

"The Valley's Dangerous Drain Problem." *San Francisco Examiner*, May 20, 1979.

Van Schilfgaarde, Jan. "Water Conservation Potential in Irrigated Agriculture." USDA Salinity Control Laboratory, Riverside, California, August 1979.

"Water resource mismanagement brewing crisis." *Oregonian*, March 20, 1983.

"What to Do When the Well Runs Dry." *Science*, November 14, 1980.

Williams, Philip. "Dam Design: Is the Technology Faulty?" *New Scientist*, February 2, 1978.

Worster, Donald. "Water and the Flow of Power." *The Ecologist*, Vol. 13, No. 5, 1983.

LETTERS, MEMORANDA, MISCELLANEOUS

Adams, Mark. Blue envelope memorandum to Floyd Dominy, Commissioner of Reclamation, "Tenor of comments by a few members of Texas Congressional Delegation concerning Reclamation program in Texas," June 12, 1962.

Gessel, Clyde, and William Culp. Memorandum to Chief, Division of River Control, Bureau of Reclamation, "Sediment Deposition—Blue Mesa Reservoir Site—Curecanti Unit—Colorado River Storage Project," March 29, 1961.

Hill, Leon. Blue envelope memorandum to Commissioner, Bureau of Reclamation, "Texas Water Situation," September 1, 1966.

Mermel, T. W. Memorandum to Commissioner Dominy, "USSR Breaks Records—Earth and Arch Dams," November 19, 1968.

Nelson, Harold. Blue envelope memorandum to Commissioner of Reclamation, "Possible Leakage, Yellowtail Reservoir," April 20, 1966.

Oka, James N. Memorandum to Chief, Special Studies Branch, Bureau of Reclamation, "Sediment Study, Morrow Point Reservoir," March 28, 1962.

EPILOGUE: **A Civilization, if You Can Keep It**

Information on the North American Water and Power Alliance is readily proffered by the Parsons Company in Pasadena, California. Background on the project was also provided by Jim Casey, Barney Bellport, C. J. Kuiper, Stanford McCasland, Derrick Sewall, Frank Moss, and Irving Fox. Derrick Sewall, Irving Fox and Moira Farrow were my main sources regarding the Canadian water situation.

BOOKS, ARTICLES, REPORTS, LETTERS,
MEMORANDA, MISCELLANEOUS

"Actual Price of High Dams Also Includes Social Costs." New York Times, July 10, 1983.

"Alaska to Mexico Canals Urged to Water Deserts." Denver Post, February 27, 1977.

"Audits Show Unbusinesslike Management of Western Basin Accounts." Environmental Policy Center, Washington, D.C.

"Billion Dollar Plot Alleged: U.S. 'Wants Canada's Water.'" Vancouver Sun, June 27, 1981.

Bocking, Richard. Canada's Water—For Sale? James Lewis and Samuel, Toronto, 1972.

"Damage to Glen Canyon Spillways Complicates Plans for Spring Runoff." Salt Lake Tribune, December 14, 1983.

Dominy, Floyd. Untitled monograph (response to U.S. Chamber of Commerce report on adverse effects of expanding government), October 25, 1957.

"Dominy Foresees Water Sharing Need After Northwest Meets Requirements." Idaho Statesman, January 22, 1966.

"Great Lakes States Seek to Keep Their Water." *New York Times*, June 13, 1982.

Kierans, Thomas W. *The Grand Canal: A Water Resources Planning Concept for North America*. Alexander Graham Bell Institute, Sydney, Nova Scotia.

King, Laura B., and Philip E. LeVeen. *Turning Off the Tap on Federal Water Subsidies* (Volume I: *The Central Valley Project: The $3.5 Billion Giveaway*). Natural Resources Defense Council, San Francisco, August 1985.

"Next B.C. Megaproject Looms." *Vancouver Sun*, September 29, 1985.

Pearce, Fred. "Fall and Rise of the Caspian Sea." *New Scientist*, December 6, 1984.

Rada, Edward L., and Richard J. Berquist. *Irrigation Efficiency in the Production of California Crop Calories and Proteins*. California Water Resources Center, Davis, California, May 1976.

Sewall, Derrick. *Water: The Emerging Crisis in Canada*. Canadian Institute for Economic Policy, Ottawa, 1981.

Snyder, Nathan. *Water from Alaska*. Ralph M. Parsons Company, October 15, 1980.

A Summary of Water Resources Projects, Plans, and Studies Relating to the Western and Midwestern United States. Senate Committee on Public Works, Washington, D.C., 1966.

Tunison, M. C. Letter to Senator William King of Utah, December 15, 1938.

"U.S. Demands Say in Stikine Dams." Telkwa Foundation *Newsletter*, May/June 1981.

Van der Leeden, Frits. *Water Resources of the World*. Water Information Center, Port Washington, New York, 1975.

Water Crisis. Ralph M. Parsons Company, Pasadena, 1980.

"Water Diversion Proposals." Idaho Water Resources Board, July 1969.

Water Use on Western Farms: An Examination of Irrigation Practices and Ways They Can Be Improved. INFORM, New York City, Spring 1982.

INDEX